beginnings of life
second edition

beginnings of life

second edition

ricki lewis
The University at Albany

WCB
Wm. C. Brown Publishers

Dubuque, IA   Bogota   Boston   Buenos Aires   Caracas   Chicago
Guilford, CT   London   Madrid   Mexico City   Sydney   Toronto

Book Team

Editor *Carol J. Mills*
Developmental Editor *Diane E. Beausoleil*
Production Editors *Carla D. Kipper and Kennie Harris*
Designer *Jeff Storm*
Art Editor *Jodi K. Banowetz*
Photo Editor *Lori Hancock*
Permissions Coordinator *Vicki Krug*

Wm. C. Brown Publishers
A Division of Wm. C. Brown Communications, Inc.

Vice President and General Manager *Beverly Kolz*
Vice President, Publisher *Kevin Kane*
Vice President, Director of Sales and Marketing *Virginia S. Moffat*
Vice President, Director of Production *Colleen A. Yonda*
National Sales Manager *Douglas J. DiNardo*
Marketing Manager *Craig Johnson*
Advertising Manager *Janelle Keeffer*
Production Editorial Manager *Renée Menne*
Publishing Services Manager *Karen J. Slaght*
Royalty/Permissions Manager *Connie Allendorf*

Wm. C. Brown Communications, Inc.

President and Chief Executive Officer *G. Franklin Lewis*
Senior Vice President, Operations *James H. Higby*
Corporate Senior Vice President, President of WCB Manufacturing *Roger Meyer*
Corporate Senior Vice President and Chief Financial Officer *Robert Chesterman*

Copyedited by Anne Cody

Cover: © SPL/Photo Researchers, Inc. Photos researched by Toni Michaels

The credits section for this book begins on page 411 and
is considered an extension of the copyright page.

Dedicated to Larry, Heather, Sarah, and Carly

"A personal library is a lifelong source of enrichment and distinction. Consider this book an investment in your future and add it to your personal library."

BRIEF CONTENTS

Beginnings of Life

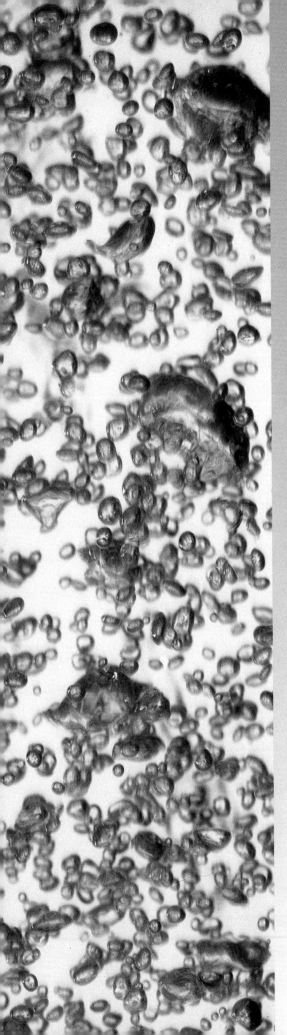

CONTENTS

UNIT 1

Overview of Biology

CONNECTIONS

Chapter 1

What Is Life?

Chapter 2

Thinking Scientifically about Life

Chapter 3

The Origin and Chemistry of Life

Contents

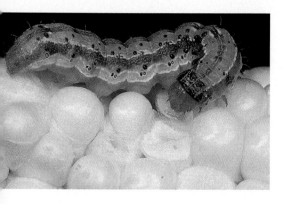

UNIT 2
Cell Biology

CONNECTIONS

Chapter 4
Cells

Chapter 5
Cellular Architecture

Chapter 6
The Energy of Life

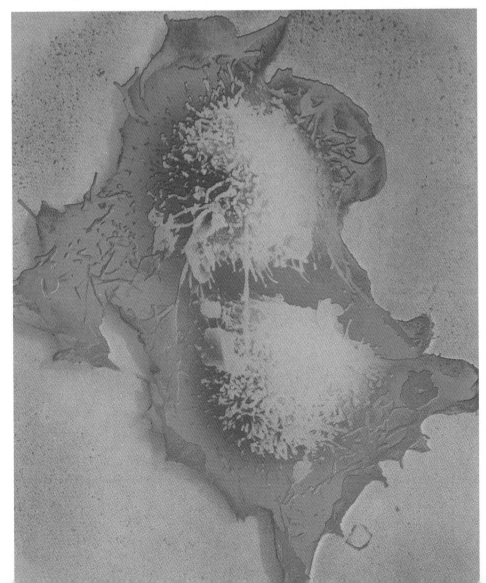

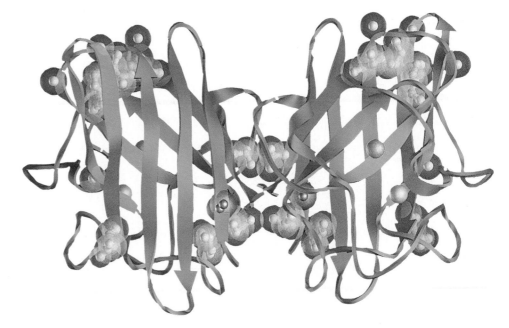

BIOLOGY IN ACTION READINGS

EXPLORATIONS

Modules from the Explorations in Human Biology software, a CD-ROM produced by Wm. C. Brown Publishers, are referenced at the ends of the following chapters.

Chapter 5 Active Transport
Chapter 9 Smoking and Cancer
Chapter 11 Life Span and Lifestyle
Chapter 14 Heredity in Families
Chapter 16 Cystic Fibrosis

ANIMATED FIGURES

Animated versions of the following illustrations are available on the Physiological Concepts of Life Science videotapes, produced by Wm. C. Brown Publishers.

Figure 3.9a Formation of Ionic Bond
Figure 3.22 Lock and Key Model of Enzyme Action
Figures 4.7 and 4.8 Journey into a Cell
Figure 4.9 Cellular Secretion
Figure 5.7 Osmosis
Figure 5.11 Active Transport Across a Cell Membrane
Figure 5.14 Endocytosis
Figure 5.20 Cyclic AMP Action
Figure 6.10 ATP As an Energy Carrier
Figure 7.5 Glycolysis
Figure 7.7 Oxidative Respiration
Figure 7.8 Electron Transport Chain and Oxidative Phosphorylation
Figures 8.8 and 8.9 The Photosynthetic Electron Transport Chain and Production of ATP
Figure 8.11 C_3 Photosynthesis (Calvin Cycle)
Figure 8.14 C_4 Photosynthesis
Figure 9.5 Mitosis
Figure 10.7 Meiosis
Figure 10.8 Crossing Over
Figure 10.12 Spermatogenesis
Figure 10.14 Oogenesis
Figure 11.20 Human Embryonic Development
Figure 12.18 A, B, O Blood Types
Figure 14.14 DNA Replication Fork
Figure 15.5 Transcription of a Gene
Figure 15.10 Protein Synthesis

THE *BEGINNINGS OF LIFE* LEARNING SYSTEM

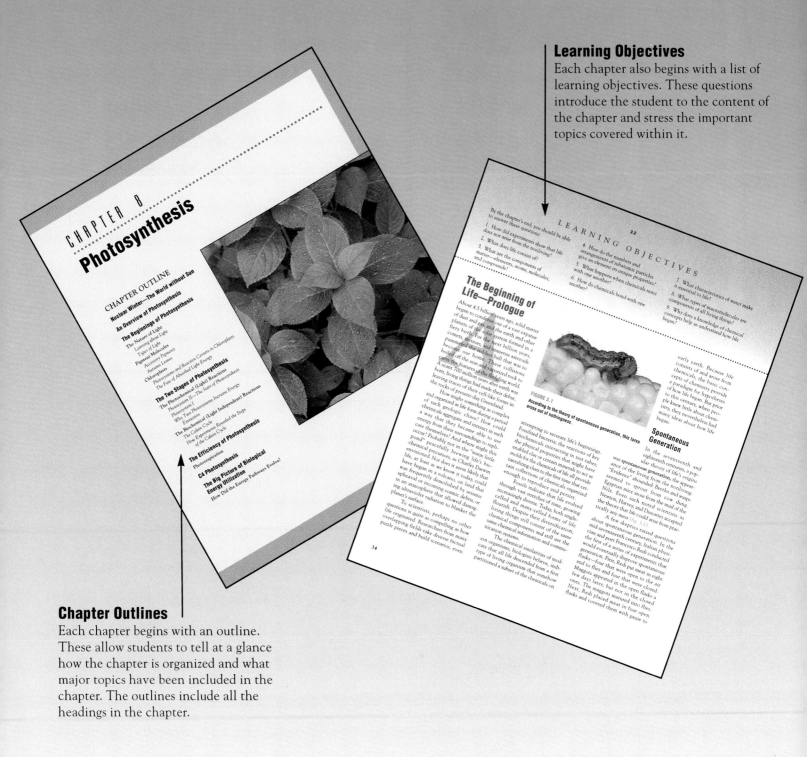

Learning Objectives
Each chapter also begins with a list of learning objectives. These questions introduce the student to the content of the chapter and stress the important topics covered within it.

Chapter Outlines
Each chapter begins with an outline. These allow students to tell at a glance how the chapter is organized and what major topics have been included in the chapter. The outlines include all the headings in the chapter.

Key Concepts

At the ends of major sections within each chapter, brief summaries highlight key concepts in the section, helping students focus their study efforts on the key material.

Tables

Numerous strategically placed tables list and summarize important information, making it readily accessible for efficient study.

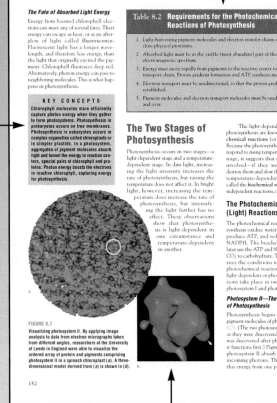

The Fate of Absorbed Light Energy

Energy from boosted chlorophyll electrons can meet any of several fates. Their energy can escape as heat, or as an afterglow of light called **fluorescence**. Fluorescent light has a longer wavelength, and therefore less energy, than the light that originally excited the pigment. Chlorophyll fluoresces deep red. Alternatively, photon energy can pass to neighboring molecules. This is what happens in photosynthesis.

KEY CONCEPTS

Chlorophyll molecules more efficiently capture photon energy when they gather to form photosystems. Photosynthesis in prokaryotes occurs on free membranes. Photosynthesis in eukaryotes occurs in complex organelles called chloroplasts or in simpler plastids. In a photosystem, aggregates of pigment molecules absorb light and funnel the energy to reaction centers, special pairs of chlorophyll and proteins. Photon energy boosts the electrons in reactive chlorophyll, capturing energy for photosynthesis.

FIGURE 8.7

Visualizing photosystem II. By applying image analysis to data from electron micrographs taken from different angles, researchers at the University of Leeds in England were able to visualize the ordered array of protein and pigments comprising photosystem II in a spinach chloroplast (a). A three-dimensional model derived from (a) is shown in (b).

152

Table 8.2 Requirements for the Photochemical Reactions of Photosynthesis

1. Light-harvesting pigment molecules and electron transfer chains must be in close physical proximity.
2. Absorbed light must be in the visible (most abundant) part of the electromagnetic spectrum.
3. Energy must move rapidly from pigments to the reactive center to the electron transport chain. Proton gradient formation and ATP synthesis must also be fast.
4. Electron transport must be unidirectional, so that the proton gradient is established.
5. Pigment molecules and electron transport molecules must be used over and over.

The Two Stages of Photosynthesis

Photosynthesis occurs in two stages—a light-dependent stage and a temperature-dependent stage. In dim light, increasing the light intensity increases the rate of photosynthesis, but raising the temperature does not affect it. In bright light, however, increasing the temperature does increase the rate of photosynthesis, but raising the light further has no effect. These observations show that photosynthesis is light-dependent in one circumstance and temperature-dependent in another.

The light-dependent reactions of photosynthesis are known as the **photochemical reactions** (or light reactions). Because the photosynthetic rate does not respond to rising temperatures during this stage, it suggests that enzymes are not involved—if they were, heat would destroy them and slow the reactions. The temperature-dependent reactions are called the **biochemical reactions**, or light-independent reactions, of photosynthesis.

The Photochemical (Light) Reactions

The photochemical reactions of photosynthesis oxidize water, release oxygen, produce ATP, and reduce NADP+ to NADPH. The biochemical reactions later use the ATP and NADPH to reduce CO_2 to carbohydrate. Table 8.2 summarizes the conditions necessary for the photochemical reactions to occur. The light-dependent or photochemical reactions take place in two photosystems: photosystem I and photosystem II.

Photosystem II—The Start of Photosynthesis

Photosynthesis begins in the cluster of pigment molecules of photosystem II (fig. 8.7). (The two photosystems were named as they were discovered; photosystem II was discovered after photosystem I, but it functions first.) Pigment molecules in photosystem II absorb the energy from incoming photons. They then transfer this energy from one pigment molecule

---

Biology in ACTION 8.2

Oxidative Stress

Approximately 3.5 billion years ago, long before plants carpeted the earth, the atmosphere held very little oxygen. As photosynthetic reactions formed and released oxygen, atmospheric levels of oxygen built up. Life had to adapt to this drastic change because oxygen is a very reactive, even toxic, substance. Some forms of the element, called free radicals, tend to combine with nearly any other molecule, driven by the tendency to complete their outermost electron shell. Free radicals are natural products of metabolism, and they also form from acid rain, pollution, and toxic metals that contaminate water.

The deleterious effects caused by activated forms of oxygen such as free radicals are examples of *oxidative stress*. In humans, oxygen damage is linked to some age-related disorders, including heart disease, emphysema, arthritis, and amyotrophic lateral sclerosis (ALS, or Lou Gehrig's disease). In plants, high oxygen levels limit growth. This happens because in warm climates, oxygen binds rubisco, the key CO_2-fixing enzyme in the Calvin cycle, and thus interferes with carbon fixation. Excess activated oxygen can also destroy chlorophyll, causing a bleaching effect in leaves.

Plants use several familiar *antioxidant* compounds to protect against oxygen damage. *Carotenoids* are vitamin A derivatives that protect chlorophyll. *Ascorbic acid* (vitamin C) is a water-soluble antioxidant believed to protect proteins on thylakoid membranes. Vitamin E, as a fat-soluble antioxidant, protects the lipid portions of thylakoid membranes, which oxygen free radicals can damage. Chloroplasts also contain antioxidant proteins, including superoxide dismutase, ascorbate peroxidase, and glutathione peroxidase. Figure 4.10b shows a peroxisome in a plant cell. A peroxisome is an organelle that contains enzymes that perform a variety of functions, including detoxification of activated forms of oxygen and cancer.

A plant's defense against oxygen damage goes to work in a particular sequence in times of stress, such as extreme weather or pestilence. During a drought, when the Calvin cycle is hampered because the stomata close to conserve water, the plant uses vitamin C reserves to forestall damage. Next, some plants increase production of vitamin E or glutathione peroxidase. These antioxidants can protect chloroplasts for about a week; after that, if the environmental threat hasn't lifted, the plant synthesizes other antioxidants such as superoxide dismutase. Using the several types of antioxidants are adaptations allowing short-term survival.

The same thing happens to your body.

Nature Made. The more you know, the better you'll feel.

FIGURE 1
The advertising industry has picked up on the role of antioxidants in promoting health.

Dramatic Visuals Program

Colorful, informative photographs and illustrations enhance the learning program of the text, as well as spark interest and discussion of important topics. This icon [■□] denotes figures that are included in the Life Science Animations videotapes/videodisc.

All figure legend titles and in-text figure references are printed in blue to facilitate students' understanding of the relationship between text and visuals.

Boldfaced Words

New terms appear in boldface print as they are introduced within the text and are immediately defined in context. They are also defined in the text glossary and are page-referenced.

Readings

Throughout *Life*, selected readings, called "Biology in Action," both elaborate and entertain. Some describe experiments; some provide health information; and others provide a closer look at specific topics. All readings are written by the author.

Chapter Summaries

At the end of each chapter is a summary. They are designed to help students more easily identify important concepts and better facilitate their learning of chapter concepts.

Review Questions

These end-of-chapter questions often continue the storytelling style of the chapter, using real anecdotes and experiments to illustrate and apply concepts.

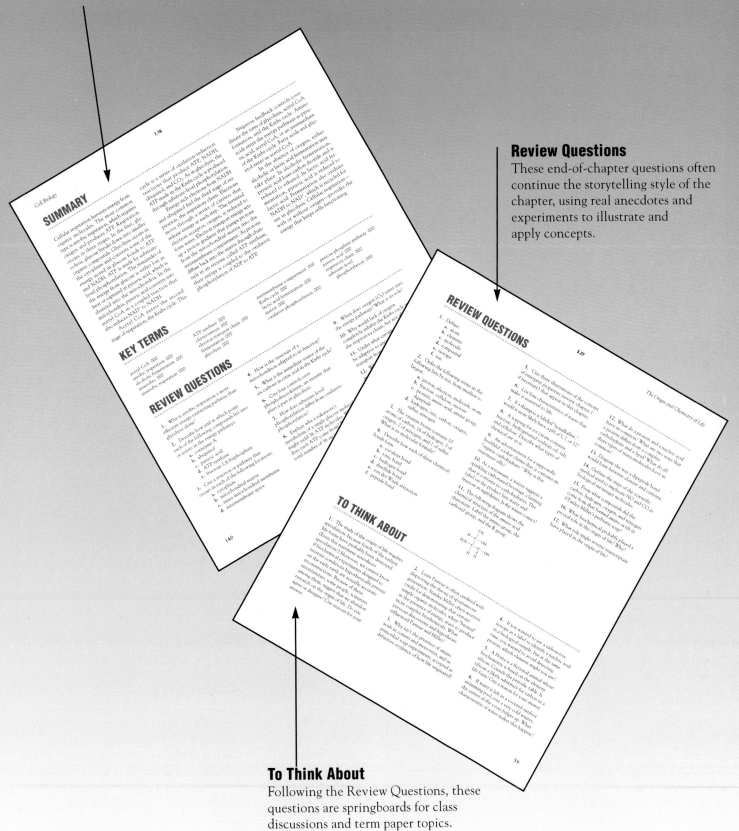

To Think About

Following the Review Questions, these questions are springboards for class discussions and term paper topics.

radiation sickness. The sudden massive doses of radiation they received affected the cells in their bodies that proliferate at the highest rates. What types of cells and tissues were most affected?

4. Cytochalasin B is a drug that blocks cytokinesis by disrupting the microfilaments in the contractile ring. What effect would this drug have on mitosis?

5. Tumor cells can often grow in culture, so that researchers can observe their response to experimental drugs. How might such a procedure benefit a cancer patient?

6. A researcher removes a tumor from a mouse and breaks it into cells. He injects each cell into a different mouse. Although all of the mice in the experiment are genetically identical and raised in the same environment, the animals develop cancers with different rates of metastasis. Some mice die quickly, some linger, others recover. What do these results indicate about the cells that comprised the original tumor?

7. Chronic myeloid leukemia is a white blood cell cancer almost always accompanied by a change in the chromosomes of these cells. The change places the tip of one type of chromosome onto another type of chromosome. This unusual arrangement moves an oncogene. Explain how the cancer likely arises.

8. Why would cancer developing in a stem cell in the basal layer of the skin's epidermis be potentially more harmful than cancer arising in a specialized cell above it?

SUGGESTED READINGS

Angier, Natalie. 1988. *Natural obsessions.* New York: Warner Books. How scientists unlocked the secrets of oncogenes.

Division of Cancer Communication, National Cancer Institute, National Institutes of Health, Bethesda, Maryland. 1–800–4–CANCER. You can obtain excellent information on all aspects of cancer by calling or writing this organization.

Gallagher, Gayl Lohse. March 1990. Evolutions: The mitotic spindle. *The Journal of NIH Research.* A short history, with beautiful illustrations, of what we know about the mitotic spindle.

Glover, David M., Cayetano Gonzalez, and Jordan W. Raff. June 1993. The centrosome. *Scientific American.* A complex interplay between microtubules and other cytoskeletal elements helps cells divide.

Harris, Curtis C. December 24, 1993. p 53: At the crossroads of molecular carcinogenesis and risk assessment. *Science,* vol. 262. *Science* magazine crowned p 53 the 1993 "molecule of the year."

Lewis, Ricki. June 1990. Wilms' tumor— The genetic plot thickens. *The Journal of NIH Research.* Pockets of embryonic tissue persisting in the kidney lead to cancer.

Lewis, Ricki. October 1990. Neurofibromatosis I revealed. *The Journal of NIH Research.* Benign tumors can result from loss of tumor suppressor genes, too.

Marx, Jean. May 7, 1993. New colon cancer gene discovered. *Science,* vol. 260. Some forms of colon cancer are caused by a cascade of gene action. Researchers have identified a gene that seems to control all the others and that may be responsible for many types of cancer.

Ohtsubo, Motoaki, and James M. Roberts. March 26, 1993. Cyclin-dependent regulation of G_1 in mammalian fibroblasts. *Science,* vol 259. The cell seems to "decide" whether to proceed through mitosis during G_1.

Pelech, Steven. July/August 1990. When cells divide. *The Sciences.* The history of our knowledge of the mitotic process and current thoughts on its pacemaker.

Science, vol. 259. January 29, 1993. This issue contains several excellent articles on breast cancer.

Travis, John. May 28, 1993. New tumor suppressor gene captured. *Science,* vol. 260. We only know of a dozen or so tumor suppressor genes. Some of them, when mutant, are responsible for a variety of cancer types.

EXPLORATIONS   Interactive Software

Smoking and Cancer

This interactive exercise allows the student to explore the relationship between smoking and cancer. The exercise presents a diagram of a human chromosome, showing the location of four genes that regulate cell growth. When their activities are disabled they actively promote growth. In this exercise, all four must be

disabled for cancer to be initiated. Students investigate the relationship between smoking and cancer by varying the amount an individual smokes and looking to see how long it takes before all four genes are mutated to a cancer-causing state.

1. What role does dose play in the probability that smoking will lead to cancer?

2. Can you discover a "safe" amount of smoking?

3. Is the 20th cigarette smoked in a day more or less dangerous than the 1st?

4. How much does smoking one pack of cigarettes a day increase the likelihood you will get cancer?

188

Suggested Readings

A list of readings at the end of each chapter suggests references that can be used for further study of topics covered in the chapter. The items listed in this section were carefully chosen by the author for readability and accessibility.

Explorations

References to the new WCB Explorations CD-ROM by George Johnson appear at the end of many chapters. Each exploration reference includes an explanation of an interactive exercise relevant to the chapter, as well as a list of questions to be answered after completing the Explorations module.

CONNECTIONS

Thoughts on Viruses

Nothing is quite as humbling as a virus. You get out of bed one morning, and even though you felt fine when you went to sleep the night before, you know instantly that all is not right. A flood of weakness washes over you as you try to stand, and you fall back into bed, aching all over. It's the flu. You know the misery will continue for four or five days.

Most of us have endured the discomfort of a cold or flu virus. For someone who has contracted a virus that causes hepatitis, the fatigue is far more long-lasting, restricting activities for weeks or even months.

What exactly is a virus, this microscopic entity that can so severely damage the human body? How can something so small harm a system so large? Understanding how viruses cripple their hosts has powerful practical implications—for example, it may help us cure AIDS. But viruses also contain clues that could unlock secrets about life itself—how it began, and exactly what it defines.

The complexity of a virus's structure depends on one's perspective. From a biologist's point of view, a virus is one type of chemical, clothed in a coat of two others. Although a virus has instructions to commandeer a host cell, it is not nearly as complex as

Yet another type of virus, the human immunodeficiency virus (HIV), devastates the human body. Researchers believe that HIV is a primary cause of acquired immune deficiency syndrome, also known as AIDS. Unlike the viruses that cause influenza and hepatitis, HIV's damage is irreversible and eventually fatal. The discomfort of the flu or hepatitis pales when compared to the psychological and physical anguish caused by AIDS.

even the simplest cell, the unit of life. Most biologists do not even consider a virus to be alive, dubbing it an "infectious agent." From a chemist's viewpoint, though, a virus is enormously more complex than a single biochemical.

Despite biologists' dubious view of viruses, the quest to understand one of them, HIV, has challenged and frustrated scientists in recent years. Investigations and experiments have ranged from molecular biologists' stripping the virus down to its component biochemical parts; to evolutionary biologists' comparing different strains of the virus to trace how and where it originated; to immunologists' probing the human body's response to HIV; to epidemiologists' racing to stop the infection's spread. AIDS research has extended well beyond the realm of science, touching sociology, law, politics, and economics.

The simple virus inspires sheer awe because of its power, yet it straddles the boundary between the living and the nonliving. Perhaps the viruses that pain us today—the influenza and hepatitis viruses, HIV, and many others—are legacies of a time long past, when chemicals carrying information first began to associate, forming the forerunners of living cells. Unlocking the secrets of the virus, and of the cell itself, is essential to understanding life.

2

Connections

These essays at the beginning of each unit tie chapter concepts together by telling a story related to the previous unit and introducing concepts that will be covered in the chapters to follow.

PREFACE

I have a confession to make—*Beginnings of Life* was *your* idea.

Back in 1990, when the first 15 chapters of my 40-chapter introductory biology text were shipped out for review, many of you thought that they stood alone as a nice, semester-sized chunk of biology. Following your lead, *Beginnings* was born, easily fitting its niche—a course built around a solid but selected core of biology, rich in concepts and applications, but not trekking through every phylum, organ system, or ecosystem.

Some of the topics in *Beginnings of Life* are those often regarded by nonscientists as the least comprehensible, if not anxiety-producing—chemistry, cells, energy, and genetics. But these topics are central to understanding and using modern biology. The presentation here will be, to many science-phobics, surprisingly familiar and engaging.

What's New

In response to numerous reviews, *Beginnings* the second time around has some changes in emphasis, but none in style.

Exciting reports on real scientists doing real science are peppered throughout, gleaned from my many interviews as a science journalist. In chapter 1, we meet a microbiologist who risks being boiled alive in a hot spring in Yellowstone National Park—all to find a particular microbe, which he indeed did in his thirteenth sample! In chapter 2, we travel with Katy Payne as she deciphers the sounds of elephants in Kenya, then tackle the medical mystery of "seabather's eruption." In chapter 15, we go for a midnight ride with Kary Mullis as he brainstorms the idea behind the polymerase

chain reaction, a way to mass-produce genes that is revolutionizing fields from anthropology to zoology.

Beginnings, 2/e, is also broader, covering a better balance of humans and other species. For example, a tour through "lost worlds" in chapter 1 contrasts life in a desolate Siberian lake to that in a busy sea bottom, a tropical rain forest, and elsewhere. But the chapter also explores the life-and-death issues of "real people." Chapter 2 underscores the similarity of life at the biochemical level, beginning with, "What do mice, sheep, humans, and bacteria have in common?"

Many "new and improved" figures also reflect this new sense of balance. Take a peek at the camouflaged adder snake of figure 1.3 or the fractured fetal skeleton in Biology in Action 13.1. The new figure 3.27 captures the important but difficult-to-visualize "RNA world" view of life's origin. In short, the illustrations are more representative and life-like throughout.

If It Ain't Broken—We Won't Fix It

The anecdotal, readable style that prompted reviewers to "invent" *Beginnings* in the first place remains the basic, distinguishing feature of the book. Simply put, a biology textbook will have no impact if it is too wordy, detailed, or dull for students, particularly nonscience majors, to read. *Beginnings* has redefined the concept of a textbook in this regard: "great stories" prevail.

Perhaps the greatest "story" of all—how life began—sandwiches the necessary but oft-dreaded chemistry of chapter 3. Another story—of gene discovery from a high school student's science fair project—began with his noticing the weird toes of

his relatives. From a woman who is both mother and grandmother to the same child, to an injured crab whose blue blood led to the development of a common medical test, to inventing ways to shuttle drugs into the brain, the stories of life will entice even the most reluctant students.

Organization

The organization of *Beginnings of Life,* 2/e, is essentially the same, but it reflects a slower, more studied approach to many major concepts.

Unit 1 Overview of Biology

Beginnings of Life, 2/e, still features two introductory chapters (chapter 1, What Is Life? and Chapter 2, Thinking Scientifically about Life), but the taxonomy from the first edition has been dropped, to better focus on the characteristics of life and the scientific method. Chapter 3, The Origin and Chemistry of Life, completes the introductory unit, as in the first edition.

Unit 2 Cell Biology

The reason behind the changes in chapter 4 (Cells) and chapter 5 (Cellular Architecture) was straightforward—update! Update! Update! Chapter 4 has greater emphasis on disease at the organelle level. Chapter 5 has a new section on cell-cell interactions, including signal transduction and cell adhesion. The evolutionary thread continues through consideration of the origin of eukaryotic cells.

Energy, the most fully-revised subject in the second edition, has been expanded to feature three chapters—The Energy of Life (chapter 6), Cellular Respiration (chapter 7), and Photosynthesis (chapter 8). Chapter 9 (Mitosis) remains a full discussion of cell division and cancer.

Unit 3 Reproduction and Development

This unit drives home the unity/diversity theme from unit 1. In chapter 10 (Meiosis) and chapter 11 (Animal Development), the reader learns about many variations on the developmental theme, examining organisms that range from aphids to zebrafish. For example, the "Two Beginnings" section of chapter 11 opens with the following example: "A millimeter-long worm, a fruit fly, and a human may not appear to have much in common. However, each organism meets the same challenges in embryonic and later development: it must develop from a single cell to an adult, with a variety of specialized cell types."

Chapter 12 is fascinating reading from a human perspective. It treats human reproduction and developmental problems, and includes many technology updates and (mainly) happy endings.

Unit 4 Genetics

Since writing *Beginnings,* I've authored another textbook, *Human Genetics: Concepts and Applications* (Wm. C. Brown Communications, 1994), seen hundreds of

patients as a genetic counselor, published numerous articles, and given many talks on genetic discoveries, technologies, and dilemmas. Reflecting the breakneck pace of this field and my own increased involvement in it, unit 4 has been reorganized into a more logical sequence of topics—one that, in fact, parallels history. The reader now proceeds from Transmission Genetics (chapter 13), to Chromosomes (chapter 14), to DNA Structure and Replication (chapter 15), to Gene Function (chapter 16), and, finally, to Genetic Technology (chapter 17); all are up-to-the-minute chapters that are easy to understand.

Pedagogy

Please compare the Review Questions and To Think About questions at the end of *Beginnings* chapters with their counterparts in other texts. Gleaned from the literature, interviews with scientists, and real clinical experiences, these questions often go well beyond the typical regurgitative recall question, asking students to apply chapter concepts to novel situations or to make informed ethical decisions. The Suggested Readings have also been chosen with care, eschewing the *Scientific American* references that litter other texts' lists in favor of references that students are likely to seek out

and read. This continues the accessibility that is the modus operandi of *Beginnings*.

Back by popular demand are Learning Objectives, Key Concepts, and Chapter Summaries. Added are Key Terms and a new feature, Connections, which conceptually links chapters within a unit to each other, and in some cases to other units, via an insightful, all-encompassing essay on an important focal topic.

Ancillaries

An exciting new ancillary is the *Student Study Art Notebook.* It is a lecture-time learning tool that contains all 183 full-color figures from the transparency package, with space for taking notes. It is packaged *free* with all new copies of this textbook.

The *Instructor's Manual,* prepared by Bob Ford of Ball State University, is helpful in preparing daily lectures or course outlines. Each chapter is correlated to *Beginnings of Life* and begins with a section called "Teaching Tips," which offers suggestions for teaching, as well as discussion questions. Following are student learning objectives, key concepts, key words, a chapter outline, answers to both sets of end-of-chapter questions, and a list of audiovisual materials. There are also an additional 25 to 40 objective questions in a *Test Item File* in the second part of the manual.

Summary of Content Changes

Unit	First Edition	Second Edition
1 Overview of Biology	Two introductory chapters: "Thinking Scientifically" and "The Diversity of Life."	Two introductory chapters: "What Is Life?" and "Thinking Scientifically about Life." Diversity has been dropped.
2 Cell Biology	One chapter on Energy—chapter 6.	Energy expanded to three complete chapters: "The Energy of Life," "Cellular Respiration," and "Photosynthesis" (chapters 6–8).
3 Reproduction and Development	Three chapters with a heavy human emphasis.	More diverse examples in "Animal Development" (chapter 11).
4 Genetics	Five chapters beginning with "Mendel's Laws" (chapter 11) and moving to "Genetic Disease" (chapter 15).	Completely updated and reorganized, beginning with "Transmission Genetics" and moving to "Chromosomes," "DNA Structure and Replication," "Gene Function," and ending with "Genetic Technology" (chapters 13–17).

A *Student Study Guide* is also available, written by Donald Breakwell and Allan Stevens of Snow College, to correlate closely with the text in offering students the opportunity to test their comprehension of basic and difficult concepts.

Instructors can also request WCB *Micro Test III*, computerized testing software that provides a data base of objective questions. Disks are available for IBM, Apple, and Macintosh PC computers and require no programming experience. If a computer is not available, instructors can choose questions from the *Test Item File* and phone/FAX their request for a printed exam.

In addition, 183 overhead *transparencies* that feature key illustrations from the text are available free to adopters.

Explorations in Human Biology CD-ROM

The Explorations in Human Biology CD-ROM is a series of hands-on, interactive activities that can be used by an instructor in lecture and/or placed in a resource center for student use or for individual student purchase. The interactive CD-ROM consists of 16 different modules that cover key topics discussed in the text, in both Macintosh and IBM Windows formats. Interactive questions at the end of many text chapters are referenced to the appropriate interactive module. The questions can be answered without the software, using critical thinking skills, but the software truly involves the student in the problemsolving process.

The Life Science Animations Videotapes/Videodisc

Illustrations taken from Wm. C. Brown Publishers' library of life science texts are brought to life through animation. Approximately 60 animations have been placed on videotape. They illustrate various physiological processes such as conduction of an action potential, synaptic transmission, electron transport and oxidative phosphorylation, and viral replication of HIV. These full-color animations enable the student to more fully and easily grasp complex physiological processes. They are free to qualified adopters.

About the Author

Ricki Lewis is Adjunct Assistant Professor at The University at Albany where she has taught human genetics, physiology, bioethics, and nutrition. An accomplished science journalist, she has had more than 1,000 articles published since 1982 and today writes for *The Scientist, BioScience, Genetic Engineering News, FDA Consumer,* and other publications. She is also a genetic counselor and author of a new Wm. C. Brown text, *Human Genetics: Concepts and Applications.* She received her Ph.D. in genetics from Indiana University in 1980.

Summary of Ancillary Package Changes

Supplement	Changes in the Second Edition
Instructor's Manual	Completely revised to include "Teaching Tips" and more extensive chapter outlines. The test item file has been almost 100% revised and contains 30–40 questions per chapter.
MicroTest III	A new computerized testing program. The test items and their answers appear in the Instructor's Manual as well as on disk available in DOS 3.5, 5.2, Mac, and Windows versions.
Student Study Guide	Completely revised. Key additions include a section called "Study Questions" containing a variety of exercises that test students' understanding. Also new to the second edition are crossword puzzles.
Transparencies	Now available are 183 full-color transparencies of key figures in the text, increased from 50 in the first edition.
Student Study Art Notebook	A new and innovative supplement that includes all 183 images from the transparency set in a notebook format. Packaged *free* with each new copy of *Beginnings* purchased from WCB.
Technology Products	Two new technology products directly tied to the text are Explorations in Human Biology CD-ROM and Life Science Animations videotapes/videodisc.

ACKNOWLEDGMENTS

A textbook, particularly a second edition, is the culmination of work by an enormous number of individuals. Most important are the adopters, readers, and reviewers of the first edition, who saw my vision and helped mold it into the even better second edition. Much credit goes to WCB vice president and publisher Kevin Kane, for hearing what reviewers were saying back at the very beginning of *Beginnings* and pursuing the idea of a non-exhaustive, short course that *Beginnings* represents. Editors Carol Mills, Diane Beausoleil, and Carla Kipper did a terrific job gestating this book. A special thanks to Anne Cody, copy editor extraordinaire, who greatly improved the clarity and logic of my writing. Special thanks also to Wayne Becker for his expertise in helping revamp the cell biology unit, to Randy Moore for his help with the energy chapters, and to Steve Miller for his help with the introductory chapters.

Finally, many, many thanks to my wonderful family and our various beasts (domestic) for supporting me through the craziness of writing a textbook.

Reviewers for Lewis *Beginnings of Life*

Many instructors have contributed to the making of this text and its teaching package. We gratefully acknowledge the assistance of the many reviewers whose constructive criticism proved invaluable to the development of this text.

Barbara J. Abraham
Hampton University

Sylvester Allred
Northern Arizona University

Jane Aloi
Saddleback College

Steven Bassett
Southeast Community College

Prem P. Batra
Wright State University

Wayne M. Becker
University of Wisconsin—Madison

Todd M. Bennethum
Purdue University

James L. Botsford
New Mexico State University

Margaret H. Bowker
Central Michigan University

John S. Choinski, Jr.
University of Central Arkansas

William S. Cohen
University of Kentucky

Christopher M. Comer
University of Illinois—Chicago

Clara Cotten
Indiana University—Bloomington

Judy M. Dacus
Frederick Community College

Marianne Dauwalder
University of Texas—Austin

Charles F. Denny
University of South Carolina—Sumter

Charles D. Drewes
Iowa State University

Larry Eckroat
Penn State University—Erie

Cory Etchberger
Penn State University—Berks Campus

Bruce Evans
Huntington College

Colleen T. Fogarty
St. John's University

Karen M. Fulford
Morris Brown College

Patricia A. Grove
College of Mount St. Vincent

Alan H. Haber
SUNY at Binghamton

Deborah D. Hettinger
Texas Lutheran College

Theresa M. Hornstein
Duluth Community College

Herbert W. House
Elon College

George A. Hudock
Indiana University

Debra A. Kirchhof-Glazier
Juniata College

Margarita Irizarry-Ramirez
Universidad Metropolitana

George M. Labanick
University of South Carolina—Spartanburg

Anton E. Lawson
Arizona State University

Mary L. Leida
Morningside College

Yue J. Lin
St. John's University

Marlene McCall
Community College of Allegheny County

Karen E. Messley
Rock Valley College

Daryl G. Miller
Broward Community College

Margaret H. Peaslee
University of Pittsburgh at Titusville

James W. Phillips
Westfield State College

Ramon Pinon, Jr.
University of California—San Diego

Stanley Rice
Southwest State University

John D. Rossmiller
Wright State University

David Sadava
The Claremont Colleges

Phillip Sheeler
California State University—Northridge

Jerry M. Skinner
Keystone Junior College

Eric Strauss
University of Massachusetts at Boston

Ross Strayer
Washtenaw Community College

Philip Stukus
Denison University

Linwood Ira Swain, Jr.
Craven Community College

Robert H. Tamarin
Boston University

Henry Tedeschi
The University at Albany

Jimmy. B. Throneberry
University of Central Arkansas

Robin W. Tyser
University of Wisconsin—LaCrosse

Jack Waber
West Chester University

Conrad Weiler
Santa Barbara City College

Roberta B. Williams
University of Nevada—Las Vegas

Kevin W. Winterling
University of Maryland—Baltimore County

UNIT

1

Overview
of
Biology

Many biologists trace their initial interest in the science of life to observing a spectacular display of nature—such as a deer against a leafy backdrop.

CONNECTIONS

Thoughts on Viruses

Nothing is quite as humbling as a virus. You get out of bed one morning, and even though you felt fine when you went to sleep the night before, you know instantly that all is not right. A flood of weakness washes over you as you try to stand, and you fall back into bed, aching all over. It's the flu. You know the misery will continue for four or five days.

Most of us have endured the discomfort of a cold or flu virus. For someone who has contracted a virus that causes hepatitis, the fatigue is far more long-lasting, restricting activities for weeks or even months.

Yet another type of virus, the human immunodeficiency virus (HIV), devastates the human body. Researchers believe that HIV is a primary cause of acquired immune deficiency syndrome, also known as AIDS. Unlike the viruses that cause influenza and hepatitis, HIV's damage is irreversible and eventually fatal. The discomfort of the flu or hepatitis pales when compared to the psychological and physical anguish caused by AIDS.

What exactly is a virus, this microscopic entity that can so severely damage the human body? How can something so small harm a system so large? Understanding how viruses cripple their hosts has powerful practical implications—for example, it may help us cure AIDS. But viruses also contain clues that could unlock secrets about life itself—how it began, and exactly what it defines.

The complexity of a virus's structure depends on one's perspective. From a biologist's point of view, a virus is one type of chemical, clothed in a coat of two others. Although a virus has instructions to commandeer a host cell, it is not nearly as complex as

even the simplest cell, the unit of life. Most biologists do not even consider a virus to be alive, dubbing it an "infectious agent." From a chemist's viewpoint, though, a virus is enormously more complex than a single biochemical.

Despite biologists' dubious view of viruses, the quest to understand one of them, HIV, has challenged and frustrated scientists in recent years. Investigations and experiments have ranged from molecular biologists' stripping the virus down to its component biochemical parts; to evolutionary biologists' comparing different strains of the virus to trace how and where it originated; to immunologists' probing the human body's response to HIV; to epidemiologists' racing to stop the infection's spread. AIDS research has extended well beyond the realm of science, touching sociology, law, politics, and economics.

The simple virus inspires sheer awe because of its power, yet it straddles the boundary between the living and the nonliving. Perhaps the viruses that pain us today—the influenza and hepatitis viruses, HIV, and many others—are legacies of a time long past, when chemicals carrying information first began to associate, forming the forerunners of living cells. Unlocking the secrets of the virus, and of the cell itself, is essential to understanding life.

CHAPTER 1

What Is Life?

Marble salamanders, in Carolina Bays.

LEARNING OBJECTIVES

By the chapter's end, you should be able to answer these questions:

1. Why is it difficult to define death?

2. What are some of the ways people have tried to define life?

3. What combination of characterisics distinguishes the living from the nonliving?

4. How can living things be very much like one another, yet diverse?

5. Why does life change over time?

6. Why is evolution so important in biology?

On the Edge of Life and Death

On a bright winter day a few years ago, young Jimmy Tontlewicz raced his sled onto the brittle surface of barely frozen Lake Michigan, plunging into its icy waters. Twenty minutes later, he was rescued. With no detectable heartbeat, pulse, or breathing, he appeared to be dead. But Jimmy received emergency care as he was rushed to the hospital, and within an hour, the characteristics of life began to reappear. The initial blast of cold water had apparently so shocked his body that his heart and lungs stopped functioning, while his other vital organs continued to function at undetectable levels. Fortunately, this was enough to keep him alive until his heartbeat and breathing were restored. Jimmy was saved by a protective biological response called the mammalian diving reflex, seen only in children and other young mammals.

The Definition of Death

The true story of Jimmy Tontlewicz illustrates how difficult it can be to define death—and, for that matter, life. Until recently, the definition of death seemed obvious and uncontroversial—life ceased when the heart stopped beating and the lungs stopped inflating and deflating. But with the development of heart transplants in the late sixties, this definition became obsolete. How could a heart donor be considered dead if his or her heart continued to beat when it was removed for transplantation into another body? A few years later, the problem of defining life's endpoint was further complicated by the invention of the respirator, a machine that artificially maintains breathing.

At present, death in humans is defined as the cessation of brain activity. Specifically, the criteria for "brain death" include the lack of observable response to any external stimulus; no spontaneous movement, such as that of the heart or breathing muscles; no reflexes (such as blinking or pupil constriction when a bright light is shone into the eye); no swallowing, yawning, or vocalization; and a "flat record" of brain-wave patterns as measured by an electroencephalograph (EEG) machine. Furthermore, the lack of movement and breathing without the aid of a respirator must be observed by more than one physician for at least an hour, and the flat brain-wave pattern must be witnessed a second time a day later.

The definition of death as cessation of brain activity is suitable for humans in a hospital setting—but is it applicable to the at least 30 million other types of living species? When, for example, is a mushroom dead? or a liverwort? Death in bacteria may be defined as the inability to reproduce, but that certainly doesn't define death in mammals. It isn't easy to establish which biological functions must cease to cause death, because to do so is really to ask the broader question, WHAT IS LIFE? How do we define it? What combination of activities distinguishes the living from the nonliving?

What Is Life?

We all have an intuitive sense of what life is: if we see a rabbit on a rock, we "know" that the rabbit is alive and the rock is not. But it's difficult to state just what it is that makes the rabbit a living thing, and the rock an inanimate object. Most scientists characterize life in terms of qualities that are uniquely shared among the living. But in the second before a newborn baby takes its first breath—or after a person dies—we may ask, as the nineteenth-century thinkers called vitalists did, whether there isn't some indefinable "essence" that renders a particular collection of structures or processes living? It is this question that lies at the heart of Jimmy Tontlewicz's near-death experience, as well as at the core of such headline-grabbing controversies as abortion rights and the right to die (fig. 1.1).

Throughout history, thinkers in many fields have been stymied by the question, "What is life?" Eighteenth-century French physician Marie Francois Xavier Bichat, credited with founding the field of histology (the study of tissues at the microscopic level), poetically but imprecisely defined life as "the ensemble of functions that resist death."

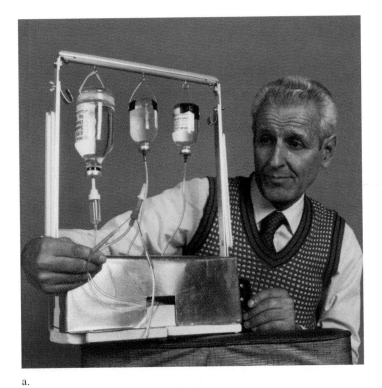

a.

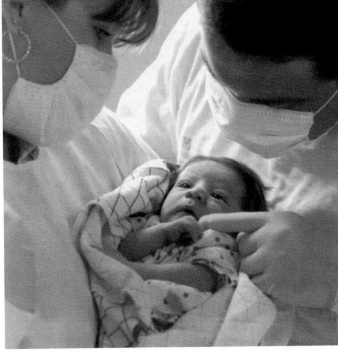

b.

FIGURE 1.1

Life and death. An issue of *Time* magazine illustrates our intense interest in the beginning and end of life. The cover story is about Dr. Jack Kevorkian, a controversial physician who helps very ill people commit suicide, pictured here with a suicide device (*a*). Inside the issue is the story of Andrew Gobea (*b*), a newborn genetically doomed to die in childhood of immune system failure. Andrew was saved by replacing the defective gene in blood cells in his umbilical cord and placing the genetically altered blood cells in his bloodstream.

Others less eloquent and still no more precise have hypothesized that life is a kind of "black box" that endows an assemblage of biochemicals with the qualities of life. Still others have tried to break life down into the smallest parts that still exhibit the characteristics of life, then to identify what it is that makes those units different from their nonliving components.

The Characteristics of Life

As mysterious as life is, most scientists agree on at least some of its basic characteristics. It is the combination of five of these qualities (listed in table 1.1) that seems to constitute life, because many of them, alone, are also characteristic of the nonliving. For example, a computer responds to stimuli, as does a self-flushing toilet, but neither is alive. A fork placed in a pot of boiling water will absorb heat energy and pass it on to the human hand grabbing it, just as a living organism receives and transfers energy, but this does not make the fork alive. And a supermarket is highly organized, but it is obviously not alive!

Organization

Living matter consists of structures arranged in a particular three-dimensional relationship, often following a pattern of structures within structures within structures. Bichat was the first to notice this pattern in the human body. During the bloody French revolution, as he performed autopsies, Bichat noticed that the body's organs were built of simpler structures, which he named **tissues** (based on the French word for "very thin"). Had he used a microscope, Bichat

Table 1.1	**Characteristics of Life**
Organization	
Energy use and metabolism	
Maintenance of internal constancy	
Irritability and adaptation	
Reproduction, growth, and development	

would have seen that tissues themselves are comprised of even smaller units called **cells.** Within the cells are structures called **organelles** that carry out specific functions. Organelles, and ultimately all living structures, are composed of chemicals (fig. 1.2). Indeed it is this "levels of organization" approach (in reverse, moving from small to large) that forms the organizational scheme of most modern biology textbooks, including this one.

A living thing, however, is more than just a collection of smaller parts. Just as a house vanishes when it has been demolished into scattered bricks, the organization of different levels of life imparts distinct characteristics. For example, individual chemicals cannot harness energy from sunlight. But groups of chemicals embedded in the membranes of an organelle within a plant cell use solar energy to synthesize nutrients. Qualities such as this, which occur as

biological complexity grows, are called **emergent properties.** These characteristics are not magical; they arise from properties common to all matter.

Organs, tissues, cells, and organelles are highly organized and carry out specific functions. At all levels, and in all organisms, structural organization is closely tied to function. Disrupt the structural plan, and proper function ceases. If, for example, a fertilized hen's egg is shaken briskly, the embryo will not develop. Likewise, if function is disrupted, a structure will eventually break down. Unused muscles, for example, begin to atrophy (waste away)—a phenomenon familiar to anyone who has ever had a broken limb immobilized for weeks in a cast. Biological function and form, then, are interdependent.

Biological organization is apparent in all life. Though they are outwardly very different, humans, eels, and evergreens are all nonetheless organized into specialized organ systems, organs, tissues, and cells. Bacteria, although less complex than animal or plant cells, are also composed of highly organized structures.

Energy Flow

The organization of life seems contrary to the natural tendency of matter described by the **second law of thermodynamics.** This law, derived from physics, states that matter tends toward randomness or disorder, a state called **entropy.** In other words, **energy** is required to maintain organization or order. A house cleaned on Sunday that grows progressively messier each day of the week is analogous to entropy. Just as a house does not regain its organization

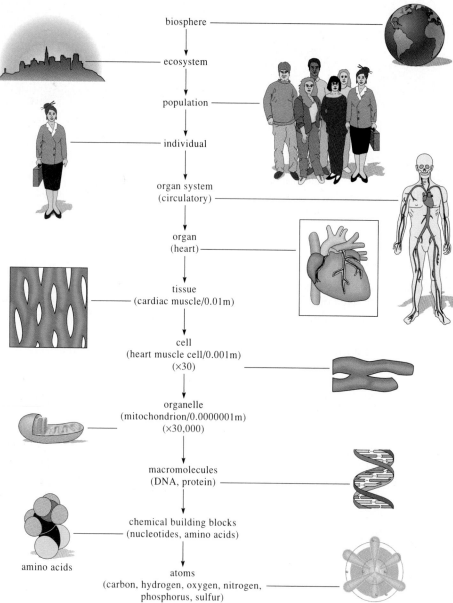

biosphere

ecosystem

population

individual

organ system
(circulatory)

organ
(heart)

tissue
(cardiac muscle/0.01m)

cell
(heart muscle cell/0.001m)
(×30)

organelle
(mitochondrion/0.0000001m)
(×30,000)

macromolecules
(DNA, protein)

chemical building blocks
(nucleotides, amino acids)

amino acids

atoms
(carbon, hydrogen, oxygen, nitrogen,
phosphorus, sulfur)

carbon

FIGURE 1.2

Levels of biological organization from the smallest components of living things, atoms, to the entire living planet, the biosphere. Many biology textbooks, including this one, follow this organization, studying life forms from small to large.

without some custodial input of energy, living things must expend energy to maintain their highly organized state.

A living system must be able to acquire and use energy to build new structures, repair or break down old ones, and reproduce. These functions occur at the whole-body or organismal level, as well as at the organ, tissue, and cellular levels. The chemical reactions within cells that direct this acquisition and use of energy, both building up and breaking down chemicals, are collectively called **metabolism.**

Maintenance of Internal Constancy

A living organism is composed of the same basic chemical elements as nonliving matter yet is obviously quite different. The chemical environment inside cells must remain within specific ranges, even in the face of drastic changes in the outside environment, to stay alive. This ability to maintain chemical constancy is called **homeostasis.** A cell must maintain a certain temperature and water balance. It must take in nutrients, rid itself of wastes, and regulate its many chemical reactions to prevent deficiencies or excesses of specific chemicals. An important characteristic of life, then, is the ability to sense and react to environmental change, keeping conditions within cells constantly compatible with being alive.

Reacting to Environmental Change—Irritability and Adaptation

Living organisms sense and respond to certain environmental stimuli while ignoring others. The tendency to respond immediately to stimuli, **irritability,** can be essential for survival. A woman touching a thorn quickly pulls her hand away; a dog lifts its ears in response to a whistle; a plant grows toward the sun. All are examples of irritability.

Whereas irritability is an immediate response to a stimulus, an **adaptation**

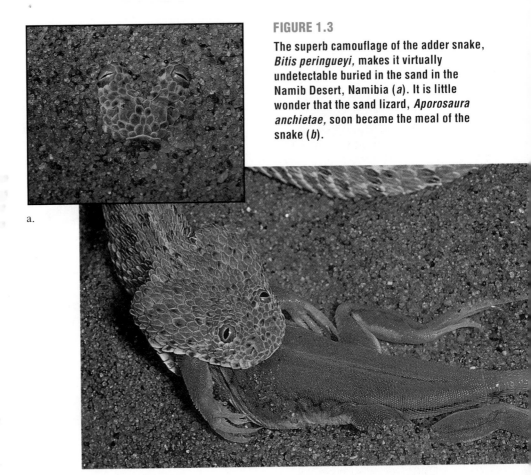

FIGURE 1.3

The superb camouflage of the adder snake, *Bitis peringueyi,* makes it virtually undetectable buried in the sand in the Namib Desert, Namibia (*a*). It is little wonder that the sand lizard, *Aporosaura anchietae,* soon became the meal of the snake (*b*).

a.

b.

is an inherited characteristic or behavior that enables an organism to successfully reproduce in a given environment. Adaptation is a response developed over time, and because of this, it is usually considered more complex than irritability. Adaptations are likely to differ from one type of organism (or species) to another, and within the same species in different environments. Both a human and a dog will pull an extremity away from fire (irritability). But on a hot summer day, the human sweats, while the dog pants (adaptations). The ability to survive a temperature extreme allows an individual to successfully reproduce.

Adaptations can be very striking. Some species of snake, for example, will literally fade into their environmental background, thanks to an adaptation that allows them to hide from their prey (fig. 1.3). But other adaptations may be

difficult for a human observer to detect. Most trees, for example, have developed several rather obvious means of surviving strong winds, including a wide base, sturdy trunk, flexible branches that sway without snapping, and strong, well-spread roots. But figure 1.4 shows another kind of adaptation to wind not readily observed in the violence of a storm. A researcher who noticed that the fronds of certain algae close into cones and cylinder shapes to survive in turbulent streams studied several types of trees during simulated "wind tunnel storms" to see if similar reactions might occur. Indeed, he found that the leaves of some types of trees do close into tighter shapes to minimize wind damage.

Adaptations accumulate in a population of organisms when individuals with certain inherited traits are more likely than others to survive in a

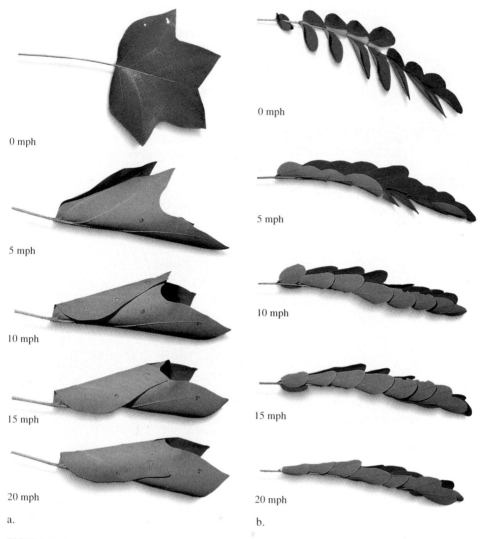

0 mph

0 mph

5 mph

5 mph

10 mph

10 mph

15 mph

15 mph

20 mph

20 mph

a.

b.

FIGURE 1.4

Leaves of many trees are adapted to minimize wind damage. Researcher Steven Vogel subjected various leaves to a wind tunnel and snapped these photos, which reveal a previously unrealized adaptation to survive a storm. Leaves characteristically fold into a more compact form—a cone for a tulip tree (*a*) and a cylinder for a black locust tree (*b*).

particular environment. Trees with leaves that curl up to resist wind are more likely to survive for the twenty or more years it takes for most trees to reproduce—even in the face of severe wind storms that strike, on average, every five years. Because trees with curling leaves survive more often than trees without this adaptation, more of them pass the advantageous trait to future generations. Trees with the adaptation eventually predominate the population—unless environmental conditions change, and protection against the wind is no longer important. This process,

when individuals with different traits survive at different rates, is called **natural selection.** It is a major force behind evolution, or biological change over time. Natural selection may explain why fewer than 1% of the life forms that have ever existed on earth are alive today.

Reproduction, Growth, and Development

Reproduction and development can be as simple as the dividing of a bacterium in two, or as complex as the germination and growth of a tree seedling or the con-

ception and birth of a mammal (fig. 1.5). Reproduction is vital if a population of organisms is to survive for more than one generation, and essential for life itself to continue.

On the organismal level, reproduction is passing on biochemical instructions (genes) for developing the defining characteristics of the organism from one generation to the next. Reproduction can be asexual or sexual. In **asexual reproduction** in single-celled organisms, cellular contents double; then the cell divides to form two new individuals genetically identical to the original parent cell. Some multicellular (many-celled) organisms reproduce asexually. For example, a potato growing on an underground stem can sprout leaves and roots that constitute an entirely new plant. Asexual reproduction also occurs in some simple animals, such as sponges and sea anemones, when a fragment of the parent animal detaches and develops into a new individual.

In **sexual reproduction,** genetic material from two individuals combines to begin the life of a third individual who has a new combination of inherited traits. Can you see why sexual reproduction would generate greater diversity than asexual reproduction?

KEY CONCEPTS

All living things share a common set of five characteristics: organization; use of energy and matter; maintenance of organization and internal chemical constancy; irritability and adaptation; and reproduction.

Evolution Explains the Unity and Diversity of Life

While all living things share common characteristics, they are also incredibly diverse. Biologists who study life's many forms—students of **biodiversity**—are continually astounded at the forbidding areas of the earth that are nevertheless home to thriving species. One such place

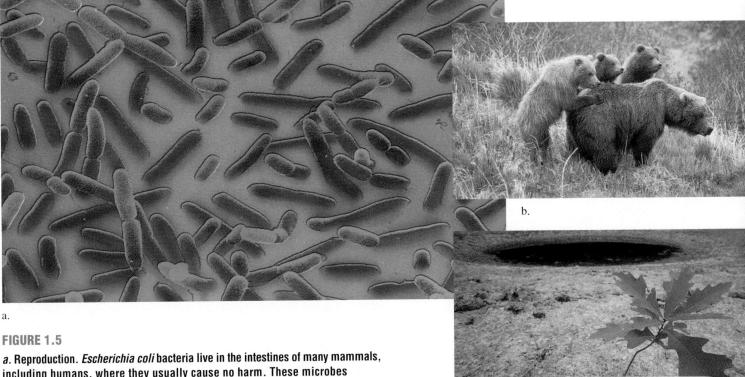

FIGURE 1.5

a. Reproduction. *Escherichia coli* bacteria live in the intestines of many mammals, including humans, where they usually cause no harm. These microbes reproduce every 20 to 30 minutes. *b.* A bear gives birth to two or three cubs during the winter. She awakens briefly from her hibernation to deliver the cubs, then goes back to sleep for two more months, while they nurse. *c.* Many more oak seedlings, such as this one, sprout than survive to grow into mature trees.

is the steaming water of a near-boiling hot spring (see Biology in Action 1.1).

Biodiversity is not only apparent in different places, but over time. For example, at the very spot on the planet you call home—currently occupied by humans, perhaps cats and dogs, certain microorganisms, birds and insects, and grass and trees—giant reptiles may have roamed 100 million years ago. Half a billion years ago, the same spot may have lain beneath an ocean, home to a very different assortment of living things.

Natural selection, in guiding the gradual accumulation of particular traits in a population, to an extent constrains evolutionary change because it weeds out the changes that do not enhance the organism's ability to successfully reproduce. But natural selection also allows for much variation, because different types of organisms develop different adaptations. Figure 1.6 illustrates different solutions or adaptations to the requirement to see.

Evolution is a continuing tale that weaves together the disparate threads of diverse adaptation and shared ancestry, connecting the biodiversity of past ages with today's. It forms the very basis of the study of life, biology. All earth's inhabitants are, in a sense, related: a tuna and a trout obviously so, because they share a fishlike ancestor. But also related are the tuna, a human, and a frog, in that they share a common vertebrate ancestor—some animal with a backbone that gave rise to all three modern vertebrates. The tuna and an oak tree are related even farther back, because they each descend from a mass of cells that was the first multicellular organism. All multicellular organisms, in turn, descend from living things that were single, complex cells. The specialized tuna, trout, human, frog, and oak, and even the single-celled amoeba, evolved as genetic changes accumulated in populations of their distant ancestors.

KEY CONCEPTS

Evolutionary change, brought about largely by natural selection, explains how all organisms past and present are related to one another.

Lost Worlds

To a biologist, perhaps nothing matches the excitement of discovering an organism never before seen by human eyes, or coming across a community of living things where life was not thought to exist. Such "lost worlds" reveal amazing adaptations to very extreme environments—amazing to us, at least. The term "lost worlds," used here to designate unfamiliar geographical areas, reflects only our human perspective; to a termite on the floor of a tropical rain forest, Los Angeles would be a lost world!

Some Like It Hot: Thermophilic Bacteria

"I still have nightmares about being boiled alive," says Francis Barany, a biochemist at Cornell University Medical College in New York City. Barany fell into a pool of boiling mud in Yellowstone Park in 1991 while searching for a bacterium. His story illustrates vividly the excitement of discovery—although his fall was frightening, Barany gained the opportunity to simultaneously explore biodiversity, clues to the origin of life, and a brand new biotechnology.

Yellowstone National Park has long held a fascination for biologists, who flock not to its tourist attractions, but to its pools of steaming water. Topped with colored mats of algae in which a few types of thermophilic (heat-loving) bacteria thrive (fig. 1), hot springs are found today only near volcanoes and lakes. But long ago, as life began, such hot environments were the norm. Thus, biologists come to places like Yellowstone, hoping that by studying how organisms live in pockets of the environment that resemble the conditions of early earth, they might discover clues to how life itself began.

Barany, and many others, study thermophilic bacteria for another reason—the microbes' biochemicals can withstand great heat. Because heat dismantles most enzymes, those from the hot springs bacteria are valuable in the laboratory, just as a detergent that launders clothes in very hot water is more useful than one that falls apart at high temperatures. The first "thermostable" enzyme used in the lab—to mass produce genetic material—came from a bacterium called *Thermus aquaticus*, which was discovered by Indiana University

FIGURE 1

The water in this hot spring in Yellowstone Park is boiling—but certain bacteria thrive in it.

microbiologist Thomas Brock in 1965. In Yellowstone's hot springs, Brock dipped glass slides several meters into the water, pulling them up a few days later. The slides were covered with bacteria. The value of using the microbe's DNA-copying enzyme was realized in 1987.

Barany ventured into the hot springs in search of an enzyme for a new gene-copying method. Here is his story:

It was evening, and we had been collecting so well that we thought we might as well finish, and the next day relax. We ended the day at Artist's Paint Pot, a hot springs and mud hole where the water had dried out, leaving hot mud. Hot springs are just beautiful, deep blue in the interior, the

surface yellow-brown, where the bacteria are. It was constantly steaming, with bubbling water.

While collecting my thirteenth sample, the ground simply gave way, collapsed under me, and my left leg sank into the hot mud. It was 70° C; that's 158° F! It blistered the skin right off. I jumped out, and hopped onto land, screaming. Someone poured cold water on my leg, ripped my jeans off. I remember screaming that I had a permit to collect, and to save that thirteenth sample!

A few months later, Barany was still hobbling about, taking pain medication—but feeling healed enough to marry the woman who pulled him out of the hot springs.

Our look at these intriguing slices of the living world sets the stage for the topic of the next chapter—how biologists study this hard-to-define but wonderful condition called life.

The Sea Bottom

Perhaps nowhere is our human idea of life as restricted as it is concerning the sea. However, life began in the sea, and this may explain why life beneath the

waves today is far more diverse than life in even heavily populated rain forests—habitats so diverse we cannot even estimate their biological wealth. The land is home to 11 **phyla** (major groups) of

FIGURE 1.6

Diversity in visual systems. All visual systems detect light and transmit the information to the brain, where an image is formed and perceived. The cameralike eyes of most vertebrates—including our own—focus light through a lens onto photoreceptor cells at the back of the eye, which detect the light. In contrast is the pinhole eye, unique to the chambered nautilus (*a*), a type of mollusc. Its eye is merely a hole that varies in diameter from 0.4 to 2.8 millimeters, allowing light to impinge upon photoreceptor cells at the back of the visual organ. The pinhole eye is rare, probably because it lacks a lens and thus does not provide a very clear image. The shrimp *Rimicaris exoculata* (*b*) inhabits deep sea thermal vents. Its "eyes" are two light-reflecting patches on its back, each consisting of 1500 clusters of photoreceptor cells packed with rhodopsin, a pigment that detects light in many types of organisms. The shrimps' eyes do not transmit images, but they sense the direction, strength, and size of a light source.

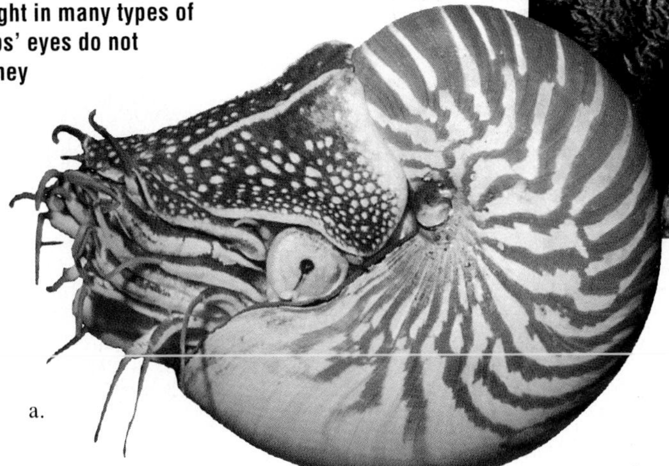

a.

b.

organisms, only one of which is restricted to the land. In contrast, the sea is home to 28 phyla, 13 of which are found only in sea water. Moreover, our view of ocean biodiversity reflects the limits of technology; biologists estimate that we have catalogued only one-tenth of one percent of the types of organisms living in the oceans!

We know more about life on land simply because the earth's surface is accessible. But life on the surface, though abundant, forms merely a film compared to that in the depths of the oceans. The residents of the sea are so much on the move that even if we sample a small portion of this vast world, we gain only a short-lived, dynamic glimpse.

Researchers at the Institute of Marine and Coastal Sciences at Rutgers University in New Jersey witnessed the incredible richness of life in the ocean when they repeatedly sampled the abyssal plain, in an experimental area the size of two tennis courts off the Atlantic coast. In the lowest portion of this marine environment, at a depth of 4,880 meters (16,000 feet), the water is

extremely cold; food from decaying organisms continually rains down. Each experiment brought to the surface a different collection of organisms, indicating tremendous movement of life in the sea. So abundant was life there that after the collections were transferred to laboratories, the researchers found they had sampled 90,677 individual organisms representing 798 species—460 of which were previously undiscovered!

A Siberian Lake

More than three million years ago, a meteorite crashing to earth gouged out White Lake in what is now the frozen tundra of Siberia. So deep was the crater that when glaciers tore through the area much later, it was not altered, as shallower lakes were. Today, for ten months of the year, several feet of ice cover White Lake; water breaking the surface is the only sign of the short summer. The lake is 183 meters (600 feet) deep and 8 miles (about 13 kilometers) in diame-

ter, with an average temperature of 1.1° C (34° F) at the surface and 2.5° C (36.5° F) below.

As harsh as White Lake may seem to us, it is home to a few cold-hardened species. These species form an **ecosystem** (an interconnected place and its residents) in which large fishes feed on smaller fishes, which eat microscopic animals called zooplankton, which eat algae. White Lake is also the occasional home of biologist Mikhail Skopets. He was first drawn to the area in 1979, after learning that other biologists had seen previously unknown fishes there—specifically, a type of salmonid called char. Skopets went fishing at the frozen lake. In 1981, he discovered a char and named it *Salvelinus elgyticus*, after the local name of the lake, E l'gygytgyn.

As part of his study, Skopets examined the contents of the fishes' stomachs. The animals had to eat something, and that something must also live in the lake. Would he find never-before-seen organisms? From previous experiments on other fishes, Skopets expected to find dined-upon zooplankton, and he did. But he also found decomposing corpses of a

a.

b.

c.

d.

smaller, unknown fish. Reasoning that the smaller fishes were bottom-dwellers, Skopets's next experiment was to travel to the lake's center by raft, let nets down to a depth of 61 meters (200 feet), and sample the life there. He hauled up a net full and happily began the task of learning all he could about an animal human eyes had never before seen.

Skopets named his find *Salvethymus svetovidovi,* meaning "long-fin char." He even determined the age of one foot-long specimen by counting the "annual rings" on stones called otoliths in its ears. This standard way to tell a fish's age is similar to dating a tree by its pattern of rings. The fish was 20 years old. The long-fin char, unknown for so long, actually was an abundant White Lake resident.

The Ndoki Region of the Northern Congo

Sandwiched between the Ndoki River to the west, swamps to the east and south, and hills to the north, lie 7.5 million acres (one million hectares) where the inhabitants have never seen humans. In May 1992, three western investigators accompanied by two pygmy guides searched the region to find two areas they had seen from aerial photographs. Even the pygmies had never ventured into this area before, because they fear a legendary dinosaurlike beast (which may be a black rhinoceros).

The visitors encountered a band of chimps, who stopped suddenly, stunned. The chimps stamped their feet, shook their arms, and chattered. But interestingly, they did not run. Chimps and

many other animals that have seen humans before run when they encounter them again.

The three explorers saw only a small portion of the abundant living community around them. Gorillas, caught offguard, reacted much as the chimps did—they stared but seemed unafraid. The visitors glimpsed them in clearings elephants had made in the thick underbrush. Wild pigs didn't bother to look up from the important business of searching for tasty roots. The humans saw buffalo and leopards as well as many kinds of monkeys who ate the same fruits elephants ate from towering trees. Smaller forms of life were everywhere, too. If the visitors took off their footwear, parasitic footworms might burrow into the soft skin of their feet, carrying bacteria with them. Equally hard to ignore were columns of driver ants who inflicted painful stings (fig. 1.7).

The researchers—thanks to their guides, who could follow clues in the terrain—found the two "lost" areas they had come to find. In the process, they gained a great appreciation for the living bounty of the planet.

e.

FIGURE 1.7

The Ndoki region of the northern Congo is home to many types of organisms, including the pangolin (*a*), the elephant (*b*), the gray parrot (*c*), driver ants (*d*), and the bongo (*e*).

KEY CONCEPTS

A look at different areas of the earth, some seemingly unfit for habitation, reveals some of the great diversity of living things.

SUMMARY

It is difficult to pinpoint when life begins and ends. Life is usually defined by describing the combination of characteristics that distinguish the living from the nonliving. First, life is organized in a sequence of increasing complexity: biochemicals, then cells, tissues, organs, organ systems, individuals, populations, ecosystems, and finally, the biosphere. As the level of organization increases, emergent properties arise as a consequence of physical and chemical laws. Second, life requires energy to maintain its organization and functions. Metabolism directs the acquisition and use of energy. A third quality of life is that living organisms must maintain an internal constancy, even in the face of a changing outside environment. A fourth characteristic is that living things respond to the environment—immediately through irritability, and in the longer term by natural selection, eliminating traits that make it difficult to successfully reproduce in particular environments. Fifth, organisms develop, grow, and reproduce, which is necessary for species survival. The combination of these five characteristics distinguish life.

Evolution through natural selection explains how life can be alike yet diverse, and how all species are related to one another by common ancestry. Biodiversity is apparent in the different types of organisms occupying different habitats. By studying the diversity of life, scientists gain an appreciation for the bounty of the biosphere and the incredible ability of life to adapt to a wide range of environments.

KEY TERMS

adaptation 7
asexual reproduction 8
biodiversity 8
cell 6
ecosystem 11
emergent property 6
energy 6
entropy 6
homeostasis 7
irritability 7
metabolism 7
natural selection 8
organelle 6
phylum 10
second law of
 thermodynamics 6
sexual reproduction 8
tissue 5

REVIEW QUESTIONS

1. Explain how biologists mentioned in the chapter explored
   a. the adaptation of leaves to strong winds.
   b. life in a small section of the ocean.
   c. the source of an enzyme that withstands very high temperatures.
   d. the types of organisms that inhabit a frigid lake.
   e. animals in the northern Congo.

2. In an episode of the television series *Star Trek*, three members of the crew of the starship USS Enterprise, exploring a planet, meet an enemy who dehydrates each one into a small box. He then rearranges the boxes and crushes one of them. He later attempts to restore the people to life, but only two of them reappear. What biological principle does this action illustrate?

3. Endothelium is a tissue consisting of cells that look like tiles adhering to one another to form a sheet. This sheet then folds to form a tiny tube called a capillary, which is the smallest type of blood vessel. In the brain, the endothelium of capillaries is so tightly packed that substances (including many drugs) cannot leak from the blood to the brain. The tight-walled capillaries are thus said to form a blood-brain barrier. What general principle discussed in the chapter is illustrated by the function of aggregated endothelium in keeping dangerous substances out of the brain?

4. What is the difference between irritability and adaptation? What do these characteristics of life have in common?

5. State three reasons why life requires energy.

6. Define the following terms, which are discussed in detail later in the text:
   a. metabolism
   b. homeostasis
   c. adaptation
   d. natural selection

7. Explain how natural selection underlies the ability of the adder snake in figure 1.3 to hide itself.

TO THINK ABOUT

1. Should biologists study geographical areas humans have never before visited? Cite a reason for your answer.

2. In 1992, a baby born at a hospital near Fort Lauderdale, Florida, was diagnosed as having anencephaly, a condition in which the brain contains only those structures that provide vital functions such as respiration and heartbeat. The parts of the brain that control reasoning ability, sensation, and perception never developed. The baby's parents wanted their daughter to be declared dead at one day of age, so that her organs could be donated for transplant to other infants before they began to deteriorate from the deficient brain function. However, the Supreme Court of the state of Florida stopped the parents' action. Justice Gerald Kogan wrote, "We find no basis to expand the common law to equate anencephaly with death."

   a. Was the Court's decision consistent with the current criteria for life and death?

   b. What further information would be helpful in making the difficult decision of whether or not this child is alive?

3. An android is a being that inhabits many science fiction stories. It is engineered from hardware and operates on software with neural networks that impressively mimic the human nervous system, giving the android intelligence, expanded memory, and self-awareness. Is an android a person or a nonliving device? Which characteristics of life does it have? Which does it lack?

4. Some people believe that human life begins with the union of sperm and egg, while others believe that it begins at birth. Some even think that human life has no start or finish—they argue that it is a continuum, because we all possess cells that have the potential to be alive when joined with another cell of the appropriate sort. What do you think? When does life begin, and why?

5. Researchers can combine genetic material of different species, increasing biological diversity. Other humans decrease diversity by destroying natural habitats, which sometimes leads to the extinction of species. What effects do you think human intervention in natural biological diversity will have on future life on this planet?

SUGGESTED READINGS

Belkin, Lisa. 1993. *First, do no harm.* New York: Fawcett Crest. A compelling look at several unforgettable young patients on the brink of death.

Brock, Thomas D. October 11, 1985. Life at high temperatures. *Science,* vol. 230. Hot springs bacteria have evolved adaptations to help them live in hot environments.

Christian, Shirley. May 29, 1992. There's a bonanza in nature for Costa Rica, but its forests too are besieged. *New York Times.* An unlikely alliance of scientists, ordinary citizens, and drug manufacturers are attempting to preserve biodiversity in the rain forest as well as better understand the biology of its inhabitants.

Dawkins, Richard. April 21, 1994. The eye in a twinkling. *Nature,* vol. 368. A computer simulation shows how eyes may have evolved.

Diamond, Jared. April 1990. The search for life on earth. *Natural History.* Discovering never-before-seen life forms is a most exciting part of biology.

Linden, Eugene. July 13, 1992. The last Eden. *Time.* A remote African rain forest is untouched by human presence.

Orstan, Aydin. Spring 1990. How to define life: A hierarchical approach. *Perspectives in Biology and Medicine.* A historical account of how different thinkers have defined life.

Skopets, Mikhail. November 1992. Secrets of Siberia's White Lake. *Natural History.* The writer identifies a new species from remains in a predator's stomach.

Vogel, Steven. September 1993. When leaves save the tree. *Natural History.* Vogel placed leaves in wind tunnels to simulate a storm and observed an adaptation.

Yoon, Carol Kaesuk. June 2, 1992. In dark seas, biologists sight a riot of life. *New York Times.* Life is far more diverse in the seas than on land, but it is also less accessible for study.

CHAPTER 2

Thinking Scientifically about Life

Biologists have learned a great deal about early animal development by studying sea urchins.

LEARNING OBJECTIVES

By the chapter's end, you should be able to answer these questions:

1. How do scientists use the scientific method to explain biological phenomena?

2. How are scientific experiments designed?

3. What are some ways of answering scientific questions?

4. What are the problems and limitations of implementing the scientific method?

5. Why is basic research important?

The Unity of Life—Of Mice and Men, and More

What do mice, sheep, humans, and bacteria have in common? Not much, you might think. But a tale of the rare inherited disorder Menkes disease vividly illustrates how these seemingly unrelated organisms are not as different as they might appear.

A newborn child with undiagnosed Menkes disease seems healthy in his first few weeks of life. But low body temperature, a slight yellowing of the skin, and sluggish feeding habits soon signal a problem. By the end of the first month, the baby is so lethargic, he must be coaxed to eat. His temperature remains low and his circulation poor. Then seizures begin, indicating more serious problems in the brain. Menkes disease is so rare, affecting only 1 to 2 of every 100,000 newborns, that physicians are usually unfamiliar with it.

Menkes disease is also called kinky hair syndrome because the hair of Menkes patients, normal at birth, soon loses its luster, then its color. It eventually stands on end and breaks off, leaving a white, kinky stubble (fig. 2.1a). Under a microscope, the hairs appear twisted and fractured.

In 1962, John Menkes of Columbia University first described the disorder, attempting to treat a New York family of Irish descent with five affected members. But it wasn't until nearly a decade later that Australian researcher David Danks, studying Menkes disease at Johns Hopkins University hospital in Baltimore, discovered the first clues to understanding it.

When Danks first saw the odd hair of his small patient, he was reminded of sheep in his native Melbourne that developed short, kinky, brittle wool after grazing on grass grown in soil deficient in the mineral copper. Might a copper deficiency cause Menkes disease in humans, he wondered? Danks had no sooner sent a sample of the patient's hair to the Australian Wool Commission for analysis of its copper content when other experiments began to support the copper deficiency explanation.

Not only did the blood of Menkes patients lack copper, but certain enzymes (proteins that control the rates of specific chemical reactions) vital for metabolism and requiring copper for their activity also lacked the metal. Connective tissue cells (fibroblasts) from Menkes disease patients grown in the laboratory revealed the basic defect—these cells could take up copper, but not release it. Other studies showed that in Menkes patients, cells lining the small intestine admit copper from food, but cannot release it into the bloodstream.

Lack of copper explains virtually all of the symptoms of Menkes disease. Copper aligns the protein rods that form hair. Without copper, hair falls apart—just as it did in the Australian sheep. The metal is also necessary for brain messenger chemicals (neurotransmitters) to function—without copper, the brain degenerates, causing lethargy, seizures, and mental retardation. An enzyme in bones that helps the body utilize vitamin C also requires copper. Without the metal, the enzyme cannot work, and the child suffers the effects of vitamin C deficiency, even if he or she consumes the vitamin. Bones bend in a manner characteristic of scurvy, a disorder caused by dietary vitamin C deficiency.

Most dangerous is the effect copper deficiency has on blood vessels throughout the body. Blood vessels consist of layers of elastin and collagen, proteins that assemble into smooth sheets; to properly assemble, they require copper. In Menkes patients, elastin and collagen in the blood vessels stretch and twist. This disrupts circulation, with wide-ranging effects that slow growth, hasten brain degeneration, and cause death before age 10.

Humans, sheep, and mice have much in common, biologically speaking, because they are all mammals (warm-blooded animals with fur and mammary glands). In fact, "mottled" mice, who have curly whiskers, colorless twisted fur, skeletal defects, altered arteries, and lethargy, are eerily reminiscent of Menkes disease patients (fig. 2.1b). But when researchers isolated the gene that causes Menkes disease in 1992, they discovered an even more astounding similarity. Using computerized databases, they compared the sequence of chemical components of the human Menkes disease gene to those of various genes in other types of organisms. A major portion of the human gene corresponds to a gene in bacteria—and that gene controls transport of copper out of the bacterium!

What is the significance of the observation that people, sheep, mice, and even bacteria all use a very similar chemical to transport copper? Perhaps they descended from a common ancestor, who had this ability and passed it on.

The story of Menkes disease illustrates the different levels at which biologists study life (table 2.1) and is an excellent example of the **scientific method.** This method of thinking underlies much of science as well as

 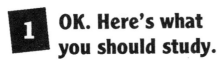

Such a deal—
cut your study time
and improve
your grades.

[**Computerized Study Guide,**
by Medi-Sim
ISBN 2X119201]

Focus your study time on only those areas where you need help. This easy-to-use software helps you identify and target those areas. Simply test yourself on each text chapter. The program not only gives you the correct answers, but creates *your own personalized study plan* based on your incorrect answers—with page references for each chapter.

This great program is available for use with IBM or compatible computers.

To order the *Computerized Study Guide,* call MEDI-SIM at 800-748-7734.

---

everyday decision making. The remainder of the chapter discusses the scientific method.

The Scientific Method

The scientific method involves making a varied series of inquiries by observing, questioning, reasoning, predicting, testing, interpreting, and concluding. Being a scientist is more a way of viewing the world than knowing a set of "facts." Figure 2.2 outlines the steps of the scientific method. Let's return to Menkes disease to see the scientific method at work.

Menkes Disease Revisited

The scientific method begins with observations:

The hair of Menkes disease patients resembles the wool of sheep that eat grass grown on soil lacking copper.

The observation leads to a question:

Why does the hair of Menkes disease patients resemble that of sheep lacking dietary copper?

The next step in the scientific method is formulating a statement to explain an observation, based on previous knowledge. This "guess," or **hypothesis,** is specific, testable, and, ideally, examines only one changeable factor, or **variable.** Sometimes a hypothesis is phrased as an "if . . . then" proposition:

If the sheep's peculiar wool is caused by copper deficiency, then Menkes patients should also show evidence of copper deficiency.

A hypothesis is more than just a hunch; it is based on some type of previous knowledge. For example, Danks knew that microscopically, the hair of Menkes disease patients greatly resembles that of copper-deficient sheep.

The next step in the scientific method is devising an **experiment,** a test designed to prove or disprove the hypothesis. Experiments and further

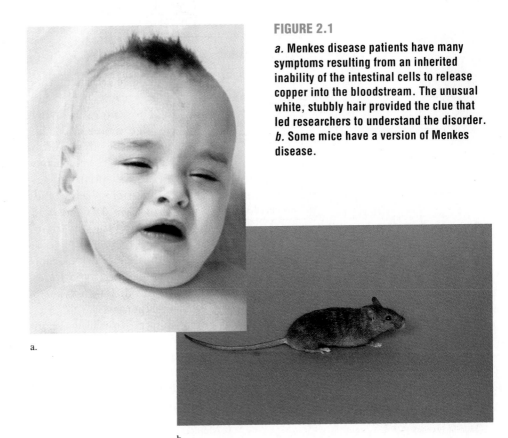

FIGURE 2.1

a. Menkes disease patients have many symptoms resulting from an inherited inability of the intestinal cells to release copper into the bloodstream. The unusual white, stubbly hair provided the clue that led researchers to understand the disorder. *b.* Some mice have a version of Menkes disease.

a.

b.

Table 2.1   Levels of Study of Menkes Disease

Science involves asking questions. Very often, scientists may approach a single problem—discovering the nature of Menkes disease, in this case—by asking different questions in different ways, at different levels. Can you add any interesting questions to this list?

Level	What to Investigate
Biochemical	How does an abnormal gene cause Menkes disease?
	How does a defective protein disrupt copper transport?
	How do defective elastin and collagen impair blood vessels, causing Menkes symptoms?
	How is the hair protein of Menkes patients abnormal?
Whole body	Which tissues and organs are affected by the disease?
Family	Which family members have Menkes disease?
	How can we predict which parents may have children with Menkes disease?
Population	Is Menkes disease more prevalent in some populations?
	Why might a disorder be more common in one population than another?
Species	Which types of organisms have disorders similar to Menkes disease?
Ecosystem	How can copper deficiency arise in different environments?

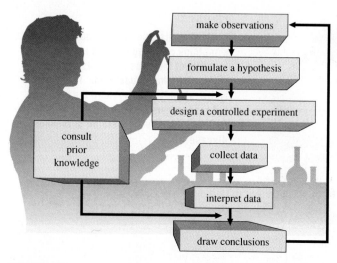

FIGURE 2.2

Steps of the scientific method.

observations build a case for or against the hypothesis. This was the case for the investigation of Menkes disease.

Tests showed that:

The blood serum of Menkes patients is low in copper.

Menkes patients have deficiencies of enzymes whose activities depend on copper.

Cells from Menkes patients take up copper but cannot release it.

Animals lacking dietary copper develop symptoms seen in human Menkes disease patients—abnormal hair, bones, blood vessels, and brain development.

Some mice have an inherited disorder that has symptoms very similar to those of Menkes disease and that involves copper deficiency.

The observations and experiments support the hypothesis that a disturbance in the body's use of copper is linked to Menkes disease. The conclusion: *Copper metabolism is abnormal in patients with Menkes disease.* But for such a conclusion to become widely accepted, the results must be repeatable—other scientists must perform the same experiments and observe the same results.

A Cycle of Inquiry

The scientific method is a cycle of inquiry. The search for knowledge never ends, because results and observations spawn new ideas, raising new questions. In the study of Menkes disease, finding the underlying role of copper led to further questions, more specific hypotheses, and different types of experiments.

Once Menkes disease was connected to copper deficiency, researchers quickly hypothesized ways of correcting the deficiency to help sick children. A more specific hypothesis leading to a new round of experimentation might be:

If Menkes disease is caused by copper deficiency, then giving a patient copper in a form that reaches the bloodstream might reduce the severity of symptoms.

Experiments to test this hypothesis began in the mid-1960s with a series of disappointments. When patients received a form of copper called copper sulfate, copper levels in their blood serum failed to rise. Other forms of copper also produced poor results. Then, in 1976, Bibudhendra Sarkar, a biochemist at the Hospital for Sick Children in Toronto, tried copper attached to an amino acid (a protein building block) called histidine on a newborn whose brother had died of Menkes disease. Not only did the copper make its way into the child's bloodstream, but today he remains the oldest survivor of Menkes disease.

At the National Institutes of Health in Bethesda, Maryland, and at Toronto's Hospital for Sick Children, several other patients given copper histidine during the first month of life developed normally. Patients treated later than this died during early child-

hood, presumably because after the age of one month, irreversible damage to the nervous system has already occurred.

The preliminary conclusion? As the hypothesis predicted, giving copper in a form that reaches the bloodstream appears to alleviate or even prevent some symptoms of Menkes disease. With the 1993 discovery of the gene that causes Menkes disease, a test can be developed to detect the condition early enough for treatment to begin within the crucial one-month period.

Even with the apparent success of copper-histidine therapy to treat Menkes disease, questions remain. Does the treatment alleviate all symptoms? Does it cause side effects? Could too much copper in the bloodstream cause a problem? Does the treatment remain effective as the child grows older?

Broader questions further engage the scientific method and widen the cycle of inquiry. Copper is only one metal. Do deficiencies or excesses of other metals cause other syndromes? Can we treat other symptoms linked to disrupted use of a metal in the body? Every line of scientific investigation, every conclusion drawn, spawns many other questions. Biology in Action 2.1 shows how the scientific method was used to solve another pressing medical problem.

KEY CONCEPTS

The scientific method consists of making observations, asking questions, formulating hypotheses, designing experiments, collecting and interpreting data, and reaching conclusions—then repeating the steps. The scientific method is a cycle. Conclusions lead to further questions.

Designing Experiments

Scientists use several approaches when they design experiments in order to make data as valid as possible. For example, researchers might examine dozens

Biology in ACTION

The Scientific Method Solves the Mystery of Seabather's Eruption

Observations

1. In the early 1990s, many ocean swimmers on New York's Long Island developed a red rash in the area covered by their bathing suits (fig. 1b). A stinging sensation preceded the rash.
2. The waters where the people swam contained millions of pink, round larvae of an unidentified sea-dwelling organism (fig. 1a).

Hypothesis

The larvae cause the rash by becoming trapped inside the swimmers' bathing suits and stinging them.

Prior Knowledge

"Seabather's eruption" was described in 1949 when swimmers in Florida acquired the uncomfortable rash. Tiny jellyfish, crab larvae, and other wormlike sea animals can cause rashes when entrapped in bathing suits. Rashes caused by algae look different from seabather's rash.

Experiments

1. Three workers in the health department on Long Island voluntarily exposed their arms and abdomens to larvae collected from a nearby beach.
2. Larvae from victims' bathing suits were grown in the laboratory.

Results

1. All of the volunteers felt stinging sensations within a few minutes and developed the rash shortly thereafter.
2. The larvae grew into adult sea anemones.

Conclusion

Seabather's eruption is caused by stings from sea anemone larvae entrapped in swimmers' bathing suits.

Further Observations

Further experiments and observations showed that the larvae usually spend the summer months inside another sea animal, the harmless comb jelly. In the fall, the larvae leave to become free-swimming. In years when seabather's eruption is prevalent, the larvae exit the comb jellies early, in midsummer.

Further Hypothesis

Does swimming in the nude or wearing a looser bathing suit prevent seabather's eruption? This hypothesis remains to be tested.

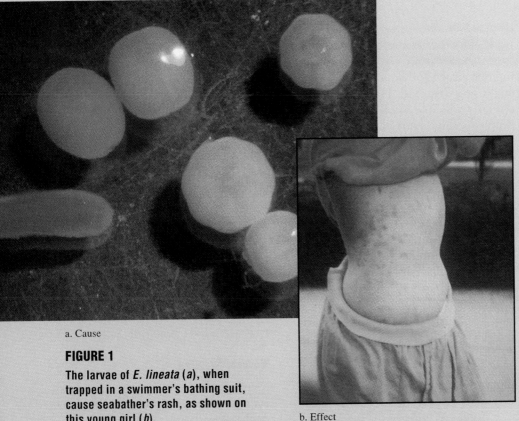

a. Cause

b. Effect

FIGURE 1

The larvae of *E. lineata* (*a*), when trapped in a swimmer's bathing suit, cause seabather's rash, as shown on this young girl (*b*).

of mice, or hundreds of cells, because a large **sample size** helps ensure meaningful results. However, this standard is not always practical. When scientists study very rare disorders, only a few patients may be available. Although scientists prefer to work with a large sample, small-scale experiments are nevertheless valuable because they may indicate whether continuing research is likely to yield valid, meaningful results. A study using a small number of experimental subjects is called a pilot study.

Experimental Controls

Logically, the unusual can only be distinguished by comparison to the usual. Ideally, scientific experiments compare a group of "normal" individuals or components to a group undergoing experimental treatment. The experimenters attempt to control the experiment so that the only factor that is different between the "normal" group and the experimental group is the one factor they want to test. This "normal" group then constitutes an **experimental control.** Designing experiments to include controls helps ensure that a single factor, or variable, truly causes an observed effect. Experimental controls may take several forms.

In studying Menkes disease, researchers might design a controlled experiment by feeding 50 mice rodent chow lacking only copper, while feeding another 50 fully nutritious food. The experimenters could then draw blood from mice on either feeding regimen and compare the blood sera of the copper-starved and normally nourished mice. If the action of taking blood samples were to affect the mice in any way, taking samples from both groups would ensure that the effects would be the same for both groups. This is an example of an experimental control that narrows the differences between the control and the experimental groups to just one variable—copper deficiency—and eliminates other possible causes of test results.

Another example of a control that ensures that the control group has the same experiences as the experimental group is called a **placebo.** A placebo is

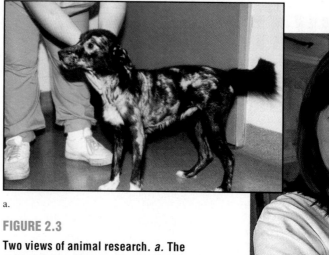

a.

FIGURE 2.3

Two views of animal research. *a.* The animal rights movement is threatening biomedical research that uses nonhuman animals. This photo is typical of those in the literature of PETA, People for the Ethical Treatment of Animals. This dog was intentionally infected with scabies mites to study the infection process. *b.* This photo depicts a happier laboratory animal.

b.

an inert substance a control group takes while the experimental group tries out the substance being tested. In evaluating new drugs, for example, a placebo might be a sugar pill. Placebos are particularly valuable in studies involving people, because some of us tend to feel better (or worse) if we know we are taking something. By using a placebo, researchers can determine whether test results are caused by the substance they are testing or by the patients' knowledge that they are taking a substance.

Another safeguard in experimental design is to make the trial **double blind,** which means that neither the researchers nor the subjects know who received the substance being evaluated and who received the placebo. The "code" is broken only after the data are tabulated. The rodent chow study is double blind because the researchers do not know which rats are receiving each type of food, and of course, neither do the rats.

When an illness is life threatening or cannot be treated, placebos are not used. It wouldn't be right to deny someone who has no options the chance to try a treatment that might be a cure. When medical researchers had evidence that the drug AZT could forestall symp-

toms of AIDS in people infected with the human immunodeficiency virus (HIV), all participants received AZT treatment.

> ### KEY CONCEPTS
> A well-designed experiment has a large sample and uses controls, so that only one variable is assessed. If the experiment uses a placebo, the trial can also be double blind.

Types of Experiments

Experimental designs vary. An experiment may take biologists to the world within a cell, to medical centers, or to remote and sometimes extreme habitats to observe life (Biology in Action 2.2).

Animal Studies

Many studies expose nonhuman organisms to treatments and situations that might provoke ethical objections if they were conducted on humans, although animal studies sometimes elicit objections, too (fig. 2.3). The results of animal studies are extrapolated to humans, or serve as the basis for further investigation. If, for example, a potential new drug causes mutations (genetic changes) in bacteria, it is discarded, because we know that chemicals that cause mutations in bacteria very often cause cancer in animals.

Animal studies can tell researchers whether or not to pursue a treatment for humans. This was true in the case of LAK cell therapy, in which white blood cells are taken from cancer patients who no longer respond to existing treatments. The cells are grown in the laboratory with a cancer-fighting immune system biochemical, interleukin 2. The cells are then injected back into the patient, where they destroy cancer cells. The ability of this treatment to reduce tumors was proven in mice before human trials took place.

Animal research has contributed to many advances in human health care and, ironically, veterinary medicine. Experimentally used animals have helped us develop:

vaccines

cancer chemotherapies

antibiotic drugs

transplants

trauma treatments

the use of insulin to treat diabetes

CAT scans

artificial joints

Sometimes animal studies identify possible risks to human health that were previously unknown. This was true of dioxin, a contaminant in the herbicide Agent Orange that was sprayed to defoliate trees in Vietnam in the 1960s. Dioxin had been known to cause skin conditions and some cancers in exposed people, and it was suspected to cause some reproductive problems, but no one recognized certain effects on reproduction until they arose in studies using rats. Researchers gave pregnant rats a single low dose of dioxin (fig. 2.4). They later noted malformations of the reproductive and urinary systems in the rats' offspring, many of which did not appear until those offspring were adults. Using rats, scientists could study the effects of dioxin

exposure over several generations in a short time—which would be impossible to do with humans. This experiment told researchers the types of health problems that might appear in the grown offspring of men exposed to dioxin in Vietnam.

Using such animals as mice and fruit flies also enables investigators to breed large numbers of genetically identical individuals so that experiments can be carefully controlled. Scientists can even genetically engineer laboratory mice to harbor the genes that cause certain human disorders, providing model systems on which to test therapies before human patients try them. Organizations that represent the people who have these disorders often underwrite the cost of developing these mice.

Many experiments use cells or tissues instead of whole animals. It is certainly more humane to observe the inflammatory effects of a cosmetic ingredient on cells growing in a dish than on the eyes of rabbits, as researchers once did. Similarly, cancer drugs are often tested on patients' cells first, to see how

they respond, rather than exposing the patients to possible toxic effects. Sometimes an experimenter can replace living matter altogether, as Jay Georgi did when he invented the "artificial dog," shown in figure 2.5, to house research fleas.

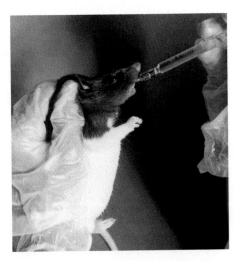

FIGURE 2.4

A rat receives dioxin to test its effects on offspring.

FIGURE 2.5

Dr. Jay R. Georgi, a Cornell University parasitologist, invented the "artificial dog" (background), a self-contained environment that produces thousands of fleas a day for pharmaceutical and biomedical research. Each of the artificial dog's flea cages (foreground) contains a body-temperature blood supply and simulated skin the insects feed through.

Biology in ACTION

Elephants Calling—The Powers of Observation

If you think only humans possess extensive language, consider an elephant's extended family. An elephant clan, led by an elder female, or matriarch, and consisting of females and young males, has quite a vocabulary. Elephant "language" includes calls in infrasound, or sounds too low for human ears to detect.

But one curious pair of human ears heard the low-pitched language of the great beasts in 1984. Cornell University researcher Katy Payne was standing near caged elephants at a zoo when she felt a "throbbing" in the air. The sensation brought back memories of singing in the church choir as a child; standing in front of the church organ pipes, she felt a similar throbbing. "In the zoo I felt the same thing again, without any sound. I guessed the elephants might be making powerful sounds like an organ's notes, but even lower in pitch," she writes in her children's book, *Elephants Calling*.[1]

To test her hypothesis, Payne and two friends used equipment that could detect infrasound to record elephants in a circus and a zoo. Infrasound was there!

Payne, already well known for having discovered the songlike language of whales, published her results, which caught the attention of elephant researchers Joyce Poole and Cynthia Moss. They invited her to join them in the wild. So Payne moved her study to the Amboseli plain, a salty, dusty stretch of land at the foot of Mount Kilimanjaro in Kenya. Her lab was a truck sitting among the elephants, who grew so used to its presence that they regarded it as a natural part of the scenery.

Payne's already highly developed powers of observation were further sharpened living among the elephants. Just as we might watch people converse in a foreign language, then discern the meanings of words from accompanying actions and expressions, Payne and her fellow elephant watchers soon became attuned to the subtle communications between mother and calf; the urges of a male ready to mate; and messages to move to find food or water. Writes Payne, "It is amazing how much you can learn about animals if you watch for a long time without disturbing them. They do odd things, which at first you don't understand. Then gradually your mind opens to what it would be like to have different eyes, different ears, and different taste; different needs, different fears, and different knowledge from ours."

One day, when Payne, Poole, and the "R" elephant family (all of which were given names starting with R) were mesmerized by two bulls fighting for dominance, the youngest family member, baby Raoul, slipped away (fig. 1). Finding an intriguing hole, Raoul stuck his inquisitive trunk inside—then leapt back, bellowing, as a very surprised warthog bounded out of his invaded burrow. Raoul's mother, Renata, responded with a roar. Payne described the scene:

Elephants in all directions answer Renata, and they answer each other with roars, screams, bellows, trumpets, and rumbles. Male and female elephants of all sizes and ages charge past each other and us with eyes wide, foreheads high, trunks, tails, and ears swinging wildly. The air throbs with infrasound made not only by elephants' voices but also by their thundering feet. Running legs and swaying bodies loom toward and above us and veer away at the last second.

The R family reunited, but all the animals were clearly shaken. Unable to resist comparing the pachyderms to people, Payne writes, "Renata does not seem angry at Raoul. Perhaps elephants don't ask for explanations."

1. Payne, Katy. 1993. *Elephants Calling*. New York: Crown.

Computer Experimentation

Many biological phenomena are informational, from the sequences of chemicals that comprise proteins and genes, to the pattern of electrical activity of a beating heart, to the comings and goings of organisms in ecosystems. This quantitative nature of life makes many aspects of biology amenable to computer analysis.

Consider protein structure. A protein is a chain of 20 types of amino acids whose sequence is encoded in genes, each of which is also comprised of a sequence of chemical units. A cell translates a gene's sequence into a protein's sequence, much as one language is translated into another. Protein chains fold into specific three-dimensional shapes that are essential to their function. In the past, a protein's structure could only be revealed by a technique called X ray crystallography. This technique is slow and requires a crystal of the protein, which isn't always possible to obtain. But if a scientist feeds the protein's amino acid sequence into a computer, along with information on how the amino acids attract or repel one another, the computer can display an image of the

The kind of observation that leads to new knowledge or understanding, Payne says, happens only rarely. "You have to be alone and undistracted. You have to be concentrating on what's there, as if it were the only thing in the world and you were a tiny child again. The observation comes the way a dream—or a poem—comes. Being ready is what brings it to you."[2]

Observation: The air near the elephant's cage at the zoo seems to vibrate.

Background knowledge: Organ pipes make similar vibrations when playing very low notes.

Hypothesis: Elephants communicate by infrasound, making sounds that are too low for the human ear to hear.

Experiment: Record the elephants with equipment that can detect infrasound.

Results: Elephants make infrasound.

Further observation: Elephants emit infrasound only in certain situations involving communication.

Conclusion: Elephants communicate with infrasound.

Further question: Do different patterns of sounds communicate different messages?

FIGURE 1

Raoul rarely ventures far from his mother, Renata.

2. Katy Payne, interview with the author, 1993.

structure of the protein. Researchers can then "enter" the site where an abnormality occurs in a gene, and the image shows where and how the protein is disrupted.

Figure 2.6 shows a computer analysis of superoxide dismutase, the protein which, when abnormal, causes amyotrophic lateral sclerosis (ALS). (ALS is also called Lou Gehrig's disease after the famous Yankees baseball player who had it.) Researchers are now trying to determine how the abnormal protein causes the symptoms of brain degeneration that characterize the illness.

Computers can solve practical problems, too. When a wild area of southern California was slated to be developed, ecologists entered into a computer all of the species known to inhabit the area. They predicted the species most vulnerable to habitat destruction and altered plans so that these species would be spared.

Epidemiological Studies

Epidemiology is the study of disease-related data from real-life situations. This can range from one-person anecdotal

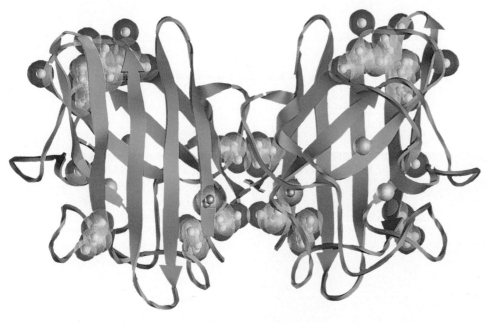

FIGURE 2.6

Computers help researchers understand the causes of disease. The disorder amyotrophic lateral sclerosis (ALS) was traced to an abnormal form of the enzyme superoxide dismutase (SOD). Using genetic data, a computer highlights the parts of the enzyme that are disrupted, showing that the defects occur at sites that destabilize the entire enzyme. The green tangle represents the enzyme, and the balls depict the sites of instability.

accounts to large-scale studies involving thousands of individuals. Such extensive studies may be retrospective, evaluating the health effects of past exposure to a chemical from an industrial accident, or they may be prospective investigations that set up a situation and then observe what happens.

One type of epidemiological investigation compares observations of two groups of people differing in one variable, such as exposure to a particular substance. Ideally, each person from one group matches a person in the other group in as many characteristics as possible, such as sex, race, age, diet, and occupation. For example, the incidence and types of birth defects in children of Taiwanese women exposed to cooking oil contaminated with polychlorinated biphenyls (PCBs) were compared to those of Taiwanese women who were not exposed. A significant number of the children exposed to PCBs were shorter and lighter than the control children and had abnormal gums, skin, teeth, nails, and lungs. On the basis of this study, these health problems were correlated to PCB exposure before birth because this was the only significant difference between the two groups of children.

Epidemiological information can only demonstrate a correlation, not a cause-and-effect relationship, between two observations. The information that runners are skinny may indicate that running is a good way to lose weight—or it may indicate only that people who are already thin tend to take up running. It is necessary to do further research to determine causes and effects. Similarly, epidemiological evidence suggests that people living near sources of electromagnetic fields are at increased risk of developing certain cancers. As a first step to test this hypothesis, the U.S. electric power industry conducted a nationwide survey to document the level of electromagnetic radiation in homes. What should the next step be?

An epidemiological study may be quite challenging to carry out. Wading through masses of information to set up experimental controls for all variables is a huge task. Moreover, biases often affect interpretations of information concerning certain events, especially if they occurred years ago. For example, a person who knows he or she was exposed to Agent Orange might attribute various health problems to that experience. A reconstruction-of-the-scene experiment, reenacting conditions in the jungle when Agent Orange was sprayed, may help resolve the issue.

Meta-Analysis

The research process includes many types of approaches—observations, animal studies, cell or tissue culture, computer simulations, or epidemiological studies (fig. 2.7). Each approach has limitations. An experiment may address only one aspect of a question; a pilot study may involve only a few people; an epidemiological investigation may lack controls. To overcome the limitations of individual experiments, researchers combine results from many studies in a **meta-analysis.**

The dioxin situation illustrates the value of a meta-analysis in evaluating data. A panel of medical researchers appointed by Congress considered 6,420 references in the medical literature to dioxin-related health effects and

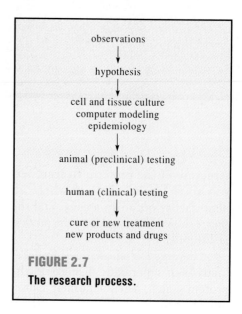

FIGURE 2.7

The research process.

230 epidemiological investigations of accidental exposures. Combining this information, the researchers identified four categories of disorders with varying likelihoods of being caused by dioxin exposure (table 2.2)—information far more extensive than would be possible from an individual study.

KEY CONCEPTS

Scientists can use many experimental methods to study life. Experiments may utilize organisms or their tissues grown in the laboratory. Computer experiments can interpret quantitative or informational aspects of biology. Epidemiological studies utilize groups of individuals who, whenever possible, differ by only the variable under study. A meta-analysis combines the results of many studies.

Applying the Scientific Method: Challenges and Limitations

Interpreting Results

The scientific method is neither foolproof, nor always easy to implement. Experimental evidence may lead to multiple interpretations or unexpected conclusions, and even the most carefully designed experiment can fail to provide a definitive answer. Consider the observation that animals fed large doses of vitamin E live significantly longer than similar animals who do not ingest the vitamin. Does vitamin E slow aging? Possibly, but excess vitamin E causes animals to lose weight, and other experiments associate weight loss with longevity. Does vitamin E extend life, or does the weight loss? The experiment of feeding animals large doses of vitamin E does not distinguish between these possibilities. Can you think of further experiments to clarify whether vitamin E or weight loss extends life?

Table 2.2	**Meta-Analysis of Dioxin-Related Health Problems**
Category	**Problems**
Definite link to dioxin exposure	Hodgkin's disease ⎫ Non-Hodgkin's lymphoma ⎬ Cancers Soft tissue sarcoma ⎭
Possible link to dioxin exposure	Porphyria cutanea tarda ⎫ Skin conditions Chloracne ⎭ Respiratory problems Prostate cancer Multiple myeloma (cancer)
Inadequate evidence to establish or disprove link to dioxin exposure	Neurological problems Reproductive problems Certain other cancers
No link to dioxin exposure	Skin cancer Cancer of digestive system Bladder cancer Brain tumors

Another limitation of implementing the scientific method is that researchers may misinterpret observations or experimental results. In the next chapter, we will see how investigators erroneously concluded from many experiments that life spontaneously arises from nonliving matter by exposing decaying meat to the air and observing the organisms that grew on it. Also discussed in chapter 3 are experiments that attempted to recreate the chemical reactions that might, on the early earth, have formed the chemicals found in life. Although the experiments produced interesting results and inspired much later work, we cannot really know if they accurately recreated conditions at the beginning of life.

Expecting the Unexpected

An investigator should try to keep an open mind towards observations, not allowing biases or expectations to cloud interpretation of the results. To do so,

scientists must expect the unexpected. But it is human nature to be cautious in accepting an observation that does not fit existing knowledge. The careful demonstration that life does not arise from decaying meat surprised many people who believed that mice were created from mud, flies from rotted beef, and beetles from cow dung. Earlier this century, researchers were so enamored of the much-studied proteins that they were hesitant to follow up suggestions that a less-understood biochemical—the nucleic acid DNA—was really the genetic material.

Perhaps no one knew better the frustration of an unexpected scientific discovery than Barbara McClintock. In the 1940s, she studied inheritance of kernel color in corn. While watching her meticulously bred plants respond, McClintock noticed that some kernels had an odd pattern of spots. After many experiments, she concluded that the spots indicated that genes (the units of inheritance) were moving, jumping into other genes, disrupting their functions. This idea seemed as preposterous at the time as the thought that the earth orbited the sun once did. Genes were thought to be immovable.

Despite McClintock's evidence and the repeated publication of her findings, many scientists did not believe her—possibly because she was a woman in a field that was then dominated by men. But in the 1970s, researchers began finding evidence of "jumping genes" in other species, including our own. McClintock was finally recognized as the perceptive researcher she always was and received the Nobel Prize.

Making Scientific Discoveries by "Accident"

Scientific discovery is not always as well planned as the scientific method may seem to suggest. Sometimes knowledge comes simply from being in the right place at the right time, or from being particularly sensitive to the unusual. Knowledge gained unexpectedly is termed serendipitous. Often a serendipitous finding can be linked with other findings or produces new information (table 2.3).

Making mental connections between accidental observations takes an especially alert and creative mind that can recognize the significance of those connections. One such mind belonged to Dr. Crawford Long, a medical student in the middle of the nineteenth century. While attending an "ether party," Long noticed that his friends who were intoxicated by the chemical appeared to feel no pain. Years later, he tested that assumption and eventually found that

he could remove tumors painlessly by placing a towel soaked in ether over his patients' faces before starting surgery (fig. 2.8).

KEY CONCEPTS

The value of the scientific method is limited by our ability to interpret experimental results. Many discoveries are not planned yet yield useful information.

Basic versus Applied Research

Scientific research seeks to understand nature. Because humans are part of nature, we sometimes tend to view scientific research, and particularly biological research, as aimed at improving the human condition. How often do you see newspaper or magazine articles that validate research results by reporting that the work "could prevent birth defects or cure cancer"? But knowledge for knowledge's sake, without any immediate application, is valuable in and of itself—

or will be. Consider one of many human proteins, discovered in 1979, thought at first to be so unimportant that it was named merely after its molecular weight—p53. By the early 1990s, many, many studies had revealed that abnormal versions of p53 lie behind half or more of all human cancers. Basic research implicated the protein in controlling cell division and cell death. Research on p53 is still mostly "basic," but many scientists are already trying to develop tests that determine which type of p53 a person has—and therefore to predict what type of cancer the individual is more likely to develop. Today's basic research often yields tomorrow's technology, in ways that cannot always be predicted.

A compelling example of the road from basic research discovery to applied technology is the development of a standard test used to detect bacterial contamination of injectable drugs and medical devices that touch the body. The story begins on the shores of Cape Cod, Massachusetts, where a horseshoe crab lay dying.

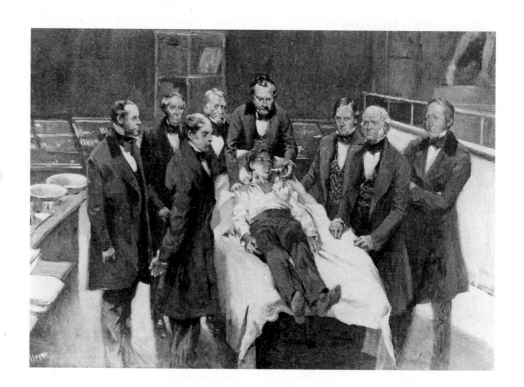

FIGURE 2.8

The discovery of ether's value as an anesthetic stems from a student's observation that friends who were using the chemical to "get high" at a party did not feel pain. This illustration depicts the use of ether as an anesthetic during surgery.

Table 2.3 From Basic Research to Technology

Discovery	Observation
Aspartame, a sweetener	Pharmaceutical researcher Jim Schlatter produced a white substance as he searched for an ulcer drug. He accidentally tasted the substance and discovered it was sweet. Further studies identified the compound as two linked protein building blocks that provide a sweet taste. Aspartame is today a widely used sweetener.
Cellulase, a whitener agent	In the 1960s, when flush toilets were not yet common in Japan, microbiologist Koki Horikoshi searched for a microbe that produced an enzyme called cellulase, which could be used to degrade human waste. By the time Horikoshi found an effective cellulase enzyme, the flush toilet had been invented. But a decade later, a soap manufacturer examined Horikoshi's cellulase and found that it could open up knots of cellulose holding stains in cotton yet leave the cellulose unharmed. Without knowing it, Horikoshi had found the perfect clothing whitener.
Cisplatin, a cancer drug	In 1969, Michigan State University biophysicist Barnett Rosenberg began studying the effects of electromagnetism on bacteria after noticing a similarity between the pattern in which iron filings align in a magnetic field and the positions of tiny fibers in cells as they divide. Bacteria in the presence of an electromagnet indeed ceased to divide, growing larger instead. Knowing that some antitumor drugs also produce big bacteria, Rosenberg hypothesized that an electromagnet could fight cancer. He was almost right. It turned out that a platinum-containing chemical formed by a chemical reaction between the electromagnet's electrodes and the chemicals they were immersed in actually halted the cell division. Rosenberg's discovery was developed into the drug cisplatin, used in cancer chemotherapy.
Penicillin	In 1928, English microbiologist Alexander Fleming noticed clear patches in a dish of bacteria exposed to the air. Instead of discarding the contaminated culture dish, Fleming hypothesized that something in the air had prevented other microbes from colonizing the exposed plate. Other scientists eventually isolated the mold that killed the microbes—it was penicillin.
Treatment for a form of mental retardation	In 1934, a mother of two mentally retarded children told a relative who was a chemist that her offsprings' diapers had an odd odor. The chemist analyzed the children's urine and found an excess of one chemical. He concluded that the biochemical that normally breaks down this substance was missing or not functioning. He then found a similar deficit in other mentally retarded children. Understanding the biochemical defect behind this disorder, PKU, led to the development of a dietary treatment to compensate for this metabolic block.

From Crab to Lab

In 1956, Frederick Bang, an infectious disease expert at Wood's Hole Marine Biological Laboratory on Cape Cod, was walking along the coast to collect specimens of *Limulus*, the horseshoe crab, when he came upon a large specimen washed ashore in the throes of an infection. Gulls had begun to pick at the animal. When Bang looked at the injured parts of the crab, he was struck by something unusual—a substance that looked like blue gelatin. Bang brought the sick crab back to his laboratory, where he isolated the bacteria causing the infection and used them to infect a healthy crab.

The second crab soon fell ill. Bang sacrificed and dissected it. The second crab's blue blood had gelled, too. What was happening?

Bang put his observation of the gelled blue blood into perspective with his previous observations and experiences:

Horseshoe crabs have persisted for a very long time and therefore probably have some immunity against infection by bacteria, which have existed for even longer.

The infectious bacteria in the crab were of a type called gram negative, known to cause disease in many types of organisms. Gram-negative bacteria contain a substance called endotoxin in their cell walls. Endotoxin stimulates the host's immune system to release chemicals that cause fever, achiness, and shock.

The crab formed blood clots in the presence of gram-negative bacteria.

Bang formulated a hypothesis to explain the cause of the clotting reaction: *If the blue blood clots in response to the presence of gram-negative bacteria, then there should be evidence of immune system action, such as the activity of protective blood cells.*

FIGURE 2.9

Horseshoe crabs are "milked" for their blood in a facility on Cape Cod, then returned, apparently unharmed, to the sea. The ability of part of the crabs' blood to clot in the presence of certain bacterial infections is the basis of a widely used medical test.

With the help of a young blood specialist named Jack Levin, Bang painstakingly separated blood components, finally finding a protein in a type of cell called an amebocyte that triggers clotting on contact with endotoxin. By wounding and infecting crabs, Bang and Levin found that the amebocytes move to the area of infection, where they initiate a localized clot that kills the bacteria. The original beached crab had died from excessive clotting.

Over the next few years, Bang, Levin, and others at Wood's Hole developed a way to purify the protein causing the clotting. They also prepared an extract of the crab blood called *Limulus amebocyte lysate*, or LAL for short. (*Limulus* refers to the horseshoe crab, *lysate* to the contents of broken cells.) Another Wood's Hole scientist, Stanley Watson, envisioned a use for their discovery. Could LAL indicate the presence of bacterial contamination?

Watson tried applying a small amount of LAL to a piece of biological membrane he was studying in another project. He discovered that if the LAL gelled, gram-negative bacteria were present. Watson began to use LAL regularly to test the purity of various parts of his projects, then realized his test might have wider uses in the pharmaceutical industry. The Wood's Hole facility eventually sold Watson the patent for using LAL. He took his idea to several drug companies, none of which showed any interest at first. Watson set up production operations in his garage, where he bled crabs, then supplied LAL to a few researchers who used it to test the purity of their equipment and cultures. The

LAL test, Watson had found, could replace a standard test for endotoxin—injecting rabbits with a contaminated substance to see if they spiked a fever.

The LAL test worked. It also saved animal lives—not only were rabbits spared, but the crabs were returned to the sea after being bled (fig. 2.9). Still, the Food and Drug Administration (FDA) and drug companies remained skeptical until 1976. That was the year of a great swine flu scare in the United States. Anticipating an epidemic of this new flu strain, then-President Gerald Ford rushed the development of a vaccine, which usually takes several years. Unfortunately, a few hundred people died from the vaccine—and the swine flu epidemic never developed.

The LAL test eventually determined the cause of the vaccine-related deaths—bacterial endotoxin contaminating the vaccine supply. In 1977, the FDA approved the LAL test. It quickly became the standard method for examining medical devices for bacterial contamination, including syringes, catheters, injectable drugs, and intravenous fluids.

Today, biologists use the LAL test to isolate and study the human immune

system biochemicals that cause the symptoms of gram-negative bacterial diseases such as septic shock (blood poisoning), toxic shock syndrome, and spinal meningitis. The story that began almost forty years ago with a seaside stroll continues, still vibrant with the possibilities scientific discoveries and their hidden applications may contain.

The Search for Scientific Knowledge Continues

As logical and elegant as the scientific method is, conclusions are tentative, because we are always learning new things about life. Complacency that we have closed the book on a question not only blinds us to new knowledge, but can be downright dangerous. The protracted effort to understand the cause of AIDS illustrates this vividly.

When discovery of the human immunodeficiency virus, or HIV, was announced in 1984, the director of the National Institutes of Health confidently predicted that a vaccine would be developed by year's end. Unfortunately, this wasn't to be. In the 1990s, even as research efforts continued to center on

HIV, evidence began accumulating that some people had AIDS symptoms without the virus, and vice versa. Scientists who proposed that HIV was not the sole cause of AIDS were ridiculed. Gradually, though, people's thinking began to change. The bulk of the evidence continued to implicate HIV, but new hypotheses had to accommodate the cases that didn't involve the virus.

One hypothesis is that HIV initially infects a person and begins a cascade of destruction in the immune system. Even if the virus eventually dies or disappears, the immune system is damaged beyond repair. This explains why some people with AIDS might have no sign of HIV. But it is the growing group of people who have lived with HIV for many years, in apparent good health, who may hold the key to eventually conquering AIDS. As researchers build hypotheses to explain the survivors' biological luck, a glimmer of understanding may finally appear. And understanding is the ultimate goal of scientific investigation.

KEY CONCEPTS

Research augments our knowledge about the living world and can lead to new technologies. Conclusions reached using the scientific method may be altered by new information gained through research.

SUMMARY

Biologists study life by applying the scientific method. The scientific method begins when a scientist makes an observation, raises questions about it, and reasons to construct a possible explanation, or hypothesis. The scientist then designs one or more experiments to test the validity of the hypothesis, including experimental controls to ensure that only one variable is examined at a time. (Placebo-controlled, double-blind experiments are used to evaluate treatments without injecting human bias.) An experiment is repeated to test the validity of the conclusions it generates. Experiments may take many forms: they may involve observations, tests on living organisms, computers, or epidemiological evidence. A meta-analysis combines the results of many types of studies to reach a consensus. Conclusions augment knowledge, but they also lead to more questions, and the cycle of scientific inquiry continues.

Even when a biologist applies the scientific method, it does not always yield a complete answer or explanation. Many experimental results are unusual or unexpected, and some are encountered by accident. New information may disprove earlier conclusions. Still, research leads to new knowledge, and sometimes to new technologies. Overall, the scientific method provides a systematic approach to exploring and understanding the living world.

KEY TERMS

double blind 20
epidemiology 23

experiment 17
experimental control 20
hypothesis 17

meta-analysis 24
placebo 20
sample size 20

scientific method 16
variable 17

REVIEW QUESTIONS

1. Suggest an appropriate control group to test

   a. conclusions drawn from the copper histidine treatments given to the children with Menkes disease mentioned in the section entitled "A Cycle of Inquiry."
   b. whether the grown children of Vietnam veterans exposed to dioxin have reproductive abnormalities.

2. Since 1980, many studies have supported the hypothesis that if pregnant women take folic acid supplements, they are much less likely to have a child with certain types of birth defects affecting the brain and spinal cord. In the early 1990s, the Centers for Disease Control and Prevention considered the results of many, many studies and reports, and introduced a cautious plan: they suggested that pregnant women who had already had an affected child should take extra folic acid if they became pregnant again. As more data accumulated, all pregnant women were advised to take extra folic acid; soon, folic acid will be added to the grain supply, so that we all can benefit from it.

   a. What type of study did the Centers for Disease Control and Prevention conduct?
   b. What further questions could be posed about the connection between folic acid and birth defects?

3. Name two ways that the experiment in which health department workers exposed themselves to the larvae causing seabather's eruption departed from ideal use of the scientific method.

4. In 1976, a chemical accident released dioxin at Seveso, Italy, near Milan, producing the highest human exposure ever known to this substance. In 1993, researchers evaluated the medical histories of people aged 20 to 74 living in and near Seveso. They examined 550 people living very near the site, 4,000 people living farther away, and 26,000 people living quite far from the site. Why were different numbers of people evaluated for these three regions?

5. A woman is more likely to develop ovarian cancer if she has a relative who has the disease. One study evaluated 500 women for the presence of ovarian cancer, focusing on whether they applied talcum powder directly to their genital areas. The study found that women who used the powder had twice the incidence of ovarian cancer as women who did not. Researchers believe that talcum powder may enter the reproductive tract and irritate tissues to cause cancer much as inhaled asbestos fibers cause a rare type of lung cancer.

   a. What information might be important to obtain before concluding that talcum powder applied to the genital area causes ovarian cancer?
   b. State the hypothesis addressed in the study.
   c. On what background information is the hypothesis based?
   d. Is the study an example of a pilot study, an epidemiological study, or a meta-analysis?

6. An informative type of experiment is placebo controlled and double blind.

   a. How is a placebo an experimental control?
   b. How does withholding information from both researchers and subjects ensure more reliable results?
   c. DMSO is a chemical believed by some people to relieve a variety of ills, but its safety and effectiveness have not yet been proven. An unpleasant side effect of rubbing DMSO on the skin is that it produces a garlicky taste as it enters the body. How might this peculiarity complicate controlled testing of DMSO?

7. A chemical manufacturing plant closes because of a toxic leak. To examine the possible health effects of the contamination, researchers ask workers and residents to fill out surveys listing their health problems and recording the extent of their exposure to the toxins. What can be learned from this type of study? What are its limitations?

TO THINK ABOUT

1. At the age of six, previously healthy and well-behaved Lorenzo Odone began having severe tantrums. Soon he lost his coordination. After many fruitless visits to doctors, he was finally diagnosed as having the genetic disorder adrenoleukodystrophy (ALD). His future was grim—gradual loss of speech, sight, swallowing, and movement as nerve cells lost their coatings of a fatty substance called myelin. The cause of the problem was a defective enzyme that disrupts the synthesis of long-chain fatty acids.

   Lorenzo's parents, Augusto and Michaela, enrolled him in an experiment to alter the course of the illness by following a special diet. The trial was to run six months—the time researchers felt was adequate to see results. But after two months, with Lorenzo worsening, the Odones took him off the diet and devised their own dietary treatment. They thought that they saw improvements in their son after months of taking their dietary supplement, later called Lorenzo's oil. Subsequent medical studies found that although the oil seems to correct the abnormal biochemistry of ALD, it does not halt symptoms once they have begun, and it is ineffective in treating a milder, adult form of the illness.

   a. Do you think that the Odones were right in not permitting their child to complete the six-month study?
   b. Should such a study include a control group? What might that control group be?
   c. What might be scientifically inaccurate about the Odones' conclusions that Lorenzo was improving after taking the oil?
   d. What further experiment might be conducted to evaluate Lorenzo's oil?

2. Animals are often used to test substances and procedures before they are tested in humans. Many people object to the suffering these animals go through. However, such objections tend to focus on furry, familiar animals, such as mice and dogs, rather than on less appealing creatures such as cockroaches and slime molds. One animal rights activist concerned about a laboratory supply facility in South Carolina withdrew her protest when she discovered that the company bred and distributed leeches, claiming, "We're interested in animal experiments." Of course, leeches *are* animals.

   a. What limits, if any, do you think should be placed on the use of nonhuman animals in experiments?
   b. Suggest two alternatives to animal experimentation.

SUGGESTED READINGS

Borman, Stu. December 13, 1993. Bioinorganic complex is possible treatment for rare genetic disease. *Chemical and Engineering News*, vol. 71. Copper histidine can treat Menkes disease, if given in time.

Forrest, Stephanie. August 13, 1993. Genetic algorithms: principles of natural selection applied to computation. *Science*, vol. 261. Natural selection via computer.

Freedman, David. August 13, 1993. Artificial intelligence helps researchers find meaning in molecules. *Science*, vol. 261. The informational molecules of life are amenable to study with artificial intelligence and computers.

Freudenthal, Anita R., Ph.D., and Paul R. Joseph, M.D. August 19, 1993. Seabather's eruption. *The New England Journal of Medicine*, vol. 329. How the scientific method nailed the cause of a mysterious rash in swimmers.

Gelb, Lenore. March 1992. Hope or hoax? *FDA Consumer*. Determining whether a study is scientifically rigorous becomes a matter of life or death when human health is concerned.

Gorman, Christine. October 26, 1992. Danger overhead. *Time*. Several studies examine hypothesized causal link between power lines and increased risk of developing cancer.

Hamer, Dean H. January 1993. "Kinky hair" disease sheds light on copper metabolism. *Nature Genetics*, vol. 3. Symptoms produced by copper deficiency in diverse organisms illustrate the common descent of life. Studies of the phenomenon illustrate the scientific method at work.

Harris, Curtis C. December 24, 1993. p 53: At the crossroads of molecular carcinogenesis and risk assessment. *Science*, vol. 262. A basic research discovery may lead to understanding (and ultimately curing) cancer.

Hileman, Bette. September 6, 1993. Dioxin toxicity research. Studies show cancer, reproductive risks. *Chemical and Engineering News*, vol. 71. Animal studies show that dioxin affects health at lower exposures than previously thought.

Kanigel, Robert. January/February 1987. Specimen no. 1913. *The Sciences*. A rat's brief life in the service of science—from its point of view. An informative and absolutely riveting account of the use of animals in experiments.

Kassirer, J. P. July 23, 1992. Clinical trials and meta-analysis—What do they do for us? *The New England Journal of Medicine*. Meta-analyses make sense out of many studies.

Larkin, Tim. June 1985. Evidence vs. nonsense: A guide to the scientific method. *FDA Consumer*. The scientific method is fine in theory; in practice, it is sometimes difficult to implement.

Lewis, Ricki. January-February 1993. Biotech devices: Replacing test animals, improving diagnoses. *FDA Consumer*. Clever new devices using biological materials are increasingly used to diagnose disease in place of laboratory-bred animals.

Long, Janice. February 1, 1993. High court ponders use of scientific evidence. *Chemical and Engineering News*, vol.71. When attorneys, judges, and juries evaluate scientific studies, a working knowledge of the scientific method helps.

Marx, Jean. August 20, 1993. Role of gene defect in hereditary ALS clarified. *Science*, vol. 261. Computers helped show how genetic misinformation causes ALS.

Raloff, J. September 4, 1993. 1976 dioxin accident leaves cancer legacy. *Science News*, vol. 144. Decades after a dioxin spill, new health effects are still cropping up.

Rizzo, W. B. September 9, 1993. Lorenzo's oil—hope and disappointment. *The New England Journal of Medicine*, vol. 329. Should parents experiment on their children, when conventional treatments no longer work, rather than cooperate with scientists conducting clinical trials? The Lorenzo's oil story is similar to research on Menkes disease.

Wynne, Brian. June 1991. Public perception and communication of risks: What do we know? *The Journal of NIH Research*, vol. 3. Every time a new risk to human health is discovered, the scientific method is used.

CHAPTER 3

The Origin and Chemistry of Life

The earth's surface was much different at the time of life's beginnings than it is today.

LEARNING OBJECTIVES

By the chapter's end, you should be able to answer these questions:

1. How did experiments show that life does not arise from the nonliving?

2. What does life consist of?

3. What are the components of matter—elements, atoms, molecules, and compounds?

4. How do the numbers and arrangements of subatomic particles give an element its unique properties?

5. What happens when chemicals react with one another?

6. How do chemicals bond with one another?

7. What characteristics of water make it essential to life?

8. What types of macromolecules are components of all living things?

9. Why does a knowledge of chemical concepts help us understand how life began?

The Beginning of Life—Prologue

About 4.5 billion years ago, solid matter began to condense out of a vast expanse of dust and gas, and the earth and other planets of the solar system formed in a fiery birth. For the next billion years, comets, meteors, and immense asteroids pummeled the red-hot ball that was to become our home. These collisions boiled off the seas and vaporized rock to carve the features of the fledgling world. A scant 700 million years after earth was born, living things had made their debut, leaving traces of their cell-like forms in the rocks of present-day Greenland.

How might something as complex and organized as life form during a period of such geologic chaos? How could chemicals aggregate and interact in such a way that they became able to use energy from their surroundings to replicate themselves? And where might this occur? Probably not in the "warm little pond" peacefully brewing life's biochemical precursors, as Charles Darwin envisioned. Nor does it seem likely that life, at least as we know it today, could have begun in a volcano, on land that was frequently demolished by seismic upheaval or incoming cosmic debris, or in an atmosphere that allowed damaging ultraviolet radiation to blanket the planet's surface.

To scientists, perhaps no other question is quite as compelling as how life originated. Researchers from many overlapping fields take diverse factual puzzle pieces and build scenarios, even

FIGURE 3.1

According to the theory of spontaneous generation, this larva arose out of nothingness.

attempting to recreate life's beginnings. Fossilized bacteria; the reactions of key biochemicals interacting in test tubes; the physical properties that might have enabled clay or certain minerals to act as molds for the chemicals of life all provide tantalizing clues to the first time that certain collections of chemicals organized enough to reproduce and persist.

Fossils indicate that life evolved through vast stretches of time, growing increasingly diverse. Today, both single-celled and many-celled forms of life flourish. Despite their diversification, living things still consist of the same chemical components and still use the same chemical information and communication systems.

The chemical similarities of modern organisms, biologists believe, indicate that all life descended from a first type of living organism that somehow partitioned a subset of the chemicals on

early earth. Because life consists of and arose from chemicals, the basic concepts of chemistry provide a paradigm for hypotheses of how life began. But prior to this century, when people knew little about chemistry, they nevertheless had many ideas about how life began.

Spontaneous Generation

In the seventeenth and eighteenth centuries, a popular theory of life's origins was **spontaneous generation,** the appearance of the living from the nonliving. "Evidence" abounded: Beetles and wasps seemed to sprout from cow dung. Egyptian mice arose from the mud of the Nile. Even such noted scientists as Newton, Harvey, and Descartes accepted the theory that life could arise from practically any mess (fig. 3.1).

A few skeptics raised questions about spontaneous generation. In the mid-seventeenth century, Italian physician and poet Francesco Redi conducted the first of a series of experiments that would eventually disprove spontaneous generation. First, Redi put meat in eight flasks—four that were open to the air and to flies and four that were closed. Maggots appeared in the open flasks a few days later, but not in the closed ones. The maggots matured into flies. Next, Redi placed meat in four open flasks and covered them with gauze to exclude flies. No maggots appeared. Redi's conclusion: flies—not some mysterious life-generating substance in the meat—were responsible for the maggots. But despite Redi's experiment, the theory of spontaneous generation persisted.

The invention of the microscope in 1675 by Anton van Leeuwenhoek made possible the study of spontaneous generation of microbial life. In 1748, English naturalist John T. Needham and French naturalist Georges Buffon boiled fresh mutton broth for a few minutes and sealed it, still hot, in a glass flask. A few hours later, on detecting microorganisms in the broth, the researchers concluded that the experiment proved spontaneous generation. But two decades later, Italian biologist Lazzaro Spallanzani showed that Needham and Buffon had not boiled their soup long enough to kill the microbes originally in it.

Spallanzani repeated their work, but he boiled the soup for an hour, then melted the mouths of the flasks shut. A few days later, the soup remained free of microbial life. However, two types of flasks did contain microorganisms: sealed flasks that contained soup that had been boiled for only a few minutes, and flasks that contained thoroughly boiled soup but that were sealed only with corks. Spallanzani's conclusion: microbes in the flasks entered from the air. Life came from preexisting life.

In 1836, German physiologist Theodor Schwann further examined how heating prevents spontaneous generation. He boiled broth in flasks, but instead of sealing the flasks, he exposed their mouths to very hot air. No microbes grew. This seemed to confirm Spallanzani's conclusions. But spontaneous generation supporters proposed that a "vital principle" essential for life was inactivated by heat. It took further experiments to finally dispel the idea that life appears, as if by magic, from nonlife.

French chemist Louis Pasteur boiled broth for a long time in flasks with long, thin S-shaped necks. Although Pasteur left the flasks unsealed, and air might presumably enter, no microbes grew. Hypothesizing that the necks of the flasks were shaped in a way that allowed air in but kept microbe-coated dust particles out, Pasteur repeated the experiment with similar flasks with broken necks. This time, microbes appeared, even though the "vital principle," if it existed, would have been "inactivated" by heat. Pasteur proved that the microbes in the flasks were generated by dust-borne microbes in the air.

Life from Space

Life drifting to the earth from space has been the theme of science fiction, scientific hypotheses, and experiments. Asteroids, meteorites, comets, and interstellar dust contain chemicals like those that appear in life on earth. Some meteorites, for example, contain traces of chemicals resembling those in biological membranes.

How might biochemicals come here from space? According to one scenario, simple organic (carbon-containing) molecules in interstellar dust clouds may have formed complex organic compounds in comets. Heat released by exploding stars may have melted ice in the comets, yielding water that enabled the chemicals to brew into a living soup. Could pieces of comets crashing to earth eons ago have seeded the planet with the molecules that became life? Although this hypothesis might seem plausible, the presence of life-related chemicals in extraterrestrial matter is not evidence of life. These chemicals appear in many unearthly rocks, and they can be synthesized by simulating conditions in outer space. Therefore, they may result from common chemical phenomena right here on earth.

Could life have come here already formed? Swedish physical chemist Svante Arrhenius suggested in his 1908 book *Worlds in the Making* that life arrived from spores that are abundant in the cosmos, a phenomenon he named **panspermia.** Two major objections arose to panspermia. First, spores would not be likely to survive exposure to radiation in space. Second, the possibility that life came to earth does not answer the question of how that life originated—it just moves life's birthplace to an extraterrestrial site. Neither spontaneous generation nor life from space offers a satisfying theory of how life began.

KEY CONCEPTS

Evidence indicates that life arose just 700 million years after the earth formed 4.5 billion years ago. Since then, though life evolved into more complex and diverse forms, it has been based consistently on the same combinations of chemicals. Biological diversity coupled with chemical constancy may reflect a single common ancestor. Experiments proving that life comes from preexisting life dispelled the theory of spontaneous generation.

Energy and Matter—A Recipe for Life

Life consists of matter (material that takes up space) and energy (the ability to do work). Living things can be viewed as sequestered subsets of chemicals that (1) include instructions to reproduce and (2) require energy to maintain their organization. Organisms must not only capture energy, they must transduce (alter) it to a usable form. They then channel it to a specific process, such as bringing a substance into a cell or duplicating genetic material.

Elements

All matter can be broken down into pure substances called **elements.** There are 92 known, naturally occurring elements and at least 14 synthetic ones. The elements are arranged into a chart, called the **periodic table,** in columns according to their properties.

It is interesting to look at the periodic table positions of the elements that comprise living things (fig. 3.2). Even without knowing much about chemistry, it is easy to see that the chemicals of life, because they appear across the

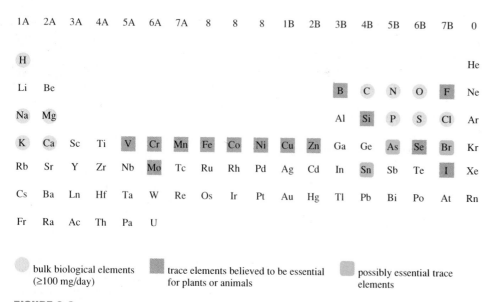

1A	2A	3A	4A	5A	6A	7A	8	8	8	1B	2B	3B	4B	5B	6B	7B	0
H																	He
Li	Be											B	C	N	O	F	Ne
Na	Mg											Al	Si	P	S	Cl	Ar
K	Ca	Sc	Ti	V	Cr	Mn	Fe	Co	Ni	Cu	Zn	Ga	Ge	As	Se	Br	Kr
Rb	Sr	Y	Zr	Nb	Mo	Tc	Ru	Rh	Pd	Ag	Cd	In	Sn	Sb	Te	I	Xe
Cs	Ba	Ln	Hf	Ta	W	Re	Os	Ir	Pt	Au	Hg	Tl	Pb	Bi	Po	At	Rn
Fr	Ra	Ac	Th	Pa	U												

○ bulk biological elements (≥100 mg/day)   ▨ trace elements believed to be essential for plants or animals   ▦ possibly essential trace elements

FIGURE 3.2

Elements of life. Robert J. P. Williams of the inorganic chemistry laboratory, Oxford University, writes: "The true chemistry of living systems is a fascinating combination of many elements from the Periodic Table used in some optimum fashion but adjusted according to species which have become stabilized during evolutionary development." The periodic table organizes the elements by their similarities in chemical reactivity. Even without knowing much chemistry, it is easy to see that the elements of life represent a diverse sampling from the chemical world.

table, represent a diverse sampling of all chemicals. Twenty-five elements are essential to life. Elements needed in large amounts—such as carbon, hydrogen, oxygen, nitrogen, sulfur, and phosphorus—are termed **bulk elements,** while those required in small amounts are called **trace elements.** Many trace elements are important parts of **enzymes,** proteins that regulate the rates of chemical reactions in living things. Some elements that are toxic in large amounts, such as arsenic, may actually be vital in very small amounts; these are called **ultratrace elements.**

KEY CONCEPTS

Living things consist of the same chemical elements, contain instructions to reproduce themselves, and require energy, which must be transduced into a usable form. The elements of life represent a diverse sampling of all elements. Bulk elements are abundant in living organisms. Trace and ultratrace elements are less plentiful.

Atoms

An **atom** is the smallest possible "piece" of an element that retains the characteristics of the element. An atom cannot be broken down by chemical means, but a great deal of energy can "split" it into **subatomic particles.**

Structure of the Atom

The three major types of subatomic particles are **protons** and **neutrons,** which form a centralized mass called the **nucleus;** and **electrons,** which surround the nucleus. Atoms of different elements are distinguished by their characteristic numbers of subatomic particles.

Protons carry a positive charge, electrons carry a negative charge, and neutrons are electrically neutral. Within an atom, the charges on protons and electrons cancel each other out. An electron has about 1/2,000 the mass of a proton or neutron. If the nucleus of an atom were the size of a meatball, then the electrons would be about a kilometer (0.62 mile) away from it!

The periodic table depicts in short-hand the structures of the atoms of each element. Each element has a symbol, which can come from the English word for that element (*He* for *helium,* for example), or from the word in another language (*Na* for *sodium,* which is *natrium* in Latin). The number in the upper left-hand corner of each element symbol is the **mass number,** which refers to the total number of protons and neutrons in the nucleus of the atom (fig. 3.3). The mass number estimates atomic mass because it does not count the tiny contributions of the electrons (which are a minute fraction of the other particles in mass). The **atomic number,** in the lower left-hand corner, shows the number of protons in the atom (which often, but not always, equals the number of electrons). The elements are arranged sequentially by atomic number (see the periodic table in the appendix). You can calculate the number of neutrons by subtracting the atomic number (the number of protons) from the mass number (the total number of protons and neutrons). Atoms of different elements may vary greatly in size. Hydrogen, the simplest type of atom, has only one proton and one electron, while an atom of uranium contains 92 protons, 146 neutrons, and 92 electrons!

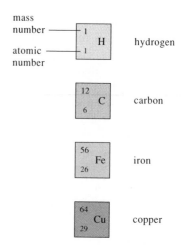

FIGURE 3.3

Four elements, as represented in the periodic table.

Isotopes Vary in Neutron Number

Elements can exist in different forms, called **isotopes,** that have different numbers of neutrons. Isotopes of an element all have the same charge, but different masses. (Weight and mass are related; weight is mass taking gravity into account.) Often one isotope of an element is very abundant, and others are rare; this is the case for oxygen (table 3.1). Similarly, 99% of carbon isotopes have six neutrons, while only 1% are isotopes with seven or eight neutrons. Tin (Sn) has the highest number of naturally occurring isotopes—10. An element's atomic mass in the periodic table is the average mass of its isotopes.

Of the 1,500 known isotopes, 1,236 are unstable, which means they have extra neutrons and tend to break down into more stable forms of the elements. (Atoms with the same numbers of protons and neutrons are more stable.) When these unstable isotopes break down, they emit radioactive energy. Every type of radioactive isotope has a characteristic **half-life,** the time it takes for half of the atoms in a sample to emit radiation, or "decay" to a different, more stable form.

Electrons Determine Reactivity

Electrons move about the nucleus near the speed of light; it is impossible to represent their positions accurately in a static diagram. Often electrons are illustrated as dots moving in concentric circles, much as planets orbit the sun (fig. 3.4). Although these **Bohr models** depict interactions between atoms, they do not realistically capture the cloudlike distribution of electrons moving constantly about the nucleus.

A more accurate way to describe electron behavior is to envision areas around the nucleus where the probability of finding an electron is the greatest. An **electron orbital** is the volume of space where a particular electron is found 90% of the time. An orbital is not a physical entity, but a way of imagining or representing electron arrangement.

Groups of electron orbitals form levels of energy called **energy shells.** Electrons tend to occupy the lowest

Table 3.1	**Isotopes of Oxygen**		
Isotope	**Atomic Shorthand**	**Abundance**	**Number of Neutrons**
Oxygen-16	$^{16}_{8}O$	99.76%	8
Oxygen-17	$^{17}_{8}O$	0.04%	9
Oxygen-18	$^{18}_{8}O$	0.20%	10

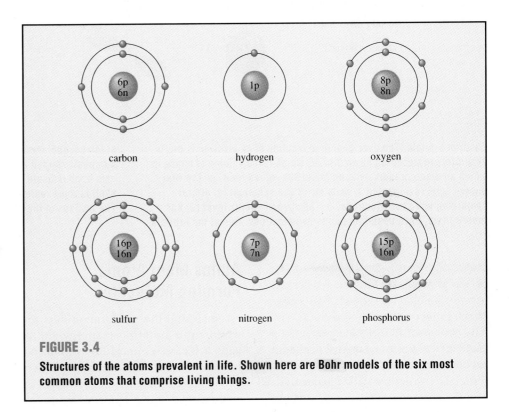

FIGURE 3.4

Structures of the atoms prevalent in life. Shown here are Bohr models of the six most common atoms that comprise living things.

possible energy shell; that is, the shell closest to the nucleus. This first energy shell can hold up to two electrons. It has one electron orbital, which is shaped like a sphere and is called an "s" orbital. The second energy shell includes one spherical orbital that is larger than the one in the first energy level, plus three dumbbell-shaped orbitals, called "p" orbitals, that are arranged as if they occupy imaginary x, y, and z axes (fig. 3.5). Each of these four orbitals in the second energy level can hold two electrons, for a total of eight electrons. Energy shells farther from the nucleus, in atoms of the larger elements, have more orbitals. The outermost shells usually contain eight or fewer electrons.

Atoms are the most stable when their outermost energy shells are filled. Usually the outermost shell is filled with eight electrons; the tendency of an atom to fill its outermost shell is thus called the **octet rule.** Atoms fulfill the octet rule and achieve stability by interacting with each other, sharing or exchanging electrons in their outermost shells until they have eight. The number of outer shell electrons, or **valence electrons,** determines the atom's chemical properties. Elements whose atoms have the same number of valence electrons have similar characteristics, and they appear in the same column in the periodic table.

Interactions in which atoms exchange or share electrons, forming new

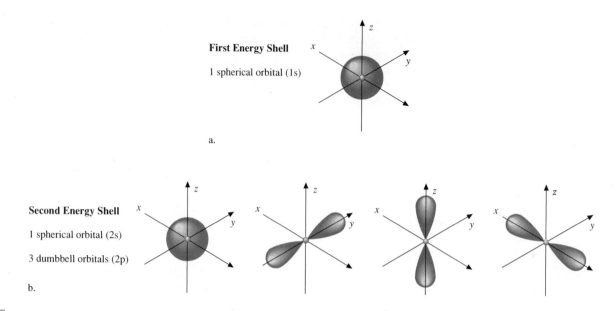

First Energy Shell

1 spherical orbital (1s)

a.

Second Energy Shell

1 spherical orbital (2s)

3 dumbbell orbitals (2p)

b.

FIGURE 3.5

Electron orbitals. The ever-changing location of an electron is more accurately represented by an orbital, which is the volume of space in which an electron is likely to be found 90% of the time. *a.* The first energy level (designated *1*) consists of one spherical (*s*) orbital containing up to two electrons. *b.* The second energy level (*2*) has four orbitals, each containing up to two electrons. One of the orbitals of the second energy level is spherical (*s*); the other three are dumbbell-shaped and arranged perpendicular to one another, as the axis lines indicate. They are designated *p.* Therefore, the first energy shell is represented as *1s,* and the second energy level as *2s* and *2p.* The nucleus of the atom is at the center, where the axes intersect.

chemicals in the process, constitute **chemical reactions.** Chemical reactions account for all of the chemical processes of the universe, including those of life. The process of sharing and exchanging electrons creates attachments, or **chemical bonds,** between atoms. Bonded atoms form larger structures called **molecules.**

KEY CONCEPTS

The smallest characteristic unit of an element, an atom, consists of neutral neutrons, positively charged protons, and negatively charged (and much smaller) electrons. The elements are organized on the periodic table according to their atomic number (number of protons). The periodic table also indicates each element's mass number—the number of protons and neutrons in one atom of that element. Isotopes are variations of atoms that have extra neutrons; isotopes may be unstable and radioactive. The farther away an electron is from the nucleus, the more likely it is to engage in a chemical reaction. Electrons are constantly moving, and the places they can be found are described as electron orbitals. Orbitals occupying the same energy levels comprise energy shells. Atoms react with each other to fill their outermost shells with electrons.

Atoms Meet Atoms, Forming Molecules

Atoms combine to form molecules when their outermost energy shells approach closely enough to overlap and to share or exchange electrons. When two or more different types of atoms—a sodium atom and a chlorine atom, for example—combine, they form a **compound.** (In a few cases, two atoms of the same element form a compound molecule. Hydrogen, oxygen, and nitrogen gases consist of such "diatomic" [two-atom] molecules.)

A compound's characteristics differ from those of the elements it contains. Consider table salt, which is the compound sodium chloride. A molecule of salt contains one atom of sodium (Na) and one atom of chlorine (Cl). Sodium is a silvery, highly reactive metal that is solid at room temperature, while chlorine is a yellow, corrosive gas. But when equal numbers of these two types of atoms combine, the resulting compound is a white crystalline solid—the familiar table salt.

Chemical Shorthand

Scientists describe molecules by writing the symbols of their constituent elements and subscripting the numbers of atoms of each element in one molecule. For example, the sugar molecule glucose, $C_6H_{12}O_6$, has 6 atoms of carbon, 12 of hydrogen, and 6 of oxygen. A coefficient indicates the number of molecules—six molecules of glucose is written $6C_6H_{12}O_6$.

Chemical reactions are depicted as equations with the starting materials, or **reactants,** on the left and the end products on the right. The total number of atoms of each element is the same on either side of the equation. For example, the equation for the formation of glucose in a plant cell is:

$$6\ CO_2 + 6\ H_2O \rightarrow C_6H_{12}O_6 + 6\ O_2$$

This means that six molecules of carbon dioxide and six molecules of water react to produce one molecule of glucose plus six molecules of diatomic oxygen. Note that each side of the equation indicates 6 total atoms of carbon, 18 of oxygen, and 12 of hydrogen. Most chemical reactions

NO and CO—Tiny Biological Messengers

Before 1992, nitric oxide (NO) had a bad reputation. It was notorious for its presence in smog, cigarettes, and acid rain. But by late 1992, researchers had implicated the small molecule in several roles in physiology, including digestion, memory, immunity, respiration, and circulation. Scientists offered one possible explanation for NO's eclectic set of roles in life—it has one unpaired electron, making it highly reactive.

NO made headlines because of its role in causing penile erection in humans. NO helps dilate blood vessels in the penis, engorging the organ with blood, so that it stiffens and rises. Elsewhere in the body, NO released by cells lining certain blood vessels signals nearby muscles to relax. This gives the vessels room to expand, lowering blood pressure. Perhaps NO's most interesting role is its action as a neurotransmitter, a type of biochemical messenger that enables nerve cells to communicate with each other.

NO was the first gas identified as a neurotransmitter, a designation usually associated with large molecules that travel slowly between cells.

If NO, a small, gaseous molecule, is a neurotransmitter, do other small molecules transmit nerve cell messages too? Solomon Snyder, a Johns Hopkins University neuroscientist who has discovered and described many neurotransmitters, suspected this might be true of carbon monoxide (CO).

Like NO, CO had a nasty reputation—as a colorless, odorless, and lethal gas—but CO also seemed to be involved in physiology. CO is found in the spleen, an organ where old red blood cells are dismantled. Laboratory studies showed that CO binds to another type of biological communicator called a second messenger. This indicated that CO might function as a neurotransmitter. Snyder hypothesized that if CO acts as a neurotransmitter, the enzyme responsible for synthesizing it would probably be found in specific brain regions, as the enzymes for other neurotransmitters are. Experiments proved his hypothesis:

CO was detected in the memory center, smell center, and lower parts of the human brain.

Nerve cells growing in culture display second-messenger activity if CO is present, but not if it is absent.

The second messenger is found in the same parts of the brain as the enzyme needed to synthesize CO.

Slices of a rat brain's memory center normally exhibit a pattern of chemical signaling between cells that indicates long-term memory being laid down. If a researcher blocks CO production, this signaling ceases, indicating that CO is involved in memory.

Neuroscientists now have an entirely new class of neurotransmitters to work with, headed by two tiny chemicals that were right under their noses. More importantly, the recently described functions of CO and NO have reaffirmed the statement made in chapter 2: when implementing the scientific method, one must always expect the unexpected.

---

proceed in two directions, as products convert back into reactants at the same time the reactants convert into products. This situation is called **chemical equilibrium;** we indicate it by drawing an arrow going both ways in the equation.

The Molecules of Life

Living things are composed mostly of water and organic (carbon-containing) molecules. Biochemicals (organic molecules found in organisms) are also rich in hydrogen, oxygen, nitrogen, phosphorus, and sulfur. Many biological molecules are so large that they are termed **macromolecules.** The plant pigment chlorophyll, for example, contains 55 carbon atoms, 68 hydrogens, 5 oxygens, 4 nitrogens, and 1 magnesium. **Molecular weight** measures a molecule's size. It is calculated by adding the atomic weights (masses) of the constituent atoms. Carbon monoxide (CO) and carbon dioxide (CO_2) are often not considered organic because of their simple structures. However, even these small molecules can have profound effects on life (see Biology in Action 3.1).

Covalent Bonds—Electron Sharing

There are several types of chemical bonds. The number of valence (outermost shell) electrons determines which type of chemical bond can form with the least amount of energy, and is therefore most likely to occur. Many of the molecules of life are held together by **covalent bonds,** which form when two atoms share valence electrons. Covalent bonds tend to form among atoms that have three, four, or five valence electrons as they seek to fill their outermost shells with eight.

A covalent bond alters the shapes of the electron orbitals involved in the bond. Consider carbon, which has four electrons in its outermost shell. It can attain the stable eight-electron configuration in its outer shell by sharing electrons with four hydrogen atoms, each of which has one electron in its only shell (fig. 3.6). When carbon bonds to the four hydrogen atoms, its energy redistributes so that the one spherical and three dumbbell-shaped orbitals in its outermost shell resemble teardrops. The resulting molecule, methane (CH_4), has a tetrahedral (square-based pyramid) shape. Methane is a swamp gas; its chemical structure can be represented in several ways (fig. 3.7).

Single electron pairs are shared in covalent bonds to form methane. But two or three electron pairs can also be shared in double and triple covalent bonds. Carbon atoms can form all three types of covalent bonds with other carbon atoms. The fact that a carbon atom may form four covalent bonds to satisfy the octet rule allows this element to assemble into long chains, intricate branches, and rings. Carbon chains and rings bonded to hydrogens are called **hydrocarbons.** When carbon bonds with other chemical groups, various other biological molecules result.

Methane illustrates a **nonpolar covalent bond,** which means that the atoms share all of the electrons in the covalent bond equally. In contrast, in a **polar covalent bond,** electrons draw more towards one atom's nucleus than the other. Water (H_2O), which consists of two hydrogen atoms bonded to an oxygen atom, is one example of a molecule held together by polar covalent bonds (fig. 3.8). The nucleus of each oxygen atom attracts the electrons on the hydrogen atoms more than the hydrogen nuclei do. As a result, the area near the oxygen carries a partial negative charge (from attracting the negatively charged electrons), and the area near the hydrogens is slightly positively charged (as the electrons draw away), similar to the way each end of a bar magnet carries an opposite charge. The tendency of an atom to attract electrons is termed **electronegativity.** Oxygen is

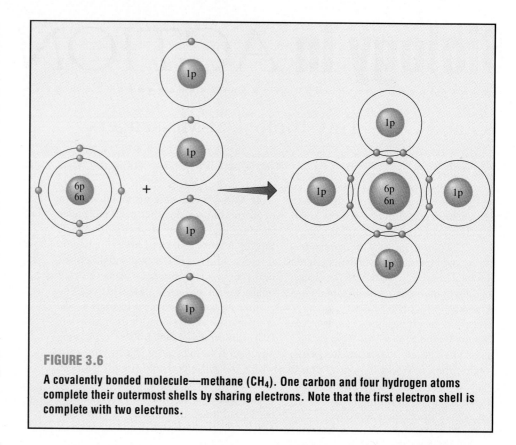

FIGURE 3.6

A covalently bonded molecule—methane (CH_4). One carbon and four hydrogen atoms complete their outermost shells by sharing electrons. Note that the first electron shell is complete with two electrons.

highly electronegative. Its ability to accept electrons is crucial in helping organisms transduce energy (as we shall see in chapter 6).

Ionic Bonds

In a covalent bond, atoms share electrons. By contrast, in an **ionic bond,** the electronegativity of one atom is so great that the atom takes an electron from the other atom. An atom with one, two, or three electrons in the outermost shell loses them to an atom with, correspondingly, seven, six, or five valence electrons. A sodium (Na) atom, for example, has one valence electron. When it donates this electron to an atom of chlorine (Cl), which has seven electrons in its outer shell, the two atoms bond ionically to form NaCl (fig. 3.9a).

Once an atom loses or gains electrons, it has an electric charge. An atom with an electric charge is called an **ion.** Atoms that lose electrons lose negative charges, and thus carry a positive charge. The process of losing an electron is called **oxidation,** and the resulting positively charged ion is called a **cation** (pronounced căt´- īon). Gaining an electron (and thus a negative charge) is called **reduction,** and a negatively charged ion is called an **anion** (pronounced ăn´-ī on). The attraction between oppositely charged ions forms the ionic bond. In NaCl, the oppositely charged ions Na^+ and Cl^- attract in such an ordered manner that a crystal results (fig. 3.9b). A large lattice built of cations and anions, like NaCl, is called a **salt.** A salt can consist of any number of ions, and it is therefore described in terms of the ratio of one type of ion to another. Ionic bonds are

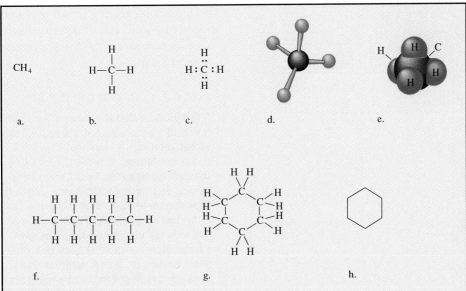

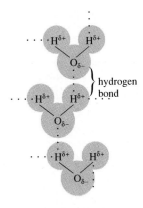

δ = partial charge

FIGURE 3.8

The two hydrogen atoms and one oxygen atom of water (H_2O) are held together by polar covalent bonds. Because the oxygen attracts the negatively charged hydrogen electrons more strongly than the hydrogen nuclei do, the oxygen atom bears a partial negative charge, and the hydrogens carry a partial positive charge. Hydrogen bonding is a result of these partial charges.

FIGURE 3.7

Different types of diagrams used to represent molecules. Consider methane, a gas present on the early earth that may have played a pivotal role in the debut of life. *a.* The molecular formula, CH_4, indicates that methane consists of one carbon atom bonded to four hydrogen atoms. *b.* The structural formula shows those bonds as single. *c.* The electron dot diagram shows the number and arrangement of shared electron pairs. *d.* A ball-and-stick model gives further information—the angles of the bonds between the hydrogens and the carbon. *e.* A space-filling model shows bond relationships as well as the overall shape of a molecule. *f.* and *g.* The organic compounds of life are built of carbon chains (*f*) and rings (*g*), with attached hydrogens, oxygens, and sometimes other elements such as nitrogen, phosphorus, and sulfur. *h.* The carbons of ring-shaped molecules are sometimes indicated only by the vertices of the bond lines. The hydrogen atoms in such structures are usually not explicitly shown.

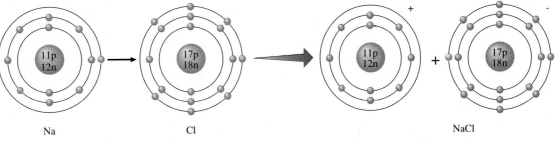

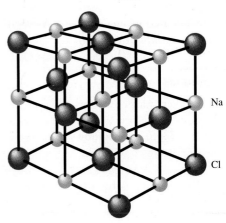

FIGURE 3.9

An ionically bonded molecule. *a.* A sodium atom (Na) can donate the one electron in its valence shell to a chlorine atom (Cl), which has seven electrons in its outermost shell. This satisfies the octet rule. The resulting ions (Na^+ and Cl^-) bond to form the compound sodium chloride (NaCl), better known as table salt. *b.* The ions that constitute NaCl are arranged in a repeating pattern.

41

Table 3.2 **Chemical Bonds**

Type	Chemical Basis	Strength	Example
Covalent	Atoms share electron pairs	Strong	Hydrocarbons
Hydrogen	Atoms with partial negative charges attract hydrogen atoms with positive charges (due to their participation in a polar covalent bond)	Weak	Water
Ionic	One atom donates an electron to another	Weak in water	Sodium chloride
Van der Waals	Oppositely charged regions between or within molecules attract one another	Weak	Protein structure
Water interactions	Hydrophobic parts of molecules avoid water; hydrophilic parts are attracted to water	Strong	DNA structure

strong in a chemical sense, but they weaken in water. Because many of the molecules of life dissolve in water, ionic bonds in biology are considered weak.

Most biochemicals are anions. Their negative charges are balanced by the many small cations abundant in living systems, such as sodium (Na^+) and potassium (K^+). Ions are important in many biological functions. Nerve transmission depends upon passage of Na^+ and K^+ in and out of cells. Muscle contraction depends upon calcium (Ca^{2+}) ions. Even the start of a new life is ionically controlled. In many types of organisms, calcium ions released by egg cells clear a path for the sperm's entry into the egg.

Weak Chemical Bonds

Chemical bonds weaker than covalent or ionic bonds are also important in the molecules of life. **Hydrogen bonds** form when hydrogen atoms covalently bonded to one atom are attracted to an atom in another molecule. A hydrogen bond is based on the hydrogen atom's partial positive charge, which results from the atom's participation in

a polar covalent bond. These bonds are abundant in water, as we shall soon see. A hydrogen bond by itself is weak, but many of them put together are much stronger. For example, numerous hydrogen bonds help provide the great strength of DNA, the genetic material. DNA consists of ladderlike chains held together at the "rungs" by hydrogen bonds.

Different parts of the large and intricately shaped chemicals of life may be temporarily charged because their electrons are always in motion. Dynamic attractions between molecules or within molecules that occur when oppositely charged regions approach one another are called **van der Waals attractions.** They help shape the molecules of life.

Interaction with water is another type of chemical attraction that shapes molecules. **Hydrophilic** (water-loving) parts of molecules are attracted to water; **hydrophobic** (water-hating) parts are repelled by water. Large molecules such as proteins contort so that their hydrophilic regions touch water, and their hydrophobic regions shun it. Interactions with water help foster

the molecular shapes that build cell membranes and their inner structures. Table 3.2 summarizes the chemical bonds important in biology.

> ### KEY CONCEPTS
> A compound's characteristics differ from those of its constituent atoms. When atoms interact in chemical reactions, bonds are formed or broken, converting reactant compounds to different compounds or products while conserving the total number of atoms of each element. In a covalent bond, atoms with three, four or five valence electrons share electrons. Carbon atoms covalently bond to hydrogen atoms to build hydrocarbons. Nonpolar covalent bonds share electrons equally, while polar covalent bonds attract electrons more towards one nucleus than the other, creating a molecule with negative and positive ends. An ionic bond forms when one, two, or three electrons transfer from the valence shell of one atom to the valence shell of another, forming ions. Ions carry electrical charges and tend to associate in an ordered network (salt). Ionic bonds in water are weak. Hydrogen bonds are also weak. They form when hydrogens covalently bonded to one atom, and thus carrying a partial positive charge, are drawn to an atom in another molecule. Van der Waals forces come into play when oppositely charged molecules or parts of molecules attract one another. These forces help create the molecular shapes essential to biological structures. Molecular shapes are also influenced by attractions to and repulsions from water.

Water—The Matrix of Life

Water has been called the "mater and matrix" of life. "Mater" means mother, and indeed life could not have begun were it not for this unusual substance. Water is also the matrix, or medium, of life, for it is vital in most biochemical reactions.

Solutions

In one second, the hydrogen bonds between a single water molecule and its nearest neighbors form and reform

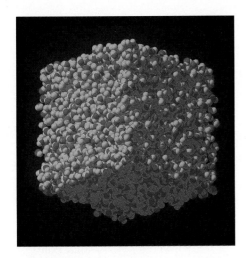

FIGURE 3.10

Using this computer-simulated "box" of 4,096 water molecules, chemists can study the intricate interactions of this fascinating and vital chemical. The blue balls represent hydrogen, the red balls oxygen.

Table 3.3 Properties of Water

Property	Example of Importance to Life
Adhesion	Biological fluids act as aqueous solutions of electrolytes
Cohesion	Water flows, forming habitats and a matrix for biological molecules
Imbibition	Embryo within seed absorbs water, swelling and bursting through seed coat
High heat capacity	Water regulates constant internal body temperature
High heat of vaporization	Water evaporation (sweating) functions as cooling mechanism

some 500 billion times (fig. 3.10)! This makes water fluid. The strong attractions between water molecules, which account for their constant rebonding, is called **cohesion.**

Water readily forms hydrogen bonds to many other compounds (fig. 3.11), a property called **adhesion.** Adhesion is important in many biological processes. For example, adhesion enables water to form body fluids that carry vital dissolved chemicals called **electrolytes.** Movement of water from a plant's roots to its highest leaves depends upon cohesion of water within the plant's water-conducting tubes and upon the adhesion of water to the walls of those tubes. The adhesiveness of water also accounts for

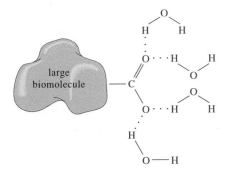

FIGURE 3.11

Water acts as a solvent for many molecules by forming hydrogen bonds.

imbibition, the tendency of substances to absorb water and swell. Rapidly imbibed water swells a seed so that it bursts through the seed coat, releasing the embryo within. Table 3.3 lists some of the important properties of water.

A **solvent** is a chemical in which other chemicals, called **solutes,** dissolve. A **solution** consists of one or more chemicals dissolved in a solvent. In an **aqueous solution,** water is the solvent. Water molecules flow between the cations and anions of a salt, and they also separate molecules held together by polar covalent bonds. (Remember, the hydrogen atoms in a water molecule carry a partial positive charge, which causes them to attract other atoms and form new hydrogen bonds. In this sense, they "pull apart" molecules held by polar covalent bonds.)

Nonpolar covalently bonded molecules such as fats and oils do not carry a charge, are not attracted to hydrogen atoms, and thus do not usually dissolve in water. This is why oil on the surface of a cup of coffee forms a visible drop; likewise, a drop of water on an oily surface beads up. Figure 3.12 illustrates how water molecules spread out on a hydrophilic substance, and cling together on a hydrophobic substance, such as a fat or oil.

Acids and Bases

In an aqueous solution, hydrogens from one water molecule tend to be pulled to another, transforming two molecules of

water ($2H_2O$) into a **hydronium ion** (H_3O^+) and a **hydroxide ion** (OH^-). (The H_3O^+, H_2O plus H^+, can be considered the chemical equivalent of H^+.) In pure water, the numbers of dissociated (free) H^+s and OH^-s are equal. However, when another chemical is dissolved in water, the balance of positive and negative charges can change. A substance that adds more H^+ to a solution is an **acid.** Common acids include hydrochloric acid (HCl), sulfuric acid (H_2SO_4), and nitric acid (HNO_3). Substances that lower the number of H^+s are **bases.** In water, bases typically dissociate (come apart) to yield OH^-, which binds with some of the H^+ present in the dissociated water to reform H_2O. A common base is sodium hydroxide (NaOH), or lye.

The **pH scale** gauges how acidic or basic a solution is in terms of its H^+ concentration. The pH scale ranges from 0 to 14, with 0 representing strong acidity (high H^+ concentration) and 14 strong basicity or alkalinity (low H^+ concentration). A neutral solution, such as pure water, has a pH of 7. Many fluids in the human body function within a narrow pH range (fig. 3.13). Illness results when pH changes beyond this range. The normal pH of blood, for example, is 7.35 to 7.45. Blood pH of 7.5 to 7.8, a condition called **alkalosis,** makes one feel agitated and dizzy. Alkalosis can be caused by breathing rapidly at high altitudes, taking too many antacids, enduring high fever, or feeling extreme anxiety. **Acidosis,** in which blood pH falls to 7.0 to 7.3,

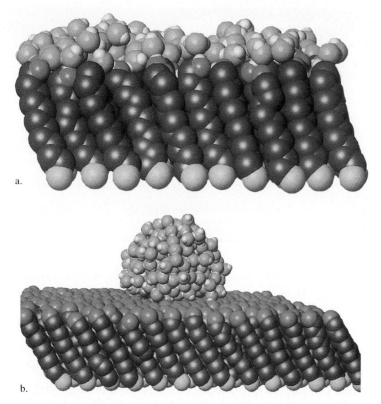

a.

b.

FIGURE 3.12

Hydrophilicity and hydrophobicity are demonstrated in this computer simulation. *a.* Water molecules form a thin film atop a hydrophilic substance. *b.* The same molecules bead up into a ball in less than a billionth of a second when placed on a hydrophobic surface. As few as a dozen or as many as a few million water molecules form these configurations.

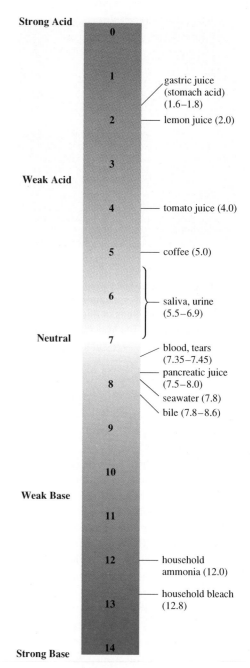

FIGURE 3.13

The pH scale is commonly used to measure the strength of acids and bases.

makes one feel disoriented and fatigued, and may impair breathing. This condition may result from severe vomiting, diabetes, brain damage, or lung or kidney disease.

Organisms have pairs of weak acids and bases, called **buffer systems,** that maintain the pH of body fluids in a comfortable range by gaining or losing hydrogens to neutralize strong acids and bases. Carbonic acid (H_2CO_3) and sodium bicarbonate ($NaHCO_3$) form one such buffer system. Carbonic acid, a weak acid, helps lower pH when sodium hydroxide (NaOH) builds up, reacting to form the weak base sodium bicarbonate:

$$HCl + NaHCO_3 \rightarrow H_2CO_3 + NaCl$$

Acid solutions are also important to life on a global scale. Sulfur and nitrogen oxides in pollution react with water in the atmosphere to return to earth as acid precipitation. Just slightly lowering the pH of a body of water can have a drastic effect on the types of organisms that can survive there. Some organisms, however, do adapt to low-pH environments (see Biology in Action 3.2).

Water and Temperature

Several temperature-related qualities of water have important implications for living organisms. Water is valuable in biological temperature control because a great deal of heat is needed to raise its temperature—a characteristic called high **heat capacity.** Because of water's high heat capacity, an organism may be exposed to considerable heat before its aqueous body fluids become dangerously warm. Water's high **heat of vaporization** means that a lot of heat is required to evaporate water. This is why evaporating sweat draws heat out from the body—again, an important factor in regulating body temperature.

Another important characteristic of water is its unusual tendency to expand upon freezing. The bonds holding ice together are farther apart than those in liquid water, and therefore ice is less dense and floats on the surface of liquid water. This characteristic presents advantages to some water-inhabiting organisms. When the air temperature drops sufficiently, a small amount of water freezes at the surface of a body of water, forming a solid cap of ice that retains heat in the water below. Many

Biology in ACTION

Acid-Loving Life Forms

A body of water with a very low (acidic) pH may not seem a very hospitable place. At first glance, the many acidic lakes and streams in Yellowstone National Park appear to lack life, except for carpets of algae and an occasional flock of Canadian geese. Nowhere to be found are the fishes, tadpoles, turtles, and snakes usually seen in freshwater habitats. But a closer perusal reveals that even at a pH as low as 2, a few hardy species thrive. Acid environments are home to bacteria, algae, protozoa (singe-celled animals), and invertebrates, particularly insects. The success of these acid-loving species, or **acidophiles,** can probably be attributed both to adaptations and to a lack of competition and predation.

Acid Environments

Some spots on earth are more acidic than others. Typically, acid environments arise when sulfur-containing compounds called **sulfides** encounter oxygen. A chemical reaction produces sulfuric acid (H_2SO_4). Sulfides are common in swamps, bogs, and volcanoes.

Yellowstone Park's sulfur-rich geothermal areas are called solfataras, named after a huge volcanic crater along Italy's Bay of Naples that is rich in elemental sulfur. The sulfur, where it is exposed, reacts with oxygen to form sulfuric acid. In the Carolinas and Georgia, and also in the Netherlands and parts of Africa and southeast Asia, sulfur-rich sediments along shorelines dry out, exposing sulfides to oxygen. As soil forms in these areas, it is very acidic; it cannot be farmed because plants will not grow at low pH.

In the eastern United States, sulfuric acid forms when mining exposes sulfur-containing sediments. Rain water carries the sulfuric acid into lakes and streams, where it dissolves iron that then forms a foam on the water's surface. When sulfur-containing oil and coal are burned and then come into contact with water, sulfuric acid forms once again. In any of these seemingly harsh environments, acidophilic organisms may be found.

The acidophiles are fascinating to biologists because they exist in an environment that normally dismantles key biochemicals, including genetic material and biochemicals that extract or store chemical energy. Somehow, these organisms adapted so that they maintain the same chemical constancy as organisms that live in less acidic places.

organisms can then survive in the depths of such shielded lakes and ponds. If water were to become denser upon freezing, ice would sink to the bottom, and the body of water would gradually turn to ice from the bottom up, entrapping the organisms that live there.

On the other hand, living things do freeze and die because water inside cells expands when it freezes, breaking apart the cells. This phenomenon is similar to filling a plastic container with liquid water, sealing it, and placing it in a freezer. The container pops its top as the freezing water expands. Some invertebrates (animals without backbones) living in the Antarctic have adapted to having icy cellular interiors. Certain springtails (wingless insects) and mites produce antifreeze biochemicals that keep body fluids liquid at temperatures low enough to freeze pure water. Tardigrades, which are relatives of spiders, dehydrate to avoid swelling and breaking their cells (fig. 3.14).

Insects called larval midges freeze, but the ice forms between their cells, not inside them.

FIGURE 3.14

Tardigrades are called waterbears. They are small (less than a millimeter long) and live on aquatic vegetation. Under environmental stress, including freezing, a tardigrade gradually dries up, its body composition changing from 85% water to about 3% water. This creates a state of suspended animation called cryptobiosis that can extend a tardigrade's life span from 1 to 60 years.

KEY CONCEPTS

Water is essential to life for many reasons. It is involved in many chemical reactions; water hydrogen bonds to itself and to many other substances, including body fluids, though not to fats and oils. Water also helps regulate pH. Acids add H^+ to water, and bases remove H^+. Buffer systems, pairs of weak acids and bases, maintain the pH of body fluids by neutralizing stronger acids and bases. Finally, water's high heat capacity and high heat of vaporization enable it to help regulate body temperature by making it difficult to heat body fluids. Water's expansion upon freezing also affects life; this quality preserves the outer environment water-dwelling organisms require, while other organisms have adapted to protect their cellular interiors from freezing and bursting.

FIGURE 3.15

Glucose + fructose −H₂O = sucrose. Dehydration synthesis between two monosaccharides yields a disaccharide.

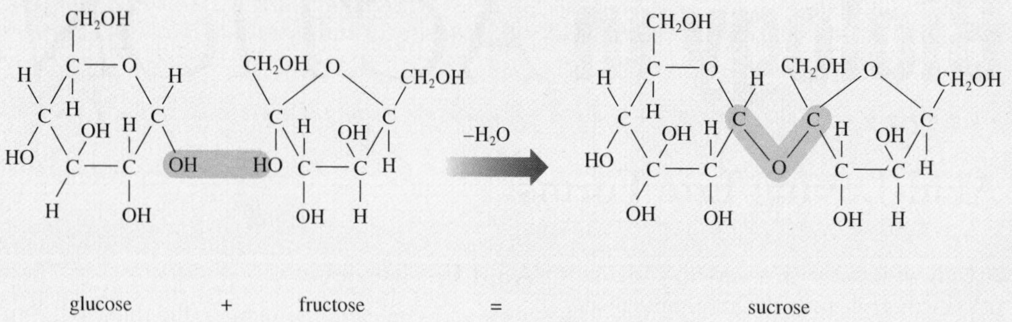

glucose   +   fructose   =   sucrose

The Organic Molecules of Life

Organisms are composed of and must take in large amounts of four types of organic compounds: carbohydrates, lipids, proteins, and nucleic acids. Vitamins are another group of biologically important organic molecules, but they are needed in smaller amounts.

Carbohydrates

Carbohydrates, which include sugars and starches, contain carbon, hydrogen, and oxygen, with twice as many hydrogen atoms as oxygen atoms in each molecule. Carbohydrates release energy when their bonds break. They also physically support cells and tissues.

Carbohydrates are classified by size. **Monosaccharides** (single sugars) contain five or six carbons. Three six-carbon sugars with the same molecular formula ($C_6H_{12}O_6$) but different chemical structures are glucose (blood sugar), galactose, and fructose (fruit sugar). **Disaccharides** (double sugars) form when two monosaccharides link by releasing a molecule of water (H_2O) (fig. 3.15). This type of chemical reaction is called **dehydration synthesis** (made by losing water). In the opposite reaction, **hydrolysis** (breaking with water), a disaccharide and water react to form two monosaccharides. Dehydration synthesis builds molecules, and hydrolysis breaks them down. Figure 3.15 shows how dehydration synthesis between glucose and fructose forms the disaccharide sucrose.

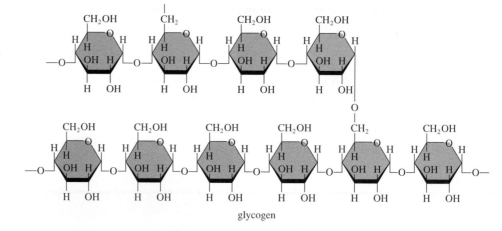

cellulose

starch

glycogen

FIGURE 3.16

The complex carbohydrates cellulose, starch, and glycogen are all chains of glucose subunits, but they differ in structure. The glucose subunits of cellulose (found in plants) link together differently than the glucose subunits of starch (also found in plants). Glycogen, a glucose polymer found in animals, is more highly branched than either cellulose or starch.

Sucrose, or table sugar, is found in many plants, including sugarcane and beets. Maltose, a disaccharide formed from two glucose molecules, provides energy in sprouting seeds and is used to make beer. Lactose, or milk sugar, is a disaccharide formed from glucose and galactose. Sugars, which are monosaccharides and disaccharides, are **simple carbohydrates.**

Complex carbohydrates, or polysaccharides (many sugars), are long chains of simple sugars linked by dehydration synthesis (fig. 3.16). Such a long

Table 3.4 The Macromolecules of Life

Type of Macromolecule	Structure	Functions	Examples
Lipid	Hydrocarbon chains plus oxygens	Membranes Energy storage	Fats, oils, cholesterol, sex hormones, triglycerides
Carbohydrate	Sugars and polymers of sugars	Energy storage Support	Sugars, starches, chitin, glycogen
Protein	Polymer of amino acids	Catalysis Receptors Transport Antibodies Control of gene expression	Digestive enzymes, hemoglobin, keratin, muscle proteins
Nucleic acid	Polymer of nucleotides	Transfer of genetic information	DNA, RNA

molecule of smaller units is called a **polymer** (many units). Each link in the chain is a **monomer** (one unit). Familiar complex carbohydrates or starches include cellulose (wood) and chitin, which forms the outer coverings of insects, crabs, and lobsters. Cellulose is the most common organic compound in nature, and chitin is second most common. Table 3.4 summarizes the functions of carbohydrates and the other macromolecules of life.

KEY CONCEPTS

Carbohydrates provide energy and support. They consist of carbon, hydrogen, and oxygen, with twice as many hydrogen atoms as oxygen atoms in each molecule. Monosaccharides (single sugars) have five or six carbons. Disaccharides (double sugars) form by dehydration synthesis from monosaccharides. Single and double sugars are simple carbohydrates; polysaccharides (starches) are complex carbohydrates.

Lipids

Lipids contain the same elements as carbohydrates, but with proportionately less oxygen. The lipids are diverse molecules that dissolve in organic solvents, but not in water. The most familiar lipids are fats and oils.

One type of fat, a **triglyceride,** consists of a 3-carbon alcohol, **glycerol,** from which three hydrocarbon chains

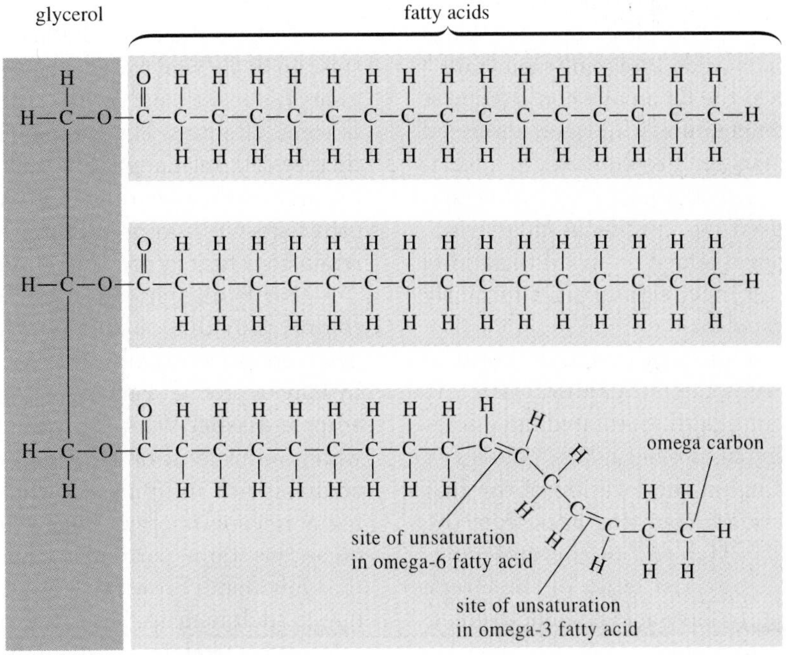

FIGURE 3.17

This lipid is an unsaturated derivative of tripalmitin, a triglyceride. Dietary fats rich in omega-3 fatty acids are healthier than those rich in omega-6 fatty acids. The double bonds bend the fatty acid tails, making the lipid oilier. The bend is shown schematically here.

called **fatty acids** extend (fig. 3.17). Single covalent bonds link the carbons of fatty acid "tails." The carbons with double bonds are bonded to one less hydrogen, because carbon can form no more than four covalent bonds.

FIGURE 3.18
Cholesterol.

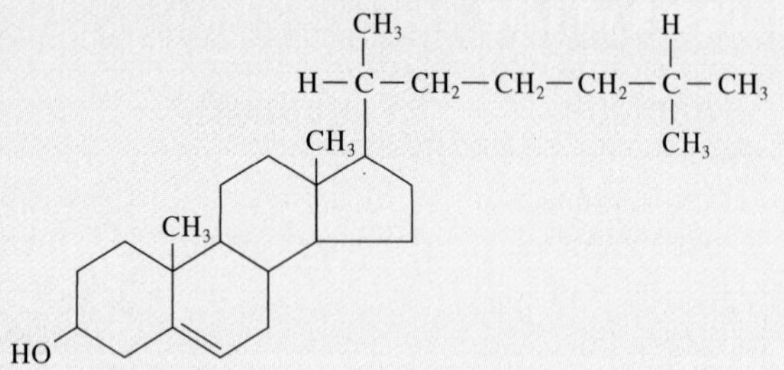

A fatty acid is **saturated** when it contains all the hydrogens it possibly can. This means that its carbons are all single bonded. A fatty acid is **unsaturated** if it has at least one double bond and **polyunsaturated** if it has more than one double bond. A **monounsaturated** fatty acid has just one double bond—olive oil is an example of a monounsaturated fat. Plant lipids are less saturated than those of animals. The sites of unsaturation, or the double bonds, cause kinks in the fatty acids, spreading their "tails." This gives the fat an oily consistency at room temperature, while more saturated animal fats tend to be solid. In a food-processing technique called hydrogenation used to produce margarine, hydrogen is added to an oil to solidify it—in essence, saturating a formerly unsaturated fat.

The site of the double bond in a dietary fat or oil can dramatically affect our health. Saturated fatty acids are designated omega-3 and omega-6, depending on the location of the first double bond, counting back from the methyl (CH₃) end of the molecule. Omega is the last letter of the Greek alphabet. In fatty-acid nomenclature, it denotes the end carbon. An omega-3 fatty acid has a double bond between the third and fourth carbons from the end, counting the carbon in the CH₃ at the end as the first carbon. In an omega-6 fatty acid, the double bond is between the sixth and seventh carbons from the end (fig. 3.17).

Most fats in red meat and dairy products are omega-6. Omega-6 fats have known effects on the heart and circulatory system and are implicated in heart disease. Greenland Eskimos, who have very low incidences of heart disease but bleed easily, eat mostly omega-3 fatty acids in their fish-rich diets. Further evidence that omega-3 fatty acids promote a healthy heart comes from studies of Dutch men who ate fish diets for 20 years. They had significantly less heart disease than Dutch men who did not each much fish.

Despite their bad dietary reputation, lipids are vital to life. They are necessary for growth and for the utilization of some vitamins. Fats slow digestion, thereby delaying hunger. Some fats make fur and feathers water-repellent, and other fats cushion organs and help to retain body heat by providing insulation.

Fat is also an excellent energy source, providing more than twice as much energy as equal weights of carbohydrate or protein. Fat cells aggregate as **adipose tissue.** White adipose tissue, which forms most of the fat in human adults, is rich in lipids. Brown adipose tissue releases energy as heat, keeping organisms warm, particularly mammals that hibernate. Brown adipose tissue is rare in adult humans, but in newborns it is layered around the neck and shoulders and along the spine.

Sterols are lipids that have four carbon rings. A very familiar sterol, **cholesterol,** is part of many cell membranes (fig. 3.18). Cholesterol is important biologically; for example, cells use cholesterol to synthesize other lipids, such as sex hormones. But because liver cells manufacture cholesterol from the breakdown products of saturated fats, excess cholesterol and saturated fats in the diet can lead to too much cholesterol in the blood. The excess cholesterol collects on the inner linings of blood vessels, impeding blood flow.

Vitamin D and cortisone are other familiar sterols. The next chapter considers phospholipids and glycolipids, the major components of biological membranes.

KEY CONCEPTS

Lipids contain carbon, hydrogen, and oxygen and do not dissolve in water. Saturated fatty acids have all single bonds and thus contain the maximal number of hydrogen atoms. Unsaturated fatty acids have at least one double bond, and polyunsaturated fatty acids have several double bonds, meaning they have far fewer hydrogen atoms. Double bonds kink the fatty acid chains, making a molecule with double bonds oily at room temperature, while a saturated fat is solid. Lipids serve many functions in the body. They store energy; make fur and feathers water-repellent; and cushion and insulate organs. Sterols, lipids with four carbon rings, are the chemical precursors of some hormones.

Proteins

Unlike carbohydrates and fats, **proteins** consist of *sequences* of smaller units—**amino acids** linked to form **polypeptide chains.** There are 20 types of amino acids in living organisms. Hundreds of these amino acids link to form a single polypeptide chain. A protein consists of one or more polypeptides. The potential variability of protein structure is therefore enormous.

Amino Acid Structure

An amino acid (fig. 3.19a) contains a central carbon atom bonded to:

1. A hydrogen atom.
2. A **carboxyl group** (COOH), which is another carbon atom double-bonded to an oxygen and single-bonded to a hydroxyl group (OH). Carboxyl groups (also called acid groups) are found in all organic acids.

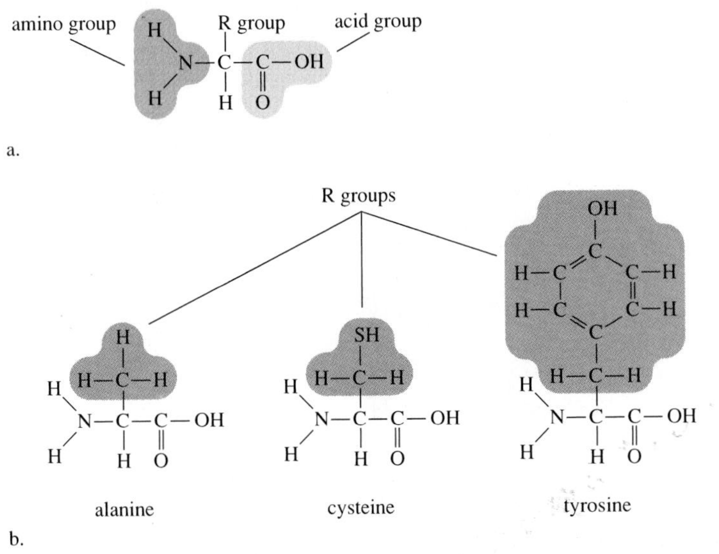

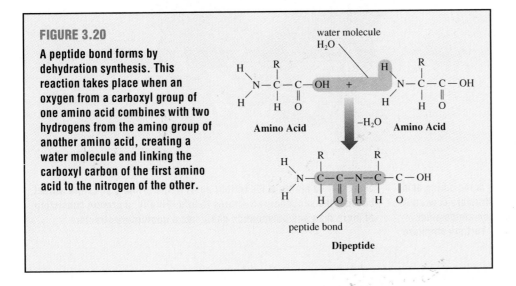

FIGURE 3.19

Amino acids, the building blocks of protein. *a.* An amino acid consists of a central carbon atom covalently bonded to an amino group, an acid (carboxyl) group, a hydrogen, and a side group designated "R." *b.* Different amino acids are distinguished by their R groups. Twenty types of amino acids are important in life; three of them are shown here.

FIGURE 3.20

A peptide bond forms by dehydration synthesis. This reaction takes place when an oxygen from a carboxyl group of one amino acid combines with two hydrogens from the amino group of another amino acid, creating a water molecule and linking the carboxyl carbon of the first amino acid to the nitrogen of the other.

3. An **amino group**—a nitrogen atom single-bonded to two hydrogen atoms (NH_2).

4. A side chain, or **R group,** that can be any of several chemical groups.

The nature of the R group distinguishes the 20 types of biological amino acids. It may be as simple as the lone hydrogen atom in glycine, or as complex as the two organic rings of tryptophan. Figure 3.19b shows three amino acids. Proteins also differ from carbohydrates and lipids in that they contain nitrogen. The R groups of two amino acids, cysteine and methionine, contain sulfur.

Protein Structure

Two adjacent amino acids join by dehydration synthesis, just as two monosac-

charides shed a water molecule to yield a disaccharide. When two amino acids bond, the acid-group carbon of one amino acid bonds to the nitrogen of the other, forming a **peptide bond** (fig. 3.20). Two joined amino acids are a dipeptide, three a tripeptide; larger chains with fewer than 100 amino acids are peptides, and finally, those with 100 or more amino acids are polypeptides. A protein breaks down into its constituent amino acids by hydrolysis (adding water).

As a protein is synthesized in a cell, it is molded into a three-dimensional structure by other proteins, by water molecules, and by the bonds that form between its atoms. Other proteins called chaperones help guide the protein into its characteristic three-dimensional shape, or **conformation.** Thousands of water molecules also surround a growing peptide, contorting it as its hydrophilic R groups are attracted to the water and its hydrophobic R groups are repelled by water. Hydrogen bonds between R groups also mold a protein's conformation. A **disulfide bond** forms between two sulfur atoms, bridging sulfur-containing parts of an amino acid chain. Disulfide bonds are abundant in keratin, the protein that forms hair.

The conformation of a protein may be described at several levels. The amino acid sequence of a polypeptide chain determines its **primary (1°) structure.** Chemical attractions between amino acids that are close together in the 1° structure fold the polypeptide chain into its **secondary (2°) structure** (fig. 3.21). Secondary structure is often created when hydrogen bonds form between amino groups. (Figure 3.23b shows two common secondary structures, alpha helices and beta-pleated sheets.) Secondary structures wind into **tertiary (3°) structures** as interactions occur between R groups and water. Finally, proteins consisting of more than one polypeptide have **quaternary (4°) structure.** For example, hemoglobin, the blood protein that carries oxygen, has

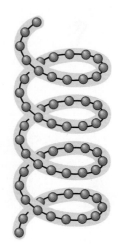

primary structure   secondary structure

tertiary structure   quaternary structure

FIGURE 3.21

Protein conformation. A protein's primary structure is the amino acid sequence of its polypeptide chain. The secondary structure is usually loops, coils, helices, barrels, or sheets created when amino acids close to each other in the primary structure attract. Tertiary structure occurs when amino acids farther apart in the primary structure attract, causing the secondary structures to fold. Finally, a protein consisting of more than one polypeptide chain has a quaternary structure.

four polypeptide chains. The liver protein ferritin has 20 identical polypeptides of 200 amino acids each. In contrast, the muscle protein myoglobin has a single polypeptide chain.

Enzymes

Among the most important of all biological molecules are enzymes. Enzymes are proteins that speed the rates of specific chemical reactions without being consumed in the process, a phenomenon called **catalysis.** Enzymes bind reactants, bringing them in contact with each other, so that less energy is needed for the reaction to proceed. Without enzymes, many biochemical reactions would proceed far too slowly to support life; some enzymes increase reaction rates a billion times. Enzymes function under specific pH and temperature conditions, and they catalyze only specific reactions.

The key to an enzyme's specificity lies in its **active site,** a region to which the reactants, also called **substrates,** bind. A substrate fits into the active site of an enzyme as the enzyme contorts slightly around it, forming a transient **enzyme-substrate complex** (figs. 3.22

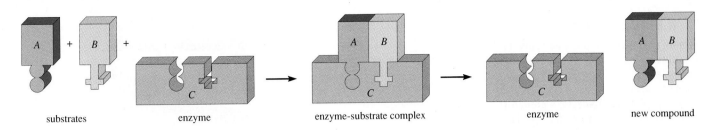

| substrates | enzyme | enzyme-substrate complex | enzyme | new compound |

FIGURE 3.22

Enzyme action. In this highly schematic depiction, substrate molecules *A* and *B* fit into the active site of enzyme *C*. An enzyme-substrate complex forms as the active site moves slightly to accommodate its occupants. A new compound, *AB*, is released, and the enzyme is recycled. Enzyme-catalyzed reactions can break down as well as build up substrate molecules. Figure 3.23 shows a more realistic depiction of enzyme action.

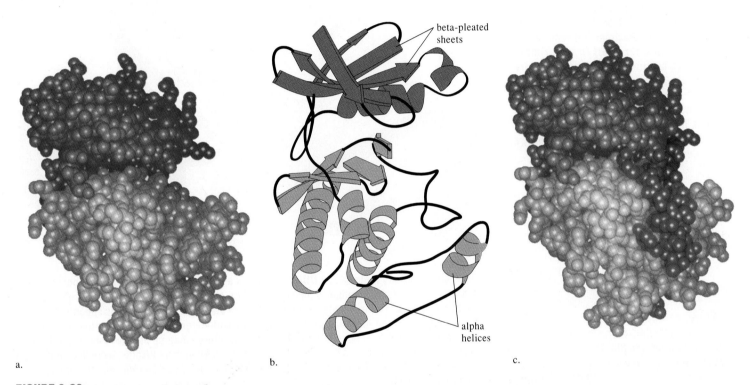

a.   b.   c.

FIGURE 3.23

The active site of a kinase-type enzyme consists of 240 amino acids that form a cleft. *a.* A kinase catalyzes the transfer of a phosphate group from a molecule called ATP, which has three phosphate groups extending from it like a tail, to another protein, the substrate. ATP binds to the smaller lobe of the (blue) kinase so that its endmost phosphate contacts the substrate protein held in the cleft of the larger (purple) lobe. The substrate fits into the cleft, as the enzyme contorts above it (*a* and *c*). *b.* A depiction of the secondary structure of the two lobes of the active site. The smaller lobe is built of amino acids associated into "beta-pleated sheets"; the larger lobe consists mostly of "alpha helices."

and 3.23). An enzyme can hold two substrate molecules that react to form one product molecule, or it can hold a single substrate that splits to yield two products. Then the complex breaks down to release the products (or product) of the reaction. The enzyme is unchanged, and its active site is empty and ready to pick up more substrate.

Parts of the enzyme other than the active site can also be vital. Some parts help maintain the active site in its proper conformation. If those parts are abnormal, the enzyme becomes unstable. For example, figure 2.6 shows an abnormal form of the enzyme superoxide dismutase.

Nucleic Acids

How does the body know how to synthesize a particular protein to perform a specific function? A protein's amino acid sequence is encoded in a sequence of chemical units of another type of biochemical, a **nucleic acid** (which will be discussed in detail in chapters 14 and 15). The nucleic acids **deoxyribonucleic acid (DNA)** and **ribonucleic acid (RNA)** are polymers of smaller units called **nucleotides.** A nucleotide consists of five-carbon sugar (deoxyribose in DNA and ribose in RNA), a phosphate group (PO_4), and one of five types of nitrogen-containing compounds called **nitrogenous bases.** The nitrogenous bases are **adenine** (A), **guanine** (G), **thymine** (T), **cytosine** (C), and **uracil** (U). DNA contains A, C, G, and T; RNA contains A, C, G, and U.

DNA and RNA contain and transmit the information that allows cells to reproduce and to synthesize the proteins necessary to life. DNA contains the information, while RNA helps transmit and translate it. Long sequences of DNA nucleotides provide information that

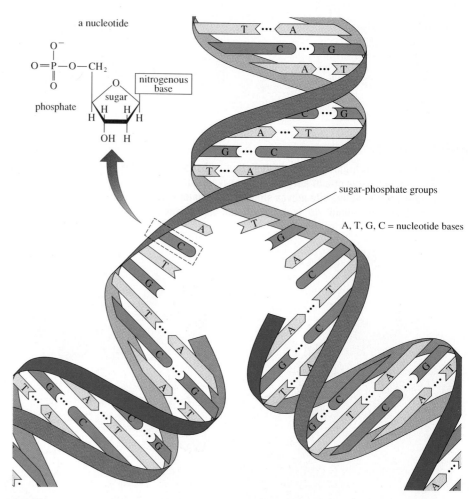

FIGURE 3.24

DNA, the hereditary molecule. A DNA molecule is a polymer of nucleotides. Each nucleotide consists of a five-carbon sugar, a phosphate group, and one of four different types of nitrogen-containing bases. The molecule resembles a spiral staircase, with the steps specific pairs of nucleotide bases and the sides chains of sugar-phosphate groups. The molecule twists into a double helix shape because the steps in the staircase are made up of only two kinds of base pairs: A–T (or T–A) and G–C (or C–G). The molecule replicates by locally unwinding so that the base pairs split apart. Each exposed base then chemically attracts a free complementary nucleotide, forming two identical DNA double helices from the original one. Hydrogen bonds hold the base pairs together.

guides other cellular components in assembling amino acids into polypeptide chains. Each group of three DNA bases in a row specifies a particular amino acid in a correspondence called the **genetic code.** A sequence of the DNA genetic code specifying a polypeptide is a **gene.** DNA is thus known as the genetic material of life. In this complex translation from gene to protein, RNA is crucial at many, if not all, steps.

DNA's structure enables it to replicate. The DNA molecule resembles a spiral staircase, with the nucleotide's sugars and phosphates combining to form

the sides of the staircase and the nitrogenous bases pairing to form steps. Pairs form only between A and T and between G and C, creating a symmetrical double helix. When DNA replicates, the spiral staircase splits down the center and unwinds, separating each base pair. Each half of the pair then chemically attracts a free nucleotide with the nitrogenous base of the proper type. The final result is two new DNA molecules, each identical to the original (fig. 3.24). DNA's ability to replicate is important for the perpetuation of life.

KEY CONCEPTS

The nucleic acids DNA and RNA are informational polymers. The building block of a nucleic acid, a nucleotide, consists of a phosphate group, a sugar, and a nitrogenous base (A, C, G, T, or U). RNA helps translate a DNA sequence into an amino acid sequence. Nucleic acids can replicate—a quality essential to the continuation of life.

The Beginning of Life—Epilogue

The molecules of life are large and complex, yet they are only a small subset of all possible molecules. The great similarity in the chemical composition of all living organisms is either remarkably coincidental or points to the descent of all living things from a single origin, a common ancestor. But how can we envision the beginnings of life, short of inventing a time machine?

In 1953, a graduate student named Stanley Miller conducted the first origin-of-life simulation experiment. Miller showed in the laboratory how simple chemicals, plus energy, might have yielded more complex chemicals—including some of those found in living things. As is true of most scientists who make scientific discoveries, Miller built his work on the ideas of scientists before him, and his conclusions have been modified by scientists since.

Evolution of an Idea— Prebiotic Earth Simulations

Back in 1929, English chemist John B. S. Haldane hypothesized that when life began, the earth's atmosphere differed markedly from today's. One of Haldane's contemporaries, Russian chemist Alexander I. Oparin, reasoned that methane (CH_4), ammonia (NH_3), water (H_2O), and hydrogen gas (H_2) might have been the raw materials for the formation of amino acids and nucleotides. (That idea was based on geological information of the time. More recent evidence points to an early atmosphere rich in carbon dioxide, CO_2.)

Oparin realized that to build large molecules from small ones requires energy. On the early earth, possible energy sources included lightning, hot springs, volcanoes, earthquakes, and unfiltered solar radiation (today, the sun's rays are filtered through an ozone layer). Heat energy would have kept water in liquid form, sped the rates of some chemical reactions, and made possible dehydration synthesis (evaporation) to form polymers.

Stanley Miller's mentor, chemist Harold C. Urey at the University of Chicago, suggested that Miller try to recreate the conditions that existed on the prebiotic (before life began) earth to see what chemicals would form. Miller set up a simple but clever apparatus. In a large glass container, he mixed the gases Oparin had suggested, exposing them to electric sparks to simulate lightning. Next, he condensed the gases in a narrow tube and passed them over an electric heater, a laboratory version of a volcano (fig. 3.25). This prebiotic soup brewed for a week. Finally, Miller examined the results, using chemical techniques. He found that he had indeed cooked up four amino acids! Because amino acids link to form proteins, and proteins are such an important part of life, Miller and others hypothesized that once amino acids formed, the other chemical reactions of life could also eventually take place.

Many researchers repeated Miller's experiment—perhaps too successfully, because several variations on Miller's theme produced more complex molecules that could be implicated in the origin of life. Perhaps ammonia and methane in the ancient atmosphere formed clouds of hydrogen cyanide (HCN), which polymerized when exposed to ultraviolet radiation from the sun. When hydrogen cyanide chains fell into the seas, they may have reacted with water to yield amino acids. Simulations of these events in laboratory glassware yield a yellow-brown, sticky substance containing six different amino acids. In still other experiments, prebiotic soup including phosphorous compounds produced nucleotides, the building blocks of nucleic acids.

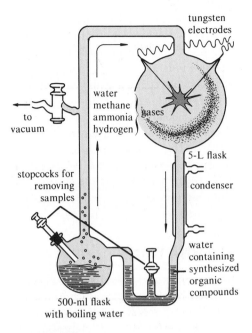

FIGURE 3.25

When Stanley Miller passed an electrical spark through heated gases, he generated amino acids—a more complex type of chemical that may have played a role in the origin of life.

Likely Stages of the Origin of Life

Prebiotic simulations suggested scenarios for the first stage of the origin of life—chemical evolution (fig. 3.26). Yet many questions remained. Exactly how did organic molecules join, so that amino acids formed peptides and nucleotides formed nucleic acids? These events constitute the second stage in the origin of life, molecular self-organization. Interestingly, the self-assembly of chemical units can also be simulated. RNA nucleotides can form polymers up to 10 nucleotides long when placed on certain types of clay. Researchers hypothesize that a repeated pattern in the crystals of the clay forms a mold to align several RNA nucleotides, easing their polymerization. A similar role has been suggested for iron pyrite, a mineral also known as "fool's gold" because of its appearance. These discoveries suggest possible mechanisms for the origin of nucleic acids such as RNA.

Finally, one more important question remains. Recall that living things are composed of chemicals that (1) include instructions to reproduce themselves and (2) use energy to maintain their organization. Which of these early molecules was able to facilitate both: to replicate as well as oversee the assembly and association—the organization—of other macromolecules? A good candidate for this pivotal role in the dawn of life is RNA; unlike DNA, which only carries genetic information, RNA plays several roles in utilizing that information to manufacture proteins, and in controlling which genes are expressed.

The RNA World

RNA comes in several types, distinguished by their sizes and functions. Before 1981, RNA was thought to be just a supporting scaffold, a biochemical bystander, to gene expression. Now we know that RNA has a unique combination of characteristics that qualify it, and probably it alone, to function as an immediate precursor of life. RNA can both replicate and organize other biochemicals. It acts as a catalyst in chemical reactions that transmit genetic information and assists in protein synthesis. Specifically, it can

> encode information in its nucleotide sequence
>
> replicate
>
> act as a scissorlike enzyme, cutting itself or DNA
>
> assist protein synthesis in various ways.

Another biochemical that may have been very important in life's beginnings is the enzyme **reverse transcriptase.** This enzyme constructs a DNA sequence from information contained in an RNA sequence. Reverse transcriptase is found today in viruses, known as retro-

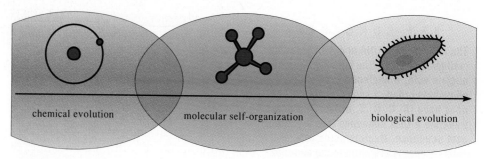

FIGURE 3.26

Stages in the origin of life. Most likely, no single, discrete event caused a nonliving biochemical—perhaps an informational polymer such as a nucleic acid or protein—to suddenly become a living thing. Rather, prebiotic and other organic compounds probably formed over a very long time (chemical evolution), with some gradually becoming able to replicate and perhaps direct or catalyze the synthesis of other compounds (molecular self-organization). At some point, these self-organizing molecules aggregated and persisted, forming the earliest viruses or even cells, and biological evolution began.

viruses, that use RNA as their genetic material. This suggests that RNA may have originally encoded sequences of amino acids, which were eventually copied into DNA.

Figure 3.27 outlines a hypothetical sequence of molecular events, involving RNA and reverse transcriptase, that could have led to the origin of life. The road to life might have begun with short RNA molecules encoding sequences of amino acids. Eventually, an RNA encoded reverse transcriptase. This enzyme then copied RNA sequences into DNA, a more permanent information store. Once information could be retained as DNA, yet translated into proteins via the informational and catalytic properties of RNA, the chemical blueprints for life were in place. These ancient collections of RNA, DNA, and protein, ancestors of living cells but not yet nearly as complex as a living cell, are termed **progenotes.** When progenotes came into existence, the stage was set for the next step—the evolution of metabolism.

The Origin of Metabolism

Metabolism is the ability of living things to acquire and use energy to maintain the organization necessary for life. Early in the road to life, another characteristic of the eclectic nucleic acids—the capacity for spontaneous change, or **mutation**—may have enabled progenotes to become increasingly self-sufficient. New variants depended more on their own internal chemistries, their ability to transduce energy into a usable form, than on environmental resources. Perhaps progenotes surrounded by lipid—a forerunner of a cell membrane—had some survival advantage in that particular environment. Similarly, perhaps progenotes best able to harness energy and use other chemicals in the environment, possibly because they had the most efficient enzymes, eventually prevailed.

Imagine a progenote that fed on a molecule, nutrient A, that was abundant in its environment. As the progenote divided and replicated, its food source must have gradually dwindled. However, because of the capacity for change inherent in all nucleic acids, the ancient seas

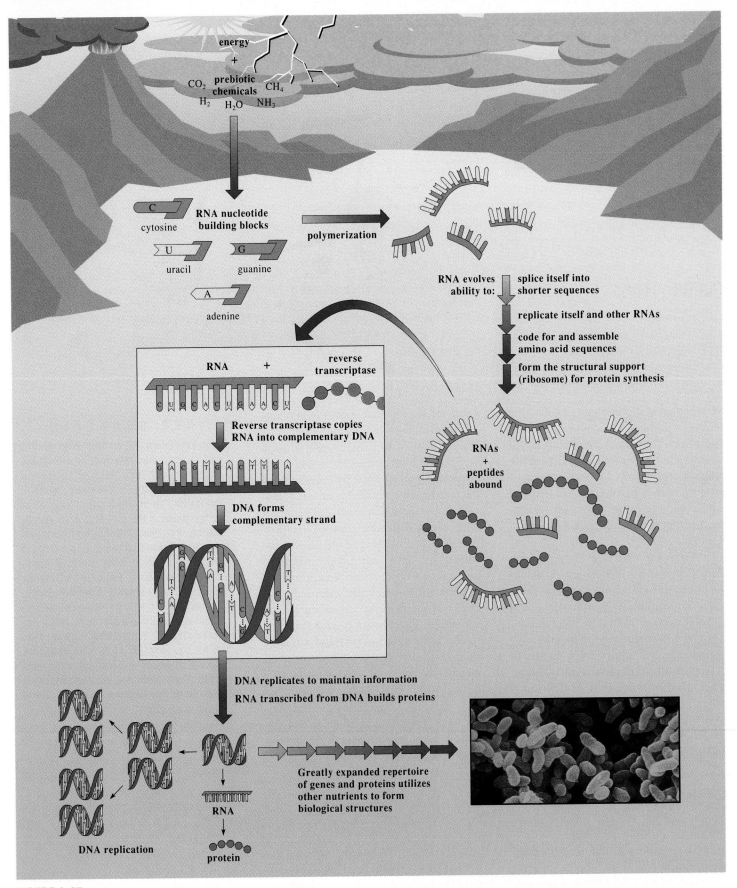

FIGURE 3.27

A scenario for life's beginnings—the RNA world. Chemical evolution in the presence of energy led to formation of nitrogen-containing nucleotide bases in an RNA-like molecule. The phosphate that tied the bases together might have been liberated from rocks when erupting volcanoes released energy. RNA evolved capabilities to replicate and encode proteins. A protein, perhaps similar to reverse transcriptase, copied the information in RNA into DNA. The DNA then polymerized a complementary strand, fashioning the first DNA double helices. As nucleic acids, proteins, carbohydrates, and lipids formed, the structures that would become parts of cells took shape. At some point, life arose from these structures.

FIGURE 3.28

Evolution of metabolic pathways. The evolution of new enzymes probably expanded the number of different nutrients certain progenotes could use, thus giving them a survival advantage over others competing for limited food. The units marked *1, 2,* and *3* are enzymes; *A* through *D* are nutrients.

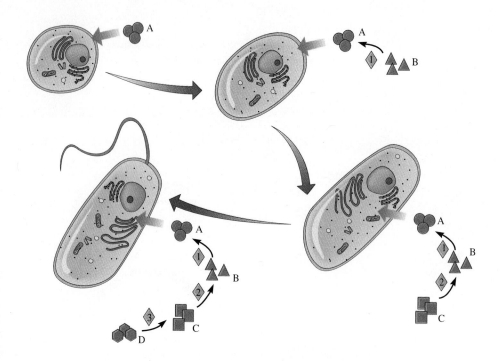

were probably home to more than one genetic variety of progenote. One type might have had an enzyme that could convert some other nutrient, B, into the original nutrient A. The new variety of progenote would have a nutritional advantage, because it could extract energy from two different food sources—B and A.

Soon the first type of progenote, totally dependent on nutrient A, would die out as its food supply vanished. For a while, the variant progenote that could convert nutrient B to A would flourish. But, in time, nutrient B would also become scarce. Perhaps a new variety of progenote would then arise, one with an additional enzyme that converted a nutrient C to B (and then, using the first enzyme, B to A). In time, this progenote would flourish; then another, and yet another. Over time, an enzyme-

catalyzed sequence of connected chemical reactions—a biochemical pathway linking D to C to B to A—would arise (fig. 3.28). More pathways would form. Intermediates of one pathway would become starting points for another, and metabolism would gradually appear.

Many different combinations of proteins and nucleic acids must have formed spontaneously. Over millions of years, with so many chance experiments of nature, it seems almost inevitable that eventually, a cell-like assemblage, with the ability to replicate provided by the nuclei acid, would persist and multiply over time. Eventually, as its internal chemistry grew more complex, this entity would become what we call a living cell. Cells would now have the ability to reproduce and to gain and use energy to maintain their organization. Biological evolution would begin.

KEY CONCEPTS

The observation that all organisms use the same energy and information molecules suggests that they descended from a common ancestor. In prebiotic chemical evolution, small monomers may have accumulated in the presence of an energy source. Polymers formed from monomer subunits. These first two possible steps in the origin of life can be studied in simulation experiments. Proteins and nucleic acids, probably organized by RNA, somehow associated with lipids and other biochemicals to form the first cells. Mutations accumulated in these simple, early cells, called progenotes. Metabolic pathways evolved as genetically variant progenotes arose. Cells could now replicate and gain and use energy to maintain organization. Biological evolution—the evolution of life—began.

SUMMARY

Life first appears in the fossil record approximately 700 million years after the earth formed 4.5 billion years ago. The chemical similarity of all living things suggests that they evolved from a common ancestor. How did that first ancestor—the first living organism—

arise? A series of experiments that prevented living organisms from contaminating flasks of broth disproved the theory of spontaneous generation. The presence of simple organic molecules in comets and meteorites supports an alternative idea—that the biochemicals

that led to life may have come from space. Some extraterrestrial environments today may resemble conditions on the prebiotic earth.

Organisms obtain energy from the environment and transduce it to a usable form. Matter can be broken down into pure substances called elements. Elements are organized in the periodic table according to their number of subatomic particles. Bulk elements, those essential to life in large quantities, include carbon, hydrogen, oxygen, nitrogen, sulfur, and phosphorus.

An atom is the smallest unit of an element. Subatomic particles include the positively charged protons and neutral neutrons that form the nucleus, and the negatively charged, much smaller electrons that circle the nucleus. An atom's mass number is its total number of protons and neutrons. Its atomic number is its number of protons. Isotopes of an element differ by the number of neutrons in the nucleus of the atom.

Electrons move constantly. The volume of space an electron most often occupies defines that electron's orbital. Electron orbitals form levels of energy called shells. An atom's tendency to fill its outermost, or valence, shell with electrons causes atoms to bond and share electrons, forming molecules. A compound usually contains different atoms that have bonded together, and it has properties different from those of any of the elements that comprise it. Chemical shorthand indicates the numbers of atoms and molecules in a compound. In a chemical reaction, though different compounds form and break, the total number of atoms of each element remains the same before and after the reaction.

There are several types of chemical bonds. Covalent bonds share electrons and tend to form between atoms that have three, four, or five valence electrons. Electron orbital configurations change as the energy redistributes itself between the bonding atoms. Carbon atoms form up to four covalent bonds. They bond with each other and hydrogen atoms to build hydrocarbons, the skeletons of many biochemicals. In nonpolar covalent bonds, all electrons are shared equally. In a polar covalent bond, the nucleus of one type of atom attracts electrons more than the nucleus of another type of atom, resulting in opposite charges in different parts of the molecule.

In an ionic bond, an atom with one, two, or three valence electrons loses them to an atom with seven, six, or five valence electrons, respectively; the resulting bonds form a lattice of cations and anions. Ionic bonds in water, and thus in living systems, are weak compared to covalent bonds. Hydrogen bonds form when a hydrogen in one molecule is drawn to part of a neighboring molecule because of unequal electrical charge distribution. Van der Waals attractions occur between parts of molecules that are temporarily carrying opposite charges. Molecular shape and interactions are affected by all these types of bonds and are also guided by attractions to and repulsions from water.

Water is essential for life; most biochemical reactions occur in an aqueous environment. Water is cohesive (attracted to itself) and adhesive (drawn to other molecules), a quality that makes many substances dissolve in it. Water helps regulate temperature in organisms because of its high heat capacity and high heat of vaporization. Water has a pH of 7, which means that the numbers of H^+ and OH^- ions in water are equal. An acid adds H^+ to a solution, lowering the pH to below 7. A base adds OH^-, which raises the pH between 7 and 14. Body fluids function at specific pH ranges, which are maintained by buffer systems consisting of weak acid-and-base pairs. The hydrogen bonds that keep water fluid are constantly forming and breaking. Water is a solvent for salts and for molecules held together by polar covalent bonds. It also expands and becomes less dense upon freezing, a characteristic that has various implications for living things.

Many of the chemicals of life are based on hydrocarbon skeletons. Carbohydrates, which provide energy and support, include the sugars and starches. They consist of carbon, hydrogen, and oxygen, with twice as many hydrogen atoms as oxygen atoms in each molecule. The simple carbohydrates include monosaccharides and disaccharides. Complex carbohydrates are polysaccharides. Two monosaccharides combine to form a disaccharide by dehydration synthesis, losing a molecule of water.

Lipids are a diverse group of organic compounds, including fats and oils, that do not dissolve in water. They contain carbon, hydrogen, and oxygen, but they have less oxygen than carbohydrates. Triglycerides consist of glycerol and three fatty acids, which may be saturated (no double bonds), unsaturated (at least one double bond), or polyunsaturated (more than one double bond). Double bonds make a lipid oily at room temperature, whereas saturated fats are more solid. Unsaturated fats are more healthful to eat than saturated fats, and omega-3 fatty acids, plentiful in fish, are healthier than the omega-6 fatty acids in meat and dairy products. Lipids provide energy, slow digestion, protect fur and feathers from water, cushion organs, and preserve body heat. Sterols are lipids containing four carbon rings, and phospholipids and glycolipids are parts of membranes.

Proteins have many functions. These informational polymers consist of 20 types of amino acids, each of which includes a hydrogen, an amino group, an acid or carboxyl group, and an R group bonded to a central carbon. Amino acids join by forming dipeptide bonds through dehydration synthesis. A protein's conformation is vital to its function. It is determined by the hydrophobicity and hydrophilicity of R groups, the amino acid sequence (primary structure), and interactions between amino acids close together (secondary structure) and far apart (tertiary structure) in the sequence. A protein with more than one polypeptide has a quaternary structure.

Nucleic acid sequences encode amino acid sequences. DNA, one type of nucleic acid, contains deoxyribose and the nitrogenous bases adenine, cytosine, guanine, and thymine. RNA contains ribose and the nitrogenous bases, except it has uracil instead of thymine. A nucleotide, the building block of a nucleic acid, consists of a phosphate, a base, and a sugar. RNA controls synthesis of an amino acid chain, using information in a DNA sequence. DNA can replicate. To do so, the DNA double helix unwinds and nucleotides with complementary bases join each half of the old strand to build two new strands, each containing one old and one new DNA molecule. DNA carries genetic information. RNA helps translate that information to enable the cell to synthesize proteins essential to life.

Simulations of prebiotic earth chemistry combine simple chemicals and energy. These simulations have produced more complex chemicals that are present in life today and that could have been important in the origin of life. A likely scenario for life's origin is that chemical evolution gradually led to molecular self-organization, which at some point evolved to replication. RNA appears to be the only molecule capable of replicating while also controlling the production and interactions of other important molecules. As RNA organized and associated with increasingly complex molecules, precursors of living cells called progenotes arose. Gradually, progenotes accumulated mutations and expanded their use of nutrients in the environment, forming metabolic pathways. At this point, cells could perform two functions necessary to life: they could replicate, and they could acquire energy and transduce it into a usable form to maintain their own organization. Biological evolution began.

KEY TERMS

acid 43
acidophile 45
acidosis 43
active site 50
adenine 52
adhesion 43
adipose tissue 48
alkalosis 43
amino acid 48
amino group 49
anion 40
aqueous solution 43
atom 36
atomic number 36
base 43
Bohr model 37
buffer system 44
bulk element 36
carbohydrate 46
carboxyl group 48
catalysis 50
cation 40
chemical bond 38
chemical equilibrium 39
chemical reaction 38
cholesterol 48
cohesion 43
complex carbohydrate 46
compound 38
conformation 49

covalent bond 39
cytosine 52
dehydration synthesis 46
deoxyribonucleic acid (DNA) 52
disaccharide 46
disulfide bond 49
electrolyte 43
electron 36
electronegativity 40
electron orbital 37
element 35
energy shell 37
enzyme 36
enzyme-substrate complex 50
fatty acid 47
gene 52
genetic code 52
glycerol 47
guanine 52
half-life 37
heat capacity 44
heat of vaporization 44
hydrocarbon 40
hydrogen bond 42
hydrolysis 46
hydronium ion 43
hydrophilic 42
hydrophobic 42
hydroxide ion 43
imbibition 43

ion 40
ionic bond 40
isotope 37
lipid 47
macromolecule 39
mass number 36
metabolism 54
molecular weight 39
molecule 38
monomer 47
monosaccharide 46
monounsaturated fat 48
mutation 54
neutron 36
nitrogenous base 52
nonpolar covalent bond 40
nucleic acid 52
nucleotide 52
nucleus 36
octet rule 37
oxidation 40
panspermia 35
peptide bond 49
periodic table 35
pH scale 43
polar covalent bond 40
polymer 47
polypeptide chain 48
polyunsaturated fat 48
primary structure 49

progenote 54
protein 48
proton 36
quaternary structure 49
reactant 38
reduction 40
reverse transcriptase 54
R group 49
ribonucleic acid (RNA) 52
salt 40
saturated fat 48
secondary structure 49
simple carbohydrate 46
solute 43
solution 43
solvent 43
spontaneous generation 34
sterol 48
subatomic particle 36
substrate 50
sulfide 45
tertiary structure 49
thymine 52
trace element 36
triglyceride 47
ultratrace element 36
unsaturated fat 48
uracil 52
valence electron 37
van der Waals attraction 42

REVIEW QUESTIONS

1. Define:
   a. atom
   b. element
   c. molecule
   d. compound
   e. isotope
   f. ion

2. Order the following terms in the following lists by size, from smallest to largest.
   a. proton, electron, molecule, atom
   b. fructose, cellulose, sucrose
   c. protein, amino acid, amine group, dipeptide
   d. hydrogen, zinc, carbon, oxygen, sulfur, nitrogen

3. The vitamin biotin contains 10 atoms of carbon, 16 of hydrogen, 3 of oxygen, 2 of nitrogen, and 1 of sulfur. What is its molecular formula?

4. Describe how each of these chemical bonds form:
   a. covalent bond
   b. ionic bond
   c. hydrogen bond
   d. disulfide bond
   e. van der Waals attraction
   f. peptide bond

5. Cite three illustrations of the concept of emergent properties (review chapter 1 if necessary) that appear in this chapter.

6. List four characteristics of water that make it essential to life.

7. If a shampoo is labeled "nonalkaline," would it more likely have a pH of 3, 7, or 12?

8. A topping for ice cream contains fructose, hydrogenated soybean oil, salt, and cellulose. What types of chemical are in it?

9. An advertisement for a supposedly healthful cookie claims that it contains an "organic carbohydrate." Why is this statement silly?

10. At a restaurant, a waiter suggests a sparkling carbonated beverage, claiming that it contains no carbohydrates. The label on the product lists water and fructose as ingredients. Is the waiter correct?

11. The following diagram shows the chemical structure of the amino acid threonine. Label the amino group, the carboxyl group, and the R group.

$$CH_3$$
$$H-C-OH$$
$$H_2N-C-C-OH$$
$$H \quad O$$

12. What do a protein and a nucleic acid have in common? What are two ways that a protein differs from a complex carbohydrate and a lipid? What do all three types of macromolecules have in common?

13. Diagram the way a dipeptide bond would form between alanine and cysteine.

14. Outline the steps of the scientific method used to implicate NO and CO as biological messenger molecules.

15. From what compounds did the carbon, hydrogen, oxygen, and nitrogen come that formed amino acids in Stanley Miller's prebiotic soup?

16. What biochemical probably played a pivotal role in the origin of life? Why?

17. What role might reverse transcriptase have played in the origin of life?

TO THINK ABOUT

1. The study of the origin of life requires speculation, because fossils of the earliest life forms have probably been destroyed. (Even this statement introduces speculation!) Moreover, we cannot know if biochemical experiments designed to recreate conditions hypothetically present on the early earth are actually accurate reconstructions. Because of these uncertainties, some people, scientists among them, suggest that we abandon research on the origin of life. Do you agree or disagree? Cite reasons for your answer.

2. Louis Pasteur is often credited with disproving the theory of spontaneous generation. Stanley Miller often receives credit for demonstrating that certain simple organic molecules, when "brewed" in the presence of energy, react to produce more complex biochemicals. What previous discoveries and hypotheses influenced Pasteur and Miller?

3. Why isn't the presence of amino acids in comets and meteorites, and in primordial soup experiments, accepted as definitive evidence of how life originated?

4. If you wanted to use a radioactive isotope as a label to identify a nucleic acid in a biological sample, but at the same time, you wanted to avoid detecting protein, which element might you use?

5. A Horta is a fictional animal whose biochemistry is based on the element silicon. Consult the periodic table. Is silicon a likely substitute for carbon in a life form? Cite a reason for your answer.

6. If water is left in a covered outdoor swimming pool over a very cold winter, the center of the cover bulges up. What characteristic of water makes this happen?

7. Amyotrophic lateral sclerosis, also known as ALS or Lou Gehrig's disease, produces a rapidly progressing paralysis. An inherited form of the illness is caused by a gene (sequence of DNA) that encodes an abnormal enzyme that contains zinc and copper. The abnormal enzyme fails to rid the body of a toxic form of oxygen. Which of the molecules mentioned in this description is a
   a. protein?
   b. nucleic acid?
   c. bulk element?
   d. trace element?

8. A man on a very low-fat diet proclaims to his friend, "I'm going to get my cholesterol down to zero!" Why is this an impossible (and undesirable) goal?

SUGGESTED READINGS

Amato, Ivan. February 14, 1992. Capturing chemical evolution in a jar. *Science*, vol. 255. Chemical reactions that may have led to life can be conducted in a laboratory today.

Amato, Ivan. June 26, 1992. A new blueprint for water's architecture. *Science*, vol. 256. A few researchers hypothesize that water's pattern of dynamic hydrogen bonds is not random, but an orderly pattern of cubes.

Barinaga, Marcia. January 15, 1993. Carbon monoxide: Killer to brain messenger in one step. *Science*, vol. 259. Carbon monoxide may be a neurotransmitter, not just a substance used to commit suicide and a pollutant.

Baum, Rudy M. June 22, 1992. RNA shown to catalyze chemical reactions required for life. *Chemical and Engineering News*, vol. 70. The first chemical of life would have had to be able to catalyze chemical reactions. Only RNA can do this as well as replicate itself.

Baum, Rudy M. January 3, 1994. Retrovirus ancestor: plasmid may link RNA-, DNA-based biology. *Chemical and Engineering News*, vol. 92. An exploration of the retrovirus's origins and the evolutionary link from RNA to DNA.

Collins, J. C. 1991. *The Matrix of Life.* East Greenbush, N.Y.: Molecular Presentations. A book devoted to explaining a fascinating molecule, water, and why it is vital to life.

Copeland, Robert A. June 1992. Proteins: Masterpieces of polymer chemistry. *Today's Chemist.* The shapes of proteins are influenced by several forces.

Culotta, Elizabeth. July 31, 1992. Forcing the evolution of an RNA enzyme in the test tube. *Science*, vol. 257. The properties of RNA that qualify it as a forerunner to life can be demonstrated in a test tube.

Culotta, Elizabeth, and Daniel E. Koshland, Jr. December 18, 1992. NO news is good news. *Science*, vol. 258. *Science* magazine names nitric oxide "molecule of the year" for 1992, in honor of its recently discovered role as a biological messenger.

Darnell, J. E., and W. F. Doolittle. March 1986. Speculations on the early course of evolution. *Proceedings of the National Academy of Sciences, USA*, vol. 83. A classic paper outlining the steps that very likely took place to bridge the evolution of a complex collection of chemicals into a living cell.

DeDuve, Christian. 1991. *Blueprint for a Cell: The Nature and Origin of Life.* Burlington, N.C.: Neil Patterson Publishers. Origin of life research melds chemistry and cell biology.

Feldman, Paul L., Owen W. Griffith, and Dennis J. Stuehr. December 20, 1993. The surprising life of nitric oxide. *Chemical and Engineering News*, vol. 71. The tiny NO molecule is important in life.

Houston, Charles S. October 1992. Mountain sickness. *Scientific American*. Breathing at high altitudes alters blood pH.

Nash, J. Madelaine. October 11, 1993. How did life begin? *Time*. There are several possible scenarios for life's beginnings.

Pace, Norman R. June 5, 1992. New horizons for RNA catalysis. *Science*, vol. 256. For many years, catalysis was thought to be the sole province of proteins.

Pennisi, Elizabeth. February 20, 1993. Water, water everywhere. *Science News*, vol. 133. Although water is perhaps the best studied molecule, we still have more to learn about it.

Pennisi, Elizabeth. July 31, 1993. Chitin craze. *Science News*, vol. 144. The second most abundant complex carbohydrate in nature has an abundance of uses.

Poole, Robert. June 29, 1990. Closing the gap between proteins and DNA. *Science*, vol. 248. Chemists are synthesizing self-replicating molecules that may have been precursors to life.

Williams, Robert J. P. 1991. The chemical elements of life. *Journal of Chemical Society, Dalton Transactions*. A detailed examination of why and how 24 elements are part of life.

Zimmer, Carl. October 1992. Wet, wild, and weird. *Discover*. Computer modeling helps biochemists learn more about water and how it interacts with the organic molecules of life.

A drop of blood contains a variety of cell types.

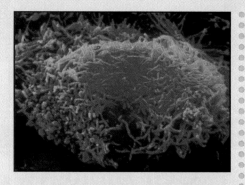

CONNECTIONS

Cancer Cells—A Personal Reflection

Although all of us fear cancer, the sheer terror of being told that this abnormality is taking place in your own body is nearly impossible even to fully imagine. During the days between diagnosis and first treatment, after the initial shock wears off, you join the ranks of those with cancer who may feel literally invaded, as if a mysterious, powerful creature has taken over some small part of their body and must be vanquished before it spreads. But this gut feeling of invasion by an unseen, alien force is not very far from what is truly happening. For at the microscopic level, certain cells have indeed strayed from biological normalcy.

Conquering cancer, and preventing it in the first place, are compelling reasons to study cell biology. By describing and understanding what is normal, we can begin to battle what is not. And cancer cells are far from normal. The chapters ahead will lead you through cell structure and function, so by the unit's final chapter, you'll understand some of the changes that constitute cancer. And they are many.

A healthy cell has a characteristic shape, with a boundary that allows entry to some substances, yet blocks others. Not so the cancer cell, which takes various shapes, with a surface more fluid and boundaries less discriminating. The cancer cell breaches the very controls that seem to keep other cells in place, as if it has a mind of its own. This renegade cell squeezes into spaces where other cells do not, secretes biochemicals that blast pathways through healthy tissue, and even creates its own personal blood supply. The cancer cell's "mind"— its genetic controls— differs from those of healthy cells and these differences are passed on when it divides. Cancer cells disregard the "rules" of normal cell division that keep other cells in check, enabling the body to develop and maintain distinct organs and other structures. To defy so many biological traditions, the cancer cell uses up tremendous amounts of energy, causing further disruptions.

Understanding the form and function of cancer cells is our most powerful weapon against them. New combinations of drugs, many derived from biological sources, are providing researchers and physicians with the means to attack cancer cells at several vulnerable points— altering cancer cell surfaces to attract the immune system; dismantling the fibers within cells that make them divide too often; bolstering the immune system's supply of cancer-fighting biochemicals. Detecting cancer cells' heightened energy use or altered surfaces permits much earlier diagnoses, when new and traditional treatments have a better chance of working.

We hear so often of those who have died of cancer. We are saddened when friends or family members are diagnosed with some variant of this group of disorders. Yet many people living perfectly normal, productive lives have cancer, or have had it. This author is one of them. Many times we can fight, and defeat, this most frightening of medical conditions. Our ability to do so rests solidly on our knowledge of the biology of cells, the units of life.

CHAPTER 4
Cells

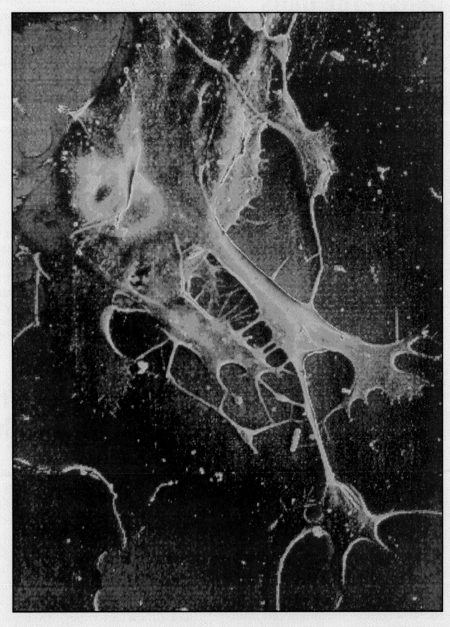

These fibroblasts are one of the more than 200 types of specialized cells in the human body.

By the chapter's end, you should be able to answer these questions:

1. What observations led to the development of the cell theory?

2. How are the two basic types of cells (prokaryotic and eukaryotic) alike and different?

3. How did compartmentalization and division of labor within the cell help to maintain functions as cells grew larger?

4. Which organelles carry out the processes of secretion, waste removal, inheritance, and energy acquisition? How are secretion and waste removal carried out in eukaryotic cells?

5. What are some examples of infectious agents? In what ways are they simpler than cells?

6. What evidence indicates that eukaryotic cells may have evolved from prokaryotic cells?

Cells—The Units of Life

A human, a hyacinth, a mushroom, and a bacterium—from outward appearances, these organisms do not seem to have much in common other than life itself. However, when we examine these organisms at a level too minute for the unaided eye to detect, astounding similarities reveal themselves. All living things consist of microscopic structures called **cells.** Within cells, highly coordinated biochemical activities carry on the basic functions of life, as well as, in some cases, highly specialized functions. The next two chapters introduce the cell, and the chapters that follow delve into specific cellular events.

Cells, the basic units of life, exhibit all of the characteristics of life. A cell requires energy, genetic information to direct biochemical activities, and structures to carry out these activities. Movement occurs within living cells, and some cells, such as the swimming sperm cell, can move about in the environment. All cells have some structures in common that allow them to perform the basic life functions of reproduction, growth, response to stimuli, and energy conversion. In short, the functions of cells are similar to the functions of whole organisms.

A cell consists of living matter, or **protoplasm,** bound by an outer membrane. **Unicellular** organisms such as bacteria consist of a single cell. **Multicellular** organisms, such as ourselves, consist of many cells. Structures within the cells of multicellular organisms and within some of the more complex unicellular organisms, called **organelles** (little organs), carry out specific functions. In multicellular organisms, different numbers of particular types of organelles endow some cells with specialized functions, such as support, contraction, or message transmission (fig. 4.1). Yet even the structurally simple cells of bacteria are organized and efficient.

KEY CONCEPTS

The basic structural unit of a living organism is the cell. Organisms are unicellular (one-cell) or multicellular (many cells). Complex cells contain specialized structures called organelles. All cells contain organelles or other structures to carry out basic life processes, but different numbers and types of organelles give cells distinct characteristics.

Viewing Cells—The Development of the Microscope

Our understanding of the structures inside cells depends upon technology, because most cells are too small for the unaided human eye to see. Today, sophisticated equipment greatly magnifies cell contents, allowing us to probe the inner workings of cells. However, the ability to make objects appear larger probably dates back to ancient times, when people noticed that pieces of glass or very smooth, clear pebbles could magnify small or distant objects. By the thirteenth century, the ability of such "lenses" to aid people with poor vision was widely recognized in the Western World.

It was more than three centuries before anyone noticed the effect of using lenses in pairs. The origin of a double-lens compound microscope can be traced to two Dutch spectacle makers, Johann and Zacharius Janssen. Actually, their children were unwittingly responsible for this important discovery. One day in 1590, a Janssen youngster was playing with two lenses, stacking them and looking through them at distant objects. Suddenly he screamed—the church spire looked as if it were coming toward him! Looking through both pieces of glass, as the elder Janssens quickly did, the faraway spire did indeed look as if it were approaching. One lens had magnified the spire, and the other lens had further enlarged the magnified image. Thus, the Janssens soon invented the first compound optical device, a telescope. Soon, similar double-lens systems were constructed to focus on objects too small to be seen by the naked human eye, and the compound microscope was born.

It was not long before such lenses were turned towards objects in nature. By 1660, an inquisitive and imaginative English physicist, Robert Hooke, melted together strands of spun glass to create lenses that were optically superior to any available before. Hooke focused his lenses on many objects, including bee stingers, fish scales, fly legs, feathers, and any type of insect he could hold still long enough to study. He was particularly fascinated by cork, which is actually bark

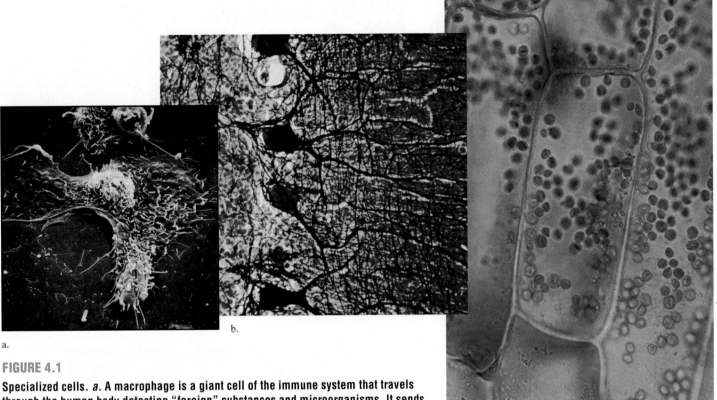

FIGURE 4.1

Specialized cells. *a.* A macrophage is a giant cell of the immune system that travels through the human body detecting "foreign" substances and microorganisms. It sends out extensions called pseudopods that capture and engulf its target, which is then drawn into the cell and destroyed by enzymes. (Magnification ×2,000.) *b.* Billions of nerve cells like these interconnect within the human brain. Note the roundish cell bodies and long extensions. (Magnification ×400). *c.* The green discs in these leaf cells of the common water plant *Elodea* are chloroplasts. Chloroplasts contain structures and biochemicals that permit the cell to utilize energy from sunlight to synthesize nutrient molecules. (Magnification ×250).

from a type of oak tree. Under the lens, the cork appeared to be divided into little boxes; although Hooke didn't know it at the time, he was viewing empty spaces left by cells that had existed when the cork was alive. Hooke called these units "cells," because they looked like the cubicles (cellae) in which monks studied and prayed. Although he did not realize the significance of his observation, Hooke was the first human to see the outlines of cells, the fundamental structural units of life.

In 1673, lenses were improved again, at the hands of Anton van Leeuwenhoek of Holland. Van Leeuwenhoek used only a single lens, but it was more effective at magnifying and produced a clearer image than most two-lens microscopes then available. One of his first objects of study was tartar scraped from his own teeth, and his words best described what he saw there:

> To my great surprise, I found that it contained many very small animalcules, the motions of which were very pleasing to behold. The motion of these little creatures, one among another, may be likened to that of a great number of gnats or flies disporting in the air.

Leeuwenhoek discovered bacteria and protozoa, and in so doing opened up a vast new world to the human eye and mind (figs. 4.2 and 4.3). However, he failed to see the single-celled "animalcules" reproduce and therefore perpetu-ated the theory of spontaneous generation, the idea that life seemingly arises from the nonliving or from nothing. Nevertheless, van Leeuwenhoek did describe with remarkable accuracy microorganisms and microscopic parts of larger organisms, although he did not recognize their great similarities.

KEY CONCEPTS

The ability of lenses to magnify objects was discovered in the late 1500s, and Robert Hooke first magnified biological objects using lenses in 1660. Leeuwenhoek improved lenses and saw and described many microbes, but he did not note the similarities among diverse types of cells.

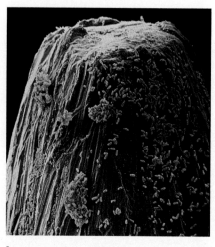
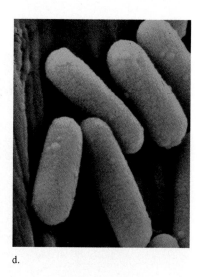

a.   b.   c.   d.

FIGURE 4.2

The living world beyond our vision. *a.* To the naked eye, this pin does not appear to be a likely site for bacterial growth. However, when the pin is examined under the scanning electron microscope at increasing magnifications (*b,c,* and *d*), rod-shaped bacteria are apparent. (Magnifications, *a.* ×7; *b.* ×35; *c.* ×178; *d.* ×4,375.)

Development of the Cell Theory

Despite the growing accumulation of microscopists' drawings of cells during the seventeenth and eighteenth centuries, the **cell theory**—the idea that the cell is the fundamental unit of all forms of life—did not emerge until the nineteenth century. Historians attribute this delay in the development of the cell theory to poor technology, not only because of the crude microscopes scientists were forced to use, but because they lacked procedures to preserve and study living cells without damaging the cells in the process. Neither the evidence itself nor early interpretations of it suggested that all living things were composed of cells. Hooke had not observed actual cells, but their absence. Leeuwenhoek made important observations, but he did not systematically describe or categorize the structures that cells had in common.

Observations Improve

Two factors converged in the nineteenth century to conceive the modern field of cell biology: more powerful microscopes, with better magnification and illumination, and the debut of the study of microbiology. In the early and mid-1830s, observers identified a darkened area, the **nucleus,** first in plant cells and then in animal cells. Many investigators also viewed a translucent, moving substance in living cells, which they called protoplasm. Slowly, scientists started to fit the pieces of the puzzle together.

In 1839, German biologists Matthias J. Schleiden and Theodor Schwann developed the cell theory. Schleiden first noted that cells were the basic units of plants, and then Schwann compared animal cells to

FIGURE 4.3

The microscope reveals another part of our living world. "Egad, I thought it was tea, but I see I've been drinking a blooming micro-zoo!" says this horrified, proper nineteenth-century London lady when she turns her microscope to her tea. People must have been shocked to learn of the active living world too small for us to see.

plant cells. After observing many different plant and animal cells, Schleiden and Schwann concluded that cells were "elementary particles of organisms, the unit of structure and function." They described the components of the cell as a cell body and nucleus contained within a surrounding membrane. Schleiden called a cell a "peculiar little organism," realizing that a cell can be a living entity on its own; but the new theory also recognized that in plants and animals, cells are part of a larger living organism.

Many cell biologists extended and elaborated upon Schleiden and Schwann's observations and ideas. German physiologist Rudolph Virchow added an important corollary to their theory, stating in 1855 that "omnis cellula e cellula"—"all cells come from cells." This statement directly contradicted the still-popular theory of spontaneous generation. Virchow's statement also challenged another concept popular at the time, the idea of free cell formation. This idea held that cells develop on their own from the inside out, the nucleus somehow forming a cell body around itself, and then the cell body growing a cell membrane, somewhat similar to the way fruit matures. Virchow's observation set the stage for descriptions of cell division in the 1870s and 1880s, discussed in detail in chapter 9.

Disease at the Organelle Level

Virchow also hypothesized **cellular pathology,** the idea that diseases attacking the whole body are ultimately caused by abnormalities within individual cells. His hypothesis was quite prophetic. Today, researchers develop new treatments for many disorders as they gain an understanding of the disease process at the cellular level.

Consider, for example, the common inherited illness cystic fibrosis, characterized by abnormal buildup of mucus in the lungs and some other organs. For decades, cystic fibrosis was treated symptomatically. Antibiotic drugs fought lung infections; daily chest-pounding exercises loosened built-up mucus and cleared breathing passages; and enzymes sprinkled onto food compensated for digestive juices trapped in a mucus-plugged pancreas, allowing the person to gain weight. Researchers finally identified the cellular defect behind cystic fibrosis in 1989, discovering abnormal channels in lung and pancreas cells. These channels trap salt within the cells. The salty cellular interiors then draw moisture in from surrounding tissue, drying out mucus until it has a sticky consistency that clogs organs. Several new treatments, including introducing a healthy gene or membrane protein into the lungs through a nasal spray, target the illness at its cellular source. (Biology in Action 4.1, later in the chapter, describes other disorders caused by defects in an organelle.)

The second half of the nineteenth century saw a flurry of activity in the field of microscopy, and cellular contents were rapidly described. By the twentieth century, subcellular function and the microscopic aspects of disease were increasingly under study. Soon, it became clear that a device was needed that could reveal structures even smaller than those seen under the compound light microscope. In the 1940s, the invention of the electron microscope met this need. (Figure 4.4 shows the range of each type of microscope.)

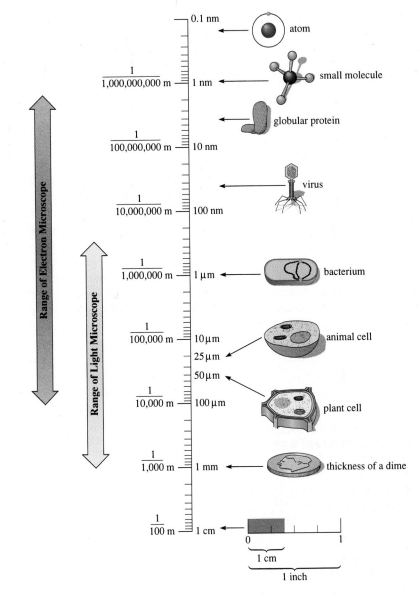

FIGURE 4.4

Ranges of the compound light and electron microscopes. Biologists use the metric system to measure size. The basic unit of length is the meter (m), which equals 39.37 inches (slightly more than a yard). Smaller metric units measure many chemical and biological structures. A centimeter (cm) is 0.01 meter (about 2/5 of an inch); a millimeter (mm) is 0.001 meter; a micrometer (μm) is 0.000001 meter; a nanometer (nm) is 0.000000001 meter; an angstrom unit (A) is 1/10 of a nanometer. Although each segment of the scale is depicted equally here, each segment actually represents only 1/10 of the length of the segment beneath it. The sizes of some chemical and biological structures are indicated next to the scale.

The cell theory is still evolving. Until just a few years ago, scientists viewed a complex cell as a structure containing a nucleus and jellylike material called cytoplasm, in which organelles were suspended in no particular organization. Researchers are now learning that organelles have precise locations in cells. Rather than drifting about randomly, organelles are transported along tracks in the cell like cars along a roller coaster track, powered by 20 or so different types of "motor molecules." An organelle is likely to be located near other structures it interacts with. So although the cell theory is still considered to be a product of the nineteenth century, we are constantly learning about the organization within cells, and about how cells interact to build larger-scale biological organization. We still have much to learn about the organization and functions within cells.

KEY CONCEPTS

The cell theory states that all living matter is composed of cells; the cell is the structural and functional unit of life; and that all cells come from preexisting cells. Human disease often reflects abnormality at the cellular level. Our knowledge of cell organization and interaction is still growing.

Cells of Two Types: Presence or Absence of a Nucleus

Cells are of two fundamental types. A **prokaryotic cell** lacks a nucleus, the dark-staining, membrane-bound body that encloses the genetic material in more complex **eukaryotic cells.** The Greek word *karyon* means "kernel," or a little less literally, "nucleus." *Eukaryote* means "true nucleus," and *prokaryote* means "before a nucleus."

Although eukaryotic cells have a greater number and variety of organelles than prokaryotic cells, the simpler prokaryotes are nevertheless organized

and efficient enough to support the biochemical reactions of life. Prokaryotes were the earliest cells to have left fossil evidence, and they flourish today, comprising the majority of living cells on earth. Both of these facts testify to the success of prokaryotic life.

Because most eukaryotic cells known today are 100 to 1,000 times the volume of a typical laboratory-cultured prokaryotic cell, biologists thought that size was also a criterion for distinguishing the two basic types of cells. However, the sizes of prokaryotes and eukaryotes can overlap. For example, a tiny single-celled eukaryote, *Nanochlorum eukaryotum,* is smaller than some bacteria. A giant bacterium discovered in 1993, the prokaryote *Epulopiscium fishelsoni,* is nearly visible to the human eye at its length of more than half a millimeter! It is a million times larger than the commonly studied *Escherichia coli* bacterium. *Epulopiscium fishelsoni* lives inside a brown surgeonfish in the Red Sea. The organism was at first thought to be a eukaryote because of its size, but the electron microscope reveals that it lacks a nucleus and has free genetic material like a bacterium. It also lacks the systems of intracellular bubbles (vesicles) that transport substances in eukaryotic cells, as well as the extensive inner scaffolding of the more complex cells. Genetic evidence ensured the organism's classification as a prokaryote—certain of its gene sequences are much more like those in prokaryotes than those in eukaryotes.

Prokaryotic Cells

All prokaryotes are either bacteria or **cyanobacteria,** organisms once known as blue-green algae because of their characteristic pigments and their similarity to the true (eukaryotic) algae. Figure 4.5 shows the major features of prokaryotic cells.

Most prokaryotic cells are surrounded by rigid **cell walls** that consist of peptidoglycans (peptide-sugars). Many antibiotic drugs, including the penicillins, cephalosporins, vancomycin, and bacitracin, halt infection by interfer-

ing with the bacterium's ability to build its cell wall. The shapes and staining properties of cell walls are used to classify bacteria as round, rod-shaped, curved, or spiral. Species whose cell walls turn purple in the presence of a dye called the Gram stain are termed gram positive; those whose cell walls turn pink are called gram negative. Cell wall characteristics and metabolic and biochemical criteria are used to distinguish bacterial species.

Beneath the prokaryote's cell wall is a **cell membrane,** or plasmalemma. The cell membrane pinches inward in places, which may indicate sites at which the cell can divide in two. Embedded in the cell membrane are enzymes that speed the rates of certain biochemical reactions, enabling the cell to obtain and utilize energy. Some types of bacteria also have internal arrays of membranes that serve as locations for specific enzymes.

Unlike bacteria, cyanobacteria contain internal membranes that are outgrowths of the surrounding cell membrane. These membranes, however, are not extensive enough to subdivide the cell into compartments, as membranes do in more complex cells. The cyanobacterium's membranes are studded with pigment molecules that absorb and extract energy from sunlight. In some prokaryotes, taillike appendages called **flagella,** which enable the cell to move, are anchored in the cell wall and underlying cell membrane. (Membrane structure and function are discussed in detail in chapter 5.)

The genetic material of a prokaryote is a single circle of DNA. A prokaryote's DNA is associated with proteins that are different than those in more complex cells. The part of a prokaryotic cell where the DNA is located is called the **nucleoid** (nucleuslike), and it sometimes appears fibrous under a microscope. Nearby are molecules of RNA and **ribosomes,** which are spherical organelles consisting of RNA and protein. Ribosomes enable the cell to utilize DNA sequence information to direct the manufacture of proteins. Because the DNA, RNA, and ribosomes in prokaryotic cells are in close contact with one another, protein synthesis in prokaryotes is rapid when compared to the process in more complex cells, whose cellular components are separated.

Cells require relatively large surface areas through which they can interact with the environment. Nutrients, water, oxygen, carbon dioxide, and waste products must enter or leave a cell through its surfaces. As a cell grows, its volume increases at a faster rate than its surface area, a phenomenon you can easily calculate if you know the size of the cell (fig. 4.6). Put another way, much of the interior of a large cell is far away from the cell's surface.

The ultimate inability of a cell's surface area to keep pace with its volume can limit the size of a cell, and this may be why prokaryotic cells are usually small. One adaptation to the cell's need for adequate surface area is for the cell membrane to fold, just as inlets and capes increase the perimeter of a shoreline.

KEY CONCEPTS

Prokaryotic cells lack nuclei, are bounded by a cell wall and cell membrane, and may contain internal membranes. They also have ribosomes and DNA in a region called the nucleoid. All prokaryotes are either bacteria or cyanobacteria.

Eukaryotic Cells

A typical eukaryotic cell is at least a thousand times the volume of a prokaryotic cell. How can a cell remain sufficiently organized to carry out the biochemical reactions of life when it is so large? Organelles accomplish this by creating specialized regions or compartments where certain biochemical reactions can occur. Saclike organelles (lysosomes, peroxisomes, and vacuoles) sequester biochemicals that might harm other cellular contents. Other organelles consist of membranes that are studded with enzymes, allowing certain chemical reactions to occur on their surfaces. On some membranes, different enzymes are organized according to the sequences in which they participate in biochemical reactions. In general, then, organelles

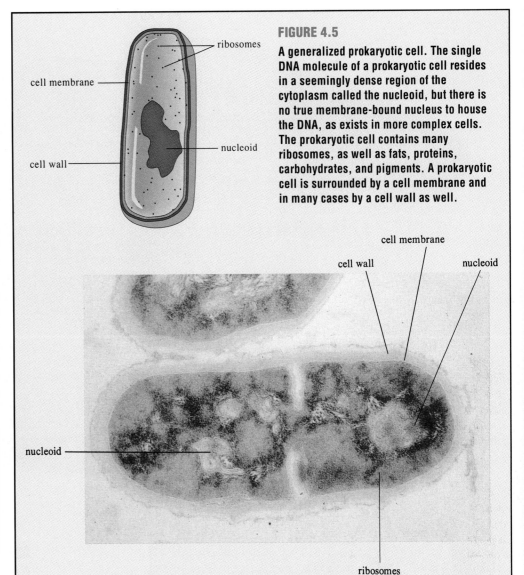

FIGURE 4.5

A generalized prokaryotic cell. The single DNA molecule of a prokaryotic cell resides in a seemingly dense region of the cytoplasm called the nucleoid, but there is no true membrane-bound nucleus to house the DNA, as exists in more complex cells. The prokaryotic cell contains many ribosomes, as well as fats, proteins, carbohydrates, and pigments. A prokaryotic cell is surrounded by a cell membrane and in many cases by a cell wall as well.

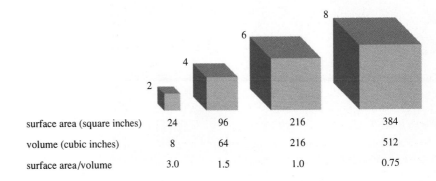

surface area (square inches)	24	96	216	384
volume (cubic inches)	8	64	216	512
surface area/volume	3.0	1.5	1.0	0.75

FIGURE 4.6

The important relationship between surface area and volume. As a cell enlarges, the amount of material inside it (its volume) increases faster than the area of its surface. The surface of a cell is vital to its functioning because communication and exchange of molecules and ions between the cell's interior and the outer environment takes place through the surface. The chemical reactions of life occur more readily when surface area is maximized. Imagine that a cell is a simple cube. Compare the surface area and volume of four increasingly larger cubes. (Surface area equals the area of each face multiplied by the number of faces; volume equals the length of a side cubed.) Can you see that volume increases faster than surface area?

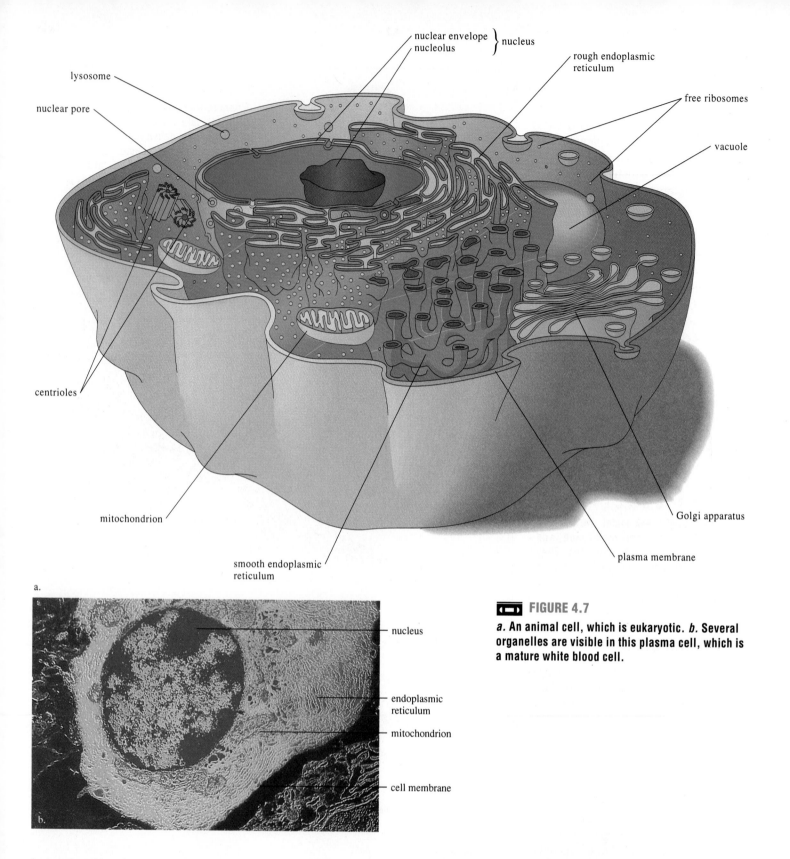

lysosome

nuclear pore

nuclear envelope
nucleolus } nucleus

rough endoplasmic
reticulum

free ribosomes

vacuole

centrioles

mitochondrion

smooth endoplasmic
reticulum

Golgi apparatus

plasma membrane

a.

nucleus

endoplasmic
reticulum

mitochondrion

cell membrane

b.

FIGURE 4.7

a. An animal cell, which is eukaryotic. *b.* Several organelles are visible in this plasma cell, which is a mature white blood cell.

keep related biochemicals and structures sufficiently close together to make them function more efficiently. The compartmentalization provided by organelles also eliminates the need for the entire cell to maintain a high concentration of a particular biochemical.

The general organization of a eukaryotic cell can be described as "bags within a bag" (figs. 4.7 and 4.8). The most prominent organelle is the nucleus, which contains the genetic material (DNA) organized with protein into rod-shaped structures called **chromosomes.**

The remainder of the cell consists of other organelles and the jellylike **cytoplasm,** the protoplasm external to the nucleus. About half of the volume of an animal cell consists of organelles; a plant cell contains about 90% water. The cytoplasm and organelles, including the

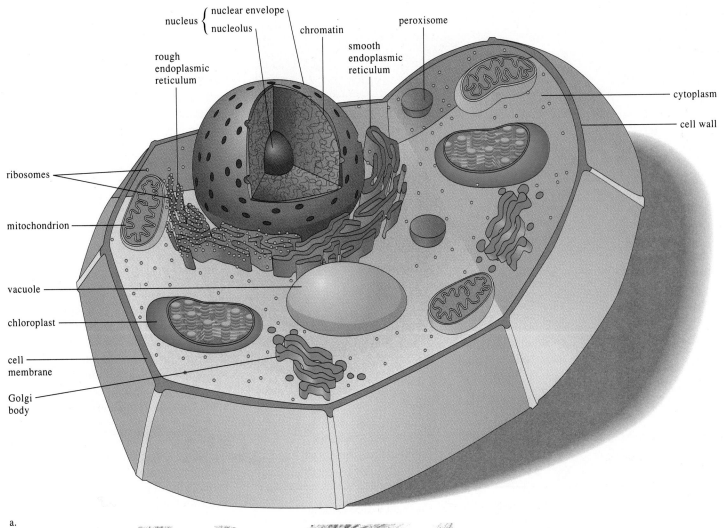

nucleus { nuclear envelope
nucleolus

chromatin

smooth
endoplasmic
reticulum

peroxisome

rough
endoplasmic
reticulum

cytoplasm

cell wall

ribosomes

mitochondrion

vacuole

chloroplast

cell
membrane

Golgi
body

a.

chloroplast

cell wall
cell
membrane

ribosomes

central
vacuole

nucleus

mitochondrion

peroxisome

b.

■□ FIGURE 4.8

a. A plant cell is eukaryotic, but unlike an animal cell, it has a cell wall, chloroplasts, and a large vacuole. A plant cell lacks the lysosomes an animal cell has. *b.* This *Coleus* leaf cell displays various organelles and structures.

Table 4.1 Comparison of Prokaryotic and Eukaryotic Cells

Characteristic	Prokaryotic Cells	Eukaryotic Cells
Organisms	Bacteria (including cyanobacteria)	Protists, fungi, plants, animals
Cell size	1–10 μm across	10–100 μm across
Oxygen required	By some	By many
Membrane-bound organelles	No	Yes
Ribosomes	Yes	Yes
DNA form	Circular	Coiled linear strands, complexed with protein
DNA location	In nucleoids	In nucleus
DNA length	Short	Long
Protein synthesis	RNA and protein synthesis are not spatially separated	RNA and protein synthesis are spatially separated
Membranes	Some	Many
Cytoskeleton	No	Yes
Cellular organization	Single cells or colonies	Some single-celled, most multicellular with differentiation of cell function

Source: From Bruce Alberts, et al., *Molecular Biology of the Cell.* Copyright © 1983 Garland Publishing Company, New York, NY. Reprinted by permission.

nucleus, are considered the living parts of the cell. Nonliving cellular components include stored proteins, fats, and carbohydrates, pigment molecules, and various inorganic chemicals. Arrays of protein rods within plant and animal cells form a framework called the **cytoskeleton,** which helps to give the cell its shape. Protein rods and tubules also form cellular appendages that enable certain cells to move, and they form structures involved in cell division. (The cell components that consist of protein rods and tubules are discussed in chapters 5 and 9.) Table 4.1 summarizes the differences between prokaryotic and eukaryotic cells.

Organelles in Action— Secretion

The different organelles within a cell interact to provide basic life functions, and also to sculpt the distinguishing characteristics of a particular cell type. The roles of several organelles can be explored in detail by examining a coordinated function, such as secretion (fig. 4.9). Consider a glandular cell in the breast of a human female. Dormant most of the time, the cell increases its metabolic activity during pregnancy and then undergoes a burst of productivity shortly after a baby is born. The ability of individual cells to manufacture the remarkably complex milk is made possi-

ble by organelles functioning together to form a secretory network.

A new mother hears her infant cry, and within minutes she may feel her breasts swell in response. When the infant suckles, the milk that it receives is a mixture of cells and biochemicals, a highly nutritious food tailored specifically to meet the needs of a human child. Deep within the mother's breasts, glandular cells rapidly produce the proteins, fats, and sugars that combine with immune system cells and biochemicals in specific proportions to form the milk.

The Nucleus

Secretion actually begins in the nucleus, where, in humans, 23 pairs of rod-shaped chromosomes contain information that other parts of the cell use to construct proteins. Each chromosome is made up of a single DNA molecule plus proteins. The DNA molecule of each chromosome consists of millions of nucleotides; long sequences of nucleotides comprise genes. DNA nucleotide sequences, or genes, encode the instructions for producing the amino acid sequences of proteins in human milk. Several types of intermediate molecules of the nucleic acid RNA transfer the gene's information to the organelles that synthesize proteins. The nucleotide sequence of a gene is transcribed into a nucleotide sequence of **messenger RNA** (mRNA). Protein synthesis also requires **transfer RNA** and **ribosomal RNA.** Ribosomes are the organelles that protein synthesis occurs on. They consist of ribosomal RNA and proteins and are assembled in a part of the nucleus called the **nucleolus.**

Different types of cells specialize in performing different functions. All cells (except red blood cells, which lack nuclei), contain a complete set of genes, which provide enough information for the cell to specialize in virtually any way by expressing (or "turning on") different subsets of genes. Proteins called **transcription factors** control which genes are turned "on" or "off" in a particular cell. As messenger RNA transcribes genetic information from sequences of DNA, it is guided from the interior of the nucleus towards its periphery by a three-dimensional network of protein fibers that form a scaffolding called the **nuclear matrix.** In the case of milk secretion, genes in the nucleus also encode the enzymes the cell needs to synthesize other components of milk, such as sugars and lipids.

The Nuclear Pore Complex

Messenger RNA exits the nucleus by passing through holes, called **nuclear pores,** in the two-layered **nuclear envelope** that separates the nucleus from the cytoplasm. Nuclear pores are not merely perforations, but channels that form as part of structures called **nuclear pore complexes** (NPCs). These complexes consist of more than 100 different proteins.

A nuclear pore complex controls passage of RNA and protein molecules between the nucleus and the cytoplasm, allowing only certain molecules through. Amazingly each of the thousands of NPCs on a nucleus completely falls apart when the cell divides. Figure 4.10 shows three views of an NPC in the egg cell of the African clawed frog. To obtain NPCs for study, researchers treat the eggs with detergent, which frees the NPCs from the nuclear envelope.

The Cytoplasm

Once it moves through the NPCs into the cytoplasm, messenger RNA encounters but does not actually enter a maze of enzyme-studded, interconnected membranous tubules and sacs that winds from the nuclear envelope to the cell membrane. This labyrinth is the **endoplasmic reticulum** (ER) (fig. 4.9). (*Endoplasmic* means "within the plasm" and *reticulum* means "network.") The portion of this membranous system nearest the nucleus is flattened and studded with ribosomes. This region is called **rough ER** because of its fuzzy appearance under the electron

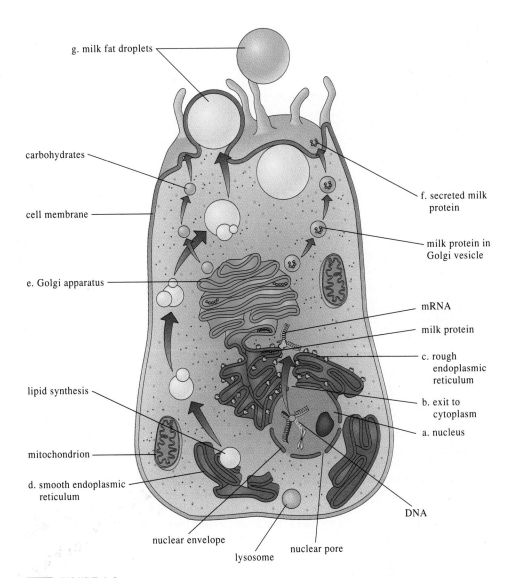

FIGURE 4.9

The process of milk secretion illustrates how organelles interact to synthesize, transport, store, and export biochemicals. Secretion begins in the nucleus (*a*), where messenger RNA (mRNA) bearing the genetic information for milk protein production is synthesized. The mRNA then exits through a nuclear pore to the cytoplasm (*b*). Many proteins are synthesized on the membranes of the rough endoplasmic reticulum or ER (*c*), using amino acids in the cytoplasm. Lipids are synthesized in the smooth ER (*d*), and sugars are synthesized, assembled, and stored in the Golgi apparatus (*e*). In an active cell in a mammary gland, milk proteins (*f*) are released from vesicles that bud from the Golgi apparatus. Fat droplets (*g*) pick up a layer of lipid from the cell membrane as they exit the cell. When the baby suckles, he or she receives a chemically complex secretion—human milk.

microscope. Messenger RNA attaches to the ribosomes on the rough ER. Amino acids from the cytoplasm are then strung together, using the instructions carried by the mRNA, to form proteins that either exit the cell or become components of a membrane. (The milk protein casein, for example, enters the tubules of the rough ER to later exit the cell.) Proteins synthesized on ribosomes not associated with ER are released into the cell's cytoplasm, where they serve specific functions.

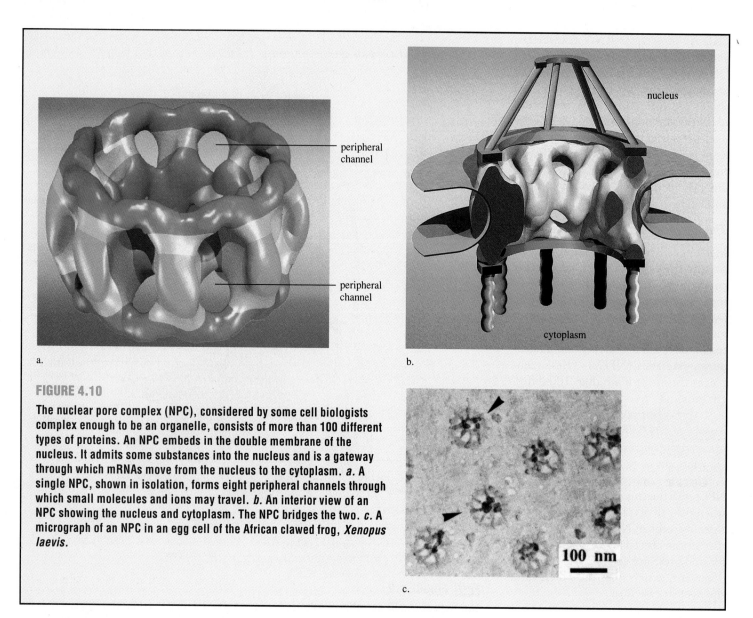

a.

b.

c.

FIGURE 4.10

The nuclear pore complex (NPC), considered by some cell biologists complex enough to be an organelle, consists of more than 100 different types of proteins. An NPC embeds in the double membrane of the nucleus. It admits some substances into the nucleus and is a gateway through which mRNAs move from the nucleus to the cytoplasm. *a.* A single NPC, shown in isolation, forms eight peripheral channels through which small molecules and ions may travel. *b.* An interior view of an NPC showing the nucleus and cytoplasm. The NPC bridges the two. *c.* A micrograph of an NPC in an egg cell of the African clawed frog, *Xenopus laevis.*

The ER acts as a quality control center for the cell. Its chemical milieu enables the protein to fold into the conformation necessary to perform its function. Misfolded proteins are pulled out of the production line and degraded in the ER.

As the rough ER winds its way outwards toward the cell membrane, the ribosomes become fewer, and the diameters of the tubules widen, forming a section called **smooth ER.** Here, lipids are synthesized and added to the proteins transported from the rough ER. The lipids and proteins travel along until the tubules of the smooth ER eventually narrow and end. The proteins and fats accumulated in the smooth ER exit in membrane-bound sacs called **vesicles** that pinch off of the tubular endings of the membrane. A loaded vesicle takes its contents to the next stop in the eukaryotic production line, a structure called the **Golgi apparatus.** The Golgi apparatus is a stack of flat, membrane-enclosed sacs that functions as a processing center. In the Golgi apparatus, sugars are synthesized and linked to form starches, or they attach to proteins to form glycoproteins or to lipids to form glycolipids. Proteins fold into their final forms in the Golgi apparatus. The components of secretions such as breast milk are temporarily stored here.

Milk proteins bud off of the Golgi apparatus in vesicles that travel outward to the cell membrane. The protein-carrying vesicles fleetingly become part of the cell membrane and then open out facing the exterior of the cell. The release of secretions out of the cell through vesicles at the cell surface, called **exocytosis,** is discussed in chapter 5. Proteins are transported along a cellular production line by a fleet of vesicles.

Fat droplets retain a layer of surrounding membrane when they leave the cell. In milk secretion, the milk constituents are stimulated to exit the cell when the baby suckles. A hormone released in the mother's system when she feels the sucking action causes the muscle cells surrounding balls of glandular cells to contract, which squeezes milk from them. The milk is released into ducts that lead to the nipple.

The composition of milk is different in different species. Human milk, for example, is rich in lipids, which suits the rapid development of the child's brain. Cow's milk, in contrast, is rich in protein, which is suited to the calf's need to build muscle rapidly.

KEY CONCEPTS

In eukaryotes, organelles partition off sets of biochemicals that function together. The nucleus contains DNA, and the rest of the cell contains cytoplasm. Protein rods form the cytoskeleton. The process of secretion illustrates the coordinated functions of organelles. mRNA carries the information for synthesizing a particular protein from the DNA, out of the nucleus through nuclear pores, and attaches to ribosomes in the rough endoplasmic reticulum (ER). Proteins are assembled on ribosomes. Proteins fold and are modified in the ER.

Lipids are added in the smooth endoplasmic reticulum, and sugars are added at the Golgi apparatus. Secretory products bud off in vesicles and exit the cell by exocytosis.

Other Organelles and Structures

Mitochondria

The activities of secretion, as well as the many chemical reactions taking place in the cytoplasm, require a steady supply of energy. In eukaryotic cells, cellular energy is provided by organelles called **mitochondria**; the energy-generating reactions of cellular respiration occur within them. The number of mitochondria in a cell can vary from a few to tens of thousands. A typical liver cell has about 1,700 mitochondria, although cells with high energy requirements, such as muscle cells, may have many more. Mitochondria appear in figures 4.7, 4.8, and 4.9.

A mitochondrion has an outer membrane similar to those of the ER and

Golgi apparatus and an intricately folded inner membrane. The folds of the inner membrane are called **cristae.** The cristae contain many of the enzymes that take part in cellular respiration. The mitochondrion is especially interesting because it contains genetic material. Another unique characteristic of mitochondria is that they are inherited from the mother only. This is because mitochondria are found in the middle regions of sperm cells but not in the head region, which is the portion that actually enters the egg to fertilize it. A class of inherited diseases whose symptoms result from abnormal mitochondria are always passed from mother to offspring. These mitochondrial illnesses usually produce extreme muscle weakness, because muscle is a highly active tissue dependent upon the functioning of many mitochondria. We will return to this fascinating organelle shortly, and we will examine it in greater detail in chapter 7.

Lysosomes

Eukaryotic cells break down molecules and structures as well as produce them. Organelles called **lysosomes** are the cell's "garbage disposals"; they chemically dismantle captured bacteria, worn-out organelles, fats, carbohydrates, and other debris. The lysosome is a sac that buds off of the ER or Golgi apparatus. In humans, each lysosome contains more than 40 different types of digestive enzymes. These enzymes can work only in a very acidic environment (pH 5); the compartmentalization provided by the lysosome membrane maintains a highly acidic region for the enzymes without harming other cellular constituents.

Material is brought to a lysosome by a vesicle, either from within the cell or from the cell membrane, which can bud inward to entrap a particle in a process called **endocytosis.** Lysosomes are particularly abundant in liver cells, perhaps because these cells break down toxins. The correct balance of enzymes within a lysosome is important to human health (Biology in Action 4.1).

Lysosomes are not found in plant cells. A similar "disposal" function is performed by enzymes in the **central vac-**

uole, a membrane-bound organelle that temporarily stores a substance or transports it within a cell.

Peroxisomes

Peroxisomes are single-membrane-bound sacs found in all eukaryotic cells, but in different numbers and sizes in different cell types. These organelles are large and abundant in liver and kidney cells in animals (fig. 4.11a). Peroxisomes, originally named microbodies, were first visualized in the 1960s as small, darkly staining bodies in animal and plant cells. The organelles were renamed peroxisomes when an enzyme called catalase was discovered within them. Catalase catalyzes the breakdown of hydrogen peroxide (H_2O_2) to water and oxygen.

The outer membrane of a peroxisome contains some 40 types of enzymes, which catalyze a variety of biochemical reactions, including:

synthesizing bile acids, which are used in fat digestion

breaking down lipids called very-long-chain fatty acids

degrading rare biochemicals

metabolizing potentially toxic compounds that form as a result of oxygen exposure

Peroxisomes reproduce by enlarging and then splitting in two.

Like the lysosomal storage disorders described in Biology in Action 4.1, abnormal peroxisomal enzymes also harm health. One such disorder was the subject of a 1992 film called *Lorenzo's Oil.*

Six-year-old Lorenzo Odone suffered from a disease called adrenoeukodystrophy (ALD). His peroxisomes lacked one of the two most abundant proteins in the outer membrane. Normally, this protein transports an enzyme into the peroxisome, where it catalyzes a reaction that helps break down a type of very-long-chain fatty acid. Without the enzyme transporter protein, the fatty acid builds up in the cells of the brain and spinal cord, eventually stripping these cells of the fatty

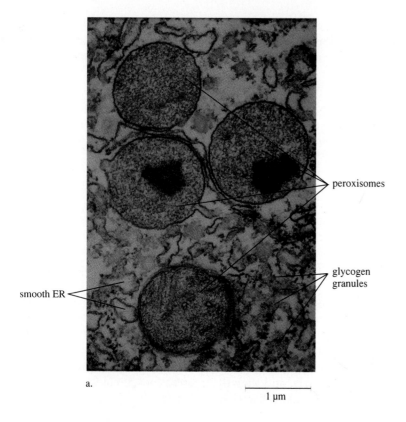

peroxisomes

glycogen granules

smooth ER

a.

1 µm

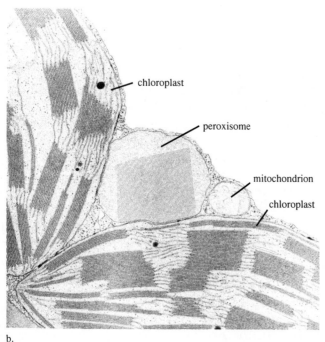

chloroplast

peroxisome

mitochondrion

chloroplast

b.

myelin sheaths necessary for nerve conduction. The early symptoms of ALD include low blood sugar, skin darkening, electrolyte imbalance, malaise, muscle weakness, dizziness, and heart arrhythmia. As the nervous system deteriorates, the patient loses control over the limbs. Death follows within a few years.

For Lorenzo and many other sufferers of ALD, ingesting a type of triglyceride from rapeseed oil prevented the buildup of the very-long-chain fatty acids for a few years, stalling symptoms. But the treatment eventually impairs blood clotting and other vital functions, and it fails to halt the ultimate progression of the illness.

In ALD, the peroxisomes appear normal, despite their protein deficit. In another condition, Zellweger syndrome, the peroxisomes themselves are absent. Thirty-five of the 40 types of peroxisomal enzymes can function in the cytoplasm, but the absence of the five enzymes that cannot causes several serious symptoms and eventually death. Affected children have abnormal development of the skull, face, eyes, ears, hands, and feet; cysts in their kidneys; and a malfunctioning liver. When their cells are grown in the laboratory, they show buildup of certain very-long-chain fatty acids. Cell biologists hypothesize that parts of the peroxisomes are present but cannot be assembled.

In plants, peroxisomes are abundant in leaf cells (fig. 4.11b). A dense area where catalase concentrates can often be visualized as a crystal in electron micrographs. Peroxisomes in plants help to metabolize or break down an organic molecule synthesized in **photosynthesis,** the process by which plants harness solar energy to synthesize simple carbohydrates.

FIGURE 4.11

a. Peroxisomes are abundant in liver cells, where they assist in detoxification. *b.* This plant peroxisome is sandwiched between two chloroplasts in a leaf cell, where it participates in a process related to photosynthesis. The square within the peroxisome is a large crystal of catalase, an enzyme found in all peroxisomes.

Biology in *ACTION*

A Closer Look at an Organelle—The Lysosome

A lysosome is a membrane-bound sac of enzymes that buds off from a Golgi apparatus. The key to the functioning of a lysosome can be summed up in one word—balance. The lysosomes in a particular type of cell contain a balanced mix of enzymes appropriate to the function of that cell. If that balance is disrupted, the cell's function can be altered—sometimes drastically.

More than 40 lysosomal enzymes are known, and most break down fats and carbohydrates (fig.1). This enzymatic digestion also breaks down worn-out organelles and membranes within the cell, as well as particles engulfed by the cell when the cell membrane buds inward. Material to be digested is carried to the lysosome in a vesicle. The membranes of the lysosome and the vesicle fuse, and the appropriate enzymes go to work. Lysosomes are sometimes called "suicide sacs" because if they rupture and release their enzymes, the entire cell is digested from within and dies. Lysosomes may actually play a role in aging by destroying cells in this way.

The absence or malfunction of just one lysosomal enzyme can be devastating to health, creating a lysosomal storage disease. In these inherited disorders, the molecule that is normally degraded by a missing or abnormal lysosomal enzyme accumulates in the lysosome. The lysosome swells with the excess waste, crowding organelles and interfering with the cell's biochemical activities. Usually only the cells that constitute a particular tissue are affected. Because there are a number of different lysosomal enzymes, several types of lysosomal storage diseases exist. Symptoms reflect the tissue with the affected cells.

Tay-Sachs disease is a lysosomal storage disease that results from a missing enzyme that normally breaks down lipids in cells surrounding nerve cells. Without the enzyme, cells of the nervous system gradually drown in lipid. Symptoms are usually noted at about six months of age, when an infant begins to lag behind in the acquisition of motor skills. On a cellular level,

however, signs of Tay-Sachs disease are present even earlier as enlarged lysosomes. Children who have inherited Tay-Sachs disease soon lose their vision and hearing and later are paralyzed. Virtually all children with the severe form of the disease die before they are four years old. A less severe form of Tay-Sachs affecting adults is also known.

Pompe disease is another childhood lysosomal storage disease. An enzyme that breaks down the complex carbohydrate glycogen into simple sugars is missing. Glycogen builds up in muscle and liver cells. The young patients usually die of heart failure, because the muscle cells of the heart swell so greatly that they can no longer function properly. Another lysosomal storage disease, Hurler disease, causes bone deformities. The electron microscope reveals that affected bone cells contain huge lysosomes swollen with mucuslike substances called mucopolysaccharides.

FIGURE 1

Lysosomal enzymes are synthesized on the endoplasmic reticulum and are transported to the Golgi apparatus. The Golgi apparatus detects and separates the specific enzymes destined for lysosomes by recognizing a particular sugar attached to these enzymes. The lysosome-specific enzymes are then packaged into vesicles that eventually become lysosomes. Lysosomes fuse with vesicles carrying debris from the outside or from within the cell, and the lysosomal enzymes then degrade the debris.

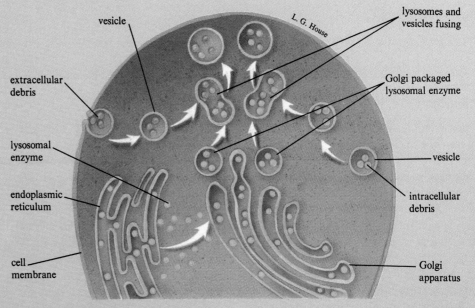

Table 4.2   Structures and Functions of Organelles

Organelle	Structure	Function
Chloroplast (plants and photosynthetic bacteria)	Plastid consisting of stacks of flattened sacs containing pigments	Photosynthesis
Endoplasmic reticulum	Membrane network: rough ER has ribosomes, smooth ER does not	Protein folding and modification; lipid synthesis; lipid attachment to protein
Golgi apparatus	Stacks of membrane-enclosed sacs	Scaffold for protein synthesis; sugar synthesis and polymerization into starches or linkage of sugars to protein or lipids
Lysosome (animals only)	Sac containing digestive enzymes	Degradation of intracellular debris, recycling of cell components
Mitochondrion	Inner membrane highly folded and studded with enzymes	Cellular respiration
Nucleus	Cell compartment containing DNA; pores in surrounding membrane	Separation of genetic material from rest of cell
Peroxisome	Membrane-bound sac containing enzymes	Catalysis of several biochemical reactions: detoxification, lipid breakdown, bile acid synthesis, metabolism of H_2O_2
Ribosome	Two associated globular subunits built of RNA and protein	Scaffold for protein synthesis
Vacuole	Membrane-bound body	Temporary storage or transport of substances

Centrioles

The organelles and structures discussed so far—the nucleus, ribosomes, endoplasmic reticulum, Golgi apparatus, mitochondria, lysosomes, and peroxisomes—are present in nearly all eukaryotic cells, although some are more abundant in certain specialized cell types than in others. A few structures are peculiar to animal or plant cells. **Centrioles** are oblong structures built of protein rods called **microtubules.** They are found in pairs in animal cells, oriented at right angles to one another near the nucleus. Centrioles appear to play a role in organizing other microtubules to pull replicated chromosomes into two groups during cell division.

Chloroplasts

An organelle unique to plants and algae, the **chloroplast,** gives these organisms their green color. Chloroplasts house the chemical reactions of photosynthesis. The inner membrane of a chloroplast is studded with enzymes and pigments necessary for photosynthesis. The membrane is organized into stacks, called **grana,** of flattened membranous disks called **thylakoids.** Like mitochondria, chloroplasts contain genetic material.

The green pigment **chlorophyll** within the chloroplasts harnesses the solar energy. An actively photosynthesizing plant cell, such as a leaf cell facing the sun, may contain more than 50 chloroplasts (fig. 4.1c). Chloroplasts are the most abundant form of a general class of pigment-containing organelles, called plastids, found in plant cells.

Mature plant cells may also be distinguished from other eukaryotic cells by the presence of a large, centrally located vacuole. Water stored in a plant cell's vacuole can amount to 90% of the cell's total volume. In addition, a rigid cell wall built of the carbohydrate *cellulose* surrounds each plant cell. The cell wall helps to support the cell and protect its contents. Animal cells do not have cell walls.

Table 4.2 summarizes organelle structures and functions.

FIGURE 4.12

A virus is a nucleic acid coated with protein. The human immunodeficiency virus (HIV), which causes AIDS, consists of RNA surrounded by several layers of proteins. Once inside a human cell (usually a T cell, part of the immune system), the virus uses an enzyme, reverse transcriptase, to synthesize a DNA copy of its RNA. The virus then inserts this copy into the host DNA. HIV damages the human body's ability to protect itself against infection and cancer by killing T cells and by using these cells to reproduce itself.
Researchers are trying to develop AIDS vaccines that target various proteins and glycoproteins of HIV.

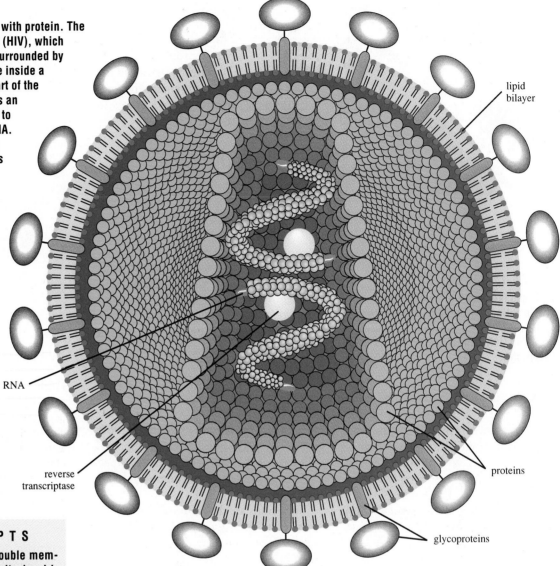

lipid bilayer

RNA

reverse transcriptase

proteins

glycoproteins

KEY CONCEPTS
Mitochondria are built of a double membrane. Along the folds of the mitochondrion's inner membrane, enzymes carry out biochemical reactions that harness the energy in food molecules. Lysosomes contain biochemicals that degrade cellular debris. Peroxisomes contain enzymes that detoxify certain substances, synthesize bile acids, break down lipids, and participate in oxygen metabolism.
Centrioles are peculiar to animal cells. Plastids, including chloroplasts, are found only in plant cells.

Simpler than Cells

The simplest form of life is a unicellular organism with no organelles, such as a bacterium. However, several types of infectious agents appear to be living while they are infecting cells but otherwise seem to be nonliving,

complex collections of chemicals. These entities include the familiar viruses and several not-so-familiar structures that straddle the border between living and nonliving.

Viruses

A **virus** is a particle, too small to be seen under a light microscope, that can only reproduce if it has access to the energy resources and protein-synthetic machinery of a host cell. A virus consists of a nucleic acid (DNA or RNA) surrounded by protein.

Many viruses, such as the human immunodeficiency virus (HIV), which causes AIDS (fig. 4.12), cannot reproduce outside a living cell. A virus reproduces by inserting its DNA or RNA

into the host cell, where it may situate itself within the host's DNA. Viral DNA sequences can probably be found within your own chromosomes.

Some viruses' genetic material is encoded in RNA rather than DNA. An RNA virus, such as HIV, is called a retrovirus. A retrovirus copies its RNA into DNA in a reaction catalyzed by reverse transcriptase, an enzyme found only in retroviruses. The DNA then inserts itself into the host cell's chromosome.

Once viral DNA integrates into the host's DNA, it can either remain there and replicate along with the host's DNA without causing harm, or it can actively take over the cell, eventually causing the cell to die. To do this, some of the virus's genes direct the host cell to replicate viral DNA rather than host

FIGURE 4.13

Plants can fall victim to an infectious particle called a viroid. Shown here are a tomato with "tomato bunchy top," caused by a viroid (left), and a healthy tomato plant (right).

DNA. As viral DNA accumulates in the cell, some of it is used to manufacture proteins. (Recall that DNA provides information from which the cell constructs proteins.) Within hours or days, the infected cell fills with viral DNA and protein. Some of the proteins wrap around the DNA to form new viral particles. Finally, the virus produces an enzyme that cuts through the host cell's outer membrane. The cell bursts, releasing new viruses to infect other cells.

Viruses are known to infect all kinds of organisms, including animals, plants, and bacteria. Any particular type of virus, however, infects only certain species, which constitute its *host range.* Biology in Action 4.2 discusses a virus that frequently inhabits human cells—herpes simplex.

Viroids

Viroids are even more streamlined and simple than viruses. They consist only of highly wound genetic material that, untwisted and stretched out, would be about 3 feet (1 meter) long. Viroids cause several exotic-sounding plant diseases,

including "avocado sun blotch," "coconut cadang cadang" and "tomato bunchy top" (fig. 4.13). They nearly destroyed the chrysanthemum industry in the United States in the early 1950s.

Prions

Like viruses, viroids infect and take over the cellular apparatus of their host. However, even viroids may not represent the smallest and simplest living things. Tinier yet are **prions,** which are only 1/100 to 1/1,000 the size of the smallest known virus. Prions are composed only of protein.

The name prion is derived from "protein infectious agent." Prions were originally described in 1966 as proteins that replicate to cause a disease called scrapie in sheep. The British researchers who originated the idea of prions were not taken very seriously. Finally, in 1982, University of California at San Francisco biologist S. B. Prusiner further described prions and implicated them in several disorders.

Prions are thought to be an abnormal form of a glycoprotein found in the cells of many mammals. The abnormal prion protein forms a gummy mass that causes different symptoms depending on its location. The buildup usually occurs in the brain, causing neurodegenerative disorders such as scrapie in sheep and goats; "mad cow disease" in English bovines; and kuru, Creutzfeldt-Jakob syndrome, and Gerstmann-Straussler-Scheinker syndrome in humans (fig. 4.14).

Experimental procedures can demonstrate prion action. When Prusiner and his coworkers introduced the abnormal prion protein into mice, the animals developed huge spongy areas in their brains. Nerve cells developed holes, gummy deposits formed, and supportive cells overgrew. Although the original source of abnormal prion protein is a genetic mutation, once the vari-

FIGURE 4.14

This child can no longer walk, stand, sit, or talk because of nervous system degeneration caused by the disease kuru, which is associated with an abnormal form of the prion protein. Deposits of gummy amyloid in his brain cause the symptoms. The child died a few months after this photo was taken.

ant prion forms, it can be transmitted infectiously. Researchers are still debating the biological status of prions.

KEY CONCEPTS

Structures even simpler than cells include viruses, viroids, and prions. A virus consists of a nucleic acid surrounded by protein. It must invade a cell to reproduce. Viral DNA inserted into a host cell may either remain dormant or replicate using the cell's organelles. Viruses attack specific species. A viroid consists only of nucleic acid (DNA or RNA) and can cause plant disease. A prion is a tiny abnormal protein that causes a variety of neurodegenerative disorders. Biologists have not definitively determined whether viruses, viroids, and prions qualify as living entities, but all three may drastically affect living cells.

Biology in ACTION

The Herpes Simplex Virus

The herpes simplex virus is far simpler in structure than even the smallest prokaryotic cell. Like all viruses, it consists only of nucleic acid surrounded by a protein coat. Viruses are too simple to be considered organisms; whether they should be considered alive is a matter of debate. Yet, despite its simplicity and its questionable biological status, it is clear that the herpes simplex virus can have drastic effects on human health.

Herpes simplex virus Type 1 (HSV-1) usually produces cold sores in or around the mouth, while Type 2 (HSV-2) usually causes genital blisters. However, each of the two viral types has been known to infect the body part the other type usually infects. Once HSV-1 or HSV-2 enters human skin, it multiplies quickly. The person initially feels tingling or itching. Within two weeks, the tingling area erupts into mouth sores or a rash of painful, clear genital blisters. The eruptions last for about three weeks; during this time, the virus can be passed on to others by skin contact.

Even after the outbreak disappears, the infected person is not free of the virus. Both viral types enter nerve cell branches in the skin (fig.1), and then the viruses follow these cells further into the nervous system. HSV-1 retreats to the brain and HSV-2 to the spinal cord. Once the virus enters the nervous system, the person's immune system is unable to attack it. The virus may resurface to cause a new eruption a week, a month, or a year later, or it may never appear again. Sufferers claim that stress seems to spark recurrences.

Herpes simplex infections are commonly spread by sexual contact. Genital herpes can be acquired by having sexual intercourse with someone who has open sores. Oral sex can spread the infection from the mouth of one partner to the genitals of the other, or vice versa. Although it is true that herpes cannot spread unless the virus is "shed" from open sores, it is possible for a person's symptoms to be so mild (particularly if the attack is not the first one, or occurs in a male), that he or she does not even know that contagion is possible. Another complication is that females sometimes have internal blisters and may be unaware of the attack.

The most seriously affected host of the herpes simplex virus is a newborn who comes in contact with open sores during the birth process. Because the infant's immune system is not yet completely functional, a herpes infection is especially devastating. Forty percent of babies exposed to active vaginal lesions become infected, and half of these infants die from it. Of those infants who are infected but survive, 25% have severe nervous system damage, and another 25% have widespread skin sores. To prevent exposing her newborn to herpes, a woman with a history of the infection can have her vaginal secretions checked periodically during pregnancy for evidence of an outbreak. If she has sores at the time of delivery, the physician performs a Caesarian section to protect the child.

At the present time, herpes simplex infections cannot be completely cured. The drug acyclovir, however, inhibits new sores, decreases healing time, and shortens the time during which a sore "sheds" live virus. The drug appears to help both new sufferers and those who suffer from recurrent attacks.

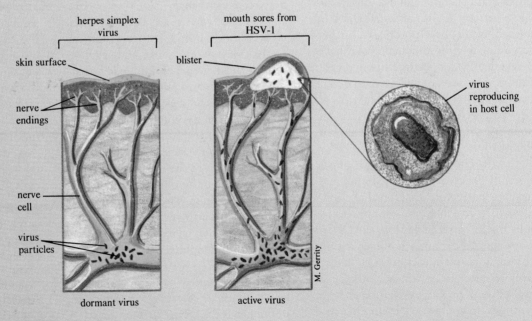

FIGURE 1

Between active outbreaks, the herpes simplex virus lies dormant in nerve cells beneath the skin's surface and causes no symptoms. When the virus moves to the skin's surface, it causes the characteristic blisters of a herpes infection flare-up.

The Origin of Eukaryotic Cells

How did the complex eukaryotic cell arise on a planet populated by the far simpler prokaryotes? The **endosymbiont theory** proposes that eukaryotic cells formed as large prokaryotic cells incorporated smaller and simpler prokaryotic cells (an endosymbiont is an organism that can live only inside another organism). The compelling evidence in support of the endosymbiont theory is that the mitochondria and chloroplasts found in eukaryotic cells bear striking similarities to prokaryotic cells.

In the 1920s, several investigators noted the resemblances between eukaryotic cell components and free-living prokaryotes. These early investigators proposed that eukaryotes evolved by incorporating prokaryotes, but their ideas were ridiculed at the time because there was little proof to support them. In the 1960s, studies with the electron microscope dramatically confirmed the similarities proposed four decades earlier between eukaryotic organelles and the prokaryotic bacteria and cyanobacteria. Then genetic studies revealed DNA in mitochondria and chloroplasts, further suggesting that these structures may have descended from once free-living organisms. These converging ideas and observations led biologist Lynn Margulis at the University of Massachusetts at Amherst to propose the endosymbiont theory in the late 1960s.

Mitochondria and chloroplasts resemble bacteria in size, shape, and membrane structure. Each organelle contains its own DNA and reproduces by splitting in two. In addition, the DNA, messenger RNA, and ribosomes within chloroplasts and mitochondria function in close association with each other, just as they do in prokaryotes during protein synthesis. In contrast, DNA in a eukaryotic cell's nucleus is physically separated from the RNA and ribosomes.

There are other similarities between prokaryotic cells and the organelles of eukaryotic cells. Pigments in the chloroplasts of eukaryotic red algae are similar to pigments cyanobacteria use to carry out photosynthesis. Sperm tails and centrioles may be descendants of ancient spiral-shaped bacteria. Mitochondria most closely resemble the aerobic (oxygen-using) purple nonsulfur bacteria, which can photosynthesize. Accordingly, the endosymbiont theory proposes that mitochondria descended from aerobic bacteria, that the chloroplasts of red algae descended from cyanobacteria, and that the chloroplasts of green plants descended from yet another type of photosynthetic microorganism. How might this hypothesized merger of organisms have taken place?

Picture a mat of bacteria and cyanobacteria, thriving in a pond some 2.5 billion years ago. Over many millions of years, the flourishing cyanobacteria pumped oxygen into the atmosphere as a by-product of photosynthesis. Eventually, only those organisms that could tolerate free oxygen survived.

Free oxygen tends to react with molecules in living things (such as nucleotides, amino acids, and sugars), turning these chemicals into oxides that can no longer carry out biological functions. One way for a large cell to survive in an oxygen-rich environment would be to engulf an aerobic bacterium in an inward-budding vesicle of its cell membrane. Eventually, the membrane of this vesicle became the outer membrane of the mitochondrion. The outer membrane of the engulfed aerobic bacterium became the inner membrane system of the mitochondrion, complete with respiratory enzymes (fig. 4.15). The smaller bacterium found a new home in the larger cell; in return, the host cell could survive in the newly oxygenated atmosphere.

Similarly, large cells that picked up cyanobacteria or other small cells capable of photosynthesis obtained the forerunners of chloroplasts and thus became the early ancestors of red algae or of green plants. Once such ancient cells had acquired their endosymbiont organelles, genetic changes impaired the ability of the captured prokaryotes to live on their own outside the host cell. The larger cells and the captured prokaryotes became dependent on one another for survival. The result of this biological interdependency, according to this theory, is the compartmentalized cells of modern eukaryotes, including our own.

Although none of us was present 2.5 billion years ago to witness the proposed endosymbiotic origin of complex cells, each of the steps to the hypothesis are visible today in symbiotic organisms. At least one modern microbe could be a descendant of a large microorganism that long ago harbored smaller prokaryotes destined to become the larger cell's organelles. Bacteria of genus *Thermoplasma* are the size of a prokaryote and lack a nucleus, but they also lack the cell wall characteristic of bacteria. However, *Thermoplasma* have proteins associated with their DNA that resemble the proteins in eukaryotes more closely than those in prokaryotes. Perhaps *Thermoplasma* represent a link in the evolutionary chain between prokaryotes and eukaryotes.

In the absence of a time machine, the elegant endosymbiont theory may be the best window we have to look into the critical juncture in cell history when cells first became more complex.

KEY CONCEPTS

The endosymbiont theory proposes that complex eukaryotic cells descended from large prokaryotes that incorporated smaller and simpler prokaryotes. The ingested prokaryotes eventually evolved to become the organelles of eukaryotic cells.

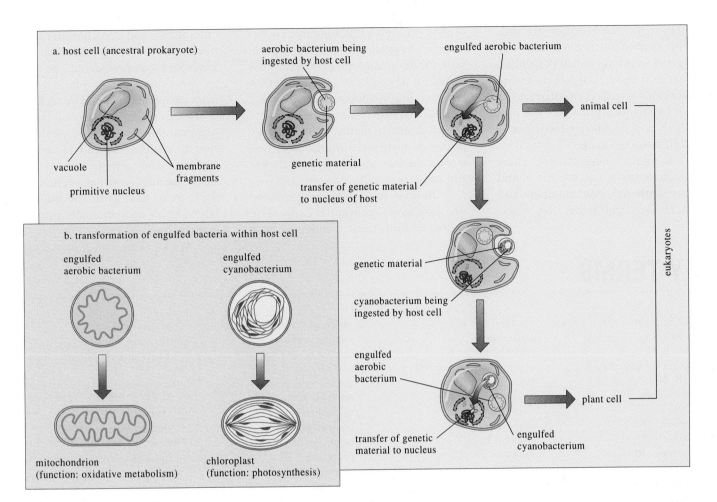

a. host cell (ancestral prokaryote)

aerobic bacterium being ingested by host cell

engulfed aerobic bacterium

animal cell

vacuole

membrane fragments

primitive nucleus

genetic material

transfer of genetic material to nucleus of host

eukaryotes

genetic material

cyanobacterium being ingested by host cell

engulfed aerobic bacterium

transfer of genetic material to nucleus

engulfed cyanobacterium

plant cell

b. transformation of engulfed bacteria within host cell

engulfed aerobic bacterium

engulfed cyanobacterium

mitochondrion (function: oxidative metabolism)

chloroplast (function: photosynthesis)

FIGURE 4.15

The endosymbiont theory. About 2.5 billion years ago, earth was home to a variety of single-celled organisms. Some of these early cells were probably larger than others and may have already contained genetic material in a structure resembling a nucleus. Pieces of membrane may have been in the cell. *a.* The large cells surrounded and engulfed some of their smaller contemporaries—aerobic bacteria and photosynthetic microorganisms. The engulfed cells, over millions of years, lost some of their genes to the nuclei of the larger cells, becoming endosymbionts of the host organisms. *b.* The photosynthetic microorganisms became chloroplasts; engulfed aerobic bacteria evolved into mitochondria.

SUMMARY

Cells, the basic units of life, are the microscopic components of all living organisms. Even the simplest cells are highly organized; complex cells may carry out very specialized functions. Organisms may be unicellular (one-cell) or multicellular (many-cells).

Cells were not observed until the late seventeenth century, when Robert Hooke focused his crude microscope on a piece of cork. Subsequent micros-copists amassed the observations that led to the development of the cell theory, which states that all life is composed of cells, that cells are the functional units of life, and that all cells come from preexisting cells. Gradually, biologists described nuclei, the cell membrane, and other parts of the cell. Virchow also hypothesized that the roots of illness lie at the organelle level. Biologists are still refining the cell theory today.

Structurally simple cells that lack nuclei are prokaryotic; all prokaryotes are either bacteria or cyanobacteria. These unicellular organisms contain genetic material, ribosomes (structures that help manufacture proteins), enzymes for obtaining energy, and various other biochemicals. The cellular contents are enclosed in a cell membrane, which is in turn usually encased by a rigid cell wall. Prokaryotes are abundant and ancient.

The more complex eukaryotic cells sequester certain biochemical activities in internal structures called organelles. This makes it unnecessary for the entire cell to maintain high concentrations of certain biochemicals; instead, the biochemicals are concentrated within the organelle that uses them.

Eukaryotic cells house their genetic material in a membrane-bound nucleus; synthesize, store, transport, and release biological molecules along a network of organelles (endoplasmic reticulum, Golgi apparatus, vesicles); degrade wastes in lysosomes; process toxins and hydrogen peroxide and perform other functions in peroxisomes; extract energy from biological compounds in mitochondria; and in plants, extract energy from sunlight in chloroplasts. A cell membrane surrounds both prokaryotic and eukaryotic cells.

Biologists have long explored the origin of eukaryotic cells. The endosymbiont theory suggests that chloroplasts and mitochondria evolved from once free-living prokaryotes swallowed by larger prokaryotes. According to the theory, chloroplasts evolved from cyanobacteria, and mitochondria evolved from aerobic bacteria. Evidence supporting the endosymbiont theory is that mitochondria and chloroplasts resemble small aerobic bacteria in size, shape, membrane structure, and reproduction, as well as in the ways their DNA, RNA, and ribosomes interact to manufacture proteins.

KEY TERMS

cell 64
cell membrane 68
cell theory 66
cellular pathology 67
cell wall 68
central vacuole 75
centriole 78
chlorophyll 78
chloroplast 78
chromosome 70
cristae 75
cyanobacteria 68
cytoplasm 70

cytoskeleton 72
endocytosis 75
endoplasmic reticulum 73
endosymbiont theory 82
eukaryotic cell 68
exocytosis 74
flagellum 68
Golgi apparatus 74
granum 78
lysosome 75
messenger RNA 72
microtubule 78
mitochondrion 75

multicellular organism 64
nuclear envelope 73
nuclear matrix 73
nuclear pore 73
nuclear pore complex 73
nucleoid 68
nucleolus 72
nucleus 66
organelle 64
peroxisome 75
photosynthesis 76
prion 80
prokaryotic cell 68
protoplasm 64

ribosomal RNA 72
ribosome 68
rough endoplasmic reticulum
  (ER) 73
smooth endoplasmic reticulum
  (ER) 74
thylakoid 78
transcription factor 72
transfer RNA 72
unicellular organism 64
vesicle 74
viroid 80
virus 79

REVIEW QUESTIONS

1. What are the functions of each of the following organelles?
   a. nucleus
   b. ribosome
   c. lysosome
   d. smooth ER
   e. rough ER
   f. chloroplast
   g. mitochondrion
   h. peroxisome
   i. Golgi apparatus

2. Proteins are very versatile molecules. List five roles proteins play that were discussed in this chapter.

3. How do the structures of a virus, bacterium, and human cell differ?

4. A fibroblast cell is shaped like a slightly irregular tile and is part of the connective tissue that cements the human body together. A nerve cell can be up to a meter in length, with many branches. A muscle cell is spindle-shaped. A white blood cell looks like a blob and can move about and engulf particles. Each of these cell types has a nucleus, a nucleolus, mitochondria, ribosomes, endoplasmic reticulum, one or more Golgi apparatuses, vesicles, and lysosomes. Yet each cell has a characteristic structure and function. How can these cells contain the same components, yet be so different from one another?

5. As a cube increases in size from 3 cm to 5 cm to 7 cm on a side, how does the surface area/volume ratio change?

6. A type of liver cell called a hepatocyte has a volume of 5,000 μm^3. Its total membrane area, including the inner membranes lining organelles as well as the outer cell membrane, fill an area of 110,000 μm^2. In comparison, a cell in the pancreas that manufactures digestive enzymes has a volume of 1,000 μm^3 and a total membrane area of 13,000 μm^2. Which cell is probably more efficient in carrying out activities that require extensive membrane surfaces? State the reason for your answer.

4.23

Cells

7. Name three structures or activities found in both prokaryotic and eukaryotic cells. List eight differences between the two cell types.

8. Cells of the green alga *Chlamydomonas reinhardtii* are grown for a few hours in a medium that contains amino acids "labeled" with radioactive hydrogen, called tritium. The *Chlamydomonas* cells are then returned to a nonradioactive medium. At various times, the cells are applied to photographic film (which radiation from the tritium exposes, producing dark silver grains). The film is then examined under an electron microscope. After 3 minutes, the film shows radioactivity in the rough ER; after 20 minutes, in the smooth ER; after 45 minutes, in the Golgi apparatus; after 90 minutes, in vesicles near one end of each of the cells. After two hours, no radioactive label is evident. What cellular process has this experiment traced? How might a similar technique be used to follow a lysosome's activity?

9. State the cell theory and the endosymbiont theory. What evidence supports each?

TO THINK ABOUT

1. Do you think a virus is alive? Cite a reason for your answer.

2. Prokaryotic organisms are far simpler in structure than organisms consisting of eukaryotic cells. Yet the "simpler" prokaryotes have probably dwelt on the earth far longer than we comparatively complex eukaryotes, and prokaryotes occupy a far more diverse range of environments than eukaryotes. Can you explain this seeming contradiction?

3. What advantages does compartmentalization confer on a large cell?

4. In an inherited condition called glycogen cardiomyopathy, teenagers develop muscle weakness, which affects the heart as well as other muscles. Samples of the affected muscle cells contain huge lysosomes, swollen with the carbohydrate glycogen. How might this condition arise?

5. The amoeba *Pelomyxa palustris* is a single-celled eukaryote with no mitochondria, but it contains symbiotic bacteria that can live in the presence of oxygen. How does this observation support or argue against the endosymbiont theory of the origin of eukaryotic cells?

6. Some of the genetic material contained in the nuclei of some species normally jumps from one chromosome to another. Do you think that movable DNA could account for the presence of DNA in mitochondria and chloroplasts? If so, what subcellular structures might participate in the transfer of genetic material from the nucleus to the mitochondria and chloroplasts?

7. Most theories about cells were developed by observing cells that could be cultured and viewed in the laboratory. However, the giant prokaryote *Epulopiscium fishelsoni* lives inside a fish and does not grow in the laboratory. Bacteria of the genus *Thermoplasma*, which may provide evidence supporting the endosymbiont theory, also will not grow in the laboratory; these bacteria are found in smoldering exposed coal veins. How might scientists alter the way they develop theories to more accurately reflect the living world?

8. Cite two ways in which the cell theory is still evolving.

SUGGESTED READINGS

Angert, Esther R., Kendall D. Clements, and Norman R. Pace. March 18, 1993. The largest bacterium. *Nature*, vol. 362. Scientists may have been wrong in assuming that any large cell is eukaryotic.

Barinaga, Marcia. April 23, 1993. Secrets of secretion revealed. *Science*, vol. 260. A fleet of vesicles carries out the process of secretion, stopping at various organelles.

de Kruif, Paul. 1966. *Microbe hunters*. New York: Harcourt Brace Jovanovich. This engrossing historical account of major discoveries in the field of microbiology highlights the effects that prokaryotic organisms can have on human health.

Edgington, Stephen M. November 1992. Rites of passage: Moving biotech proteins through the ER. *Bio/Technology*, vol. 10. Because the ER folds and modifies proteins, it is crucial for protein engineers to understand its workings.

Hoffman, Michelle. June 26, 1992. Motor molecules on the move. *Science*, vol. 25. Three classes of proteins provide much of the movement within cells.

Hoffman, Michelle. February 26, 1993. The cell's nucleus shapes up. *Science*, vol. 259. The nucleus is more organized than previously thought.

Kaplan, Donald R., and Wolfgang Hagemann. November 1991. The relationship of cell and organism in vascular plants. *BioScience*, vol. 41. These botanists argue that the cell theory does not always apply to plants.

Margulis, Lynn, and Mark McMenamin. September/October 1990. Marriage of convenience. *The Sciences*. An excellent review of developments leading to Margulis's endosymbiont theory.

Panté, Welly, and Veli Aebi. September 1993. The nuclear pore complex. The *Journal of Cell Biology*, vol. 122, no. 5. Nuclear pores are much more than simple holes.

Payne, Claire M. January 1985. The ultrastructural pathology of cell organelles. *Biology Digest*. Many disorders can be traced to abnormal organelles.

Prusiner, S. B. April 9, 1982. Novel proteinaceous infectious particles cause scrapie. *Science*, vol. 216. Protein can be infectious.

Raff, Rudolf A., and Henry R. Mahler. August 18, 1972. The nonsymbiotic origin of mitochondria. *Science*, vol. 177. An alternative to the endosymbiont theory.

Science, vol. 258, November 6, 1992. Frontiers in biology. This section features eight excellent articles on cell biology advances.

Sogin, Mitchell L. March 18, 1993. Giants among prokaryotes. *Nature*, vol. 362. Putting the giant bacterium into perspective.

Thomas, Lewis. 1974. *The lives of a cell*. New York: Viking Press. Essays entitled "The Lives of a Cell" and "Organelles as Organisms" discuss the nature of organelles and the endosymbiont theory.

Touchette, Nancy. May 1991. Scrapie prion: Teaching an old dogma new tricks. The *Journal of NIH Research*, vol. 3. Debate continues on the nature of the agent causing scrapie.

Valle, David, and Jutta Gartner. February 25, 1993. Penetrating the peroxisome. *Nature*, vol. 361. Adrenoleukodystrophy is caused by a missing or abnormal peroxisomal protein—but not the one researchers had suspected.

CHAPTER 5

Cellular Architecture

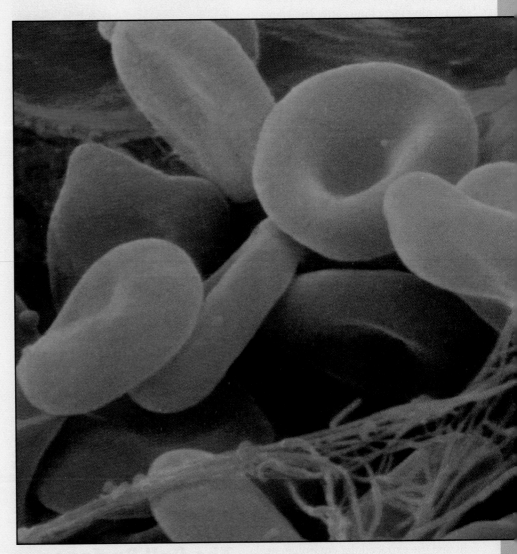

The cellular architecture molds the distinctive shapes of these red blood cells.

LEARNING OBJECTIVES

By the chapter's end, you should be able to answer these questions:

1. What structures comprise the cellular architecture?

2. Why are a cell's surface molecules and their locations and movements important?

3. What are the components of a cell membrane, and how are they organized?

4. How do substances cross a cell membrane?

5. What are the functions and components of the cytoskeleton?

6. What role does the cellular architecture play in a cell's external communications and cell-cell interactions?

The two-year-old's health appeared to be returning mere hours after the transplant. Her new liver, needed to replace a degenerated one, came from a child just killed in an automobile accident. The patient's skin was losing its usual sickly yellow pallor, and she was becoming much more alert. Yet already, as the new organ began taking over the jobs the old one had abandoned, immune system cells detected the new organ and, interpreting it as an infection, began to produce chemicals to attack it. Even though the donor's liver was carefully "matched" to the little girl—meaning that the arrangement of molecules on the surfaces of its cells appeared to be very similar to those on the cells of her own liver—the match was not perfect. A rejection reaction was in progress, and the transplanted organ would soon cease to function. The little girl would need another transplant to survive.

The rejection of a transplanted organ illustrates the importance of cell surfaces in the coordinated functioning of a multicellular organism. The cell surface is one component of the cellular architecture—structures that give a cell its particular three-dimensional shape and topography, help determine the locations and movements of organelles and biochemicals within the cell, and participate in the cell's interactions with other cells and the extracellular environment.

The cellular architecture consists of surface molecules embedded in the cell membrane, the outer covering of a cell. Just beneath the cell membrane are protein fibers that are part of the cell's interior scaffolding, or **cytoskeleton.**

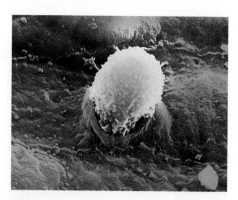

FIGURE 5.1

Cellular architecture. A white blood cell's inner skeleton gives it the spheroid shape that eases its journey through the circulatory system. But it also enables the cell to squeeze between the cells of blood vessel walls and enter the tissues. Specializations of the white blood cell's surface enable it to recognize and adhere to part of a blood vessel wall near an injury and to move between the lining cells. This is the cellular basis of the inflammatory response, discussed at the end of the chapter.

Together, the cell surface and cytoskeleton form a dynamic structural framework that helps to distinguish one cell from another (fig. 5.1). The various components of the cellular architecture must communicate and interact for the cell to carry out the processes of life.

The Cell Surface— Cellular Name Tags

At a conference where most of the participants do not know one another, name tags are often used to establish identities. Cells also have name tags in the form of sugars, lipids, and proteins that protrude from their surfaces. These cellular name tags are found on all cells, from the single cell of a bacterium to the trillions of cells of the human body. Some surface molecules distinguish cells of different species, like company affiliations on name tags. Other surface structures distinguish individuals within a species. A multicellular organism must be able to determine which cells are part of the body and which are not. Within the human body, the immune system performs this function. The immune system is a huge collection of white blood cells and the biochemicals they produce. These cells recognize the surfaces of one person's cells as "self" and all other cell surfaces as "nonself." When the immune system encounters the nonself surfaces of some bacteria or other foreign agents that have entered the body, it launches an attack, calling forth other cells and biochemicals to fight the invasion. Unfortunately for the child who received the liver transplant, her immune system recognized and attacked the cells of her transplanted liver as nonself.

Surface structures also distinctively mark cells of different tissues within an individual, so that a bone cell's surface is different from the surface of a nerve cell or a muscle cell. The human body has more than 200 cell types, and each has its own distinctive surface. These surface differences between cell types are particularly important during the development of the embryo, when different cells sort to grow into specific tissues and

Biology in *ACTION*

Cell Surfaces and Health Predictions—The HLA System

Would you want to know which medical conditions you are likely to suffer from later in life, and perhaps even die from? In just a few years, a blood test performed at birth, coupled with computer analysis of disease-incidence statistics, might make it possible to construct at least a partial health-prediction profile for any newborn baby. This may become possible because of a group of proteins called the human leukocyte antigens (HLA) that dot the surfaces of human white blood cells (leukocytes).

Individuals are distinguished by unique combinations of HLA molecules. Certain HLA combinations are statistically associated with a higher-than-normal probability of developing a particular disease. So far, about 50 medical conditions have been linked to specific HLA types. As table 1 shows, the HLA-associated disorders produce symptoms in a range of organ systems, but all of the disorders seem to stem from immune system abnormalities. HLA-linked diseases also tend to run in families, but not in any predictable manner.

An example of an HLA-linked disease is ankylosing spondylitis, a disease characterized by inflammation and deformation of the vertebrae. If a person has an antigen called B27 on his or her white blood cells, then the chances of developing the condition are 100 times greater than those of a person who lacks the B27 antigen. This HLA prediction is based on the observation that more than 90% of the people who suffer from ankylosing spondylitis possess the B27 antigen, although this antigen is present in only 5% of the general population. Of course, these statistics suggest that predictions of disease based on HLA tests should be approached with caution; the statistics show that 10% of people who have ankylosing spondylitis do *not* have the B27 antigen and that some people who have the antigen never develop the disease.

So far, the precise nature of the relationship between HLA types and human health is not well understood. A particular HLA type may indicate a genetic susceptibility to a disease or perhaps a sensitivity to certain viruses. Even though our present knowledge of the HLA system is not complete, the statistical associations may indeed prove clinically useful by providing warnings to individuals at high risk of developing certain conditions.

Table 1 **Medical Problems Linked to Specific HLA Types in Humans**	
Condition	**Description**
Ankylosing spondylitis	Inflammation and deformation of vertebrae
Reiter syndrome	Inflammation of joints, eyes, and urinary tract
Rheumatoid arthritis	Inflammation of joints
Psoriasis	Scaly skin lesions
Dermatitis herpetiformia	Burning, itchy skin lesions
Systemic lupus erythematosis	Rash on face; destruction of heart, brain, and kidney cells; very high, persistent fever
Addison disease	Malfunction of adrenal glands producing anemia, discolored skin, diarrhea, low blood pressure, and stomachache
Grave disease	Malfunction of thyroid gland, producing goiter
Juvenile-onset diabetes	Defect in beta cells of pancreas, disrupting sugar metabolism
Multiple sclerosis	Degenerative disease of brain or spinal cord producing weakness and poor coordination
Myasthenia gravis	Progressive paralysis
Celiac disease	Childhood diarrhea
Gluten-sensitive enteropathy	Sensitivity of intestine to wheat
Chronic active hepatitis	Inflammation of liver

organs. Cell surfaces also change over time; the number of surface molecules declines as a cell ages.

Individual patterns of cell surface features may one day be used to predict the likelihood that a person will develop certain diseases. The **human leukocyte antigens** (HLA) are cell surface molecules that appear in different patterns in different people. (The name comes from the fact that these cell surface molecules, or **antigens,** are found on leukocytes, the white blood cells.) A person who has certain HLA cell surface molecules may have a much greater chance of developing a certain disease than a person with a different HLA makeup (Biology in Action 5.1).

The Cell Membrane— Cellular Gates

Just as the character of a community is molded by the people who enter and leave it, the special characteristics of different cell types are shaped in part by the substances that enter and leave them. The cell membrane, a selective barrier that completely surrounds the cell, monitors the movement of molecules in and out of the cell. In eukaryotes, membranes are also found within cells, where they compartmentalize structures and form many organelles (fig. 5.2)

The Protein-Phospholipid Bilayer

The chemical characteristics and the arrangement of the molecules that comprise a membrane determine which substances that membrane will allow to cross. The structure of a biological membrane is possible because of a chemical property of the phospholipid molecules that comprise it—one end of such a molecule is attracted to water, while the other end is repelled by it (fig. 5.3). **Phospholipids** are lipid molecules with attached phosphate groups (PO_4, a phosphorus atom bonded to four oxygen atoms).

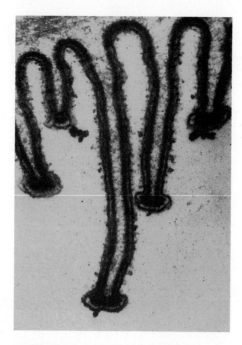

FIGURE 5.2

Biological membranes define the outer limits of the cell, and form vesicles and organelles within the cell. This is a cell membrane of a protozoan, a single-celled animal.

The phosphate end of a phospholipid molecule, which is attracted to water, is **hydrophilic** (water-loving); the other end, consisting of two fatty acid chains, moves away from water because it is **hydrophobic** (water-fearing). Because of these water preferences, phospholipid molecules in water spontaneously arrange into a **phospholipid bilayer** (fig. 5.4). This two-layered, sandwichlike structure forms so that its hydrophilic surfaces are on the outsides of the "sandwich," exposed to the watery medium outside and inside the cell. Its hydrophobic surfaces face each other on the inside of the "sandwich," and thus are unexposed to water. The phospholipid bilayer forms the structural backbone of a biological membrane. Phospholipid bilayers are also used in the pharmaceutical industry to construct microscopic bubbles called liposomes, which are used to encapsulate drugs (Biology in Action 5.2).

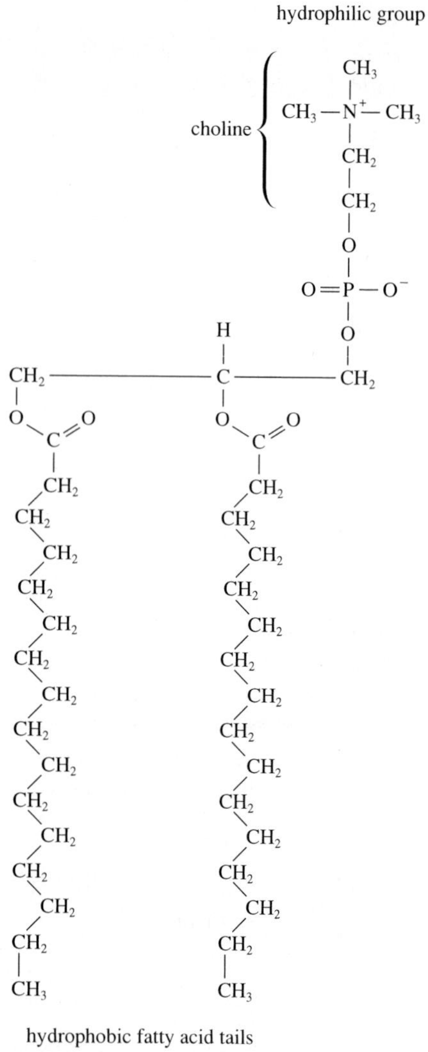

FIGURE 5.3

The two faces of membrane lipids. A phospholipid is literally a two-faced molecule, with one end attracted to water (hydrophilic, or water-loving) and the other repelled by it (hydrophobic, or water-fearing).

The phospholipid bilayer's hydrophobic interior presents a barrier to most substances dissolved in water. However, proteins embedded throughout the bilayer form passageways for water-soluble molecules and ions to pass through. The membranes of living

FIGURE 5.4

In the presence of water, phospholipids form bilayers. When the bilayers are exposed to intense sound waves, they fold up, forming spheres called liposomes. Liposomes are used to encapsulate drugs, which can then be delivered in the human body.

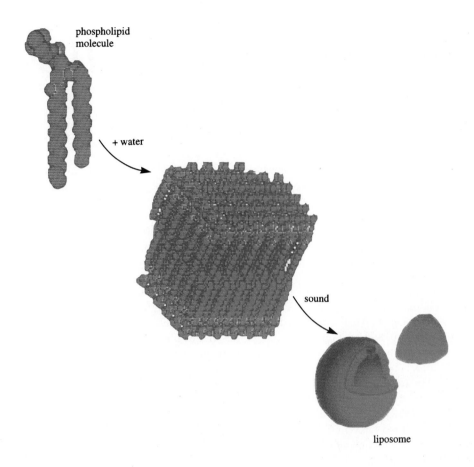

phospholipid molecule

+ water

sound

liposome

cells, then, consist of phospholipid bilayers and the proteins embedded in them (fig. 5.5).

Membrane proteins are diverse. Some lie completely within the phospholipid bilayer. Others traverse the membrane to extend out of one or both sides, including the protein which, when abnormal, causes the inherited disorder cystic fibrosis (fig. 5.6).

Peripheral proteins are membrane proteins that are attached to one face of the membrane, but do not traverse the phospholipid bilayer and are not totally embedded in it. In animal cells, membrane proteins often attach to branchlike sugar molecules to form **glycoproteins,** which protrude from the membrane's outer surface.

The proteins and glycoproteins that jut from the cell membrane create the surface characteristics that are so important to a cell's interactions with other cells. Proteins within the oily lipid sandwich can move about, sometimes at remarkable speed. Because of this, the protein-phospholipid bilayer is often called a **fluid mosaic.** A pigment protein found in the retina of a human eye, for example, moves to different depths within the phospholipid bilayer of the cell membrane depending upon the intensity of the incoming light. The lipid molecules can also move laterally in the plane of the layer they are a part of.

The number or distribution of cell membrane proteins distinguishes different cell types and organelle functions. Recall the mitochondrion, the organelle made of membranes on which many of the reactions of cellular metabolism occur.

The outer mitochondrial membrane has 1.2 times more protein than lipid. The inner mitochondrial membrane, in which the enzymes required to catalyze the energy reactions are embedded, has 3.6 times as much protein as lipid. These protein distributions differ because of the different functions of the membranes.

Disruptions in the protein/lipid ratio of a membrane can have devastating effects on health. Consider the cell membrane of a Schwann cell, 75% of which is made up of a lipid called myelin. Schwann cells enfold nerve cells in tight layers, somewhat like plastic wrap repeatedly wrapped around a hot dog. This insulation greatly speeds the electrical transmission of nerve messages. If Schwann cell membranes contain too much or too little lipid, illness can result. In multiple sclerosis, the lipids in Schwann cell membranes are destroyed, leading to visual impairment, numbness, tremor, and difficulty moving. In Tay–Sachs disease, Schwann cell membranes accumulate too much lipid, and the nerve cells beneath them cannot transmit messages to muscle cells. Symptoms of Tay–Sachs disease include paralysis and loss of sight and hearing.

KEY CONCEPTS

Because phospholipids are hydrophobic at one end and hydrophilic at the other, they readily form phospholipid bilayers. A cell membrane consists of proteins, glycoproteins, lipoproteins, and glycolipids embedded in a phospholipid bilayer. The structure of the bilayer is a fluid mosaic in which the embedded protein and lipid molecules can move. Different types of specialized cells have different characteristic proportions of lipid and protein in their cell membranes.

Biology in *ACTION*

Liposomes—New Drug Carriers

In 1961, English investigator Alec Bangham poured water into a flask containing a film of phospholipid molecules as part of his research on blood clotting. He was surprised to see that the lipid turned milky. Looking at the material under a microscope, Bangham saw that the phospholipid film had broken into thousands of tiny bubbles, each surrounding some of the water. The bubbles ranged in diameter from 25 nanometers to several micrometers. They were named liposomes, meaning "bodies of lipid."

What Bangham had discovered were microscopic spheres made of a simple phospholipid bilayer, identical to the structure that forms the basis of cell membranes (fig. 1). Some of the lipids even had more than one phospholipid bilayer coat, a little like an onion skin. Throughout the 1960s, liposomes were used by cell biologists as models of cell membranes.

In 1981, a young biologist named Marc Ostro realized the potential of liposomes as drug carriers. Water-soluble drugs can be packaged in the watery interior of liposomes; fat-soluble drugs can be lodged within the phospholipid bilayer itself. The advantages of packing a drug in a liposome are twofold—the drug can be released slowly from the liposome, and if researchers can find a way to direct the liposome to diseased cells, drug delivery can be targeted. This would solve a major drawback of conventional drug treatment—that is, getting enough drug to the site of disease to be effective, yet keeping it away from healthy cells, where it can cause side effects.

The natural reaction of the human immune system to liposomes provides a way to move them to disease sites. Large scavenger cells of the immune system, called macrophages, are attracted to liposomes and rapidly engulf them. The "swallowed" liposome is sent to a lysosome, where enzymes take the liposome apart. If the liposome contains a drug, some of it seeps into the cytoplasm and perhaps eventually out of the cell. The success of liposome-carried drugs lies in the function of macrophages, which normally congregate in parts of the liver, spleen, bone marrow, and lymph nodes but move to sites of inflammation or infection. Macrophages engulf liposomes containing antibiotic drugs and transport them to the infection site—precisely where they are needed. Liposomes harboring antiinflammatory drugs, for example, travel via macrophages to inflamed arthritic joints.

Liposomes are also useful for delivering cancer drugs. When these drugs are given in "free" form, they cause severe side effects because they kill healthy cells as well as cancer cells. When such toxic drugs are encapsulated in liposomes and injected, they accumulate in macrophage-laden areas, where they leak out slowly and steadily enough to destroy the cancer cells but not the healthy ones.

Liposome-enclosed drugs used to treat infection, inflammation, and cancer are injected because they are not absorbed well in the digestive tract. The tiny bubbles are useful, however, in some topical applications. "Artificial tears," for example, are liposomes packed with tear components (water, salts, lipid, and a mucuslike substance) that soothe the irritating condition called dry eye. Another liposome-enclosed drug treats fungal infections in the female reproductive tract, and a hair-growing drug encased in liposomes is used to treat pattern baldness. Because of their slow release, liposome-enclosed topical drugs are needed less often and are therefore less irritating.

FIGURE 1

Liposomes are microscopic bubbles composed of phospholipid bilayers. They form spontaneously when certain concentrations of fatty molecules are mixed with water. The diameters of these liposomes are about 0.15 micrometer.

Movement across Membranes

A muscle cell can contract and a nerve cell can conduct a message only if certain molecules and ions are maintained at proper levels inside and outside the cell. In all cells, the cell membrane oversees these vital concentration differences. Before considering how cells control which substances enter and leave, it is helpful to define some terms.

An **aqueous solution** is a homogeneous mixture of a substance (the **solute**) dissolved in water (the **solvent**). Lemonade made from a powdered mix illustrates the relationship between solute and solvent: the solvent is water, the solute is the powdered mix, and the solution is lemonade. Concentration refers to the relative number of one kind of molecule compared to the total number of molecules present, and it is usually

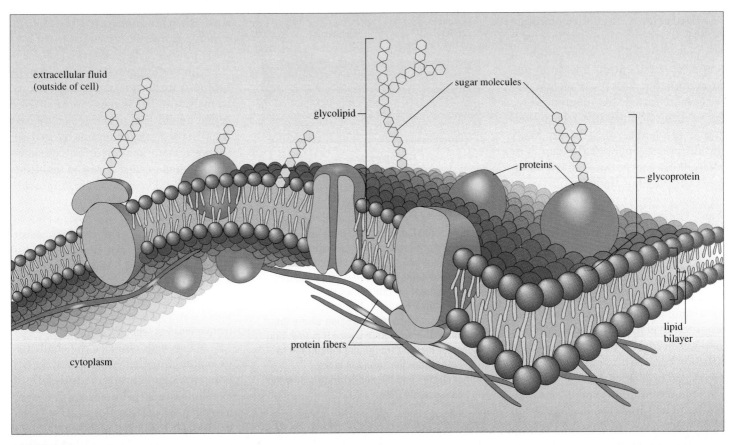

FIGURE 5.5

In a cell membrane, mobile proteins are embedded throughout a phospholipid bilayer, which is also mobile, to produce a somewhat fluid structure. An underlying meshwork of protein fibers supports the cell membrane. Jutting from the membrane's outer face are sugar molecules attached to proteins (glycoproteins) and lipids (glycolipids). Because a membrane is made of several different kinds of molecules, and proteins move within the fluid phospholipid bilayer, the structural organization of a membrane is described as a fluid mosaic.

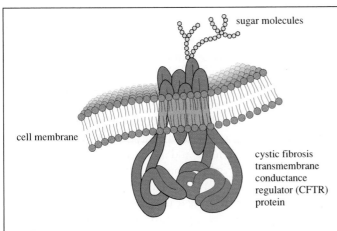

FIGURE 5.6

The CFTR (cystic fibrosis transmembrane conductance regulator) protein that, when defective, causes cystic fibrosis (CF) normally forms a channel through the cell membrane that allows chloride ions to exit the cells lining the lungs and pancreas. In CF, the protein channels do not form properly. The chloride ions trapped inside these cells attract water, diluting the salty cellular interiors. This influx of water into the cells deprives other cells of the fluids they need to produce normal secretions. The secretions dry out, impairing respiration and digestion.

given in terms of the solute. When solute concentration is high, the proportion of solvent (water) present is low, and the solution is concentrated. When solute concentration is low, solvent is proportionately high, and the solution is dilute.

Passive Diffusion

The cell membrane is selectively permeable—that is, some molecules are able to pass freely through the membrane (either between the molecules of the phospholipid bilayer or through special protein-lined channels), while others are not. Molecules of oxygen (O_2), carbon dioxide (CO_2), and water (H_2O) are among those that freely cross biological membranes, and they do so by the process of diffusion.

Diffusion is the movement of a substance from a region where it is very concentrated to a region where it is not very concentrated without any input of energy. This phenomenon may be demonstrated by placing a tea bag in a cup of hot water. Compounds in the tea leaves dissolve gradually and diffuse throughout the cup. The tea is at first concentrated in the vicinity of the bag, but the brownish color eventually spreads to create a uniform brew. The natural tendency of a substance to move from where it is highly concentrated to where it is less so is called "moving down" or "following" its concentration gradient.

When molecules move through a membrane because of this natural tendency to travel from high concentration to low concentration, we call it **simple diffusion** because it requires no input of energy and no carrier molecule.

Simple diffusion eventually reaches a point at which the concentration of the substance is the same on both sides of the membrane. After this, molecules of the substance continue to flow randomly back and forth across the membrane at the same rate, so that the concentration remains the same on both sides.

A Special Case of Diffusion—The Movement of Water

The fluids that continually bathe our cells consist of molecules dissolved in water. Because cells are constantly exposed to water, it is important to understand how the entry of water into a cell is regulated. If too much water enters a cell, it swells; if water leaves, it shrinks. In either case, the cell's functions may be hampered. Water moves across biological membranes by a form of simple diffusion called **osmosis,** which is influenced by the concentration of dissolved substances inside and outside the cell (fig. 5.7).

When the solute concentrations on each side of a membrane differ, water moves across the membrane. According

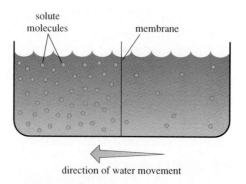

solute molecules   membrane

direction of water movement

FIGURE 5.7

Osmosis—a biologically important type of simple diffusion. Either a synthetic nonliving material (called an artificial membrane) or a biological membrane can be used to demonstrate osmosis, which is the movement of water down its concentration gradient. In osmosis, water molecules diffuse from a region where their concentration is high to where it is low; that is, water moves through the membrane to the side where the solute is more concentrated and the concentration of water is consequently lower.

to the general tendency of diffusion, water moves in the direction that dilutes the solute. That is, water moves to where the solute is more concentrated. Most cells are **isotonic** to the surrounding fluid—that is, solute concentrations are the same within and outside the cell, so that there is no net flow of water.

If a cell's isotonic state is disrupted, its shape changes as water rushes in or leaks out. If a cell is placed in a solution in which the concentration of solute is lower than it is inside the cell (a **hypotonic** solution), water enters the cell to dilute the higher solute concentration there. The cell swells. In the opposite situation, if a cell is placed in a solution in which the solute concentration is higher than it is inside the cell (a **hypertonic** solution), water leaves the cell to dilute the higher solute concentration outside. This cell shrinks. (Hypotonic and hypertonic are relative terms in that they can be applied both to the surrounding solution and to the solution inside the cell. It may help to remember that hyper means "over" and hypo means "under.") The effects of immersing a cell in a hypertonic or hypotonic solution can be demon-

strated with a human red blood cell, which is normally suspended in an isotonic solution called plasma (fig. 5.8).

Living things have evolved interesting strategies for regulating osmosis to maintain cell shapes. Some single-celled inhabitants of the ocean are isotonic to their salty environment, so their shapes remain unaltered. The paramecium, a single-celled organism that lives in ponds, takes a different approach to maintain its footprint-shaped form. A paramecium contains more concentrated solutes than the pond, so water tends to flow into the organism faster than it flows out. The paramecium has a special organelle, a **contractile vacuole,** which enables it to pump extra water out, much as a bucket bails water from a leaky rowboat (fig. 5.9).

Plant cells also face the challenge of maintaining their shapes even with a concentrated interior. Instead of expelling the extra water that rushes in, as the paramecium does, plant cells expand until their cell walls restrain their cell membranes. The resulting rigidity, caused by the force of water against the cell wall, is called **turgor pressure** (fig. 5.10). This can be demonstrated with a piece of wilted lettuce. When placed in water, it becomes crisp, as the individual cells expand like inflated balloons.

Osmosis influences how frequently we urinate. Kidney tubules have very active membranes that return valuable substances to the blood and excrete unneeded substances in the urine. One chemical recycled in the kidney is water. The amount of water returned to the blood is influenced by water intake as well as by other chemicals, such as caffeine and alcohol. Have you ever drunk a huge mug of coffee or tea and then had to frequently run to the bathroom over the next few hours? The caffeine in coffee and tea is a diuretic; it causes frequent passage of watery urine by decreasing the movement of water from the kidney tubules back into the blood. Beer has a similar effect, caused both by its high water content and by the diuretic action of alcohol. The rare disorder diabetes insipidus is caused by the absence of a hormone that controls

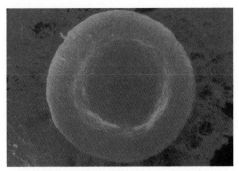

a. isotonic (no net change in water movement)

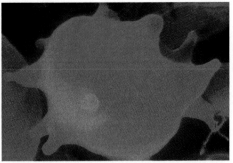

b. blood cells in hypertonic solution (water diffuses outward)

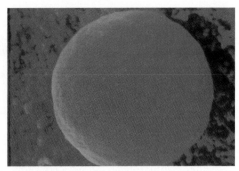

c. blood cells in hypotonic solution (water diffuses inward)

FIGURE 5.8

A red blood cell changes shape in response to changing plasma solute concentrations. *a.* A human red blood cell is normally shaped like a doughnut without a hole. When the surrounding blood plasma is isotonic, water enters and leaves the cell at the same rate and the cell maintains its characteristic shape. *b.* When the salt concentration of the plasma increases, so that the surrounding fluid is hypertonic to the cell, water leaves the cells to dilute the outside solute faster than water enters the cell. The cell shrinks. *c.* When the salt concentration of the plasma decreases, so that the surrounding fluid is hypotonic to the cell, water flows into the cell faster than it leaves. The cell swells and may even burst.

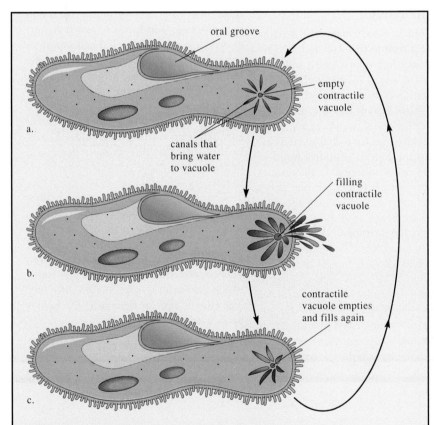

FIGURE 5.9

How paramecia keep their shapes. Because solutes within paramecia are more concentrated than they are in the surrounding water (making the inner solution hypertonic), water tends to enter these unicellular organisms faster than it leaves. Organelles called contractile vacuoles fill and then pump excess water out of the cells by way of the cell membranes. When filled, the contractile vacuole resides in the cell's interior (*a*). It moves near the cell membrane as it fills with water (*b*), then releases the water to the outside. The organelle then resumes its empty shape (*c*) and moves away from the cell membrane back to the interior of the cell.

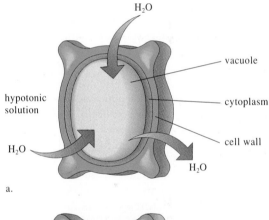

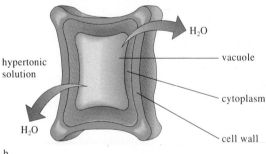

FIGURE 5.10

How plant cells keep their shapes. Like paramecia, plant cells usually contain more concentrated solutes than their surroundings, causing water to flow into the cell. *a.* In a hypotonic solution, water enters the cell and collects in vacuoles. The cell swells against its rigid, restraining cell wall, generating turgor pressure. *b.* When a plant cell is placed in a hypertonic environment (so that solutes are more concentrated outside the cells), water flows out of the vacuole, and the cell shrinks. Turgor pressure is low.

Breaching the Blood-Brain Barrier

Winding through the human brain are 400 miles of tiny tubules—a network of blood vessels, many just a single cell in thickness, called capillaries. Capillaries elsewhere in the body have tiny holes and are glued together at a few places called tight junctions. Brain capillaries, in contrast, adhere as smoothly as bricks cemented together, knit into a tight, curved sheet by a continuous set of tight junctions. This anatomical arrangement, called the blood-brain barrier, allows only a few substances to pass from the blood to the brain, protecting the nerve cells from large biochemical fluctuations. However, many useful drugs cannot gain entry.

A Selective Barrier

As they gain an understanding of how substances cross the blood-brain barrier, researchers are developing new ways to send useful drugs into the brain to treat illness. The barrier readily admits lipid-soluble drugs, because the cell membranes of the endothelial cells that form the capillaries are lipid-rich. Heroin, valium, nicotine, cocaine, and alcohol are lipid soluble and can cross the barrier, which is why the effects of these substances are felt so rapidly. Oxygen also enters the brain by directly crossing through the endothelial cell membrane. However, many useful drugs cannot gain entry.

Water-soluble molecules must take other routes across the blood-brain barrier. Insulin, for example, enters a vesicle by endocytosis on the blood side, travels through the cells, and arrives on the brain side by exocytosis. Glucose, amino acids, and iron each use their own carrier molecules. These biochemicals are vital. If the brain is starved of glucose for just a few seconds, the person loses consciousness. Amino acids are constantly needed to build protein neurotransmitters, and iron is needed for metabolism.

Delivering Drugs across the Barrier

Until recently, penetrating the barrier required drilling into the skull and delivering a substance by a tube called a catheter. A less drastic way is to inject the sugar-alcohol mannitol into the carotid artery in the neck. This dehydrates the endothelium, loosening the tight junctions for 15 seconds, enough time for a drug given intravenously to sneak past. Mannitol infusion can deliver drugs to battle brain tumors when other treatments fail, but one drawback is that other molecules may also rush in, causing seizures.

Some biochemicals open the barrier. A substance called RMP–7 loosens tight junctions enough to let small molecules pass through. RMP–7 is used to enable the drug amphotericin-B to treat a fungal brain infection in AIDS patients (fig. 1). Delivering this drug directly to the brain may make lower dosages as effective as higher dosages delivered by conventional routes, easing the severe nausea that earns the drug its nickname among AIDS patients: "amphoterrible."

All these routes across the blood-brain barrier admit whatever happens to be near the temporarily compromised barrier into the brain. A more targeted approach involves carrier-mediated endocytosis. This approach uses a receptor for a protein called transferrin that transports iron from the blood to the brain. Each cell of the blood-brain barrier may contain up to 100,000 transferrin receptors. A transferrin molecule carrying iron binds to the receptor, moves into the endothelial cell, migrates through the cytoplasm, and departs on the brain side of the cell, where the iron is released. A drug linked to the brain's iron transport system may thus be shuttled to the usually out-of-reach brain without admitting unwanted molecules. Experiments with this system have been used to treat cancer and Alzheimer's disease in rats.

The ability to target drugs to the brain is coming at a good time. Some common degenerative neurological disorders, such as Alzheimer disease and Parkinson disease, are on the rise, as are AIDS-associated brain infections. Moreover, we now know which genes cause several brain disorders. The time is ripe for breaching the blood-brain barrier.

---

water resorption in the kidneys. People with this disorder urinate profusely, up to 7 gallons (26 liters) a day! In the opposite situation, in which water returns to the blood too readily, urine volume is low and the solutes in the urine are concentrated.

Facilitated Diffusion

Many polar or ionic biochemicals that move in and out of cells cannot simply diffuse across the cell membrane because they are repelled by its hydrophobic portion. One way for these molecules to cross the cell membrane is by moving through channels within membrane proteins called **carrier proteins.** Whether the carrier molecule moves into or out of the cell depends upon its concentration on both sides of the membrane. The process of crossing a membrane down a concentration gradient with the aid of a carrier protein is termed **facilitated diffusion.**

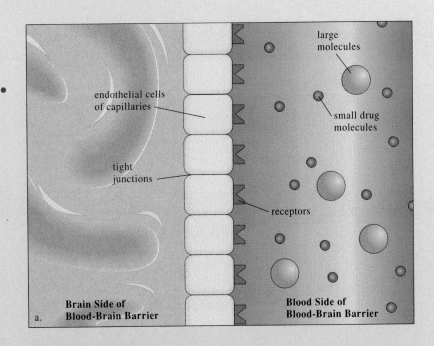

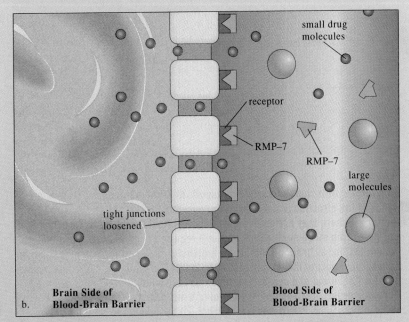

FIGURE 1

The blood-brain barrier. *a.* The barrier is a continuous sheet of capillaries whose cells are tightly knit together by tight junctions. *b.* When RMP–7 binds to the receptors on the endothelial cells of a brain capillary, the tight junctions are fleetingly loosened, allowing time for a drug to sneak through.

Glucose, amino acids, and iron travel from the blood to the brain by facilitated diffusion. Facilitated diffusion and simple diffusion are forms of **passive transport** because they do not require an input of energy. Biology in Action 5.3 explains how researchers are using natural transport mechanisms, such as facilitated dif-

fusion, to help molecules cross from the blood to the brain.

Active Transport

Life is, in a sense, a matter of partitioning and maintaining certain concentrations of specific biochemicals within

cells. Often these concentrations cannot be maintained by passive transport mechanisms, which involve movement down a concentration gradient without any input of energy. How does a cell import additional quantities of a needed substance—perhaps a nutrient—that is already more concentrated inside the cell

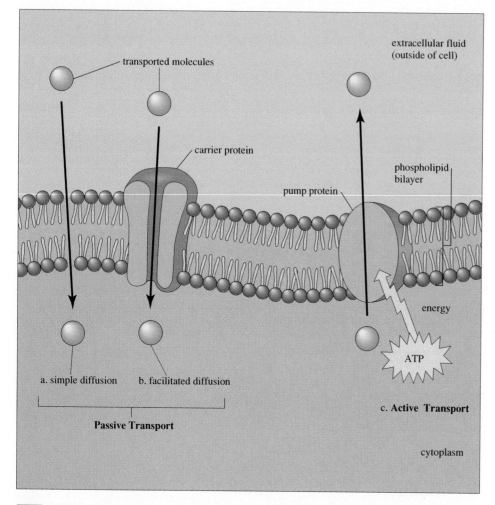

extracellular fluid (outside of cell)

transported molecules

carrier protein

pump protein

phospholipid bilayer

energy

ATP

a. simple diffusion

b. facilitated diffusion

Passive Transport

c. **Active Transport**

cytoplasm

FIGURE 5.11

Simple diffusion, facilitated diffusion, and active transport. In passive transport, molecules cross cell membranes in the direction dictated by their concentration gradient: they move from the side of the membrane where they are more highly concentrated to the side where they are less highly concentrated. This movement, which does not require energy, occurs by either simple or facilitated diffusion. *a.* In simple diffusion, molecules move through spaces between the lipid and protein components of the membrane. *b.* In facilitated diffusion, molecules move across the membrane with the aid of a carrier protein embedded within the membrane. *c.* Active transport, by contrast, involves the movement of a molecule across a membrane against its concentration gradient. This process is carried out by membrane proteins called "pumps" and requires energy. The energy is provided by ATP or by a gradient of protons or other ions across the membrane.

than outside? How would a cell export a metabolic by-product that is already more concentrated outside the cell?

Movement of substances against their concentration gradients is possible, but the process requires energy. The molecule that often provides biological energy is called **adenosine triphosphate** (ATP), which is discussed in the next chapter. When a phosphate group splits from ATP, energy is released; that energy can then be harnessed to help drive a cellular function. The movement of a substance against a concentration gradient with the aid of a carrier protein and energy is called **active transport.** Figure 5.11 compares simple diffusion, facilitated diffusion, and active transport.

The first active transport system discovered was a membrane protein called the **sodium-potassium pump.** This pump resides in the cell membranes of most animal cells, which indicates that it is extremely important. Cells must contain high concentrations of potassium ions (K$^+$) and low intracellular levels of sodium (Na$^+$) to perform such basic functions as maintaining their volume and synthesizing protein, as well as to conduct more specific activities, such as transmitting nerve impulses and enabling the lungs and kidneys to function.

The sodium-potassium pump is actually a carrier protein in the cell membrane. The pump first binds Na$^+$ from inside the cell (fig. 5.12). When ATP splits, the freed phosphate also binds to the pump. This changes the shape of the pump, allowing Na$^+$ to move out of the cell and K$^+$ to move into the cell. The sodium-potassium pump thus uses energy from ATP to help keep the proper concentrations of Na$^+$, K$^+$, and other substances in and outside of the cell.

KEY CONCEPTS

Molecules cross membranes by various mechanisms. In simple diffusion, a substance moves down its concentration gradient from areas where the substance is highly concentrated to areas where it is less so. Simple diffusion of water is called osmosis, and osmosis influences a cell's shape. In facilitated diffusion, a substance moves down its concentration gradient with the aid of a carrier protein. In active transport, a substance moves against its concentration gradient, often using energy from split ATP.

Large-Scale Membrane Transport— Exocytosis and Endocytosis

Most molecules dissolved in water are small, and they can easily cross cell membranes by simple diffusion, facilitated diffusion, or active transport. Large molecules (and even bacteria) can also get into and out of cells with the help of the cell membrane.

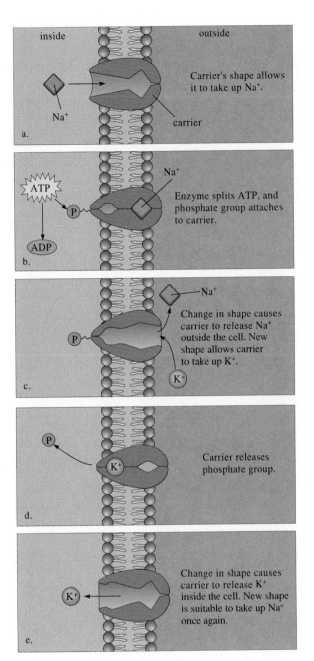

FIGURE 5.12

The sodium-potassium pump. This "pump," actually a carrier protein embedded in the cell membrane, uses energy to move potassium ions (K^+) into the cell and sodium ions (Na^+) out of the cell. The pump first binds Na^+ on the inside face of the membrane (*a*). ATP is split to yield ADP and a phosphate group, and the liberated phosphate binds to the carrier or pump (*b*). This binding alters the conformation of the pump and causes it to release the Na^+ to the outside. The altered pump can now take up K^+ from outside the cell (*c*). Next, the pump releases the bound phosphate (*d*), which alters the conformation of the carrier protein once again. This time, the change in shape releases the K^+ to the cell's interior (*e*). The pump is also back in the proper shape to bind intracellular Na^+.

Exocytosis

We have already encountered **exocytosis,** the process by which large particles leave cells, in chapter 4. Inside the cell, a vesicle (a bubblelike structure) made of a phospholipid bilayer surrounds the structure to be transported out of the cell—a droplet of a secretion such as milk, for example. The vesicle moves to the cell membrane and joins with it, and the transported substance is released on the outside of the membrane (fig. 5.13). Nerve transmission relies on exocytosis for the release of chemical messengers called neurotransmitters. Table 5.1 summarizes the transport mechanisms that move substances across membranes.

Endocytosis

Large particles enter cells by **endocytosis,** in which the cell membrane in a localized region buds inward to surround and enclose the particle (fig. 5.14). The pocket of membrane then pinches off from the interior of the membrane, producing a vesicle containing the particle. In animal cells, the vesicle is released into the cytoplasm, where it eventually fuses with a lysosome. (See Biology in Action articles 4.1 and 5.2.) Digestive enzymes within the lysosome dismantle the foreign particle. Endocytosis enables a white blood cell to engulf a bacterium, which it then destroys with lysosomal enzymes.

FIGURE 5.13

Exocytosis transports substances out of the cell. Biochemicals or particles can exit cells by exocytosis. Phospholipid bilayer spheres within the cell surround the structures to be exported. These structures or particles travel to the cell membrane and merge with it. The particles are then released to the outside.

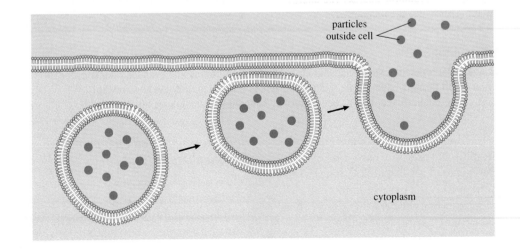

Table 5.1   **Movement across Membranes**

Mechanism	Characteristics	Example
Simple diffusion	Follows concentration gradient	Diffusion of oxygen from lung into capillaries
Osmosis	Simple diffusion of water	Reabsorption of water from kidney tubules
Facilitated diffusion	Follows concentration gradient, assisted by carrier protein	Diffusion of glucose into red blood cells
Active transport	Moves against concentration gradient, assisted by carrier protein and energy, usually from ATP	Reabsorption of salts from kidney tubules
Endocytosis	Membrane engulfs substance and draws it into cell in membrane-bound vesicle	Ingestion of bacterium by white blood cells
Exocytosis	Membrane-bound vesicle fuses with cell membrane, releasing its contents outside of cell	Release of neurotransmitters by nerve cells

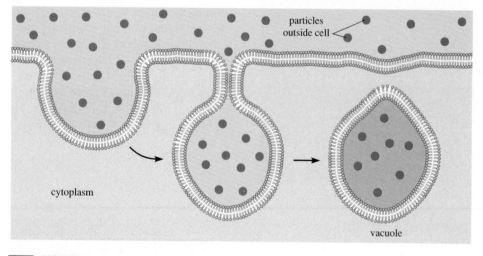

particles outside cell

cytoplasm

vacuole

FIGURE 5.14

Endocytosis transports substances into the cell. Large particles and even whole bacteria can enter a cell through endocytosis. A small portion of the cell membrane buds inward, entrapping the particles, and a vacuole forms, bringing the substances into the cell.

Some substances that enter cells by endocytosis are used rather than destroyed. This is true for some neurotransmitters. These biochemicals are recycled when the cells that secrete them take them back in by endocytosis.

Receptor-Mediated Endocytosis—Handling Cholesterol

When biologists first viewed endocytosis in white blood cells in the 1930s, they thought it might be a general route of entry into a cell. They believed a cell would gulp in anything at its surface, as if a theatre manager were letting anyone in rather than just ticket holders. We now recognize a more specific form of endocytosis called **receptor-mediated endocytosis.** In this process, a **receptor** protein on a cell's surface binds a biochemical, called a **ligand;** the cell membrane then surrounds the ligand and conducts it into the cell. One example of receptor-mediated endocytosis is liver cells' intake of dietary cholesterol.

Cholesterol is carried in various types of lipoprotein particles. One such cholesterol carrier is called a **low-density lipoprotein,** or LDL. The protein portions of LDL particles, called **apolipoproteins,** bind to LDL receptors clustered in protein-lined pits in the surface of a liver cell (fig. 5.15). The membrane buds off internally to form a vesicle around the LDL, and the loaded vesicle travels to a lysosome. Recall from chapter 4 that the lysosome is a sac of enzymes that degrade certain biochemicals. Within the lysosome, an enzyme liberates the cholesterol from its LDL carrier. The cholesterol then enters the endoplasmic reticulum, and the receptor recycles to the cell surface.

The presence of cholesterol in the cell has interesting regulatory effects, shown in figure 5.15. Taking in dietary cholesterol:

inhibits the cell's production of an enzyme needed to synthesize cholesterol

stimulates synthesis of another enzyme that catalyzes the conversion of cholesterol in the endoplasmic reticulum to a storage form (cholesteryl oleate)

inhibits production of new LDL receptors

Can you see how these three actions would tend to lower the cell's cholesterol level? Why, given this built-in feedback system, do so many people have overly high levels of serum cholesterol? The problem is a fatty diet. When the blood is flooded with dietary fat, LDL receptors become saturated with cholesterol. Even though the influx of cholesterol-

Liver Cell

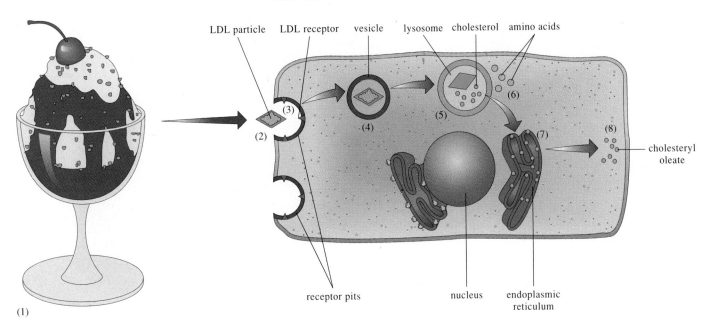

FIGURE 5.15

The intracellular effects of dietary cholesterol. Dietary cholesterol (1) is carried in an LDL particle (2) which enters a liver cell by receptor-mediated endocytosis (3). A vesicle (4) transports the engulfed LDL particle to a lysosome (5), where the apolipoprotein portion of the LDL particle is dismantled into amino acids (6), and the cholesterol is transported to the endoplasmic reticulum (7). Here, cholesterol is converted into a storage form called cholesteryl oleate (8). The presence of freed cholesterol in the cell inhibits it from synthesizing more cholesterol and making new LDL receptors.

loaded LDL receptors inhibits cholesterol synthesis *inside* the cell, the amount of cholesterol in the blood awaiting pickup by LDL receptors becomes overwhelming. The excess cholesterol clings to artery walls, forming a sticky plaque that blocks blood circulation and sowing the seeds for heart disease.

People who inherit a disease called familial hypercholesterolemia (FH) have too few LDL receptors. In the milder form of the illness, individuals with one abnormal copy of the FH gene have approximately half the normal number of LDL receptors. These individuals tend to suffer heart attacks as young adults. Stormie Jones inherited two copies of the FH gene, the more severe form of the illness. She lacked LDL receptors altogether, and as a result had a cholesterol level 6 times the normal. Before she died at age 10, Stormie suffered several heart attacks and underwent two cardiac bypass surgeries, several heart valve replacements, and finally a heart-liver transplant that reduced her blood cholesterol to near-normal levels. But ultimately, her body could not survive all the trauma.

Vesicle Trafficking

Phospholipid bilayers provide a transportation route from the cell membrane to and from various organelles, much as a monorail system links the various theme parks at Disney World. Figure 4.9 illustrated this intracellular transportation system for milk secretion. Endocytosis and exocytosis occur at the points of origin and destination for the intracellular transport route. The carriers between the cell membrane and organelles are vesicles, the bubblelike structures formed from phospholipid bilayers. How do vesicles "know" where to take their cargo? As is often the case in living systems, proteins provide the specificity needed in a generalized process.

A vesicle "docks" at its target membrane using a series of proteins as cues and then anchors (receptors). Some of these proteins are located within the phospholipid bilayers of the organelle and the approaching vesicle, while other participating proteins are initially free in the cytoplasm. Certain combinations of proteins come together, forming a complex that helps guide the vesicle to its destination (fig. 5.16).

KEY CONCEPTS

In exocytosis and endocytosis, the cell membrane surrounds large particles of a substance, transporting them out of or into the cell, respectively. Receptor-mediated endocytosis is a more discriminating form of endocytosis in which the substance entering a cell first binds to specific cell-surface receptors. A fleet of vesicles transports substances between organelles and the cell membrane. Proteins provide the specificity needed to ensure that substances arrive at the appropriate organelle.

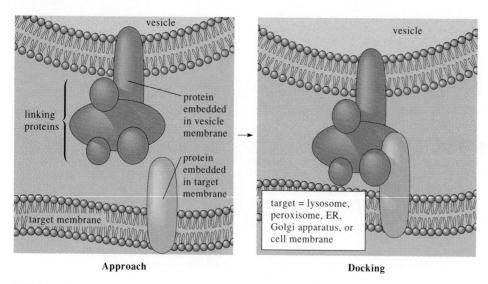

FIGURE 5.16

Vesicles recognize and dock at the appropriate organelle by means of a series of linking proteins.

The Cytoskeleton— Cellular Support

Until about 20 years ago, scientists thought eukaryotic cells were little more than sacs containing various structures that floated randomly in the jellylike cytoplasm. Today, thanks to modern methods of viewing cells, we know that cells are highly organized. Much of this organization, as well as the distinctive shapes of specialized cell types, is determined by the cell's cytoskeleton, a meshlike network of fine rods and tiny tubes of protein.

The protein girders of the cytoskeleton are dynamic structures that constantly break down and rebuild as the cell performs specific activities. Some cytoskeletal elements function as rails, forming conduits for cellular contents to travel along; other components of the cytoskeleton, the **motor molecules,** power the movements of organelles along these rails. Our knowledge of motor molecule proteins is relatively young. In 1986, we knew of only one family of motor molecule proteins—a group called the myosins that participate in muscle contraction. Today, we know of more than 20 different types of

motor molecules that fall into three categories—the myosins, the kinesins, and the dyneins.

The motor molecules, fibers, and hollow tubules that comprise the cytoskeleton carry out a variety of cellular tasks. Specifically, the cytoskeleton:

moves chromosomes apart during cell division

controls vesicle trafficking

builds organelles by helping to move components

accounts for the wandering movements of macrophages

enables cilia (hairlike structures that form a fringe on some cells) to wave

builds the cell walls of plants and

secretes and takes up neurotransmitters in nerve cells

Let's take a closer look at the components of the cytoskeleton.

Microtubules

All eukaryotic cells contain long, hollow tubes, called **microtubules,** that are responsible for many cellular movements. A microtubule is only about a

FIGURE 5.17

Microtubules self-assemble. Microtubules assemble when they gain tubulin dimers at one end and tear down when they lose tubulin dimers from the other end.

millionth of an inch (about 25 nanometers) in diameter. Its walls consist of pairs, or dimers, of the protein **tubulin.**

Cells contain both formed microtubules and individual tubulin molecules. When the cell requires microtubules to carry out a specific function—dividing in two, for example—the free tubulin dimers self-assemble into more of the tiny tubes. After the cell divides, some of the microtubules dissociate into individual tubulin dimers, replenishing the cell's supply of building blocks (fig. 5.17). Cells, then, are in a perpetual state of flux, building up and breaking down microtubules to carry out particular functions. Some drugs used to fight cancer (in which certain cells divide too frequently) work by dismantling the

FIGURE 5.18

Cilia are hairlike appendages found on many eukaryotic cells. *a.* On cells lining the human respiratory tract, beating cilia move dust particles upward so the lungs can exhale them. *b.* A cilium consists of nine pairs of microtubules surrounding a central pair. The microtubules bend when ATP molecules break down. The bending action causes the cilium to wave.

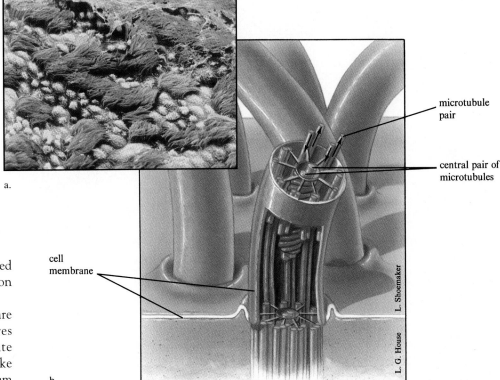

microtubule pair

central pair of microtubules

cell membrane

L. Shoemaker

L. G. House

a.

b.

microtubules that pull a cell's duplicated chromosomes apart, halting cell division in the cancer cells.

Many types of eukaryotic cells are fringed with cilia, hairlike structures built of microtubules that coordinate their movements to produce a wavelike motion (fig. 5.18). An individual cilium is constructed of nine microtubule pairs surrounding a central pair, and the entire structure is studded with another type of protein, the motor molecule **dynein.** Cilia move using energy obtained from ATP, the biological energy molecule.

Cilia perform many vital functions in animal cells, including beating particles up and out of the human respiratory system and waving an egg cell down the human female's reproductive tract for a possible meeting with a sperm cell. (A sperm's "tail," or flagellum, contains microtubules that slide past one another to generate movement.) Single-celled organisms may have thousands of individual cilia to enable them to move through water.

Microfilaments

Another building block of the cytoskeleton is the **microfilament,** a tiny rod made of the protein actin. Microfilaments are about one-third as thick as microtubules, and they are solid rather than hollow. In muscles, actin microfilaments are interspersed with larger filaments of myosin. When the actin microfilaments and larger myosin rods slide past one another,

muscle cells contract. Microfilaments are also important in endocytosis by white blood cells in animals. The microfilaments align beneath the cell membrane and propel portions of the membrane outward to entrap particles. You can actually see microfilaments in action by watching the blood clot that forms on a scraped knee. The retraction of the clot as the wounded area grows new skin is carried out by microfilaments and other proteins.

Intermediate Filaments

When cells are treated with harsh substances such as detergents, microtubules and microfilaments break down; but another type of filament, intermediate in size between the other two, remains. These remaining filaments, called **intermediate filaments** because of their size, are thought to provide scaffolding in the cell because they are seen in parts of the cell under structural tension. Intermediate filaments may, for example, hold the nucleus in place. Different types of intermediate filaments exist in different cell types.

KEY CONCEPTS

A cell's shape is largely determined by its cytoskeleton, a network of protein rods and tubes. Microtubules consist of tubulin, are involved in cell division, and form cilia. Microfilaments consist of actin and form part of muscle.

Intermediate filaments are diverse structures, intermediate in size between microtubules and microfilaments, that may help maintain cell shape.

Coordination of Cellular Architecture

Cellular architecture is a coordinated collection of interacting structures and molecules. The interface between the cell membrane and the underlying cytoskeleton is particularly important because the membrane directly connects the cell to the environment.

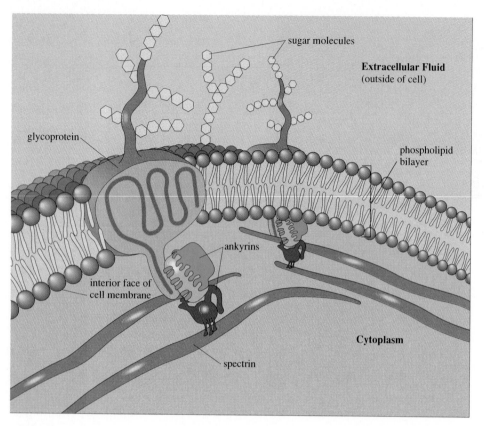

FIGURE 5.19

Proteins called ankyrins bind spectrin from the cytoskeleton to the interior face of the cell membrane. On its other end, ankyrin binds a large glycoprotein that helps transport ions across the cell membrane. In hereditary spherocytosis, abnormal ankyrin causes the cell membrane to collapse—a problem for a cell whose function depends upon its shape.

The Integrity of the Red Blood Cell Membrane

Much of what we know about the cohesion between the cell membrane and the cytoskeleton within it comes from studies of human red blood cells. The doughnut shapes of these cell fragments, which lack nuclei, enable them to squeeze through the narrowest blood vessels on their 300-mile, 120-day journey through the circulatory system.

A red blood cell's strength derives from rodlike **spectrin** proteins that form a meshwork beneath the cell membrane (fig. 5.19). The spectrin rods are attached to the membrane by proteins called **ankyrins.** Spectrin molecules are like the steel girders of the red blood cell architecture, and the ankyrins are like the nuts and bolts. If either is absent, the cell's structural support collapses.

Cell collapse is precisely what happens in an illness called hereditary spherocytosis, characterized by abnormal red blood cell ankyrins. Parts of the red blood cell membrane disintegrate, causing the cell to lose its characteristic shape and balloon out. The bloated red blood cells obstruct narrow blood vessels in the spleen, dying and producing anemia as the spleen fails to maintain the red blood cell supply. Doctors treat hereditary spherocytosis by removing the spleen. Researchers expect to find defective ankyrin behind problems in other cell types, because it seems to be a key protein in establishing contact between the components of the cell's architecture.

Cell-Cell Interactions

In prokaryotes and single-celled eukaryotes, the cell is the entire organism. In multicellular organisms, cells communicate and interact.

Cell biologists are just beginning to understand some of the biochemical steps that orchestrate the cell-cell interactions that make multicellular life possible. We look now at two broad types of interactions between cells—signal transduction and cell adhesion.

Signal Transduction

A cell's organelles, cell membrane, and cytoskeleton are dynamic structures that must communicate with each other and with the external environment to maintain life. A key participant in this biological communication network is the cell membrane, which helps assess and transmit incoming messages in a process called **signal transduction.**

Signal transduction involves a cascade of protein action. It begins when receptor molecules jutting from cell membrane surfaces bind to messenger molecules, such as **hormones** and **growth factors,** known as **first messengers.** Hormones are chemicals carried in the blood to a specific site, where they exert an action. Growth factors act more locally. When a first messenger molecule docks at a cell surface, it binds to specific receptor molecules that are part of the membrane. The receptors contort, which triggers changes in other nearby molecules. A series of molecules in the cell membrane and cytoplasm respond, passing the signal for a particular cellular response in a biochemical version of a sequence of dominoes toppling. The culmination of all this signaling is a specific cellular response, such as secretion or cell division, involving various organelles. Figure 5.20 illustrates how signal transduction triggers ATP splitting inside the cell, releasing energy that is then used for any of several cellular functions. In this illustration, cyclic AMP acts as the **second messenger** that transmits the signal within the cell.

Many disorders can be traced to disrupted signal transduction. One such condition is the inherited neurofibromatosis type I (NF1), in which tumors grow in nervous tissue beneath the skin and in the central nervous system. Normally, a protein called neurofibromin blocks a growth factor docked at a cell surface receptor from activating

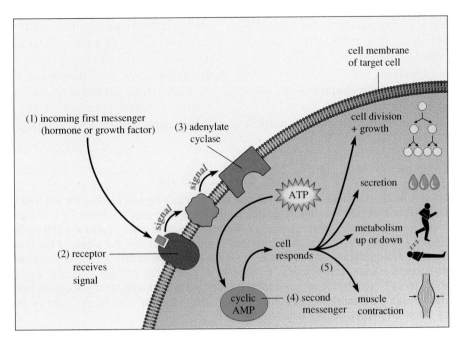

FIGURE 5.20

Signal transduction. A first messenger (1) binds a receptor (2), triggering a cascade of biochemical activity at the cell's surface. An enzyme, adenylate cyclase (3), catalyzes a reaction inside the cell. In this reaction, ATP loses phosphates, becoming cyclic AMP, the second messenger (4). The energy released by splitting ATP is channeled into various cellular activities (5).

cell division. In NF1, abnormal neurofibromin cannot stifle cell division, and a tumor forms.

Cell Adhesion—The Area Code Hypothesis

A multicellular organism is a dynamic structure. Its component parts—cells—interact through signals carried by messengers such as hormones or growth factors, or through physical contact with one another. Cells touch each other not at random, but through adhesion—a precise sequence of interactions between proteins that jut from one cell to meet receptors on other cells. Inflammation is a familiar process that provides an example of the complexity of cell-cell interactions.

Inflammation is the painful, red swelling surrounding a site of injury or infection, such as the area surrounding a jagged sliver of wood embedded in the palm of your hand. At the cellular level, inflammation occurs when a contingent of white blood cells (leukocytes) rush through the bloodstream to the endan-

gered part of the body. Here, the white blood cells squeeze between the cells of the **endothelium,** the curved layer of single cells that forms the walls of the smallest blood vessels (capillaries). Once the white blood cells have crossed from the blood into the tissues, they are chemically attracted to the injury, where they either attack infecting microbes or secrete biochemicals that do so. This migration of white blood cells from the bloodstream to the injury site is called **leukocyte trafficking.** A series of proteins, called **cellular adhesion molecules,** or CAMs, help guide white blood cells to the injured area. How do they do so?

The "area code hypothesis" suggests that white blood cells respond to three types of proteins activated in a sequence, as a three-digit area code directs a telephone call to a certain geographic area. The code stops the white blood cell and directs it to leave the blood vessel at the site of the injury. Figure 5.21 depicts the steps in this process.

The first "digit" in the area code is a type of CAM called **selectin** that coats

the white blood cell's tiny projections. The selectins provide traction, slowing the white blood cell into a roll, braking it from its usual turbulent rush by binding to carbohydrates on endothelial cell surfaces. The second digit in the area code is a specific type of chemical attractant, released by clotting blood, bacteria, or decaying tissue, whose presence signals the white blood cell to stop. The final digit in the code, called an **integrin,** is activated by the chemoattractant. The integrin latches onto yet another type of protein, an adhesion receptor protein, that protrudes from the capillary wall at the site of the damage. The adhesion receptor protein is thought to touch the cytoskeleton beneath the endothelial cell membrane, forming an anchor that pulls the white blood cell between the tilelike cells of the endothelium so that it exits on the other side. Figure 5.21 shows an illustration and a photograph of leukocyte trafficking—white blood cells rolling, adhering to, and crossing the blood vessel wall to reach the site of an injury.

What happens if a person's leukocyte-trafficking system fails? A Dallas teenager named Brooke Blanton knows the answer all too well. Her parents and doctors first suspected she had a problem when, as a baby, her teething sores did not heal. Oddly, these and other small wounds never accumulated pus, which consists of bacteria, cellular debris, and white blood cells. Eventually, doctors discovered that Brooke has a never-before-recognized disorder named leukocyte-adhesion deficiency. Her body lacks the CAMs that enable white blood cells to stick to blood vessel walls. The blood cells zip along in the bloodstream right past the wound, unable to slow down, stick, and cross the vessel wall. Brooke must be very careful to avoid injury and infection, and she receives antiinfective treatments for even the slightest wound.

Cell adhesion is critical to many other functions. CAMs guide cells surrounding an embryo to grow toward maternal cells, forming the beginnings of the placenta, the supportive organ linking a pregnant woman to the fetus. Sequences of CAM also establish connections between the nerve cells that underlie learning and memory.

Because cell adhesion is a general process involving several different types of proteins, adhesion disruptions can cause a variety of disorders. The spread of cancer cells, for example, is at least partly caused by a lack of cell adhesion that allows cancer cells to travel easily from one part of the body to another. Arthritis may occur when white blood cells are reined in by the wrong "area code" of adhesion molecules, inflaming a joint where there isn't an injury. Even the prevalence of the common cold and flu may be attributed to the ability of certain viruses to temporarily alter integrins,

thereby gaining access to our tissues. A major focus of recent drug development is to find substances that alter cell adhesion molecules.

The last two chapters have illustrated the complexity of a cell. Although the study of these tiny units of

life has a long and impressive history, we understand only a few cellular functions at the molecular level. In the next three chapters, we will narrow our focus to view three organelles in some detail—the chloroplast, mitochondrion, and nucleus.

KEY CONCEPTS

Cells in multicellular organisms must interact and communicate with one another and with the extracellular environment. One means of communication is signal transduction. In signal transduction, a first messenger (perhaps a hormone or growth factor) docks at a cell surface receptor and triggers a series of biochemical changes to transmit the signal to the inside face of the cell membrane. Inside the cell, a second messenger carries the signal. Eventually, this cascade of chemical responses culminates in a particular cellular response. Another vehicle for cell-cell interaction is cell adhesion. Cell adhesion molecules (CAMs) guide leukocytes through the bloodstream and blood vessel walls to the site of an injury. To do so, they use a sequence of cell-protein interactions.

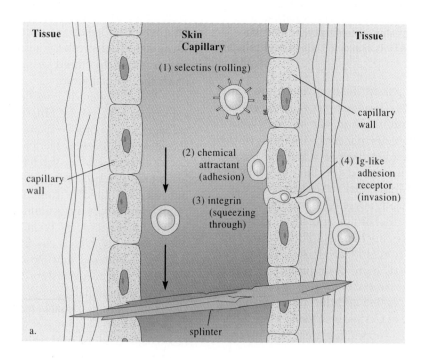

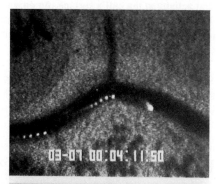

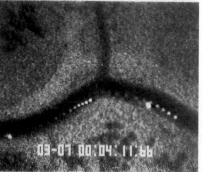

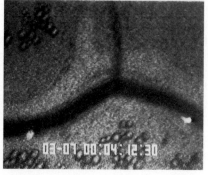

FIGURE 5.21

Leukocyte trafficking. *a.* (1) A traveling white blood cell slows into a more leisurely roll by binding selectins, slowing from 2500 micrometers per second to 50 micrometers per second. (2) Next, specific chemical attractants stop the white blood cell at the site of an injury. (3) An integrin then anchors the white blood cell in place, and an adhesion receptor (4) helps the cell squeeze through the junctures of the tight lining. *b.* Cell adhesion molecules (CAMs) assist white blood cells in their movements. These micrographs trace the movement of fluorescently tagged white blood cells past a venule, in a rabbit.

SUMMARY

Many cellular functions depend upon a cell's surface and cytoskeleton, which together can be thought of as the cellular architecture. Different cell types can be distinguished by the characteristics of their cellular architectures, and abnormalities in the cellular architecture are often associated with disease.

The features of a cell's surface identify it as belonging to a particular species, to a particular individual, and to a particular tissue within that individual. The surface consists of intricate patterns of sugars, proteins, and other molecules jutting from the cell's membrane. The human immune system recognizes the surfaces of body cells as "self" and the surfaces of other cells or particles as "nonself." Cell surfaces are particularly important in sorting out embryonic cells to form tissues and organs.

Cell membranes control which substances enter or leave a cell, and this, in turn, affects the cell's functions. A biological membrane consists of a phospholipid bilayer embedded with movable proteins. The percentage and distribution of proteins varies in the outer membranes of different cell types and in organelles.

Substances pass through cell membranes in several ways. A molecule that passes through openings in a membrane by following its concentration gradient moves by simple diffusion. Osmosis is the simple diffusion of water. Molecules that move through a membrane by following the concentration gradient with the aid of a carrier protein move by facilitated diffusion. Molecules that pass through a membrane by going against the concentration gradient, requiring both a carrier protein and energy supplied by ATP, move by active transport. Vesicles inside the cell can carry molecules and bacteria to the cell membrane, where they fuse with the membrane and release the cargo outside the cell. This process is called exocytosis. Vesicles that form on the cell membrane envelop and transport molecules into the cell by endocytosis. Receptor-mediated endocytosis is more specific because a substance must first bind to a cell surface receptor before the cell draws it in. Phospholipid bilayers (which comprise the cell membrane), various organelles, and vesicles form a continuous network along which substances are transported into and out of a cell.

The cytoskeleton is a network of rods and tubes that provides cells with form, support, and the ability to move. Microtubules self-assemble from hollow tubulin subunits to become such cellular constituents as flagella, cilia, and the spindle fibers that separate one cell into two during cell division. Microfilaments, structures smaller than microtubules, are solid, are composed of the protein actin, and provide contractile motion. Tough intermediate filaments provide scaffolding.

The integrity of the red blood cell membrane, signal transduction, and cell adhesion all illustrate the interaction of components of the cellular architecture. The red blood cell membrane retains its shape through a meshwork of spectrin protein rods bolted to the cell membrane by proteins called ankyrins. Without these proteins, the cell would collapse and be unable to survive. In signal transduction, the cell membrane transmits messages from the cell's exterior to the organelles within to regulate certain cellular functions. Cell adhesion facilitates cell-cell interactions by causing the proteins on one cell's surface to attach to receptors on other cells. All these cellular activities depend on the coordination of the cell surface and components of the cytoskeleton—in other words, on cellular architecture.

KEY TERMS

active transport 98
adenosine triphosphate 98
ankyrin 104
antigen 89
apolipoprotein 100
aqueous solution 92
carrier protein 96
cellular adhesion molecule (CAM) 105
contractile vacuole 94
cytoskeleton 88
diffusion 94
dynein 103
endocytosis 99

endothelium 105
exocytosis 99
facilitated diffusion 96
first messenger 104
fluid mosaic 91
glycoprotein 91
growth factor 104
hormone 104
human leukocyte antigen 89
hydrophilic 90
hydrophobic 90
hypertonic 94
hypotonic 94
inflammation 105

integrin 105
intermediate filament 103
isotonic 94
leukocyte trafficking 105
ligand 100
low-density lipoprotein 100
microfilament 103
microtubule 102
motor molecule 102
osmosis 94
passive transport 97
phospholipid 90
phospholipid bilayer 90

receptor 100
receptor-mediated endocytosis 100
second messenger 104
selectin 105
signal transduction 104
simple diffusion 94
sodium-potassium pump 98
solute 92
solvent 92
spectrin 104
tubulin 102
turgor pressure 94

REVIEW QUESTIONS

1. Why don't all human cells have the same pattern of surface features?

2. Cite an illustration of how the cell membrane is a dynamic rather than static structure.

3. How is a liposome similar to a cell membrane? How is it different?

4. Explain the differences between simple diffusion, facilitated diffusion, and active transport.

5. Endocytosis provides a general route for a substance to enter a cell. How can this process be more selective?

6. How do the cell membrane and cytoskeleton utilize ATP?

7. List five functions of the cytoskeleton.

8. What roles do receptors play in

   a. transporting substances across a cell membrane?
   b. signal transduction?
   c. vesicle trafficking?
   d. cell adhesion in inflammation?

9. Would a substance that destroys the integrity of the blood-brain barrier be dangerous? Why or why not?

10. List the functions of the following proteins.

    a. tubulin
    b. dynein
    c. actin
    d. ankyrin
    e. spectrin
    f. cell adhesion molecules

11. Describe how components of the cellular architecture are affected in each of the following illnesses.

    a. familial hypercholesterolemia
    b. diabetes insipidus
    c. hereditary spherocytosis
    d. neurofibromatosis type I
    e. leukocyte-adhesion deficiency
    f. arthritis
    g. spread of cancer

12. How does each of the following processes illustrate the interaction of components of the cellular architecture?

    a. maintaining the integrity of the red blood cell membrane
    b. signal transduction
    c. cell adhesion in leukocyte trafficking

TO THINK ABOUT

1. How would you test the hypothesis that a particular medical condition is associated with a particular HLA type?

2. In many species, including our own, sperm cells have surface proteins that bind to receptor molecules on the surface of the egg cell. Use this information to suggest two possible methods of birth control.

3. A widely used drug, lovastatin, works by lowering cellular production of an enzyme necessary for liver cells to synthesize cholesterol. What medical condition does this drug treat?

4. Liver cells are packed with glucose. What mechanism could be used to transport more glucose into a liver cell? Why would only this mode of transport work?

5. At the tip of a human sperm cell is a small elevation called the acrosome, which contains enzymes that allow the sperm cell to enter an egg cell. The acrosome contains many vesicles beneath its membrane. How do you suppose the enzymes are released from the tip of the sperm cell?

6. A drop of a 5% salt (NaCl) solution is added to a leaf of the aquatic plant *Elodea*. When the leaf is viewed under a microscope, colorless regions appear at the edges of each cell as the cell membranes shrink from the cell walls. What is happening to these cells?

SUGGESTED READINGS

Barinaga, Marcia. April 23, 1993. Secrets of secretion revealed. *Science*, vol. 260. A vast intracellular transport system moves molecules.

Bollag, Gideon, and Frank McCormick. April 23, 1992. Neurofibromatosis is enough of GAP. *Nature*, vol. 356. NF1 disrupts signal transduction.

Brown, Michael S., and Joseph L. Goldstein. April 6, 1986. A receptor-mediated pathway for cholesterol homeostasis. *Science*, vol. 232. The way cells in the liver handle cholesterol from the diet illustrates many interactions of the cellular architecture.

Collins, Francis S. May 8, 1992. Cystic fibrosis: Molecular biology and therapeutic implications. *Science*, vol. 256. Collins led one of the teams that discovered the CF gene. Here, he explains how the mutation causes the symptoms.

Edgington, Stephen M. April 1992. How sweet it is: Selectin-mediating drugs. *Bio/Technology*, vol. 10. Researchers are deciphering the sequence of protein-carbohydrate interactions involved in cell adhesion.

Ezzell, Carol. June 13, 1992. Sticky situations: Picking apart the molecules that glue cells together. *Science News*. Many mechanisms bind cells.

Fackelmann, K. A. August 24, 1991. Energy duo takes on cystic fibrosis's chloride defect. *Science News*. The clogged lungs of CF patients are caused by an abnormally shaped channel in the membranes of cells lining the respiratory tract.

Hoffman, Michelle. March 8, 1991. New clue found to growth factor action. *Science*, vol. 252. Growth factors are often the first step in signal transduction pathways through the cell.

Hoffman, Michelle. June 26, 1992. Motor molecules on the move. *Science*, vol. 256. Several types of proteins help move cellular components.

Hooper, Celia. June 1990. Ankyrins aweigh! *The Journal of NIH Research*, vol. 2. As its name and this article title imply, ankyrin helps anchor proteins of the cytoskeleton to the cell membrane.

Lewis, Ricki. October 1990. Neurofibromatosis I revealed. *The Journal of NIH Research*, vol. 2. The tumors and spots of NF1 result from a defect in signal transduction, causing abnormal cell division.

Lewis, Ricki. March 1994. Gateway to the brain. *BioScience*, vol. 44. Understanding the function of the blood-brain barrier leads to development of new drug delivery systems.

Lewis, Ricki. April 15, 1994. Biotechnology companies formulate the next generation of liposomes. *Genetic Engineering News*, vol. 14. Liposomes are excellent drug carriers.

Maddox, John. March 18, 1993. Why microtubules grow and shrink. *Nature*, vol. 362. Microtubules are constantly undergoing remodeling.

Thompson, Dick. September 28, 1992. The glue of life. *Time*. People missing adhesion molecules live dangerously.

 # EXPLORATIONS   **Interactive Software**

Active Transport

This interactive exercise allows students to explore how substances are transported across membranes against a concentration gradient (that is, towards a region of higher concentration). The exercise presents a diagram of a coupled channel within a membrane through which amino acids are pumped into the cell. By altering ATP concentrations the student can speed or slow the operation of the ATP-driven sodium-potassium pump and explore the consequences for amino acid transport. Similarly, the student can alter the cellular or extracellular levels of amino acid, and investigate the effect on cellular expenditure of ATP. Because the amino acid transport channel is coupled to the ATP-driven sodium-potassium pump, students will discover that both ATP and amino acid levels have important influences.

1. How does the level of ATP influence the operation of the sodium-potassium pump?

2. Does a fall in cellular levels of ATP inhibit cellular uptake of amino acids?

3. Does an increase in extracellular levels of amino acid always lead to an increased expenditure of ATP?

4. As long as ATP is readily available, is there any condition under which an increase in extracellular levels of amino acids does not result in increased transport of amino acids into the cell?

In bioluminescence, fireflies convert chemical energy into light energy.

LEARNING OBJECTIVES

By the chapter's end, you should be able to answer these questions:

1. What is energy?

2. Why is energy important in living systems?

3. What sources provide energy for life?

4. What is the difference between potential energy and kinetic energy? How are the two related?

5. How do the laws of thermodynamics underlie biological energy acquisition and use?

6. What energy transformations do cells use to synthesize and break down molecules?

7. How do chemical reactions alter the energy available to do work?

8. What kinds of reactions constitute metabolism?

9. Which molecules help cells use energy?

10. How do chemical interactions control metabolism?

The Energy of Life

A tortoiseshell butterfly expends a great deal of energy to stay alive. Not only does it fly from flower to flower, but it migrates seasonally over long distances. A less obvious way the insect uses energy is to power the many biochemical reactions of its metabolism. The butterfly obtains this energy by feeding on nectar produced by a flower (fig. 6.1); the flowering plant gets its energy from the sun. All animals ultimately extract life-powering energy from plants and from certain microorganisms that capture the energy in sunlight and convert it into chemical energy.

Evidence of energy pervades our lives. We experience a burst of energy soon after eating a candy bar. Thunderstorms, volcanoes, and earthquakes violently display energy. A vibrating guitar string, a sunbeam, and a flying hockey puck each demonstrate energy. All aspects of life require energy, from replicated chromosomes dividing and distributing into two cells, to cells migrating in an early embryo to form tissues, to a flower blossoming. Energy is required to actively transport substances across membranes against their concentration gradients, and to transport secreted products through the cellular production line, and into and out of organelles, as they make their way out of the cell. Because energy is so basic to life, we need to understand just what it is and how living things use it.

FIGURE 6.1

A butterfly expends great amounts of energy flying from flower to flower and migrating sometimes thousands of miles. This tortoiseshell butterfly is feeding on a flower.

What Is Energy?

The term *energy* was coined about two centuries ago, when the Industrial Revolution redefined familiar ideas of energy from the power behind horse-drawn carriages and falling water to the power of the internal combustion engine. As people began to think more about harnessing energy, biologists realized that understanding energy could not only improve the quality of life, but reveal how life itself is possible. **Bioenergetics** is the study of how living organisms use energy to perform the activities of life.

Energy is the ability to do work—that is, to change or move matter against an opposing force, such as gravity or friction. Because energy is an ability, it is not as tangible as matter, which has mass and takes up space. One way to measure energy is in calories. A calorie (cal) is the amount of energy required to raise the temperature of one gram of water 1°C. The most common unit for measuring the energy content of food and the heat output of organisms is the Calorie (Cal), or kilocalorie (kcal), which is the energy required to raise the temperature of a kilogram of water 1°C.

Sources of Energy

Energy on earth comes from various sources:

  the sun

  fossil fuels such as coal, oil, and gas

  wind

  organisms (biomass)

  nuclear power from cosmic events that occurred before our solar system formed

  the pull of the moon (tidal power)

  geothermal energy from the earth's core

The Sun

Most organisms obtain their energy from the sun. Deep within the sun, temperatures of 10,000,000°C fuse hydrogen atoms, forming helium and releasing gamma rays. These rays then emit electrons and **photons,** particles of light energy that travel in waves. Life depends upon the ability to transform this energy, contained in photons of sunlight, into chemical energy before it converts to heat—the eventual fate of all energy.

The sun's total energy output is about 3.8 sextillion megawatts of electricity. Earth intercepts only about two-billionths of this; although this is the equivalent of burning about 200 trillion tons of coal a year, most of this energy doesn't reach life. Nearly a third is reflected back to space, and another half is absorbed by the planet, converted to heat, and returned to space. Another 19% of incoming solar radiation powers wind and water and drives **photosynthesis,** the process by which plants harness solar energy. Of this 19%, only 0.05 to 1.5% is incorporated into plant material—and only about a tenth of that makes its way into the bodies of animals that eat plants. Far less reaches animals that consume plant-eating animals (fig. 6.2). Obviously, an extremely tiny portion of solar energy is captured to support life on this planet. To obtain that energy, we are entirely dependent upon photosynthetic organisms.

The Earth

While most living organisms depend, directly or indirectly, on photosynthetic plants to acquire solar energy, a few species have adapted to obtain and utilize geothermal energy. Crabs, clams, worms, and bacteria thrive around dark cracks in the ocean floor where hot molten rock seeps through, creating an intensely hot environment. The residents of these deep-sea hydrothermal vents obtain energy from inorganic chemicals surrounding them. Bacteria, for example, extract energy from the chemical bonds of hydrogen sulfide (H_2S) and then add the freed hydrogens to carbon dioxide to synthesize organic compounds. Organisms like these, which use inorganic chemicals to manufacture nutrient molecules, are called **chemoautotrophs** (fig. 6.3). This term comes from

FIGURE 6.2

Biological energy. The lion gets its energy by eating other animals, such as this giraffe. The giraffe, in turn, got its energy from eating leaves. At each step in a food chain, 90% of usable energy is dissipated as heat.

the more general term **autotroph,** which refers to organisms that synthesize their own nutrients.

Potential and Kinetic Energy

To stay alive, an organism must continually take in energy through eating or from the sun or the earth and convert it into a usable form. Every aspect of life centers on converting energy from one form to another. Two basic types of energy constantly interconvert: **potential energy** and **kinetic energy.**

Potential energy is stored energy available to do work. When matter assumes certain positions or arrangements, it contains potential energy. A teaspoon of sugar, a snake about to pounce on its prey, a child at the top of a slide, and a baseball player about to slam a ball (fig. 6.4) illustrate potential energy. In organisms, potential energy is stored in the chemical bonds of nutrient molecules, such as carbohydrates, lipids, and proteins.

FIGURE 6.3

Chemoautotrophic bacteria support life in deep sea hydrothermal vents. The bacteria derive energy from inorganic chemicals such as hydrogen sulfide.

Kinetic energy is energy being used to do work. Burning sugar, the snake pouncing, the child riding down the slide, and the soaring baseball all demonstrate kinetic energy. A chameleon shooting out its sticky tongue to capture

FIGURE 6.4

The ball this pitcher is about to throw has potential energy.

a butterfly (fig. 6.5) uses kinetic energy, as does an elephant trumpeting and a Venus's-flytrap closing its leaves around an insect. The adder snake in figure 1.3a demonstrates potential energy as it awaits the approach of a sand lizard. Somewhere between figures 1.3a and 1.3b, the snake strikes using kinetic energy to grab its lizard meal.

Kinetic energy transfers motion to matter. The movement in the batter's arm transfers energy to the ball, which then takes flight. Similarly, flowing water turns a turbine, and a growing root can push aside concrete to break through sidewalk. Put another way, kinetic energy moves objects. Heat and sound are types of kinetic energy because they result from the movement of molecules.

Potential energy is translated into the kinetic energy of motion, and that burst of energy can be used to do work. Under most conditions, potential and kinetic energy are readily interconvertible, although not at 100% efficiency.

K E Y   C O N C E P T S

Energy is the ability to do work. Potential energy is stored energy available to do work, and kinetic energy is energy actually being used to do work. Under most conditions, kinetic energy and potential energy are freely interconvertible, though not at 100% efficiency.

The Laws of Thermodynamics

The energy conversions vital for life, as well as those that occur in the nonliving world, are regulated by rules called the laws of **thermodynamics** (fig. 6.6). These laws operate on a system and its surroundings; the system is the collection of matter under consideration, and the surroundings are the rest of the universe. An open system exchanges energy with its surroundings, while a closed system is isolated from its surroundings—it does not exchange energy with anything outside the system. For example, consider a thermos bottle and its contents. The uncapped thermos, an open system, loses steam from the heated food to the surrounding air, while a closed thermos prevents such an exchange of energy. The laws of thermodynamics are unbreakable and apply to all energy transformations—gasoline combustion in a car's engine, or glucose breakdown in a cell.

FIGURE 6.5

The living world abounds with illustrations of kinetic energy, the energy of action. Here, a chameleon's tongue whips out to ensnare a butterfly.

The First Law of Thermodynamics

Although *thermodynamics* may seem a complicated term, the laws actually make much common sense. The first law of thermodynamics is the law of energy conservation. It states that energy cannot be created or destroyed, but only converted to other forms. This means that the total amount of energy in a system and its surroundings remains constant; thus, on a grander scale, the amount of energy in the universe is constant.

In a practical sense, the first law of thermodynamics explains why we can't

---

First Law of Thermodynamics

Energy cannot be created or destroyed, but only converted to other forms.

Second Law of Thermodynamics

All energy transformations are inefficient because every reaction results in an increase in entropy and the loss of usable energy as heat.

FIGURE 6.6

The laws of thermodynamics underlie energy use in organisms.

get something for nothing. The energy released when the baseball hurtles towards the outfield doesn't appear out of nowhere—it comes from the batter's muscles. Likewise, green plants do not manufacture energy from nothingness, they trap the energy in sunlight and convert it into chemical bonds. The energy in sunlight, in turn, comes from chemical reactions in the sun's matter.

According to the first law of thermodynamics, the amount of energy your body uses cannot exceed the amount of energy you take in through the chemical bonds contained in the nutrient molecules (food) that you eat. Even if you are fasting, your body cannot use more energy than your own tissues already contain. Similarly, the energy produced by leaves during photosynthesis cannot exceed the amount of energy in the absorbed light. No system can use or release more energy than it takes in.

The Second Law of Thermodynamics

The second law of thermodynamics states that all energy transformations are inefficient because every reaction results in increased entropy and loses some usable energy to the surroundings as heat. Unlike other forms of energy, heat energy results from random molecule movements. Any other form of energy can convert completely to heat, but heat cannot completely convert to any other form of energy. Because all energy eventually becomes heat, and heat is disordered, all energy transformations head towards increasing disorder, or **entropy.** The concept of entropy is illustrated in figure 6.7. In general, the more disordered a system is, the higher its entropy.

Because of the second law of thermodynamics, events tend to be irreversible unless energy is input to reverse them. Processes that occur without an energy input are spontaneous. A house becomes messier—not neater—as the week progresses unless someone expends energy to clean up. In natural processes, irreversibility results from the loss of

a.

b.

FIGURE 6.7

A teenager's messy room symbolizes entropy—extreme disorder (*a*). Cleaning the room (*b*) requires an input of energy.

usable energy as heat during energy transformation. It is impossible to reorder the molecules that have dispersed as heat.

Cell energetics are also governed by the second law of thermodynamics. Cells derive energy from nutrient molecules and use it to perform such activities as growth, repair, and reproduction. The chemical reactions that free this energy are (as the law predicts) inefficient and release much heat. The cells of most organisms are able to extract and use only about half of the energy in their food. Although organisms can transform energy—storing it in tissues or using it to repair a wound, for example—ultimately, much of the energy is dissipated as heat.

Because organisms are highly organized, they may seem to defy the second law of thermodynamics—but only when you consider them alone, as closed systems. Organisms remain organized because they are *not* closed systems. They use incoming energy and matter, from sources such as sunlight and food, in a

constant effort to maintain their organization and stay alive. Although the entropy of one system, such as an organism or a cell, may decrease as the system becomes more organized, the organization is temporary; eventually the system will die. The entropy of the universe as a whole is always increasing.

We're most familiar with energy transformations that release large amounts of energy at once—explosions, lightning, or the sight of a plane taking off. Cellular energy transformations, by contrast, release energy in tiny increments. Cells extract energy from glucose in small amounts at a time via the biochemical pathways of **cellular respiration.** If all the energy in the glucose chemical bonds was released at once, it would be converted mostly to heat, producing deadly high temperatures. So instead, cells extract energy from glucose and other molecules by slowly stripping electrons from them in a series of chemical reactions. Each reaction lowers the potential energy of the reactant molecule. Some of this energy is lost as heat, but much of it is trapped in the chemical bonds of other molecules in the cell. Then the energy is either stored or utilized to power the activities of life.

Metabolism

Metabolism, Greek for "change," consists of the chemical reactions that change or transform energy in cells. The reactions of metabolism are organized into step-by-step sequences called **metabolic pathways** in which the product of one reaction becomes the starting point, or substrate, of another (fig. 6.8). Metabolism in its entirety is an enormously complex network of interrelated biochemical reactions organized into chains and cycles.

Building Up and Breaking Down—Anabolism and Catabolism

Metabolism includes both the construction and breakdown of molecules within cells. The energy-requiring pathways

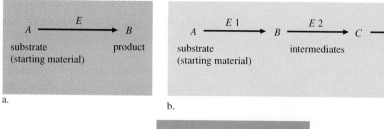

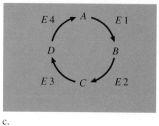

FIGURE 6.8

The chemical reactions of life form chains and cycles of metabolic pathways. Depicted are (*a*) an enzyme-catalyzed chemical reaction, (*b*) a biochemical pathway, and (*c*) a biochemical cycle. In each case, *E* represents the enzyme that catalyzes a specific reaction. Note that in a cycle, the product of the last reaction is also the starting material of the first. Cycles release by-products that are important in other biochemical pathways.

Table 6.1	Differences between Catabolism and Anabolism	
Catabolism	**Anabolism**
Breaks down large molecules	Synthesizes large molecules
Energy released	Energy required
Reactions converge	Reactions diverge

that construct large molecules from small ones comprise **anabolism,** or biosynthesis. In anabolism, a small number of chemical subunits join in different ways to produce many different types of large molecules. The pathways of anabolism diverge (branch out) because a few types of precursor molecules combine to yield many different types of products. For example, the 20 different types of amino acids link together in varied sequences to synthesize thousands of different kinds of proteins.

Metabolic pathways that break down large molecules into smaller ones constitute **catabolism,** or degradation. These pathways release energy. Catabolic pathways converge (come together) because many different types of large molecules degrade to yield fewer types of small molecules. For example, an animal may

eat many different types of foods, consisting of large and diverse molecules of carbohydrates, fats, and proteins. These major nutrients ultimately break down into just a few types of compounds—glucose, glycerol and fatty acids, and amino acids, respectively. Table 6.1 summarizes the basic differences between anabolism and catabolism.

KEY CONCEPTS

Metabolism is the sum of the chemical and energy transformations that take place in cells. Metabolism occurs in stepwise sequences called metabolic pathways. In anabolism, the cell uses a few starting compounds to synthesize a larger number of products, consuming energy in the process. In catabolism, the cell breaks down a large number and variety of starting materials to a smaller variety of smaller units, releasing energy in the process.

Free Energy

Each reaction in a metabolic pathway rearranges atoms into new compounds, and each reaction either absorbs or releases energy. According to the first law of thermodynamics, it takes the same amount of energy to break a particular kind of bond as it does to form that same bond—energy is not created or destroyed. The amount of energy stored in a chemical bond is called its **bond energy.** Some chemical bonds are stronger than others—that is, they have greater bond energies. When a stronger bond breaks, it releases more energy. Conversely, forming a stronger bond requires a greater input of energy.

The potential energy of a compound, then, is contained in its chemical bonds. When these bonds break, some of the energy released can be used to do work—for example, to form other bonds. The amount of stored energy potentially available to form new bonds is called the **free energy** of the molecule. It is represented by a G for its discoverer, Yale physicist Josiah Gibbs.

Chemical reactions change the amount of free energy stored in a bond and potentially available to do work. This change in free energy, symbolized ΔG, is the most fundamental property of a chemical reaction. It tells us the difference between the free energies of the reactants and the products. ΔG can be positive or negative. In a reaction with a negative ΔG,

the products contain less free energy than the reactants,
energy is released, and
entropy increases.

Such a reaction is spontaneous because it occurs without an input of energy. It is also **exergonic** (energy outward), which means that entropy increases and energy is released.

However, not all reactions are spontaneous—some require an input of energy rather than release it outward. For example, combining the monosaccharides glucose and fructose to form water and sucrose (table sugar) requires energy. In such a reaction, the ΔG is positive, meaning that the products have more energy than the reactants. Therefore, this reaction requires an absorption of energy from the surroundings and is not spontaneous. Reactions that require a net input of energy are called **endergonic** (energy inward). Any reaction with a positive ΔG is endergonic, and will not occur spontaneously. In fact, the reverse reaction will tend to occur.

The change in free energy of a reaction dictates how much work the reaction can perform. Consider the breakdown of glucose ($C_6H_{12}O_6$) to carbon dioxide (CO_2) and water (H_2O). The ΔG for this reaction is negative; therefore, this reaction is spontaneous and exergonic. The carbon dioxide and water products store less energy than glucose; some of the energy released can do work, and the rest is lost as heat.

Free Energy and Chemical Equilibrium

Most chemical reactions can proceed in both directions—that is, when enough product forms, some of it converts back to reactants. **Chemical equilibrium** is the point at which a reaction proceeds in both directions at the same rate. When this happens, the change in free energy (ΔG) is zero. When a reaction departs from equilibrium (that is, when either reactants or products accumulate), the change in free energy increases.

Because all activities of life require energy, cells must remain far from equilibrium to obtain the energy required to stay alive. They do this by continually preventing the accumulation of any metabolic pathway reactants. For example, the large difference in free energy between glucose and its breakdown products, carbon dioxide and water, propels cellular metabolism strongly in one direction—as soon as glucose enters the cell, it is quickly broken down to release energy. This energy allows the cell to stave off equilibrium by continuously forming new products needed for life.

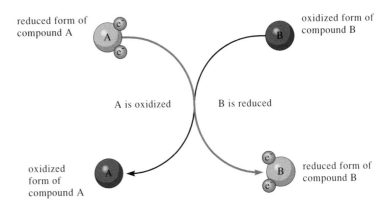

FIGURE 6.9

Oxidation and reduction reactions are paired. Here, compound A is initially in the reduced (electron-rich) form. It becomes oxidized in the reaction shown, losing a pair of electrons to compound B, which is reduced in the process.

KEY CONCEPTS

Each reaction of a metabolic pathway rearranges atoms into new compounds. Chemical reactions change the amount of free energy available to do work (for example, to form other bonds) by releasing the energy contained in chemical bonds. If the change in free energy (ΔG) is negative, the reaction is spontaneous. If the ΔG is positive, the reaction is not spontaneous; energy is required.

Oxidation, Reduction, and Energy Content

Most energy transformations in organisms occur in chemical reactions called **oxidations** and **reductions.** Oxidation occurs when a molecule loses electrons. The name comes from the observation that many reactions in which molecules lose electrons involve oxygen. Oxidation is the equivalent of adding oxygen because oxygen strongly attracts electrons away from the original atom. Oxidation reactions, such as the breakdown of glucose to carbon dioxide and water, are catabolic. They degrade molecules into simpler products as they release energy.

Reduction occurs when a molecule gains electrons. Reduction changes the chemical properties of a molecule, but not necessarily its size. Electrons removed from one molecule during oxidation join another molecule and reduce it (fig. 6.9).

Reduction reactions, such as the formation of lipids, are usually anabolic. They require a net input of energy. Oxidation and reduction reactions always occur simultaneously; if one molecule is reduced (gains electrons), another must be oxidized (loses electrons).

Many energy transformations in living systems involve carbon oxidations and reductions. Reduced carbon contains more energy than oxidized carbon. This is why reduced molecules such as methane (CH_4) are explosive, while oxidized molecules such as carbon dioxide (CO_2) are not. The same principle applies to other compounds: the more reduced they are, the more energy they contain. Anyone who has ever dieted is at least intuitively familiar with this concept of energy content. Saturated fats are highly reduced, and they contain more than twice as many kcal by weight as proteins or carbohydrates. Thus, a fat-laden meal of a bacon double cheeseburger with fries and a shake many contain the same number of kcal (and thus the same amount of energy) as a bathtub full of celery sticks.

KEY CONCEPTS

Many energy transformations in organisms occur in oxidation and reduction reactions. Oxidation occurs when a molecule loses electrons, and reduction occurs when a molecule gains electrons. Oxidation and reduction reactions always occur simultaneously.

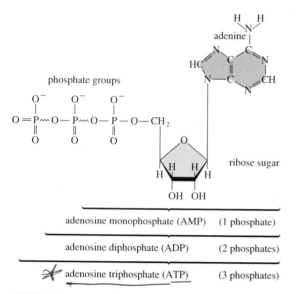

FIGURE 6.10

ATP—energy currency of the cell. Energy is required to form bonds between negatively charged phosphate (PO_4) groups. The squiggly lines that represent these bonds indicate that a great deal of energy is released when they break down.

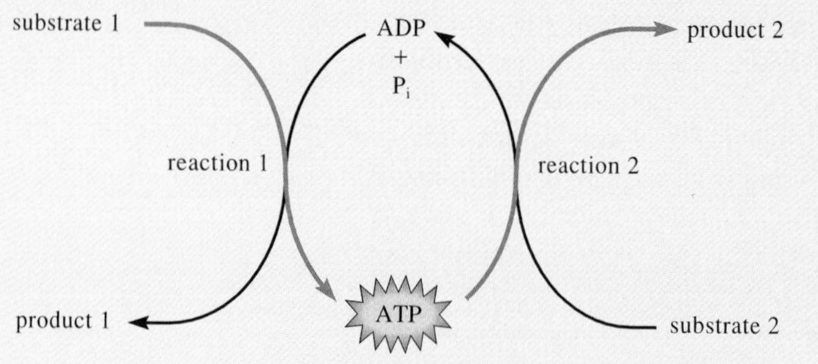

FIGURE 6.11

ATP breakdown and formation link biochemical reactions. Here, reaction 1 releases energy, which is used to phosphorylate ADP to ATP. Reaction 2 requires energy, which is released when ATP splits to yield ADP.

ATP: Cellular Energy Currency

Organisms store potential energy in the chemical bonds of nutrient molecules. It takes energy to make these bonds, and energy is released when bonds of these molecules break. Much of the released energy of life is stored temporarily in the covalent bonds of **adenosine triphosphate,** a compound more commonly known as **ATP.**

ATP is composed of the nitrogen-containing base adenine, the 5-carbon sugar ribose, and three phosphate groups (each group including a phosphorus atom bonded to four oxygen atoms) (fig. 6.10). ATP is one of the building blocks of the nucleic acids DNA and RNA. The covalent bonds that join ATP's phosphate groups can break, releasing a large amount of energy. Because they are able to release so much energy, these phosphate bonds are sometimes called "high-energy" bonds, and they are represented graphically by squiggly lines. However, ATP's "high energy" is not contained entirely in its phosphate

bonds, but in the complex interaction of atoms that comprise the entire molecule.

When the endmost phosphate group of an ATP molecule detaches, energy is released and the molecule becomes adenosine diphosphate (ADP, which has two phosphate groups rather than three). Still another phosphate bond can break to yield adenosine monophosphate (AMP) and another release of energy. ATP thus provides an energy currency for the cell. When a cell needs energy for an activity, it "spends" ATP by converting it to ADP, inorganic phosphate (symbolized Pi), and energy (fig. 6.11). This reaction is represented as:

$$ATP + H_2O \rightarrow$$
$$ADP + Pi + energy$$

The ΔG of this reaction is negative; energy is released. Biology in Action 6.1 discusses the role of ATP in bioluminescence, the reaction that gives fireflies a "glow."

ATP is an effective biological energy currency for several reasons. First, converting ATP to ADP + Pi releases about twice the amount of energy needed to drive most intracellular reactions. The extra energy is dissipated as heat. Second, ATP is readily available. The large amounts of energy in the bonds of fats and starches are not—they must first convert to ATP before they can be used. Finally,

ATP's terminal (end) phosphate bond, unlike the covalent bonds between carbon and hydrogen in organic molecules, is unstable—so it breaks and releases energy easily.

All cells use ATP for energy transformations. Just as you can use currency to purchase a huge variety of different products, cells use ATP in all kinds of different chemical reactions to do different kinds of work. If you ran out of ATP, you would die instantly. Life requires huge amounts of ATP. A typical adult uses the equivalent of 2 billion ATP molecules a minute just to stay alive, or about 200 kilograms (about 440 pounds) of ATP a day. However, ATP is used so rapidly that only a few grams are available at any given instant. Organisms recycle ATP at a furious pace, adding phosphate groups to ADP to reconstitute ATP, using the ATP to drive reactions, and turning over the entire supply every minute or so. The reaction that converts ADP to ATP is represented as:

$$ADP + PO_4 + energy \rightarrow ATP + heat$$

The ΔG for ATP formation is positive, which means it requires an input of energy. This energy comes from molecules broken down in other reactions that occur at the same time and in the same place as ATP synthesis. Such complementary, simultaneous reactions are called **coupled reactions.**

Biology in ACTION

Firefly Bioluminescence

Photosynthesis transforms light energy into chemical energy. In some forms of a process called **bioluminescence,** the reverse happens—chemical energy converts into light energy.

Bioluminescence is a familiar phenomenon—we see it every summer in the glow of a firefly's abdomen. More than 1,900 species of firefly are known, and each uses its own distinctive repertoire of light signals to attract a mate. Typically, flying males emit a pattern of flashes. Wingless females, called glowworms, are usually found on leaves, and they glow in response to the male. In one species, *Photuris versicolor*, the female emits the mating signal of another species and then eats the hapless male who approaches her. Some frogs consume so many delectable fireflies that they glow! Some summer evening, try mimicking a firefly's pattern of flashes with a flashlight. Can you entice a glowworm to respond?

Johns Hopkins University researchers William McElroy and Marlene DeLuca, who enlisted the aid of Baltimore schoolchidren to bring them jars of fireflies, deciphered the firefly luminescence reaction in the 1960s. McElroy and DeLuca found that the light is emitted when an organic acid called *luciferin* reacts with ATP, yielding the intermediate compound *luciferyl adenylate* (fig. 1). The enzyme *luciferase* then acts upon this intermediate in the presence of molecular oxygen (O_2) to yield the compound *oxyluciferin*—and a flash of light. The oxyluciferin is then reduced to luciferin, and the cycle is ready to start over again.

Chemical companies sell luciferin and luciferase, which researchers use to detect ATP. When ATP appears in a sample of any substance, it indicates contamination by a living thing. For example, the manufacturers of Coca-Cola use firefly luciferin and luciferase to detect bacteria in syrups used to produce the beverages. Contaminated syrups glow in the presence of luciferin and luciferase because the ATP in the bacteria sets the bioluminescence reaction into

motion. Firefly luciferin and luciferase were also aboard the Viking spacecraft sent to Mars. Scientists sent the compounds to detect possible life—a method that would only succeed if Martians use ATP.

Although we understand the biochemistry of the firefly's glow, the ways animals use their bioluminescence are still very much a mystery. This is particularly true for the bioluminescence phenomenon of synchrony seen in the trees. When night falls, first one firefly, then another, then more, begin flashing from the tree. Soon the tree twinkles like a Christmas tree. But then, order slowly descends. In small parts of the tree, the lights begin to blink on and off together. The synchrony spreads. A half hour later, the entire tree seems to blink on and off every second. Biologists studying animal behavior have joined mathematicians studying order to try to figure out just what the fireflies are doing—or saying—when they synchronize their glow.

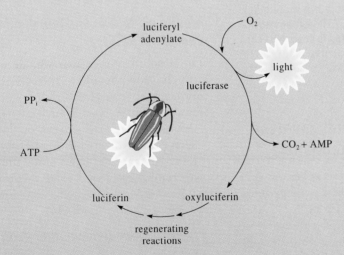

FIGURE 1

Luciferin reacts with ATP to produce luciferyl adenylate and two phosphate groups (designated PPi). Luciferyl adenylate then reacts with oxygen in a luciferase-catalyzed reaction to release CO_2, AMP (adenosine monophosphate), and light. Luciferin is then regenerated by a series of reactions that reduce oxyluciferin.

FIGURE 2

Bioluminescence from fireflies lights the air around this Iowa farmhouse.

Table 6.2 Some Molecules Involved in Cellular Energy Transformations

Molecule	Mechanism
ATP	Compound that phosphorylates molecules, energizing them to participate in other reactions. Conversions between ATP and ADP link many biochemical reactions.
Cofactors	Substances, usually ions, that transfer chemical groups such as phosphates between molecules.
Coenzymes	Vitamin-derived cofactors that transfer electrons or protons.
NAD^+	Coenzyme that transfers electrons; used to synthesize ATP.
$NADP^+$	Coenzyme that supplies hydrogen to reduce CO_2 in photosynthesis.
FAD	Coenzyme that transfers electrons; used to synthesize ATP in cellular respiration.
Cytochromes	Iron-containing molecules that transfer electrons in metabolic pathways.

Coupled Reactions

Cells couple the breakdown of nutrients to ATP production, and they couple the breakdown of ATP to other reactions that occur at the same time and place in the cell. Coupled reactions, as their name implies, are reactions that occur in pairs. One reaction drives the other, which does work or synthesizes new molecules. Consider, for example, the formation of sucrose (table sugar) and water from glucose and fructose. Because the ΔG for this reaction is greater than zero, the reaction is not spontaneous; it requires a net input of energy. When ATP provides additional energy, the reaction has a negative ΔG (the reactants now have more energy available than the products). The reaction proceeds, with ATP breakdown coupled to sucrose synthesis.

ATP accomplishes most of its work by transferring its terminal (end) phosphate group to another molecule in a process called **phosphorylation.** In muscle cells, for example, ATP transfers phosphate groups to contractile proteins. Similarly, plants use ATP energy to build cell walls. Phosphorylation energizes the molecules receiving the phosphate groups so that they can be used in later reactions. The original energy "cost" of this phosphorylation is thus returned in subsequent reactions.

KEY CONCEPTS

All cells couple ATP breakdown to other reactions to do work. ATP releases about twice the energy needed to drive most cellular reactions. Molecules that receive phosphate groups from ATP phosphorylation are energized.

Other Compounds Involved in Energy Metabolism

Several other compounds besides ATP affect metabolism, or the energy transformations in cells (table 6.2). Non-protein helpers called **cofactors** aid metabolism by acting with other substances to cause certain reactions. Cofactors are often ions. Mg^{2+}, for example, is a cofactor required to transfer phosphate groups between molecules.

Organic cofactors, which are called **coenzymes,** usually carry protons or electrons. Coenzymes are often nucleotides, as is ATP. But a coenzyme's energy content, unlike ATP's, depends on its oxidation state, not on the presence or absence of a particular phosphate bond. Vitamins act as coenzymes in cells, helping to drive metabolic reactions. NAD^+, $NADP^+$, and FAD are molecules that function as coenzymes.

Nicotinamide Adenine Dinucleotide (NAD+)

Nicotinamide adenine dinucleotide (NAD^+), like ATP, consists of adenine, ribose, and phosphate groups; but NAD^+ also has a nitrogen-containing ring, called nicotinamide, which is derived from niacin (vitamin B_3). The nicotinamide is the active part of the molecule. NAD^+ is reduced when it accepts two electrons and two protons from a substrate. Both electrons and one proton actually join the NAD^+, leaving a proton (H^+). This reaction is written as:

$$NAD^+ + 2H^+ + 2e^- \rightarrow NADH + H^+$$

The ΔG of this reaction is negative. $NADH + H^+$ is fully reduced and is therefore packed with potential energy. The cell uses it to synthesize ATP and to reduce other compounds. Other biochemical pathways also tap the energy in $NADH + H^+$ to produce ATP.

Nicotinamide Adenine Dinucleotide Phosphate (NADP+)

The structure of **nicotinamide adenine dinucleotide phosphate ($NADP^+$)** is similar to that of NAD^+, but with an added phosphate group. $NADP^+$ supplies the hydrogen that reduces carbon dioxide to carbohydrate during photosynthesis. This process "fixes" atmospheric carbon into organic molecules that then serve as nutrients. The energy that drives this process comes from ATP.

Flavin Adenine Dinucleotide (FAD)

Flavin adenine dinucleotide (FAD) is derived from riboflavin (vitamin B_2). FAD, like NAD, carries two electrons. However, it also accepts both protons to become $FADH_2$. In cells, FAD helps break glucose down to carbon dioxide and water.

Cytochromes

The **cytochromes** are iron-containing molecules that transfer electrons in metabolic pathways. When oxidized, the iron in cytochromes is in the Fe^{3+} form. When the iron accepts an electron, it is reduced to Fe^{2+}. There are several types of cytochromes, all of which carry electrons in cells. In metabolic pathways, several cytochromes align to form **electron transport chains,** with each molecule accepting an electron from the one before it and passing it to the next. Figure 6.12 shows such a chain; we will see more of them in the next two chapters. Small amounts of energy are released at each step of an electron transport chain, and the cell uses this energy in other reactions. Cytochromes take part in all energy transformations in life, suggesting that the cellular strategy for energy transformation is quite ancient.

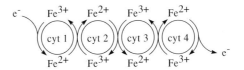

FIGURE 6.12

Different cytochromes align, forming electron transport chains. Energy is released in small increments as electrons are passed down the chain in a series of oxidation-reduction reactions. Electrons are passed between iron molecules attached to the cytochromes. Iron in ferric form (Fe^{3+}) gains an electron, becoming reduced to ferrous (Fe^{2+}) form. The cytochromes are embedded in membranes.

KEY CONCEPTS

Several other compounds besides ATP help transform energy within cells. Cofactors, which are usually ions, transfer chemical groups from molecule to molecule. Vitamin-derived coenzymes, including NAD^+, $NADP^+$, and FAD, transfer protons or electrons. NAD^+ is reduced to become rich in potential energy, which it then transfers to synthesize ATP and to reduce other compounds. $NADP^+$ supplies hydrogen to fix atmospheric carbon into useful organic molecules, and FAD^+ is an electron carrier important in cellular respiration. Iron-containing molecules called cytochromes transfer electrons in metabolism.

Enzymes and Energy

Many chemical reactions require an initial boost of energy. This first burst of energy, called the **energy of activation,** often comes from heat in the environment. A piece of wood must be heated to provide the energy of activation needed to ignite it. In cells, the amount of heat needed to activate most metabolic reactions would swiftly be lethal. Instead, cells use enzymes to lower the required energy of activation.

Recall from chapter 3 that an enzyme is a protein than speeds certain chemical reactions, without taking part in them, by decreasing the energy of activation. An enzyme binds the substrate in a way that enables the reaction to proceed. Without enzymes, many biochemical reactions would not occur fast enough to support life.

Higher temperatures increase enzyme activity—to a point. The activity of an enzyme doubles for every increase of $10°C$, but beyond about $60°C$, entropy grows too great, and the enzyme is denatured (falls apart). At this point, the reaction halts. Although the enzymes of a few organisms—such as the hot springs bacteria discussed in chapter 1—can tolerate high temperatures, most enzymes work best at much lower temperatures. Most of the enzymes in the human body function optimally near body temperature.

Control of Metabolism

A cell is in a constant state of flux, as it rapidly dismantles some molecules and synthesizes others. Disturbances in the balance between these two actions threaten the life of the cell, as we learned from the lysosomal storage disorders described in Biology in Action 4.1. Several biological mechanisms regulate cellular metabolism and preserve this delicate balance.

Certain enzymes control metabolism by functioning as pacesetters at important junctures in biochemical pathways. The enzyme whose reaction precedes the slowest reaction sets the pace for the pathway's productivity, just as the slowest runner on a relay team limits the overall pace for the whole team. This is because each subsequent reaction in the metabolic pathway (like each subsequent relay runner) requires the product of the preceding reaction to continue. The reaction catalyzed by this enzyme is called the rate-limiting step; the enzyme is a **regulatory enzyme** because it regulates the pathway's pace and productivity.

Inhibition of a Metabolic Pathway

Often a regulatory enzyme is highly sensitive to a chemical cue. When this cue is the end product of the pathway the enzyme takes part in, the enzyme may "turn off" for a time, and the regulation is called **negative feedback,** or feedback inhibition. Negative feedback prevents too much of one substance from accumulating—an excess of a particular biochemical effectively shuts down its own synthesis until its levels fall. At that point, the pathway resumes its activity. Negative feedback, then, is somewhat like a thermostat—when the temperature in your home reaches a certain level, the thermostat shuts the heat off for a while. When the temperature falls, it cues the thermostat to ignite the furnace again.

Exactly how does a product inhibit its own synthesis? A product may bind to the regulatory enzyme's active site, preventing it from binding substrate and temporarily shutting down the pathway. This is called **competitive inhibition** because a substance other than the substrate competes to occupy the active site. In **allosteric inhibition,** product molecules bind to the regulatory enzyme, but not to its active site. This alters the conformation of the enzyme so that it can no longer bind substrate. Allosteric inhibition is also called **noncompetitive**

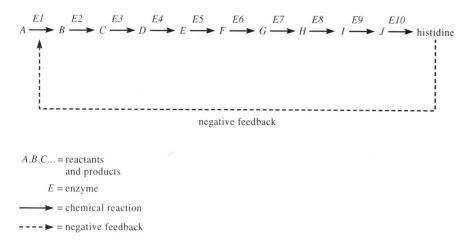

A,B,C... = reactants
         and products

E = enzyme

———➤ = chemical reaction

----➤ = negative feedback

FIGURE 6.13

Negative feedback (feedback inhibition). Like other organisms, *E. coli* bacteria require the amino acid histidine to synthesize proteins. This diagram isolates the reactions leading to histidine production. In the 10-reaction biochemical pathway for histidine synthesis in *E. coli*, each step involves a chemical reaction catalyzed by a different enzyme. When sufficient histidine accumulates, the histidine allosterically inhibits the activity of E1, the first enzyme in the pathway. Further production of histidine is temporarily limited.

KEY CONCEPTS

Enzymes catalyze biochemical reactions by lowering the energy of activation. Enzymes ensure that spontaneous reactions proceed fast enough to support life; regulatory enzymes control the pace and production of a biochemical pathway. In feedback inhibition, or negative feedback, the accumulation of a pathway's product temporarily shuts the pathway down by disrupting the function of the regulatory enzyme. The inhibitor can bind to the active site (competitive inhibition) or to another region of the enzyme, changing the enzyme's conformation so that the substrate cannot bind to it (noncompetitive inhibition). Certain drugs and poisons can inhibit metabolic pathways. Positive feedback is much more rare than negative feedback. In positive feedback, a pathway's product activates the pathway.

inhibition because the inhibitor does not directly compete to occupy the active site. Both competitive and noncompetitive inhibition are forms of negative feedback.

In allosteric inhibition, the product of a pathway usually binds to the junction between two subunits in a multi-subunit protein. This temporarily destabilizes the enzyme and impairs catalysis. One example of allosteric inhibition is the 10-step pathway to synthesize the amino acid histidine (fig. 6.13).

Substances from outside the body, such as drugs and poisons, can also inhibit enzyme function. A molecule of a foreign substance similar in conformation to the substrate may fit into the enzyme's active site much like a dummy key fits into a lock, preventing the substrate from binding. The antibiotic drug sulfanilamide, for example, competitively inhibits certain enzymes in bacteria. Sometimes, increasing the concentration of the substrate can overcome such a competitive inhibitor.

Drugs that bind noncompetitively, to a part of the enzyme other than the active site, also inhibit enzyme function. Nerve gas, for example, inhibits the activity of the enzyme acetylcholinesterase, which normally recycles the neurotransmitter acetylcholine back into nerve cells after it sends its message. When the nerve gas binds to acetylcholinesterase, it is unable to bind its substrate, and nerve transmission ceases. Penicillin acts in a similar way, binding irreversibly to an enzyme that bacteria need to build cell walls. When the cell walls fall apart, the bacteria die, and the infection is squelched.

Activation of a Metabolic Pathway

Much rarer than negative feedback in living systems is **positive feedback,** in which a sequence of events is activated by the presence of its product. Blood clotting, for example, begins when a biochemical pathway synthesizes fibrin, a threadlike protein. The products of the later reactions in the clotting pathway stimulate the enzymes that activate earlier reactions. As a result, fibrin accumulates faster and faster, until there is enough to stem the blood flow. When the clot forms and the blood flow stops, the clotting pathway shuts down— an example of negative feedback.

The Major Energy Transformations— Respiration and Photosynthesis

The energy transformations that sustain life are similar in all organisms. The most important of these pathways are respiration and photosynthesis, the topics of the next two chapters. As we shall see, these pathways are intimately related.

The *energy-requiring* stage of biological energy acquisition and utilization is photosynthesis. During this process, chloroplasts in leaves absorb light energy. They use this energy to release oxygen and reduce carbon dioxide (a low-energy compound) to carbohydrate (a high-energy compound). Carbohydrate, in turn, fuels the activities of the plant and ultimately all other organisms. During cellular respiration, the *energy-releasing* stage of the biological energy process, energy-rich carbohydrate molecules are oxidized to carbon dioxide and water. The energy released by respiration provides the energy of life.

SUMMARY

Energy is the ability to do work. Potential energy is stored energy potentially available to do work, while kinetic energy is energy actually being used to do work. All of life's activities are fueled by energy transforming from one form to another. These energy transformations are governed by the laws of thermodynamics. The first law, concerning the conservation of energy, states that energy cannot be created or destroyed, but only converted to other forms. The second law states that all energy transformations are inefficient, because every reaction results in increased entropy and the loss of usable energy as heat. Entropy measures the disorder of a system.

Metabolism is the sum of the energy and matter conversions in a cell. It consists of enzyme-catalyzed reactions organized into often interconnected pathways and cycles. Anabolic reactions synthesize or build molecules, diverge from a few reactants to many products, and require energy. Catabolic reactions break down molecules, converge from many reactants to a few products, and release energy. Each reaction of a metabolic pathway rearranges atoms into new compounds. These rearrangements, or chemical reactions, change the amount of free energy available to do work. The change in free energy (ΔG) is the amount of energy potentially available to form other bonds.

Many energy transformations in organisms occur via oxidation and reduction reactions. Oxidation is the loss of electrons from a molecule; reduction is the addition of electrons. Oxidation and reduction reactions occur simultaneously, in pairs.

ATP, the energy currency of cells, readily releases more energy than cells require, allowing for entropy and heat loss. Many energy transformations involve coupled reactions; in such reactions, the cell uses the energy released by ATP to drive another reaction. Other compounds that take part in cellular metabolism include coenzymes and cytochromes.

Enzymes speed spontaneous biochemical reactions to a biological useful rate by lowering the energy of activation. Negative feedback, through either competitive or noncompetitive (allosteric) inhibition, disables and thus regulates enzyme function. Positive feedback also regulates enzyme function, but it activates rather than inhibits the enzyme. Positive feedback is rare in living systems.

Photosynthesis and respiration are the major energy transformation pathways in organisms. In photosynthesis, plants absorb light energy to reduce carbon dioxide and water to carbohydrate, while in respiration, cells convert carbohydrates to carbon dioxide and water, releasing energy to form ATP. Together, photosynthesis and respiration comprise a system that allows individual organisms, and ultimately, ecosystems, to acquire and use energy to maintain life.

KEY TERMS

adenosine triphosphate (ATP) 118
allosteric inhibition 121
anabolism 116
autotroph 113
bioenergetics 112
bioluminescence 119
bond energy 116
catabolism 116
cellular respiration 115
chemical equilibrium 117
chemoautotroph 113

coenzyme 120
cofactor 120
competitive inhibition 121
coupled reactions 118
cytochromes 121
electron transport chains 121
endergonic 117
energy 112
energy of activation 121
entropy 115
exergonic 116

flavin adenine dinucleotide (FAD) 120
free energy 116
geothermal energy 112
kinetic energy 113
metabolic pathway 115
negative feedback 121
nicotinamide adenine dinucleotide (NAD$^+$) 120
nicotinamide adenine dinucleotide phosphate (NADP$^+$) 120

noncompetitive inhibition 121
oxidation 117
phosphorylation 120
photon 113
photosynthesis 113
positive feedback 122
potential energy 113
reduction 117
regulatory enzyme 121
thermodynamics 114

REVIEW QUESTIONS

1. Give one example of potential energy and one of kinetic energy.

2. Cite everyday illustrations of the first and second laws of thermodynamics. How do these principles underlie every organism's ability to function?

3. Explain what a negative change in free energy (ΔG) means. What does a negative ΔG tell you about the energy of reactants as compared to products of a chemical reaction?

4. Give an example of entropy.

5. Why isn't heat usable energy?

6. When ΔG is zero, what is happening in the chemical reaction?

7. Cite three reasons why ATP is an excellent source of biological energy.

8. Look at figure 6.11. Which reaction is exergonic, and which is endergonic?

9. List three differences between anabolism and catabolism.

10. What do ATP, NAD^+, $NADP^+$, and FAD have in common chemically?

11. What evidence supports the idea that nucleotides do not function only as parts of nucleic acids?

12. How does an enzyme speed a chemical reaction?

13. How are competitive enzyme inhibition and allosteric (noncompetitive) inhibition alike? How do they differ?

TO THINK ABOUT

1. Some people argue that life's high degree of organization defies the laws of thermodynamics. How is the organization of life actually consistent with the principles of thermodynamics?

2. How might you explain the fact that even species as diverse as humans and yeast utilize the same biochemical pathways to extract energy from nutrient molecules?

3. Cytochrome c is an electron carrier that is nearly identical in all species. No known disorders affecting cytochromes exist. What does this suggest about the importance of this molecule?

4. Figure 5.15 depicts the cellular fate of dietary cholesterol. A series of steps shuts off cellular synthesis of cholesterol. Is this an example of negative or positive feedback? Why?

5. In 1991, a large volcanic eruption in the Philippines threw great dust clouds into the atmosphere, lowering temperatures around the world for many months. How might this event have affected energy transformations in living things?

SUGGESTED READINGS

Baum, Rudy M. February 22, 1993. Views on biological, long-range electron transfer stir debate. *Chemical and Engineering News*, vol. 73. Chemists are isolating parts of the chain of electron carriers in cellular respiration.

Becker, Wayne M., and David W. Deamer. 1991. *The World of the Cell*. Redwood City, Calif.: Benjamin/Cummings. Part 3 is an in-depth look at the cell's energy transformation and utilization.

Herring, Peter J. May 13, 1993. Light genes will out. *Nature*, vol. 363. The bioluminescent bacteria that live in anglerfish and flashlight fish are very unusual.

Peterson, Ivars. August 31, 1991. Step in Time. *Science News*. How do fireflies flash in synchrony?

Storey, Richard D. March 1992. Textbook errors and misconceptions in biology: Cell energetics. *The American Biology Teacher*. Understanding the laws of thermodynamics helps us understand the energy pathways in a cell.

CHAPTER 7

Cellular Respiration

A panda bear must eat bamboo nearly constantly to extract enough energy to function.

LEARNING OBJECTIVES

By the chapter's end, you should be able to answer these questions:

1. Which biochemical reactions and pathways do cells use to extract energy from the bonds of nutrient molecules?

2. Which molecules carry transferred energy?

3. How does electron transfer in the respiratory chain set up the conditions that lead to ATP synthesis?

4. What role does oxygen play in extracting energy from nutrients?

5. How efficient is aerobic respiration?

6. What energy reactions occur in the absence of oxygen?

7. How can energy pathway disruption affect health?

Eating for Energy

The chipmunk in figure 7.1 is ingesting a bonanza of nutrients in its nutty meal. As the nut passes through the rodent's digestive system, it gradually breaks apart into clumps of cells. These clumps of cells then degrade into nutrient molecules of proteins, carbohydrates, and lipids. After the chipmunk has digested these macromolecules into their component amino acids, monosaccharides, and fatty acids and glycerol, they are small enough to enter the chipmunk's blood and lymphatic systems. When these smaller nutrient molecules enter the animal's cells, they break down further, and some may be converted to glucose. When glucose and other nutrients break down, energy is released in the form of ATP, which the chipmunk's cells then capture and use. This chapter describes exactly how cells extract energy from nutrients.

The Three Stages of Cellular Respiration

All organisms harvest energy from stored chemicals in much the same way. Many use the metabolic pathways of cellular respiration. This process oxidizes glucose into carbon dioxide (CO_2), water (H_2O), and energy.

If glucose were burned to release energy outside a living thing, it would create a small fire lasting several minutes. In a cell, respiration controls this process, so that the heat does not destroy living tissue. Instead, part of the energy

FIGURE 7.1

Eating and digestion are the first steps in procuring biological energy. Nutrient molecules must be absorbed into the cells lining the intestines, then transferred to the bloodstream, where they are transported to cells. Once in the cells, nutrients are broken down in biochemical reactions, pathways, and cycles that release the energy in their chemical bonds.

from glucose becomes trapped in ATP. The rest of the released energy—the heat lost during respiration—never causes an organism to combust because its release is spread over many biochemical reactions. The carbon liberated from stored chemicals is recycled into other organic molecules.

There are several kinds of cellular respiration. The most efficient, requiring oxygen and producing ATP, is called aerobic respiration. Other types of respiration make little or no ATP, but lose heat energy or transfer energy into organic molecules other than ATP.

Cellular respiration harvests energy in three stages. The first is **glycolysis,** in which glucose is split into smaller molecules in the cytoplasm. The second stage, the **Krebs cycle,** completely metabolizes the products of glycolysis and captures some of the energy. In the third stage, several electron carriers form an **electron transport chain** to harvest energy from the movement of electrons (fig. 7.2).

Into the Mitochondrion

Two of the stages of aerobic respiration, the Krebs cycle and the electron transport chain, occur in the mitochondria in eukaryotes. These organelles are particularly plentiful in cells that require a lot of energy, such as muscle cells. A mitochondrion consists of an outer membrane and a highly folded inner membrane, an arrangement that creates two compartments (fig. 7.3). The innermost compartment, called the **matrix,** is where the first reactions of aerobic respiration occur. A second area between the two membranes is called the **intermembrane compartment.**

The inner mitochondrial membrane is folded into numerous projections called cristae, which are studded with enzymes and electron carrier molecules. This extensive folding greatly increases the surface area on which reactions can occur. In aerobic prokaryotes, which lack mitochondria, respiratory enzymes are embedded in the cell membrane.

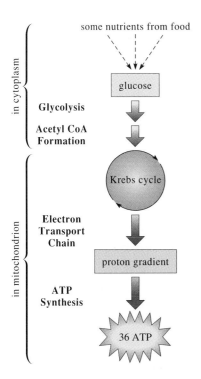

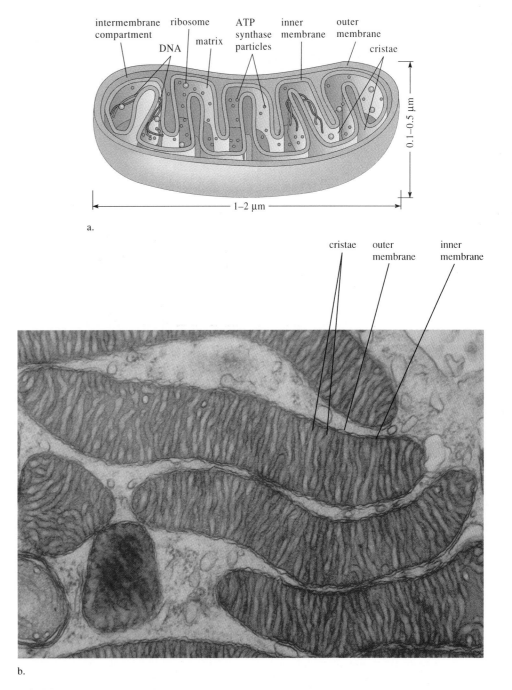

FIGURE 7.2

An overview of cellular respiration. This highly schematic overview shows the stages of energy procurement that constitute cellular respiration. As the chapter progresses, the illustrations of biochemical pathways will become more detailed. Look to the insets that repeat this diagram with different sections highlighted to orient yourself to the part of the overall pathway under discussion.

Energy Transfer from Glucose to ATP

All three stages of respiration produce ATP. The overall result of these pathways is to use energy stored in organic molecules to phosphorylate ADP to ATP. Cellular respiration is quite efficient, with one glucose molecule theoretically yielding 36 ATPs in eukaryotes and 38 in prokaryotes. Just as we use ordinary currency, the cell can use its energy currency—ATP—in several ways.

KEY CONCEPTS

The cell retrieves the energy stored in glucose through cellular respiration. This occurs in three stages: glycolysis, the Krebs cycle, and the electron transport chain. Part of the energy released converts into ATP, which the cell can then use to do work.

FIGURE 7.3

Two depictions of the anatomy of a mitochondrion. The mitochondrion's interior is a highly folded membrane studded with enzymes and electron carriers important in cellular respiration. An outer membrane surrounds the organelle. Energy from an electron transport chain sets up a proton gradient between the inner membrane and the space between the two membranes, called the intermembrane compartment or space. Protons moving down their concentration gradient through channels in an enzyme (ATP synthase) cause ADP to be phosphorylated to ATP.

Retrieving Glucose from Other Molecules

In most organisms, the reactions of cellular respiration begin with glucose. But before this, other carbohydrates such as sucrose (table sugar) or fructose (fruit sugar) must break down or convert to glucose. Other organisms, including plants, also carry out cellular respiration. In leaves, respiration begins with sucrose, but in other parts of the same plant, such as roots and stems, it may begin with starch. Later in the chapter, we will see how the other major nutrients—proteins and lipids—enter the energy pathways. But for now, we will focus on glucose.

Routes to ATP Formation

The general equation for glucose respiration is:

glucose + oxygen →
$C_6H_{12}O_6$ + O_2

carbon dioxide + water + energy
$6CO_2$ + $6H_2O$

Some of the energy product ends up in ATP; the rest is in CO_2 or H_2O, or is lost as heat.

Substrate-Level Phosphorylation

ATP synthesis occurs in two ways. In the first, **substrate-level phosphorylation,** a phosphate (PO_4) transfers from organic compounds (substrates) to ADP, forming ATP. (*Phosphorylation* is the process of taking up a phosphorus-containing group, which energizes the molecule receiving it.) An enzyme that binds both the substrate and ADP catalyzes the transfer (fig. 7.4). The energy of substrate-level phosphorylation comes from the phosphate bond of the substrate. Some energy escapes as heat when the phosphate group transfers.

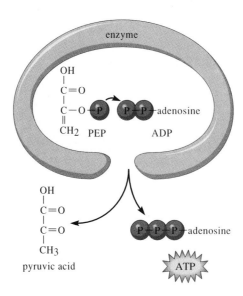

FIGURE 7.4

Substrate-level phosphorylation. A phosphate group on an organic substrate is transferred to ADP, phosphorylating it to ATP. The reaction is enzyme-catalyzed. This particular reaction is the final step in glycolysis. A molecule of phosphoenolpyruvate (PEP) loses a phosphate to ADP, forming pyruvic acid and ATP.

Oxidative Phosphorylation

The second way respiration generates ATP is through **oxidative phosphorylation,** so named because it depends on oxidation-reduction reactions. Electrons from NADH pass along a chain of electron carriers through a series of oxidation-reduction reactions. The energy from this electron movement sets up differing proton concentrations on each side of the inner mitochondrial membrane, establishing a gradient. When protons move down their concentration gradient (traveling from where they are highly concentrated to where they are not), they physically contact an enzyme at one site, triggering phosphorylation of ADP to ATP at a different site on the same enzyme. This process is called **chemiosmosis** because it entails chemical reactions as well as transport across membranes (osmosis). We will discuss chemiosmosis in more detail later in the chapter.

Substrate-level phosphorylation is the simpler and more direct mechanism for producing ATP, but it accounts for only a small percentage of ATP synthesis—the ATP produced from glycolysis and the Krebs cycle. In the electron transport chain that follows these reactions, ATP forms through oxidative phosphorylation and chemiosmosis. This represents the bulk of the ATP generated in cellular respiration.

KEY CONCEPTS

The reactions of cellular respiration break down glucose in the presence of oxygen to yield CO_2, H_2O, and energy. Some of this energy ends up in ATP. ATP synthesis occurs in two ways: ATP is synthesized directly by substrate-level phosphorylation or indirectly by oxidative phosphorylation, a form of chemiosmosis.

Glycolysis— Glucose Breaks Down to Pyruvic Acid

Glucose contains considerable bond energy, but cells recover only a small portion of it during glycolysis. Furthermore, cells gather this energy in small increments, so that excess heat does not damage the cell. During glycolysis, glucose is split into two three-carbon compounds. The entire process requires ten steps, all of which occur in the cytoplasm (fig. 7.5). The first half of the pathway activates glucose so that energy can be extracted from its bonds. The second half of the pathway actually extracts the energy.

The First Half of Glycolysis— Glucose Activation

The first step of glycolysis uses one molecule of ATP to phosphorylate glucose. Phosphorylation activates glucose so that the appropriate enzyme can carry out the next step. Since phosphate is negatively charged, and since charged compounds

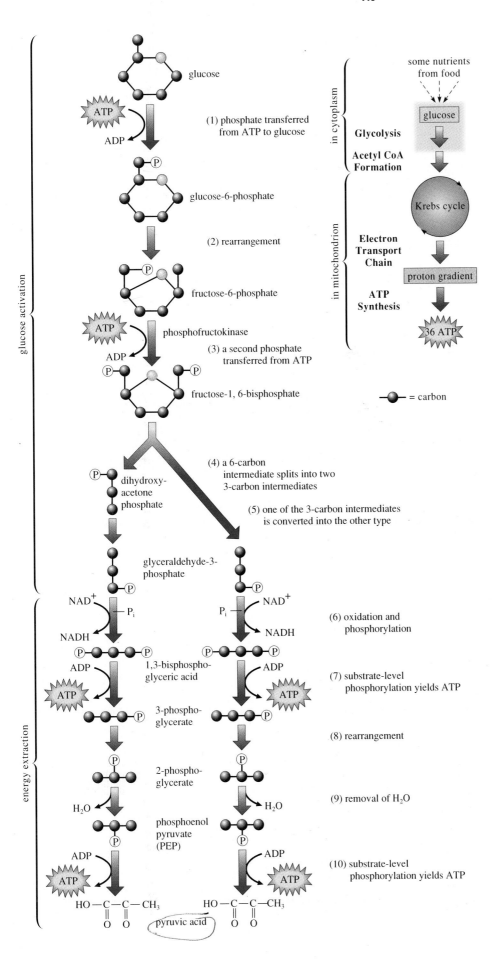

(◖◗) **FIGURE 7.5**

In the glycolysis reactions, glucose is rearranged and split into two three-carbon intermediates, each of which is rearranged further to eventually yield two molecules of pyruvic acid. Along the way, four ATPs and two NADHs are produced. Two ATPs are consumed in activating glucose, so the net ATP yield is two ATP molecules per molecule of glucose.

cannot easily cross the cell membrane, phosphorylation also traps glucose in the cell. In step 2, the atoms of phosphorylated glucose are rearranged; in step 3 the new molecule (fructose-6-phosphate) is phosphorylated again by a second ATP, forming fructose-1, 6-bisphosphate. This compound is split in steps 4 and 5, and each of the three-carbon products has one phosphate. This ensures that none of the carbons from the original glucose can escape from the cell. One of the products, glyceraldehyde-3-phosphate, is further metabolized in glycolysis. The other product, dihydroxyacetone phosphate, is converted to glyceraldehyde-3-phosphate and then is catabolized along with the other glyceraldehyde-3-phosphate. These products represent the halfway point of glycolysis (steps 1 through 5 in fig. 7.5). So far, energy in the form of ATP has been invested, but no ATP has been produced.

The Second Half of Glycolysis—Energy Extraction

The first energy-obtaining step of the glycolysis pathway occurs in the second half when NAD^+ is reduced to NADH through the coupled oxidation of glyceraldehyde-3-phosphate (step 6). After this step, some of the energy from glucose is stored in the energy-rich electrons of NADH. This oxidation also releases enough energy to add a second phosphate group to glyceraldehyde-3-phosphate, making it 1,3-bisphosphoglyceric acid. Finally, the cell is ready to make ATP. Substrate-level phosphorylation occurs when one of the phosphates of 1,3-bisphosphoglyceric acid transfers to

ADP (step 7). The three-carbon molecule that remains is then rearranged to form PEP or phosphoenolpyruvic acid (step 8), which becomes **pyruvic acid** when it loses water (step 9) and donates its phosphate to a second ADP (step 10). Each glyceraldehyde-3-phosphate from the first half of glycolysis progresses through this pathway to make two ATPs and one pyruvic acid in the second half.

Because one molecule of glucose yields two molecules of glyceraldehyde-3-phosphate, and each molecule of glyceraldehyde-3-phosphate yields two molecules of ATP and one of pyruvic acid, each glucose that undergoes glycolysis yields four ATPs and two pyruvic acids. However, because the first half of glycolysis requires two ATPs, the net gain of glycolysis is two ATPs per molecule of glucose.

At the end of glycolysis, a small amount of the chemical energy that started out in glucose ends up in ATP and NADH. However, most of the energy of glucose remains in pyruvic acid. The energy in the pyruvic acid bonds helps synthesize ATP in the mitochondrion during the second stage of cellular respiration—the Krebs cycle.

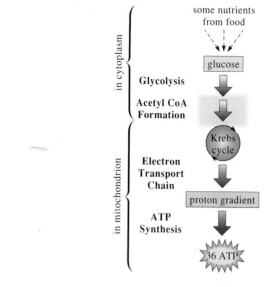

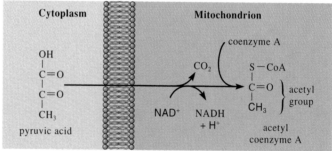

FIGURE 7.6

Acetyl CoA formation is a bridge between glycolysis and the Krebs cycle. After pyruvic acid enters the mitochondrion, crossing both membranes, it loses CO_2, reduces NAD^+ to NADH, and combines with coenzyme A to yield acetyl CoA.

KEY CONCEPTS

During the first five steps of glycolysis, ATP phosphorylates glucose twice, and the glucose then splits into two three-carbon compounds. The three-carbon compounds are oxidized to yield pyruvic acid during the second five steps of glycolysis. This oxidation process yields ATP and NADH.

Acetyl CoA Formation Bridges Glycolysis and the Krebs Cycle

The pyruvic acid transported into the mitochondrial matrix is not directly used in the Krebs cycle. First it loses a molecule of carbon dioxide as NAD^+ is reduced to NADH. The remaining molecule, called an acetyl group, attaches to

a coenzyme to form **acetyl coenzyme A,** abbreviated **acetyl CoA** (fig. 7.6).

The conversion of pyruvic acid to acetyl CoA links glycolysis and the Krebs cycle. Pyruvic acid is the final product of glycolysis, and acetyl CoA is the compound that enters the Krebs cycle.

This step is critically important in the cell. It is irreversible in the sense that once carbons from pyruvic acid are used to form CO_2 and acetyl CoA, the carbons may not be used to regenerate glucose.

The Krebs Cycle

The Krebs cycle is a cycle because the last step regenerates the reactants of the first step (fig. 7.7). Seven of the eight steps occur in the mitochondrial membrane. In addition to continuing the catabolism of glucose, the Krebs cycle

forms intermediate compounds. The cell then uses the carbon skeletons of these compounds to manufacture other organic molecules, such as amino acids.

In the first step of the Krebs cycle, coenzyme A is cleaved from acetyl CoA, and the acetyl group attaches to four-carbon oxaloacetic acid. The resulting six-carbon compound is citric acid. (The Krebs cycle is also called the citric acid cycle.) In the next step, citric acid rearranges to isocitric acid, which becomes the substrate for two oxidation steps that together remove two molecules of carbon dioxide. These steps also reduce two molecules of NAD^+ to NADH.

In the first oxidation, carbon dioxide removal yields alphaketoglutaric acid. In the second oxidation, carbon dioxide is removed from the alphaketoglutaric acid, producing succinic

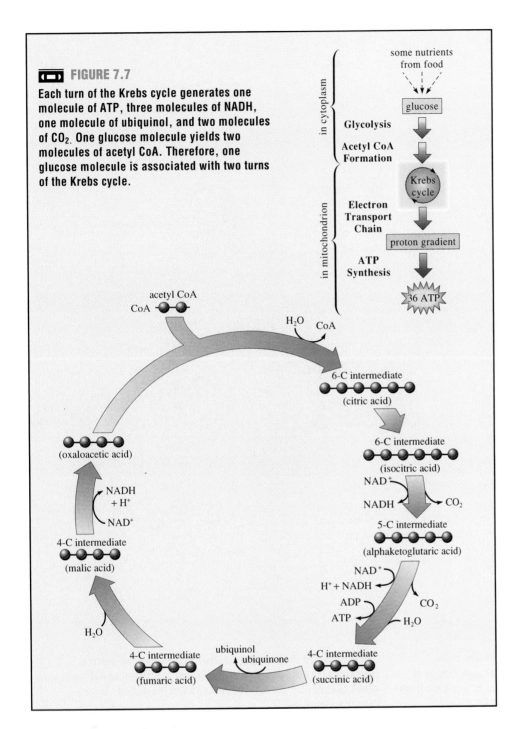

FIGURE 7.7

Each turn of the Krebs cycle generates one molecule of ATP, three molecules of NADH, one molecule of ubiquinol, and two molecules of CO_2. One glucose molecule yields two molecules of acetyl CoA. Therefore, one glucose molecule is associated with two turns of the Krebs cycle.

To summarize, each molecule of acetyl CoA entering the Krebs cycle produces only one ATP through substrate-level phosphorylation. Most of the energy derived from the oxidative steps of the Krebs cycle is stored in the high-energy electrons of NADH and ubiquinol. The cell harvests the energy in these molecules in the third stage of respiration—oxidative phosphorylation.

KEY CONCEPTS

Before the pyruvic acid produced by glycolysis enters the Krebs cycle, it moves into mitochondria and converts into acetyl CoA. Acetyl CoA then enters the Krebs cycle. Each turn of the cycle uses one molecule of acetyl CoA to produce one ATP, one ubiquinol, and three NADHs. The acetyl carbons are released as carbon dioxide.

Oxidative Phosphorylation and Electron Transport

Most of the ATP generated in cellular respiration comes from oxidative phosphorylation. However, the cell does not directly use the high-energy electrons of NADH and ubiquinol, derived from the Krebs cycle, to synthesize ATP. Rather, these electrons initiate a series of oxidation-reduction reactions that move electrons through several carrier molecules, forming an electron transport chain also called the **respiratory chain** (fig. 7.8).

The Respiratory Chain

The respiratory chain is like a series of tiny, successively stronger magnets; each carrier has a greater ability to accept and donate electrons than the previous molecule. Each carrier thus pulls electrons from a weaker neighbor, and gives them up to a stronger one. The strongest and final carrier in the chain is oxygen. Once reduced, oxygen combines with protons in the mitochondrial matrix to yield water. Without oxygen, the electron flow would halt, and ATP would not be generated.

acid. This reaction provides energy for the substrate-level phosphorylation of ADP to ATP and reduces a molecule of NAD^+ to NADH.

Three more oxidations occur after the formation of succinic acid: succinic acid is oxidized to fumaric acid, fumaric acid is oxidized to malic acid, and malic acid is oxidized to oxaloacetic acid. The oxidation of succinic acid also reduces the electron carrier

molecules ubiquinone to ubiquinol. (This is where the reduction of FAD to $FADH_2$, mentioned in the last chapter, occurs. FAD and $FADH_2$ are part of the enzyme that catalyzes the reduction of ubiquinone to ubiquinol. FAD and $FADH_2$ are considered intermediates of the reaction, and ubiquinol is the product.) Finally, oxidation of malic acid to oxaloacetic acid reduces a third NAD^+ to NADH.

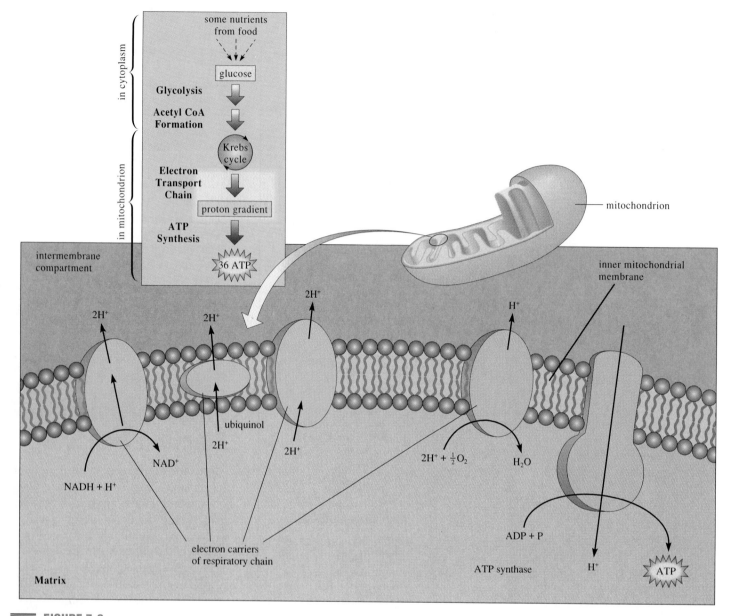

The respiratory chain and chemiosmosis. The respiratory chain consists of several electron carrier molecules that accept protons on the matrix side of the inner mitochondrial membrane and release them in the intermembrane compartment. This establishes a proton gradient, which causes some protons to leak back across the membrane. In doing so, they pass through a channel in ATP synthase, which causes ADP bound to another part of the enzyme to be phosphorylated to ATP.

The respiratory chain is so important that blocking it with a poison is deadly (Biology in Action 7.1). Inherited inborn errors of metabolism can affect the respiratory chain. A group of disorders in humans collectively called Leber hereditary optic neuropathy, for example, affects subunits of the first or third electron carriers in the respiratory chain. The major symptom of this disease is sudden loss of vision at about age 20 because of damage to the optic nerve. No one knows why the eye is affected.

Energy from the electrons passing down the respiratory chain fuels the formation of the proton gradient (the differing proton concentrations on each side of a membrane), which in turn drives ATP synthesis. In the final step, oxygen is reduced. Why don't NADH and ubiquinol simply donate electrons directly to oxygen instead of passing them down the chain? Such a single-step reaction would probably release a damaging amount of heat. It would also eliminate the proton pumping needed to drive ATP synthesis.

Chemiosmosis and ATP Synthesis

How do the protons pumped by energy from the respiratory chain power ATP synthesis? Many scientists hypothesized that the respiratory chain forms high-

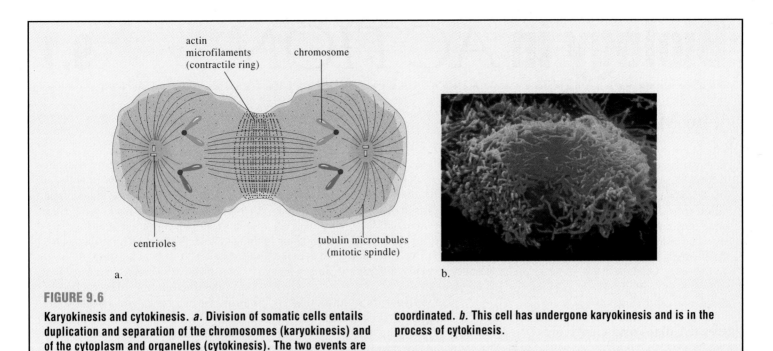

FIGURE 9.6

Karyokinesis and cytokinesis. *a.* Division of somatic cells entails duplication and separation of the chromosomes (karyokinesis) and of the cytoplasm and organelles (cytokinesis). The two events are coordinated. *b.* This cell has undergone karyokinesis and is in the process of cytokinesis.

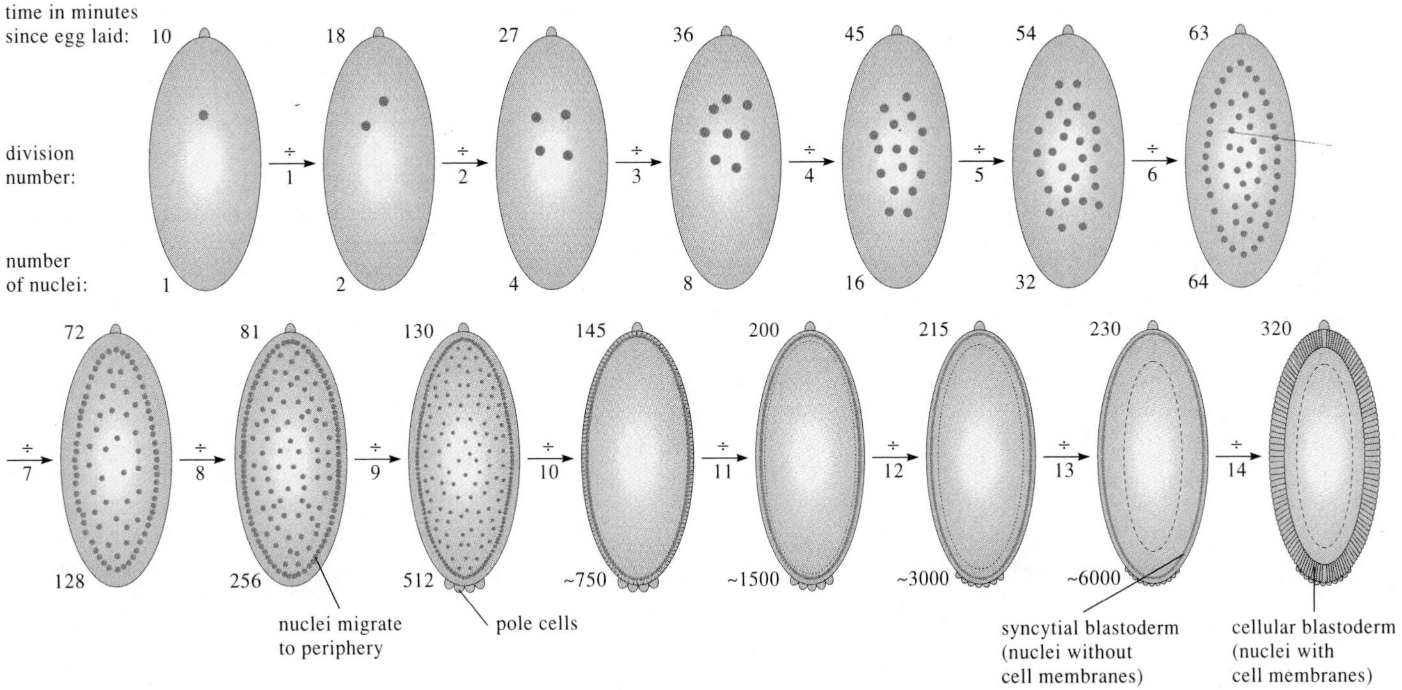

time in minutes since egg laid: 10   18   27   36   45   54   63

division number: ÷1 ÷2 ÷3 ÷4 ÷5 ÷6

number of nuclei: 1   2   4   8   16   32   64

72   81   130   145   200   215   230   320

÷7 ÷8 ÷9 ÷10 ÷11 ÷12 ÷13 ÷14

128   256   512   ~750   ~1500   ~3000   ~6000

nuclei migrate to periphery

pole cells

syncytial blastoderm (nuclei without cell membranes)

cellular blastoderm (nuclei with cell membranes)

FIGURE 9.7

An egg of the fruit fly *Drosophila melanogaster* is a barely visible, tiny white oval structure. The size of the egg does not change as many rapid mitotic cell divisions occur in the cells within it. Four hours (240 minutes) after a fertilized egg is laid, after the thirteenth division, it consists of a one-cell-thick layer of nuclei that lack cell membranes, although cytoskeletal elements form basketlike structures around them. At the fourteenth division, cell membranes form. The embryo is called a syncytial blastoderm just before the fourteenth division, and a cellular blastoderm after.

The Mitotic Spindle

Scientists were able to view the mitotic spindle long before they understood its role in separating replicated chromosomes. Illustrations in textbooks from the 1890s show microtubules emanating from two sets of chromosomes. These depictions do not differ greatly from views of dividing cells in today's texts. Our knowledge of the mitotic spindle began with the discovery of centrosomes and centrioles.

Centrosomes and Centrioles

The 1880s was a golden age for the study of cells, as dyes that revealed many cellular constituents became available. In 1887, German cytologist Theodor H. Boveri saw an area of cytoplasm near the cell nucleus—a structure that appeared to orchestrate mitosis. He named it a **centrosome.** Belgian cytologist Edouard Joseph Louis-Marie van Beneden independently made the same observation. Both men saw centrosomes in the eggs of the roundworm *Ascaris.* They noted that a nondividing cell had one centrosome, but a dividing cell had two centrosomes that moved away from each other as the division proceeded.

During the early part of the 20th century, researchers saw centrosomes in a variety of species and cell types. The invention of the electron microscope permitted a closer look at the intriguing structures. Biologists discovered that in animal cells, centrosomes contained microtubules arranged in a pattern called a **centriole**—nine rods, each consisting of three microtubules fused along their length. (Plant cells do not have centrioles, but may have different organizations of microtubules.)

The dynamic nature of microtubule growth enables the mitotic spindle to assemble rapidly during the cell cycle. Microtubules extend when one end adds alpha and beta tubulin subunits faster than tubulin disassembles at the other end (see fig. 5.17). Tubulin growth may be coordinated by a third type of tubulin, called gamma tubulin, that was discovered in 1989.

The Role of the Mitotic Spindle in Mitosis

During the first half of mitosis, microtubules extend, and when they bump into chromosomes, attach to them. In the interphase preceding mitosis, the centrosome lies near the nucleus. As prophase begins, arrays of microtubules throughout the cell fall apart. At the same time, in animal cells, the centrosome replicates by forming a second centriole at a 90-degree angle to the first. The two centrosomes then migrate from the nucleus towards opposite ends, or poles, of the cell (fig. 1). Microtubules extend from the centrosomes, then back away, then extend again, like fingers groping for something to hold. When a microtubule touches a centromere, it stops elongating.

Once each chromosome is attached to a microtubule extending from each centromere, the mitotic spindle is in place and metaphase begins. During metaphase, the chromosomes align down the cell's equator; in anaphase, the centromeres part, and motor molecules (discussed in chapter 5) power the separation of the two sets of chromosomes. By the time the mitotic spindle disassembles during telophase, one cell is well on the way to becoming two.

FIGURE 1

The mitotic spindle. The apparatus that pulls apart replicated chromosomes is intricate. The spindle consists of highly organized microtubules that form the fibers that grow outwards from two structures called centrioles. The centrioles, also built of microtubules, occupy opposite ends of the cell. Short microtubules extending from the centrioles are called astral rays; they may be the starlike projections that Flemming first saw in salamander cells in 1879, which led him to discover and describe the stages of mitosis three years later. Kinetochores are points on chromosomes where the chromosomes attach to spindle fibers.

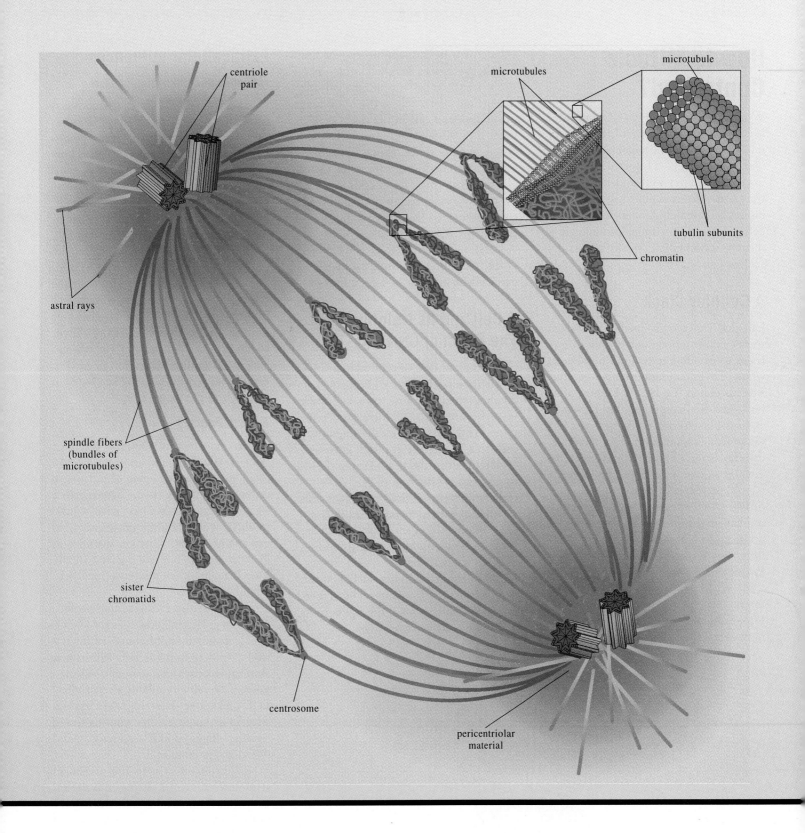

centriole
pair

microtubules

microtubule

astral rays

tubulin subunits

chromatin

spindle fibers
(bundles of
microtubules)

sister
chromatids

centrosome

pericentriolar
material

How Is Mitosis Controlled?

What causes a somatic cell to divide? Mechanisms that control the rate of mitosis operate at the molecular, cellular, and tissue levels. These vital mechanisms are not yet well understood, and they constitute a major focus of current biological research. Understanding how cells control mitosis may provide clues to how abnormalities of the cell cycle, such as cancer, arise.

A Cellular Clock

Laboratory-grown mammalian cells seem to obey an internal "clock" that allows them to divide a maximum number of times. This maximum number of divisions is called the Hayflick limit, after its 1960s discoverer, Leonard Hayflick. A fibroblast (connective tissue cell) taken from a human fetus, for example, divides from 35 to 63 times, with an average of about 50 times. However, a fibroblast taken from an adult divides only 14 to 29 times; the younger the individual, the greater the number of cell divisions. It is as if the cells "know" how long they have existed and how much longer they have to exist.

Within an organism, however, different cell types undergo mitosis at different characteristic rates. These rates may exceed the division limit seen in cells grown in the laboratory—or not even come close to it. A cell lining a person's small intestine might divide throughout life; a cell in the brain may never divide; a cell in the deepest skin layer of a 90-year-old could divide a dozen or more times if the person lives long enough. In fact, by the time a very elderly person dies, many cells may not even have begun to tap their proliferative potential—that is, to wind down their built-in mitotic clock.

Several factors affect the cellular clock, including crowding and various extracellular and intracellular influences.

A Cell and Its Neighbors— The Effect of Crowding

Crowding can slow or even halt mitosis. Normal cells growing in culture stop dividing when they form a one-cell-thick layer (a monolayer) lining their container. If the layer tears, the remaining cells that border the tear will grow and divide to fill in the gap but stop dividing once it is filled (fig. 9.8). This control of mitosis in culture by adjacent cells is called **density-dependent inhibition.**

Many cells in the adult human are inhibited from dividing by surrounding cells. In some tissues, only cells in certain positions can divide, such as those deep in the skin. In plants, cells in meristems, special regions at the tips of roots and shoots, divide frequently.

Extracellular Influences

Hormones

Certain animal cells divide frequently at some times, infrequently or not at all at others. This difference in rate is due to the influence of biochemicals called **hormones.** In an animal, a hormone is manufactured in a gland and travels in the bloodstream to another part of the body, where it exerts a specific effect. For example, at a particular time in the monthly hormonal cycle in the human female, peak levels of the hormone estrogen stimulate the cells lining the uterus to divide, building tissue a fertilized egg can implant in. If an egg is not fertilized, another hormonal shift breaks the lining down, resulting in menstruation. Hormones also control the cell proliferation needed to rapidly convert a fatty breast into an active milk-producing gland.

Growth Factors

When an animal is wounded, different cells in the damaged area must begin or increase the frequency of mitosis in order to build new tissue. The increase in mitotic rate associated with wound healing is mediated by proteins called **growth factors.** Unlike hormones, growth factors are not carried in the bloodstream—they act more locally to stimulate mitosis.

One of the most studied growth factors is epidermal growth factor (EGF), which stimulates epithelium (lining tissue) to undergo mitosis. EGF fills in new skin underneath the scab of a skinned knee. The salivary glands also produce EGF. Can you see how this might explain why an animal licks its wounds to aid healing? Ingested EGF also helps mend digestive system ulcers.

Fibroblast growth factor (FGF) stimulates division of the endothelial cells that form the one-cell-thick walls of the tiniest blood vessels (capillaries) and the inner linings of the larger blood vessels. FGF also provokes mitosis in fibroblasts, connective tissue cells which secrete collagen, a protein that also helps build blood vessels. The formation of new blood vessels, called **angiogenesis,** is important in wound healing because

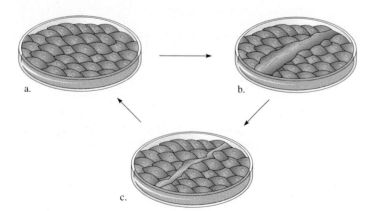

FIGURE 9.8

a. **Normal cells in culture divide until they line their container in a one-cell-thick sheet (called a monolayer). If the monolayer is torn (*b*), the cells at the wound site grow and divide (*c*) to fill in the gap.**

the increased blood supply brings oxygen and nutrients to nourish new tissue.

Large cells that give rise to blood platelets synthesize platelet-derived growth factor (PDGF). Platelets break apart at the site of a wound, where they release biochemicals that cause the blood to clot locally. PDGF provokes fibroblast mitosis in the area of a wound; this, like EGF's stimulation of epithelial mitosis, helps fill in the damaged area beneath the clot.

Scientists can produce growth factors in the laboratory using genetic engineering techniques (discussed in chapter 14) and subsequently use them as drugs. EGF, for example, can speed healing of a corneal transplant. The cornea is a one-cell-thick layer covering the eye, and normally its cells do not divide. However, a torn or transplanted cornea treated with EGF will divide to restore a complete cell layer. EGF is also being tested as a treatment for skin grafts and for nonhealing skin ulcers that occur as a frequent complication of diabetes. FGF may help heal surgical incisions by filling in fibroblasts. It may even help repair heart tissue damaged by a heart attack by building a new blood vessel network.

Intracellular Influences— Kinases and Cyclins

Not only do extracellular influences affect the rate of cell division, certain substances within cells also stimulate them to divide. Teams of proteins called **kinases** and **cyclins** activate genes whose products carry out mitosis.

For many years, researchers suspected that some type of trigger molecules set the events of mitosis into motion. In the early 1970s, researchers at Yale University and Argonne National Laboratory identified, but could not isolate, a signal substance in the cytoplasm of a cell about to divide. When the substance was transferred to a cell not yet ready to divide, that cell would begin dividing. The researchers named the biochemical **maturation-promoting factor (MPF)**.

MPF was first identified in immature frog egg cells and then in a variety of somatic cell types in several species. MPF levels seemed to rise and fall in synchrony with the cell cycle. What controlled MPF production?

In 1988, Manfred Lohka and James Maller, at the University of Colorado Medical School, isolated the long-sought MPF. They discovered that it is a two-protein complex. One protein is a regulator protein already known, from yeast experiments, to control the cell cycle. This protein is called **cdc2 kinase** (cdc stands for cell division cycle, and a kinase is a type of enzyme that activates other proteins by adding a chemical group called a phosphate to them). The cdc2 kinase controls the cell cycle in all organisms composed of eukaryotic cells. Yeast is the simplest of these organisms.

The discovery that cdc2 kinase is one of the two proteins in MPF wasn't the whole story behind cell cycle control. Although cdc2 kinase is present throughout the cell cycle, MPF is not. What turns MPF on and off? The answer is another type of protein called a cyclin. Cyclin was discovered at the Marine Biological Laboratory in Woods Hole, Massachusetts, in the early 1980s. Biologists first identified it in fertilized eggs undergoing their characteristic rapid cell division, its levels rising and falling to match mitoses. Oddly, the researchers found that cyclin is produced continuously, but at one point in the cell cycle, its levels plummet as if something is rapidly degrading it.

Figure 9.9 illustrates how kinases and cyclins control the cell cycle. First, cdc2 kinase binds to cyclin, which has

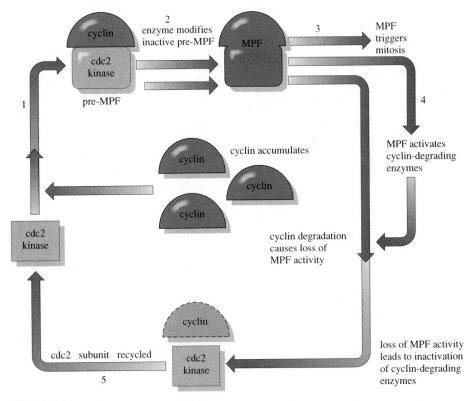

FIGURE 9.9

Cyclins and kinases regulate the cell cycle. The enzyme cdc2 kinase is always present in the cell. It combines with cyclin that has built up during interphase to form pre-MPF (step 1), which is activated by an as-yet unknown enzyme (step 2). Activated MPF triggers mitosis (step 3) and also stimulates production of enzymes that degrade cyclin (step 4). Once the lack of cyclin causes MPF to fall apart (step 5), the cycle begins anew as more cyclin accumulates during the next interphase.

accumulated during the previous interphase, to form a pre-MPF molecule. An enzyme activates pre-MPF to become mature MPF. The presence of the activated MPF makes mitosis inevitable. MPF also stimulates enzymes that break down cyclin, shutting down MPF's own activity in a negative feedback mechanism. With no active MPF, levels of cyclin-degrading enzymes drop, and cyclin accumulates again. When enough accumulates that cyclin again combines with the always-present kinase, division begins anew.

KEY CONCEPTS

Various factors control mitosis. A cellular "clock" limits the number of times a cell can divide. Outside influences on the rate of division include crowding, hormones, and growth factors. Within cells, kinases and cyclins activate the genes whose products carry out mitosis.

How Mitosis Maintains the Cellular Composition of a Tissue

Precise control of mitosis is necessary for normal growth and development. The balance between cell death and cell reproduction maintains the organization of tissues forming organs in a growing individual. This is why the livers of a newborn human and an adult human differ dramatically in size, but still are recognizable as liver. In many tissues and organs, cells are continually dying and being replaced by mitosis.

Many tissues contain a few cells, called **stem cells,** that divide often. When a stem cell divides to yield two daughter cells, one remains a stem cell, ready to divide again, while the other specializes to perform certain functions. In this way, the tissue can maintain its specialized nature as well as its ability to generate new cells.

Figure 9.10 shows stem cells in the basal (bottommost) layer of the epidermal skin layer in a human. These basal

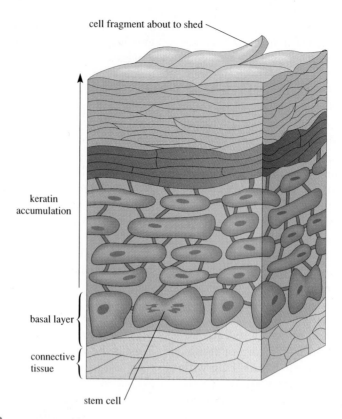

FIGURE 9.10

In some tissues, only cells in certain positions divide. The outer layer of human skin, the epidermis, has actively dividing stem cells in its basal (deepest) layer that push most of their daughter cells upward, yet maintain a certain number of stem cells. The cells pushed upward accumulate so much of the protein keratin that eventually the nuclei and organelles are squeezed aside and degenerate. The cell remnant is flattened and shed from the skin's outer surface.

cells undergo mitosis, but the cells above them, which divided from the basal stem cells, do not. Another example when stem cells push more specialized cells upward may be found in the folds of the small intestinal lining. Rapidly dividing cells, such as those in the skin and small intestine, are the cells most easily damaged by some drugs and forms of radiation. This is why radiation exposure from bombs and cancer therapy, as well as the drugs used in cancer chemotherapy, cause skin and digestive problems.

Cell Populations

Cell populations indicate the percentage of cells in particular stages of the cell cycle in a tissue. In a renewal cell population, the cells are actively dividing. Renewal cell populations maintain linings within animal bodies that are continually shed and rebuilt from dividing cells, such as the cell layers that line the inside of the digestive tract. In the human body, renewal cell populations replace many trillions of cells each day.

In an expanding cell population, up to 3% of the cells are dividing. The remaining cells of the expanding population are not actively dividing, but they can enter mitosis when a tissue is injured and new cells are required to repair it. The fast-growing tissues of young organisms, as well as adult kidney, liver, pancreas, and bone marrow tissues, consist of expanding cell populations.

Cells that are highly specialized and no longer divide comprise static cell populations. Nerve and muscle cells form static cell populations, remaining in the first gap phase (G_1). These cells grow by enlarging rather than dividing. A single nerve cell may grow to a meter in length, but it cannot divide.

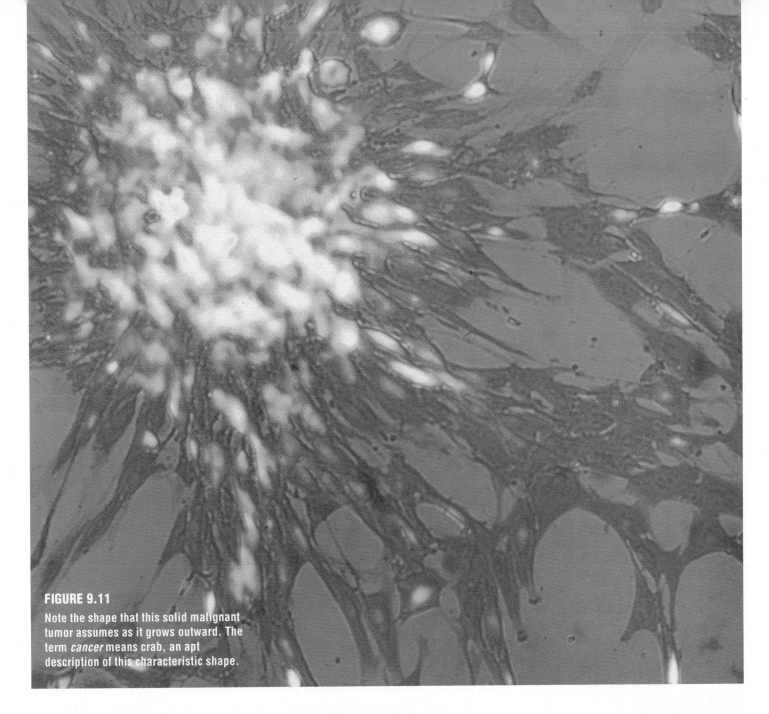

FIGURE 9.11
Note the shape that this solid malignant tumor assumes as it grows outward. The term *cancer* means crab, an apt description of this characteristic shape.

KEY CONCEPTS

Mitosis helps maintain the cellular composition of tissues. In multicellular organisms, stem cells divide often to maintain the specialized function of the tissue and to generate new cells. Cell populations are certain proportions of cells in a particular stage of the cell cycle in a tissue. Renewal cell populations are actively dividing. Up to 3% of the cells in expanding cell populations are dividing. Static cell populations are inactive and do not contain dividing cells.

Cancer—When the Cell Cycle Goes Awry

One out of three of us will develop cancer, a group of disorders in which certain body cells lose normal control over both mitotic rate and the number of divisions they undergo. Cancer begins with a single cell, which divides to produce others like itself, eventually either growing into a mass called a cancerous or **malignant tumor** or traveling in the blood.

Not all tumors are cancerous. A malignant tumor differs from a **benign** (gentle) **tumor.** A benign tumor is usually round and appears distinct from surrounding tissues. It does not travel to other locations in the body. If a benign tumor remains small, shrinks, or is surgically removed, it poses no health threat. Fibroids, which are often found in the uterus, are one example of a benign tumor.

An invasive malignant tumor, in contrast to a benign tumor, grows irregularly, sending tentacles in all directions (fig. 9.11). In fact, the word *cancer,*

179

FIGURE 9.12

When a plant is wounded, a common soil bacterium, *Agrobacterium tumefaciens,* can infect the wound and produce crown galls—plant tumors with many of the characteristics of animal cancers. This crown gall is growing on a tobacco plant.

Table 9.1	Time to Complete One Cell Cycle in Some Human Cells	
Cell Type		**Hours**
Normal Cells		
Bone marrow precursor cells		18
Lining cells of large intestine		39
Lining cells of rectum		48
Fertilized ovum		36–60
Cancer Cells		
Stomach		72
Acute myeloblastic leukemia		80–84
Chronic myeloid leukemia		120
Lung (bronchus carcinoma)		196–260

which means "crab" in Greek, comes from the resemblance between malignant tumors and crabs first noticed by Hippocrates, the Greek physician, in the fifth century B.C.

Cancer cells probably arise from time to time in everyone because mitosis occurs so frequently that an occasional cell is bound to escape from the mechanisms that normally control the process. In multicellular animals, certain cells, sometimes organized into immune systems, destroy cancer cells. Plants have an abnormal sort of growth, a crown gall, that is similar to cancer (fig. 9.12).

Characteristics of Cancer Cells

Cancer cells can divide uncontrollably and eternally, given sufficient nutrients and space. These characteristics are vividly illustrated by the cervical cancer cells of a woman named Henrietta Lacks, who died in 1951. Her cells persist today as standard cultures in many research laboratories (Biology in Action 9.2).

Although we say that cancer cells divide frequently, their rate of mitosis is a relative matter. Some normal cells divide frequently, others rarely. Even the fastest-dividing cancer cells, which complete mitosis every 18 to 24 hours, do not divide as often as some normal human embryo cells (table 9.1). Therefore, a cancer cell divides faster than the normal cell type it arose from, or it divides continuously at the normal rate. In a sense, a cancer cell is an immature cell that doesn't "know" when to stop dividing. Once a tumor forms, it may grow faster than surrounding tissue simply because a larger proportion of its cells are actively undergoing mitosis.

Cancers can grow at a very fast rate. The smallest detectable fast-growing tumor is half a centimeter in diameter and can contain a billion cells, dividing at a rate that produces a million or so new cells an hour. If 99.9% of the tumor's cells are destroyed, a million would still be left to proliferate. Other cancers are very slow to develop and may not be noticed for several years. Lung cancer may take three to four decades to develop. However, any tumor's growth rate is slower at first because fewer cells are dividing. By the time the tumor is the size of a pea—when it is usually detectable—billions of cells are actively dividing.

When a cell becomes cancerous, it passes its abnormal division characteristics on to its descendants. That is, when a cancer cell divides, both daughter cells inherit the altered cell cycle control and are also cancerous. Therefore, cancer is heritable, in the sense that it is passed from cell to daughter cell (table 9.2).

A cancer cell is also transplantable. If a cancer cell is injected into a healthy animal, the disease spreads as the cell divides, creating more cancerous cells. A cancer cell also looks different from a normal cell. It is rounder, possibly because it is less adhesive to surrounding normal cells than usual, and because the cell membrane is more fluid than normal, allowing different substances across.

Along with being heritable and transplantable, cancer cells often undergo genetic changes. Detecting gene-sequence differences between cancer cells and normal cells can help distinguish between the two and establish a diagnosis (fig. 9.13).

Another characteristic of a cancer cell is that it is dedifferentiated, exhibiting less specialization than the normal cell type it derives from. A skin cancer cell is rounder and softer than the flattened, scaly, healthy skin cells above it in the epidermis. Cancer cells also act much differently than normal cells. Whereas normal cells placed in a container divide to form a monolayer, cancer cells pile up on one another—they lack

Biology in ACTION

Enticing Cells to Divide in the Laboratory

Because it is difficult to observe cells dividing in a living organism, much of our knowledge of mitosis comes from cells growing in glass containers, a technique called **cell culture.** Often, cultured cells multiply into very large numbers, and their secreted products can be extracted and used as drugs. Cell culturing is as much an art as it is a science. Cultured cells must be given nutrients, hormones, and growth factors, and their wastes must be removed. Some types of cells, such as blood cells, float freely in containers of liquid. Most cell types, though, adhere in single layers to surfaces. Researchers have devised glassware with intricate nooks and crannies to encourage maximal cell growth in minimal space. For example, a 4-liter vessel containing extensive surface area in its bumpy interior holds 40 billion cells—a number that formerly required 1,300 smooth-surfaced bottles! Another ingenious innovation is to sculpt labware surfaces that resemble the contours of naturally occurring proteins. This apparently creates an environment more like the one a cell would normally grow in, and the cultures flourish.

A major hurdle in early attempts to culture cells was the fact that most cells of vertebrate animals divide in culture only 50 times, and then they die. The reason for this natural limit on laboratory-induced mitosis is not known, but it produces some startling results. For example, if a human cell divides 20 times, is frozen for a few years, and then is thawed, it will usually resume dividing until it has reached 50 divisions. To be of use in biomedical experiments that follow cellular changes over long periods of time, cell cultures need to remain mitotically active much longer. One way to overcome the 50-division limit is to enlist the aid of

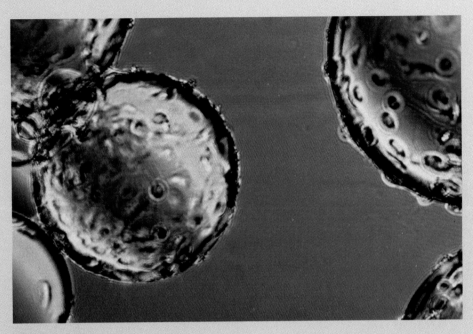

FIGURE 1

Growing cells on tiny oil beads helps culture the large numbers of cancer cells needed for research.

cancer cells, which divide well beyond 50 times in their lives. Scientists can extract cancer cells from an animal tumor, or expose normal cells in culture to cancer-causing viruses; and then grow the cells in cultures in the lab (fig. 1). The "immortality" of cancer cells is a boon to cell culture.

Perhaps the greatest help to cell culture technology has come from a seemingly unlikely source—a woman named Henrietta Lacks who died of cervical cancer in 1951. Before she died, she donated a few of her cancerous cells to a research laboratory at Johns Hopkins University. Although the cancer killed Henrietta within eight months, her cells lived on. These "HeLa"

cells, named for their donor, were the first human cells to be cultured successfully, and they remain in widespread use in biological research projects today. Ironically, the very characteristics that have made HeLa cells so valuable to research—their unrelenting mitosis and their seeming immortality—have also created a problem. The cells grow so well, dividing nearly once a day in any environment, that they rapidly proliferate in any culture of non-HeLa cells that they come into direct contact with. Within a few days, a HeLa-contaminated culture contains nearly all HeLa cells. So far, some 90 cell types thought to be of non-HeLa origin have turned out to be HeLa cells!

Table 9.2 Characteristics of Cancer Cells

Loss of cell cycle control

Heritability

Transplantability

Genetic mutability

Dedifferentiation

Loss of density-dependent inhibition

Ability to induce local blood vessel formation (angiogenesis)

Invasiveness

Ability to spread (metastasis)

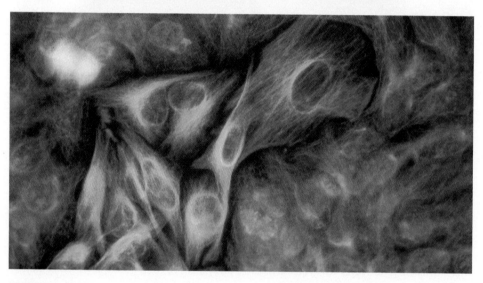

FIGURE 9.13

The orange cell, a melanoma (skin cancer cell), invades normal skin. Cancer cells, when stained for the presence of gene variants characteristic of cancer cells only, look very different from surrounding healthy tissue.

A benign tumor is not cancerous because it is not crab-shaped like a cancerous tumor, and it does not invade other tissues.

KEY CONCEPTS

Cancer occurs when cells divide faster or more times than the cells they descended from. Cancer cells are heritable, transplantable, genetically mutable, and dedifferentiated; they also lack density-dependent inhibition. Cancer cells cut through tissue boundaries to invade neighboring tissues, and they eventually metastasize to other parts of the body. Although benign tumors can crowd healthy tissue, they do not invade surrounding tissue or metastasize as cancerous tumors do.

The Causes of Cancer

Cancer can be viewed as a normal process—mitosis—that is mistimed or misplaced. If cells in an adult liver divide at the rate or to the extent that embryo liver cells divide, the resulting overgrowth may lead to liver cancer. Because proteins such as growth factors, kinases, and cyclins partially control the pace of mitosis, and because these proteins are constructed using genetic information, genes play a role in causing cancer.

Two major classes of genes contribute to causing cancer. **Oncogenes** must be activated to cause cancer. **Tumor suppressors,** which normally hold mitosis in check, must be inactivated or removed to eliminate control of the cell cycle and initiate cancer.

Oncogenes

Oncogenes (*onco* means "cancer") are genes that normally activate cell division in specific situations. When an oncogene activates cell division in the wrong time or place, cancer may result. For example, one human oncogene normally active at the site of a wound stimulates production of growth factors, which prompt mitosis to heal the damage. When the oncogene is active at a site other than a wound, it still hikes

density-dependent inhibition. In an organism, this pileup would produce a tumor.

Cancerous cells have surface structures that enable them to squeeze into any available space, a property known as invasiveness. They anchor themselves to tissue boundaries and secrete chemicals that cut paths through healthy tissue (fig. 9.14).

Eventually, the malignant cells of a tumor reach the bloodstream, and they travel through it to other parts of the body to establish new secondary tumors. The traveling cells then secrete chemicals in their new locations that stimulate

the production of tiny blood vessels to nourish the rapidly accumulating cells. The cancer has spread, or **metastasized** (from the Greek for "beyond standing still"). Once a cancer spreads, it is very difficult to treat because the cells of secondary tumors often undergo genetic changes that make them different from the original tumor cells. A drug that shrinks a malignant stomach tumor may have no effect at all on a secondary tumor in the liver. However, with some cancers, the type of genetic change that occurs in a secondary tumor provides clues about which treatment is likely to work.

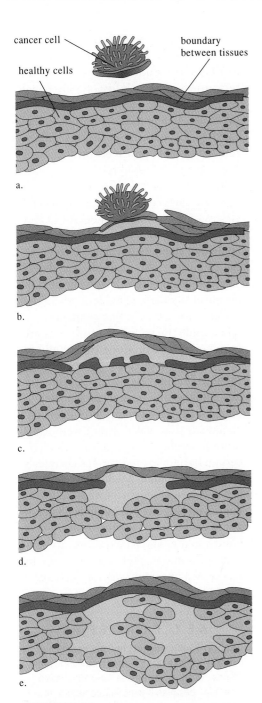

FIGURE 9.14

A cancer's spread takes many steps. *a.* A cancer cell adheres to normal cells that lie next to a boundary between two sections of tissue. *b.* The cancer cell secretes substances that cause the neighboring normal cells to move away, so that the cancer cell can now attach directly to the boundary. *c.* Next, the cancer cell secretes enzymes that allow it to penetrate the boundary and (*d*) to invade the tissue in the adjacent compartment. *e.* The cancer cell continues its migration and divides, starting a new tumor.

growth factor production and therefore stimulates mitosis, but, because there is no damaged tissue to replace, the new cells form a tumor.

How are oncogenes "turned on" to cause cancer? Researchers think that when an oncogene is placed next to a gene it is not normally next to, it boosts the oncogene's expression. A virus infecting a cell, for example, may insert its genetic material next to an oncogene. When the viral DNA begins its characteristic rapid reproduction, it also stimulates unusually rapid expression of the oncogene next to it. The heightened activation of the oncogene causes it to produce more of the growth factor or other protein it controls, which promotes inappropriate mitosis.

Oncogenes can also be activated when they move from their normal location on a particular chromosome and end up next to a gene that is normally very active. This appears to be the case in Burkitt's lymphoma, a cancer of a type of white blood cell that normally manufactures immune system chemicals called antibodies. In the cancer, a virus triggers chromosome breakage, which places an oncogene next to an antibody gene. In this abnormal location, the oncogene is more highly expressed than normal because the antibody genes in such a cell are highly expressed. In Africa, where Burkitt's lymphoma is common, patients exhibit a specific chromosome breakage pattern (fig. 9.15). Infection with the Epstein-Barr virus appears to trigger the chromosome breakage, which places an oncogene in a position near an antibody gene, causing an overexpression of the oncogene.

Tumor Suppressors

While some cancers occur when oncogenes are activated, others result when a gene fails to perform its normal function—suppressing tumor formation. For example, a childhood kidney cancer called Wilms tumor is caused by the absence of a gene that normally halts mitosis in the rapidly developing kidney tubules in the fetus. If the gene is missing, mitosis does not cease on schedule,

and the child's kidney retains pockets of cells that divide as frequently as if they were still in the fetus. In the child, these cells form tumors.

The role genes play in causing cancer is just now being deciphered. Rarely is it a straightforward or simple cause and effect. For example, some forms of colon cancer occur because of a sequence of genetic abnormalities involving both oncogene activation and tumor suppressor inactivation as well as possible environmental influences, such as diet. In other cases, a single gene can trigger a variety of cancer types. The loss or inactivation of one such gene, a tumor suppressor called p53, can cause cancers of the colon, breast, bladder, lung, liver, blood, brain, thyroid, esophagus, or skin (fig. 9.16).

Mutations in p53 seem to permit environmental insults to cause cancer. In one type of liver cancer prevalent in southern Africa and Qidong, China, most of the patients have a particular variant of the p53 gene. People in these areas are commonly exposed to the hepatitis B virus and to a food contaminant called aflatoxin B1. Somehow, the gene, virus, and/or food contaminant interact to send liver cells on the pathway to cancer. Similarly, the one in eight heavy smokers who develops lung cancer may owe his or her misfortune to a p53 gene variant that predisposes an individual to cancer.

Discovering the normal function of the p53 gene is a top priority in biomedical research. So far, the normal form of the gene is implicated in cell cycle control, in a cell's expression of other genes, and in several metabolic processes. Understanding p53 will open the door to understanding many types of cancer.

Carcinogens

While oncogenes and tumor suppressors play a part in causing cancer, other possible causes include certain chemicals, nutrient deficiencies, and radiation. Many toxins produced by plants also cause cancer in tests on cells in culture. Associations between many of these factors and cancer often derive from

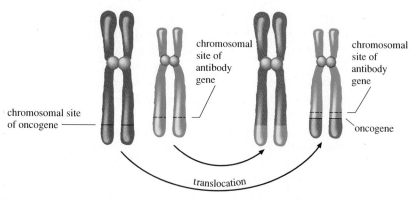

FIGURE 9.15

The cause of Burkitt's lymphoma appears to be the movement of an oncogene on the 8th largest human chromosome to a specific site on the 14th largest chromosome. The oncogene is placed near an immune system gene that is normally highly expressed. Overexpression of the oncogene in certain cells sets into motion the biochemical changes of cancer.

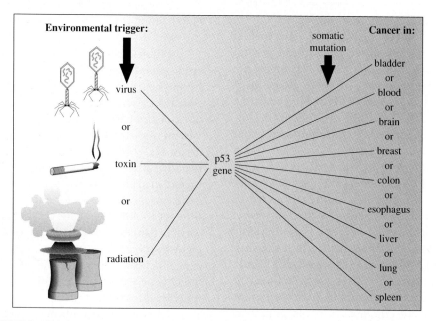

FIGURE 9.16

Environmental factors trigger genetic changes that lead to cancer. The p53 gene is one of several that seem to control the interactions between environmental and inherited risk factors for cancer.

population data. For example, people who smoke cigarettes are 40 times more likely to develop lung cancer than people who do not smoke. Therefore, it is reasonable to conclude that something in cigarette smoke contributes to causing cancer.

Chemical carcinogens (cancer-causing chemicals) were recognized as long ago as 1775, when British physician Sir Percival Potts suggested that the high rate of skin cancer in the scrotum among chimney sweeps was due to their

exposure to a chemical in soot. Today, carcinogens are identified both by epidemiological studies and by the Ames test, in which the ability of a chemical to cause genetic change in bacteria is taken as a strong indication that it may cause cancer in other organisms. A substance that tests positively on the Ames test is typically tested further on multicellular animals.

Researchers recognize three general classes of chemical carcinogens. Direct carcinogens cause cancer when they are applied to standard fibroblast cells growing in culture. An example of a direct carcinogen is benzene, a solvent used in chemical laboratories and formerly in the dry-cleaning industry. Procarcinogens are safe outside the body, but inside they are metabolized to produce intermediate compounds that cause cancer. They include certain organic dyes, cigarette tars, and the nitrites and nitrates used to preserve processed meats. Promoters are chemicals that make other carcinogens more powerful. They include alcohol, certain hormones, and various chemicals in cigarette smoke. Medical researchers are developing ways to integrate information on heredity (an individual's oncogenes and tumor suppressors), dietary habits, and chemical exposures to estimate personal cancer risks. Biology in Action 9.3 offers information on detecting and treating cancer.

KEY CONCEPTS

Cancer-causing agents include oncogenes, tumor-suppressing genes, and carcinogens. If an oncogene is overexpressed, perhaps because it is activated near a viral sequence or a very active gene, cancer may result. Conversely, when a tumor-suppressing gene is absent or inactive, cancer may also occur. Carcinogens—certain chemicals, nutrient excesses or deficiencies, and radiation—are substances that cause cancer. Usually, cancer results from some combination of these causes.

The War on Cancer

Researchers continually develop new weapons in the battle against cancer. The fight is taking place on three fronts: prevention, diagnosis, and cancer treatment.

Prevention

You can do a great deal to prevent cancer. The National Cancer Institute claims that 20,000 cancer deaths could be prevented each year if Americans were to follow these suggestions:

1. Avoid obesity and eat less fat.
2. Eat more fruits and vegetables, especially those rich in vitamins A and C. Cabbage and cauliflower are also valuable because they contain minerals that stimulate production of enzymes that break down carcinogens.
3. Limit consumption of foods containing mutagens or carcinogens, such as smoked, nitrated, or charred meats.
4. Limit alcohol intake.

You can also reduce your cancer risk by not smoking cigarettes, following health and safety rules in the workplace, avoiding X rays whenever possible, and protecting the skin from ultraviolet radiation (from the sun or artificial tanning facilities).

Learn to recognize cancer's warning signs and report them to a physician:

1. A change in bowel or bladder habits
2. A sore that does not heal
3. Unusual bleeding or discharge
4. A lump
5. Persistent indigestion
6. Difficulty in swallowing
7. A change in the appearance of a wart or mole
8. Chronic cough or hoarseness

Diagnosis

The key to treating cancer successfully is early detection, so that the spread of cancer cells can be stopped. Health care professionals place increasing emphasis on simple examinations and tests that can be performed at home. Women are encouraged, beginning at age 20, to examine their breasts at the same time each month for lumps that could be tumors. A breast exam, which takes about 10 minutes, involves palpating each breast in widening concentric circles to feel all the tissue. As many as 56% of all breast cancers are initially detected by women themselves.

A new genetic test can identify the 1 in 200 women who have inherited a particular gene that gives them an 85% chance of developing breast cancer. Before this test was developed, a woman with many affected female relatives could have her breasts removed, a drastic measure to prevent cancer.

Males between the ages of 15 and 35 should periodically examine their testicles for lumps. Starting at age 35, men should be regularly examined by a physician for prostate cancer. The physician checks the prostate directly with a rectal exam, and may run a blood test to try to detect a molecule called prostate specific antigen, or PSA, that is more abundant when a man has prostate cancer.

Colon cancer can be spotted early with an at-home *hemoccult* (hidden blood) test. To perform this test, a person sends a stool sample to a laboratory, where technicians check the sample for blood—a sign of colorectal cancer 50% of the time. Adult males, in particular, should have an annual hemoccult, especially if they have family members who have had intestinal growths called polyps. A new test will help identify individuals who have inherited genes that put them at very high risk of developing colon cancer.

Doctors routinely perform other low-cost tests to screen for certain cancers. Most women have a yearly *Pap smear* to detect cervical cancer. Many women over the age of 40 undergo yearly *mammography,* an X-ray technique that detects breast tumors too small to detect by touch (fig. 1). For women in this age group, the

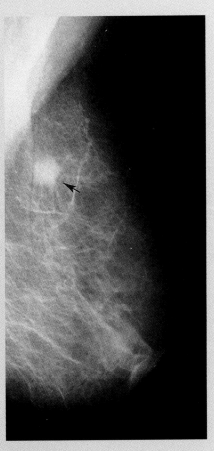

FIGURE 1

This mammogram shows a nonpalpable cancer (arrow) in the right breast.

risk of developing cancer from the X rays in the test is less than the risk of dying from undetected breast cancer.

A type of blood test to diagnose cancer in its early stages detects certain molecules that occur in larger numbers on cancer cells than on healthy cells, such as PSA. These tests are based on highly pure preparations of immune system proteins called **monoclonal antibodies** that have specialized binding sites that fit certain cancer cell surface molecules as a key fits a lock. If above normal levels of antibodies bind to a blood sample, then cancer may be present (fig. 2). *Continued*

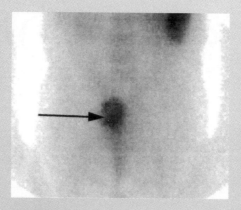
SUMMARY

Mitosis, the division of somatic cells, is responsible for growth, development, and repair of damaged tissue. A cell reproduces by first duplicating its contents, including the genetic material, and then dividing into two "daughter" cells. The cell cycle is the sequence of preparing for division and dividing. The cycle consists of interphase, when genetic material, macromolecules, and organelles are duplicated, and mitosis, when the cellular constituents divide into two daughter cells.

Interphase is further broken down into two gap periods, G_1 and G_2, when the cell makes proteins, carbohydrates, and lipids, and a synthesis period (S), when it replicates genetic material. Each replicated chromosome consists of two complete sets of genetic information, called chromatids, attached by a section of DNA called a centromere.

Mitosis consists of four stages. In prophase, the chromosomes condense and become visible when stained, the nuclear membrane disassembles, and the mitotic spindle builds up from microtubules. In metaphase, the replicated chromosomes attach to spindle fibers, which align the chromosomes down the center of the cell. In anaphase, the chromatids of each replicated chromosome separate, so that a complete copy of the genetic material segregates to each end of the cell. In telophase, the spindle breaks down and a nuclear membrane forms around each of the two sets of chromosomes. The division of the two sets of genetic material into daughter cells is called karyokinesis. Cytokinesis also begins during telophase. Cytokinesis is the distribution of cytoplasm, organelles, and macromolecules into the daughter cells. Karyokinesis and cytokinesis are closely coordinated.

The rate, timing, and number of mitotic divisions among the various specialized cells of a multicellular organism are highly regulated. Although biologists do not fully understand cell-cycle control, external factors such as hormones, growth factors, and proximity to other cells appear to regulate the cell cycle. Cyclins and kinases are intracellular biochemicals that control the cell cycle. Healthy human cells generally do not divide more than 50 times when grown in culture; in the body, the presence of other cells also limits their division. Different cell populations are defined as the proportion of cells in different stages of the cell cycle.

A loss of cell cycle control can result in growth of a tumor. A tumor cell divides more frequently or more times than cells surrounding it. It also has altered surface properties, lacks the specializations of the cell type it arose from, and produces daughter cells like itself. A benign tumor is rounded in shape and localized, whereas a malignant tumor infiltrates nearby tissues and can spread, or metastasize, to other locations. Malignant tumors secrete enzymes that open a route to the bloodstream. When cancer cells enter the bloodstream, they can establish secondary tumors elsewhere in the body.

Some cancers are caused by oncogenes, genes whose protein products control mitotic rate in certain cells at certain times in development. An oncogene activated in the wrong place or at the wrong time can alter mitosis in a way that causes cancer. Oncogenes can be activated by being moved to a region of very active DNA. Loss or abnormality of tumor suppressor genes can also cause cancer, as can carcinogens, including some chemicals, nutrient deficiencies, and radiation. Genetic factors interact with exposure to carcinogens, viruses, and diet to determine an individual's cancer risk.

KEY TERMS

anaphase 171
angiogenesis 176
aster 171
benign tumor 179
cdc2 kinase 177
cell culture 181
cell cycle 168
cell population 178
centriole 174
centromere 170

centrosome 174
chromatid 170
chromatin 168
cleavage furrow 171
cyclin 177
cytokinesis 171
density-dependent inhibition 176
gap phase 168
growth factor 176
G_1 phase 168
G_2 phase 170

hormone 176
interphase 168
karyokinesis 171
kinase 177
kinetochore 174
malignant tumor 179
maturation-promoting factor
   (MPF) 177
metaphase 171
metastasis 182
mitosis 168

mitotic spindle 170
monoclonal antibody 185
M phase 170
oncogene 182
prophase 171
S phase 170
stem cell 178
syncytium 171
synthesis phase 168
telophase 171
tumor suppressor 182

REVIEW QUESTIONS

1. Describe the events that take place during mitosis.

2. What is a cell doing during interphase?

3. Give an example of how overly frequent or extensive mitosis can harm health. Give an example of how too infrequent or too little mitosis can harm health.

4. A cell is taken from a newborn human, allowed to divide 19 times, and then frozen for 10 years. Upon thawing, what is the cell likely to do in terms of mitosis?

5. If a layer of cells is torn, mitosis sometimes fills in the missing cells. Cite two examples of this phenomenon—one observed in the laboratory and one in the human body.

6. In what ways do cancer cells differ from normal cells?

7. How do biochemicals from inside and outside the cell control the cell cycle?

8. How can a virus cause cancer?

9. A young boy had a tumor in his stomach. Because the doctors said the tumor was benign, his parents, thinking it was not dangerous, refused to allow them to surgically remove it. The boy died. Why?

10. How can one cancer be caused by gene activation, yet another be caused by a gene's absence?

11. Cite two methods used to distinguish colon cancer cells from surrounding healthy cells.

12. Suppose some cells growing in culture have a mutant cyclin that cannot be degraded. They proceed through the cell cycle until metaphase and then halt. Why?

TO THINK ABOUT

1. A researcher classifies mouse tissue cells according to what stage of the cell cycle they are in. She finds that 3% of the cells are dividing. Of the cells in interphase, 50% are in the G_1 phase, 40% in the S phase, and 10% in the G_2 phase. Based on this information, the investigator concludes that, for the cells in this tissue, the G_1 phase is of the longest duration, followed by the S phase.

The cells spend the least amount of time in G_2. What is the basis for this interpretation?

2. A device called a fluorescence-activated cell analyzer measures the DNA content in cells of a large cell population. The device distinguishes three groups of cells. Group A is large, and its cells have a certain amount of DNA. Group B is also large, and its cells have twice as much

DNA as the cells in group A. The third group, C, is very small, and each of its cells has four times as much DNA as a cell from group A. Which stages of the cell cycle correspond to the cells in groups A, B, and C?

3. When the United States dropped atomic bombs on Japan in World War II, many people were not immediately killed but suffered slow, agonizing deaths from radiation sickness. The sudden, massive doses of radiation they received affected the cells in their bodies that proliferate at the highest rates. What types of cells and tissues were most affected?

4. Cytochalasin B is a drug that blocks cytokinesis by disrupting the microfilaments in the contractile ring. What effect would this drug have on mitosis?

5. Tumor cells can often grow in culture, so that researchers can observe their response to experimental drugs. How might such a procedure benefit a cancer patient?

6. A researcher removes a tumor from a mouse and breaks it into cells. He injects each cell into a different mouse. Although all of the mice in the experiment are genetically identical and raised in the same environment, the animals develop cancers with different rates of metastasis. Some mice die quickly, some linger, others recover. What do these results indicate about the cells that comprised the original tumor?

7. Chronic myeloid leukemia is a white blood cell cancer almost always accompanied by a change in the chromosomes of these cells. The change places the tip of one type of chromosome onto another type of chromosome. This unusual arrangement moves an oncogene. Explain how the cancer likely arises.

8. Why would cancer developing in a stem cell in the basal layer of the skin's epidermis be potentially more harmful than cancer arising in a specialized cell above it?

SUGGESTED READINGS

Angier, Natalie. 1988. *Natural obsessions.* New York: Warner Books. How scientists unlocked the secrets of oncogenes.

Division of Cancer Communication, National Cancer Institute, National Institutes of Health, Bethesda, Maryland. 1–800–4–CANCER. You can obtain excellent information on all aspects of cancer by calling or writing this organization.

Gallagher, Gayl Lohse. March 1990. Evolutions: The mitotic spindle. *The Journal of NIH Research.* A short history, with beautiful illustrations, of what we know about the mitotic spindle.

Glover, David M., Cayetano Gonzalez, and Jordan W. Raff. June 1993. The centrosome. *Scientific American.* A complex interplay between microtubules and other cytoskeletal elements helps cells divide.

Harris, Curtis C. December 24, 1993. p53: At the crossroads of molecular carcinogenesis and risk assessment. *Science,* vol. 262. *Science* magazine crowned p53 the 1993 "molecule of the year."

Lewis, Ricki. June 1990. Wilms' tumor— The genetic plot thickens. *The Journal of NIH Research.* Pockets of embryonic tissue persisting in the kidney lead to cancer.

Lewis, Ricki. October 1990. Neurofibromatosis I revealed. *The Journal of NIH Research.* Benign tumors can result from loss of tumor suppressor genes, too.

Marshall, Eliot. June 10, 1994. Tamoxifen: hanging in the balance. *Science,* vol. 264. Should healthy women take a drug to prevent breast cancer?

Marx, Jean. May 7, 1993. New colon cancer gene discovered. *Science,* vol. 260. Some forms of colon cancer are caused by a cascade of gene action. Researchers have identified a gene that seems to control all the others and that may be responsible for many types of cancer.

Ohtsubo, Motoaki, and James M. Roberts. March 26, 1993. Cyclin-dependent regulation of G_1 in mammalian fibroblasts. *Science,* vol. 259. The cell seems to "decide" whether to proceed through mitosis during G_1.

Pelech, Steven. July/August 1990. When cells divide. *The Sciences.* The history of our knowledge of the mitotic process and current thoughts on its pacemaker.

Science, vol. 259. January 29, 1993. This issue contains several excellent articles on breast cancer.

Travis, John. May 28, 1993. New tumor suppressor gene captured. *Science,* vol. 260. We only know of a dozen or so tumor suppressor genes. Some of them, when mutant, are responsible for a variety of cancer types.

EXPLORATIONS   Interactive Software

Smoking and Cancer

This interactive exercise allows the student to explore the relationship between smoking and cancer. The exercise presents a diagram of a human chromosome, showing the location of four genes that regulate cell growth. When their activities are disabled they actively promote growth. In this exercise, all four must be disabled for cancer to be initiated. Students investigate the relationship between smoking and cancer by varying the amount an individual smokes and looking to see how long it takes before all four genes are mutated to a cancer-causing state.

1. What role does dose play in the probability that smoking will lead to cancer?

2. Can you discover a "safe" amount of smoking?

3. Is the 20th cigarette smoked in a day more or less dangerous than the 1st?

4. How much does smoking one pack of cigarettes a day increase the likelihood you will get cancer?

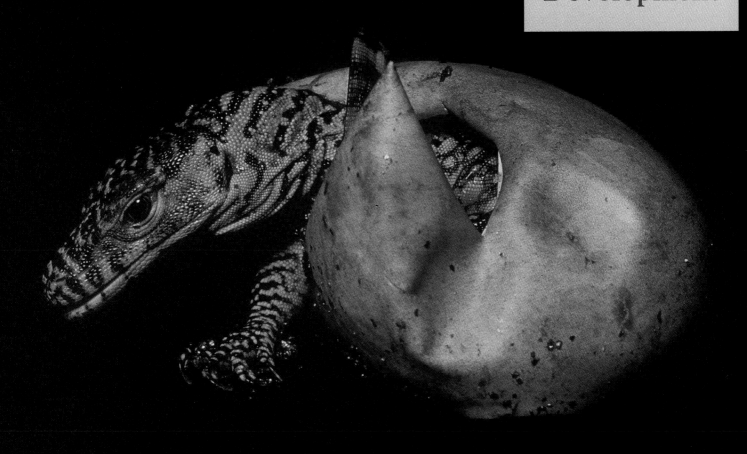

U N I T
3

Reproduction
and
Development

A Komodo dragon begins life.

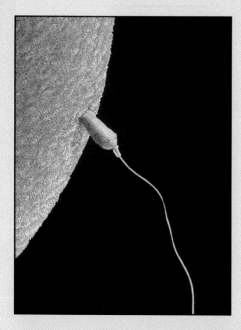

CONNECTIONS

Guinea Pig Love

The fourth graders gathered around Watson and Crick, their pet guinea pigs, watching and waiting. The pets had arrived in the classroom shortly after the start of the school year in September and now, in December, Crick was huge with her unborn babies. She lay in a corner on a bed of alfalfa, barely moving, while Watson skittered about at the other end of the cage.

One sunny Thursday, as the students lined up to go to lunch, a single screech came from the vicinity of the guinea pig cage. The first few students to reach the cage watched as three babies emerged quickly from Crick. The children barely noticed the blood as they stared, open-mouthed, at the three furry, tiny, brown and white babies staggering about. By the time the news had flashed around the school that Room 210's guinea pigs had finally become parents, the babies were clean and moving about quite well.

Adult humans are no less enthralled at witnessing a birth. It is an astonishing process, capping a remarkable period of growth and development. Today, the excitement begins soon after conception, when ever-earlier tests detect the pregnancy hormone. Still, a woman may not feel very different, even if she experiences nausea or other discomforts of pregnancy. An ultrasound exam before the 8-week mark of prenatal development doesn't reveal much, either. On the screen, the embryo looks like a lima bean with a blip in its middle. But if the developing human were more visible, the viewers would see astounding detail. All major structures and organs are present, at least in rudimentary forms.

An ultrasound exam four months later is literally a different picture. An excited parent-to-be can see the fetus's heart beat and his or her legs extend, and maybe even see the fetus suck a thumb. The doctor sees even more, comparing leg lengths and "crown to rump" measurements to see if the fetus is developing on schedule.

As the pregnancy advances, sophisticated scans are hardly needed to reveal the presence of the fetus within. The mother grows so accustomed to the fetus's pokes and jabs, and then whole-body contortions, that she is usually quite aware, and sometimes worried, if they cease for more than a few minutes. The father can feel, and even see, his child moving through the skin stretching over the mother's growing uterus.

Finally the big day (or night) arrives. No matter how extensively the couple has read, no matter how many classes they have taken or birth stories they have listened to, nothing can quite prepare them for the indescribable joy of seeing their child come into the world. Biology is perhaps at its best at the point when prenatal development blossoms into childhood. These chapters will allow you to glimpse that journey.

CHAPTER 10

Meiosis

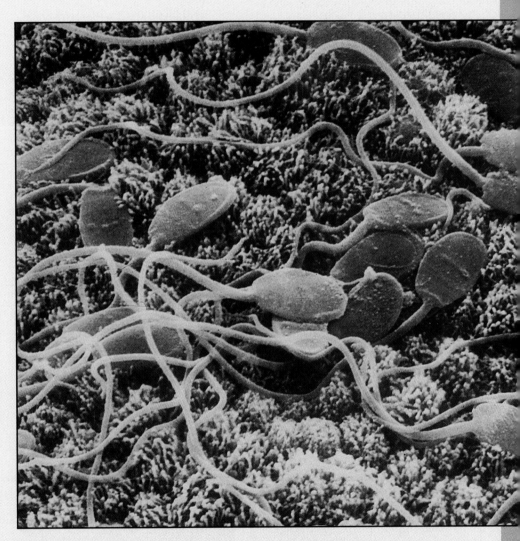

Sperm cells have such distinctive shapes and movements that they are easily recognized. These cells are rabbit sperm on an oviduct, the tube in the female that transports the eggs.

LEARNING OBJECTIVES

By the chapter's end, you should be able to answer these questions:

1. How does asexual reproduction differ from sexual reproduction?

2. How can the same type of organism reproduce both asexually and sexually?

3. What are some ancient reproductive strategies that might have been forerunners to sexual reproduction?

4. What structures form the human male and female reproductive systems?

5. Why is it necessary for germ cells to have half the number of chromosomes found in other cells?

6. What steps halve the chromosome number during germ cell formation?

7. How does germ cell formation increase genetic variability?

8. How do the male and female germ cells differentiate their specialized characteristics?

9. Where in the reproductive system of each sex does each germ cell formation stage take place?

10. How does germ cell formation differ in the two sexes?

Why Reproduce?

Organisms must reproduce—they must generate other individuals like themselves—for a species to survive. The amoeba in figure 10.1a demonstrates a straightforward strategy for meeting the challenge of reproduction—replicating genetic material, then splitting in two, redistributing the cellular structures of one individual into two. This ancient form of reproduction, called **binary fission,** is still common among single-celled organisms.

In an unchanging environment, the mass production of identical individuals, as in binary fission, makes sense. However, external conditions are rarely constant in the real world. If all organisms in a species were well suited to a hot, dry climate, the entire species might perish when exposed to a frost. A population of individuals of the same species, but with a diversity of characteristics, provides some insurance of species survival in a changeable environment, even if only a few unusual individuals survive. Binary fission cannot create or maintain this diversity; sexual reproduction can.

Sexual reproduction provides protective genetic variability by producing novel combinations of traits in new individuals. The young organisms in figure 10.1b are visibly different from each other because they were conceived by a sexual route. The amoebae in figure 10.1a, the products of **asexual reproduction,** or reproduction without sex, are identical to each other.

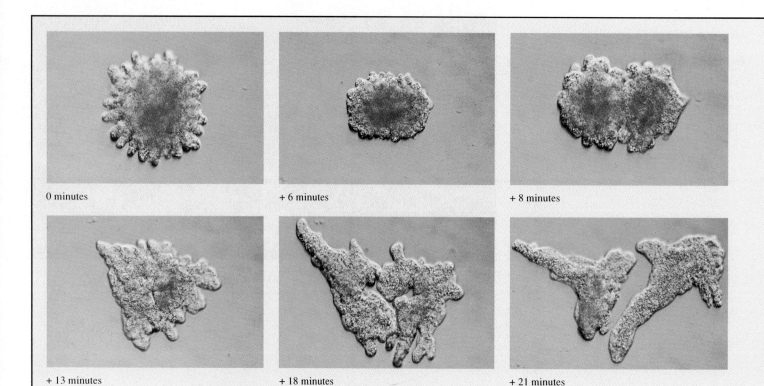

0 minutes

+ 6 minutes

+ 8 minutes

+ 13 minutes

a.

+ 18 minutes

+ 21 minutes

The Evolution of Sexual Reproduction

Sexual reproduction, a two-sex system, is the most common reproductive strategy among complex multicellular organisms. (A few organisms have more than two sexes—see To Think About question 1.) Sexual reproduction has two essential qualities: it introduces new combinations of genes from different individuals, and it increases the number of individuals by producing offspring. When a biological phenomenon such as two-sex reproduction persists over time and appears in a variety of organisms, that phenomenon probably offers an evolutionary advantage over other approaches to solving the same problem. The advantage that two-sex reproduction offers is the variety of combinations of possible traits that arise when cells from two individuals join to form a third individual. This diversity offers the species a greater chance to survive in an ever-changing environment.

Reproductive Diversity

Despite the success of sexual reproduction, living things reproduce in many ways. Some organisms exhibit both sexual and asexual phases—for example, the single-celled eukaryote yeast, *Saccharomyces cerevisiae.* A single **diploid cell** (a cell with two sets of chromosomes) can replicate its genetic material and "bud," yielding genetically identical diploid daughter cells. Or, yeast can form a structure called an ascus, which gives rise to specialized **haploid cells** containing a single set of chromosomes. Two of these haploid cells can then fuse, restoring the diploid chromosome number. Many plants reproduce asexually when a "cutting" from a larger plant grows into a new organism; they reproduce sexually through flowers and fruits. Biology in Action 10.1 describes an insect with a complex combination of asexual and sexual reproduction.

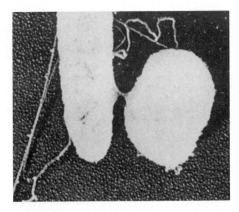

The bacterium *Escherichia coli* usually reproduces asexually by binary fission. It can also transfer genetic material to another bacterium; it is drawn to the other by a projection from its cell membrane called a sex pilus. The transfer of DNA between bacterial cells, called conjugation, is similar to sexual reproduction in that it produces a new combination of genes. But it is not reproduction because a new individual does not form.

Conjugation—Sex without Reproduction

Learning how diverse organisms reproduce and exchange genetic material can provide clues to how sexual reproduction may have evolved. The earliest sexual process appeared about 3.5 billion years ago, according to fossil evidence, in a form of bacterial gene transfer called **conjugation.** An outgrowth on one bacterial cell, called a **sex pilus,** enabled that cell to transfer genetic material to another (fig. 10.2). Bacterial conjugation, still seen today, is sex without reproduction, for it alters the genetic makeup of an already existing organism rather than resulting in a new individual.

Paramecium is a unicellular eukaryote that clearly separates reproduction and sex. Paramecia reproduce asexually by binary fission, yet they conjugate by aligning their oral cavities and forming a bridge of cytoplasm between them. The nuclei transfer across this bridge, resulting in new gene combinations but no new offspring.

b.

FIGURE 10.1

a. The single-celled *Amoeba proteus* follows an asexual lifestyle by splitting in two (binary fission). *b.* These siblings are obviously not exactly alike. The reason—each receives different combinations of the parents' genes, courtesy of meiosis.

The Complicated Sex Life of the Aphid

A sumac gall aphid reproduces through any of several strategies. These small insects inhabit diverse types of vegetation, including trees, mosses, ornamental flowers, and garden vegetables. Their unusual reproductive cycle begins in June, when small females called fundatrices hatch on tree bark and move to the leaves. There they burrow into the leaves, nestling into sacs called galls (fig. 1). As each fundatrix grows and matures within a gall, embryos begin to develop in her abdomen, without the aid of a male aphid. Female reproduction without male fertilization is called **parthenogenesis.**

By August, the gall that at first housed a single fundatrix has swollen to accommodate a thriving matriarchy of the original female and her daughters, all pale and wingless, and some granddaughters (fig. 2). By autumn, the granddaughters emerge from the gall, mature, winged, and dark. The granddaughters take flight, landing on mosses to live out their last few days. Sacs of embryos develop in the granddaughters, again without male input. As death nears, the granddaughters deposit groups of young on the moss.

Individuals of this fourth generation reproduce by parthenogenesis, grow, and live with their young protected beneath the waxy coat of the moss. Come springtime of that year, or sometimes the next, the young aphids fly up from the mosses. But the aphid's curious multigenerational reproductive cycle isn't over yet.

The winged aphids alight on sumac trees, where they release offspring—this time of two types, male and female. The males and females do what males and females do: they mate. This rare sexual generation introduces genetic variability; individuals from different colonies, and therefore descended from different fundatrices, mate and pass their genes to the next generation in new combinations. The sexual females deposit fertilized eggs on sumac bark. Some time later, the eggs hatch as a new generation of fundatrices, and the complex sequence of reproductive events begins anew.

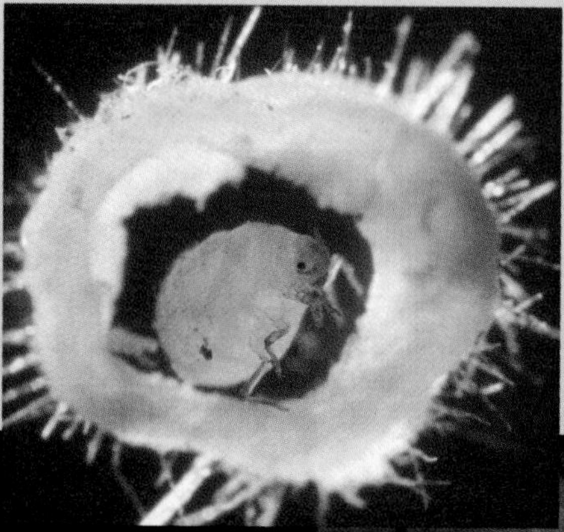

FIGURE 1

In this cross section of a protective gall on a sumac tree, you can see a fundatrix, a female aphid who will reproduce clones of herself asexually, by parthenogenesis.

FIGURE 2

The fundatrix's daughters look like her—pale and wingless. Her granddaughters, destined to migrate, are dark and winged. Curiously, this colony is a clone—all of the individuals are genetically identical. Obviously, the third generation expresses a different set of genes than the previous two generations.

Cell Fusion in *Chlamydomonas*

Conjugation is sexual in the sense that it exchanges genetic material. The two participating cells are of a single type—not necessarily genetically identical, but not as glaringly different as the sperm and egg cells of multicellular organisms. The first evolutionary inklings of different cell types engaging in sexual reproduction to produce new individuals may have occurred about 1.5 billion years ago, when two types of the same single-celled organism fused. Modern unicellular green algae of the genus *Chlamydomonas* exhibit a form of sexual reproduction that may echo that of its ancestors.

In the more prevalent nonreproductive phase, *Chlamydomonas* cells are haploid (fig. 10.3a). The cells are of two mating types, designated plus and minus. Under certain environmental conditions, a plus and minus cell join, forming a single diploid cell. This cell then undergoes a special type of cell division called **meiosis** that halves the chromosome number, yielding four haploid cells—two plus and two minus. When *Chlamydomonas* plus and minus cells join, it fulfills the two requirements of sexual reproduction— the number of individuals increases, and the genetic contributions of the two parents are mixed and reapportioned in the offspring.

When reproduction involves the meeting and fusion of two (or more) types of cells, it is critical that the different types recognize each other. The two types of *Chlamydomonas* cells look identical, but they are distinguished by the molecules on their surfaces, and perhaps in other ways that are obvious to them (fig. 10.3b). Sexual reproduction in *Chlamydomonas* begins when molecules on the tail-like flagella of each type attract, drawing the plus and minus cells

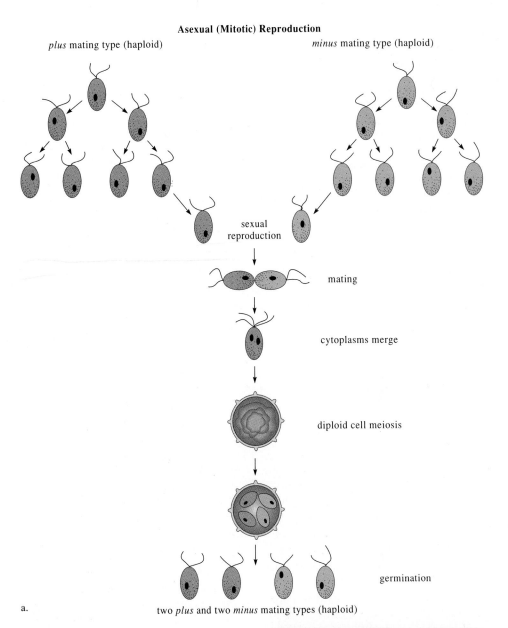

Asexual (Mitotic) Reproduction

plus mating type (haploid)   *minus* mating type (haploid)

sexual reproduction

mating

cytoplasms merge

diploid cell meiosis

germination

a.

two *plus* and two *minus* mating types (haploid)

FIGURE 10.3

a. The unicellular alga *Chlamydomonas* has two mating types, each haploid, that can reproduce asexually when they are separated, or sexually when together. In some circumstances, cells of different mating types merge, forming a diploid cell that undergoes meiosis, mixing up combinations of traits. Meiosis yields four haploid cells, two of each mating type.
b. *Chlamydomonas* cells.

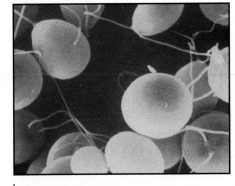

b.

FIGURE 10.4

The human male reproductive system. Sperm cells are manufactured within the seminiferous tubules, which are tightly wound within the testes, which descend into the scrotum. Sperm mature and are stored in the epididymis and exit through the vas deferens. The paired vasa deferentia join in the urethra, through which seminal fluids exit the body. Secretions are added to the sperm cells from the prostate gland, the seminal vesicles, and the bulbourethral gland.

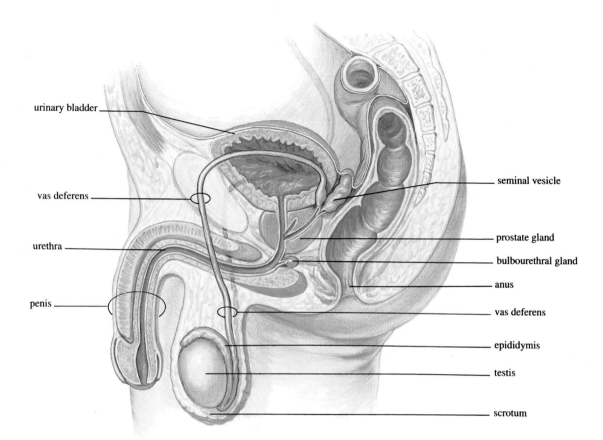

urinary bladder

vas deferens

urethra

penis

seminal vesicle

prostate gland

bulbourethral gland

anus

vas deferens

epididymis

testis

scrotum

into physical contact. When certain regions of the two cell membranes touch, the cytoplasms become continuous, like two soap bubbles coalescing. The plus cell then extends a fingerlike appendage toward the minus cell, establishing a bridge of cytoplasm that further weds the two cells.

In more complex organisms, the two types of mating cells differ more in appearance. Still, despite the differences—such as the enormity of eggs compared to sperm—each cell plays the same role as an emissary carrying genetic material to the next generation.

An essential characteristic of sexual reproduction is the halving of the chromosome number, which reduces two sets to one set. If the genetic material did not divide in half, the amount of genetic material would double as two parental cells come together. Meiosis is the process that halves the genetic material. During meiosis, much of the cytoplasm and most of the organelles are stripped from the forming sperm, yet retained around the developing egg. This renders distinctive products of meiosis—sperm and egg cells, collectively called

germ cells or **gametes**—that contain half the normal amount of genetic material. The remainder of the chapter considers meiosis and the maturation of gametes in humans.

KEY CONCEPTS

Organisms must reproduce to ensure species survival. In asexual reproduction, an organism generates another genetically identical to itself; in binary fission, one cell replicates its genetic material and then splits in two. In sexual reproduction, the genetic traits of two individuals combine in a third, allowing reproductive diversity that helps a species to survive. Some organisms exhibit both asexual and sexual reproduction.

Conjugation may represent an ancient precursor to sexual reproduction. In conjugation, genetic material is transferred, but no new individual is produced. Cell fusion in *Chlamydomonas* may be an ancient example of actual sexual reproduction; it involves the exchange of genetic material and results in an increase in the number of individuals. Meiosis is the form of cell division that halves the chromosome number each sex contributes, forming gametes that combine into new individuals.

Where Gametes Form in the Human Body

The reproductive systems of the human male and female are similarly organized. Each system has paired structures, called **gonads,** in which the **sperm** and **ova** are manufactured; a network of tubes to transport these cells; and various hormones and glandular secretions that control the entire process.

The Human Male Reproductive System

Sperm cells are manufactured within a 125-meter-long network of tubes called **seminiferous tubules,** which are packed into paired, oval organs called **testes** (sometimes called testicles) (fig. 10.4). The testes are the male gonads. They lie outside the abdomen within a sac called the **scrotum.** Their location outside of the abdominal cavity allows the testes to maintain a lower temperature than the rest of the body, which is necessary

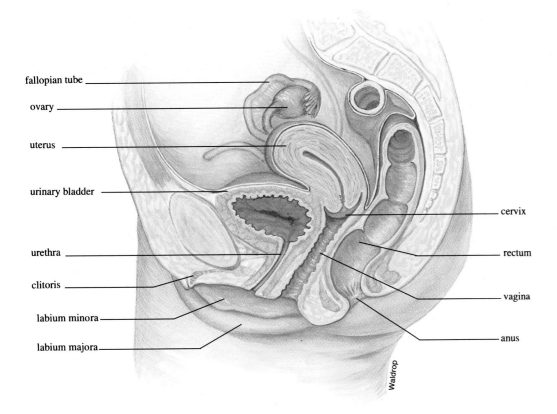

fallopian tube

ovary

uterus

urinary bladder

urethra

clitoris

labium minora

labium majora

cervix

rectum

vagina

anus

Waldrop

FIGURE 10.5

The human female reproductive system. Immature egg cells are packed into the paired ovaries. Once a month, one oocyte is released from an ovary and is drawn into the fingerlike projections of a nearby fallopian tube by ciliary movement. If the oocyte is fertilized by a sperm cell in the fallopian tube, it continues into the uterus, where it is nurtured for nine months as it develops into a new individual. If the ovum is not fertilized, it is expelled along with the built-up uterine lining through the cervix and then through the vagina. The external genitalia consist of inner and outer labia and the clitoris.

for the sperm cells to develop properly. Leading from each testis is a tightly coiled tube, the **epididymis,** in which sperm cells mature and are stored. Each epididymis continues into another tube, the **vas deferens.** Each vas deferens bends behind the bladder and joins the **urethra,** the tube that also carries urine out through the **penis.**

Along the sperm's path, three glands produce secretions. The vasa deferentia pass through the prostate gland, which produces a thin, milky, alkaline fluid that activates the sperm to swim. Opening into the vas deferens is a duct from the **seminal vesicles,** which secrete the sugar fructose the sperm need for energy, plus hormonelike prostaglandins, which may stimulate contractions in the female reproductive tract that help sperm and ovum to meet. The **bulbourethral glands** (also called Cowper's glands), each about the size of a pea, are attached to the urethra where it passes through the body wall. They secrete alkaline mucus, which coats the urethra before sperm are released. All of these secretions combine to form the **seminal fluid** that the sperm cells travel in.

During sexual arousal, the penis becomes erect so that it can penetrate the vagina and deposit sperm in the female reproductive tract. At the peak of sexual stimulation, a pleasurable sensation called **orgasm** occurs, accompanied by rhythmic muscular contractions that eject the sperm from the vas deferens through the urethra and out the penis. The discharge of sperm from the penis is called **ejaculation.** One human ejaculation typically delivers about 100 million sperm cells.

KEY CONCEPTS

The human male and female reproductive systems each have paired structures called gonads that house reproductive cells, tubes for transporting these cells, and glands whose secretions enable the cells to function. Sperm develop in the male's seminiferous tubules, which wind inside the testes, and the testes, the male gonads, reside in the saclike scrotum. Sperm cells mature and collect in each epididymis, which lead from each testis into the vasa deferentia. The vasa deferentia join at the urethra in the penis. Three glands contribute secretions to the semen: The prostate gland adds an alkaline fluid, the seminal vesicles add fructose and prostaglandins, and the bulbourethral glands secrete mucus. About 100 million mature sperm cells are ejaculated during orgasm.

The Human Female Reproductive System

The female sex cells develop within paired organs in the abdomen called the **ovaries** (fig. 10.5). The ovaries are the female gonads. Within each ovary of a newborn female are about a million **oocytes,** or immature egg cells. An individual oocyte is surrounded by nourishing **follicle cells.** In the adult, the ovary houses oocytes in different stages of development. Approximately once a month, one ovary releases the most mature oocyte. Beating cilia sweep the mature oocyte into the fingerlike projections of one of two **fallopian tubes.** The tube carries the oocyte into a muscular saclike organ, the **uterus** or womb.

Once released from the ovary, an oocyte can live for about 72 hours, but it can be fertilized only during the first 24 hours of this period—possibly even less. If the oocyte encounters a sperm cell in the fallopian tube and the cells combine and their nuclei fuse, the oocyte completes its development and becomes a **fertilized ovum.** It then travels into the uterus and becomes implanted in the thick, blood-rich uterine lining that has built up over the preceding few weeks—

Meiosis

Germ cells or gametes (reproductive cells) are formed from certain somatic cells, called **germ-line cells.** Meiosis, which halves the chromosome number, and maturation, which sculpts the distinctive characteristics of sperm and egg, combine to constitute **gametogenesis** (making gametes).

Stages of Meiosis

Meiosis entails two divisions of the genetic material. The first division is called **reduction division** (or meiosis I) because it reduces the number of chromosomes—in humans, from 46 to 23. The second division, called the **equational division** (or meiosis II), produces four cells from the two formed in the first division (fig. 10.6). The cells undergoing meiosis are diploid, abbreviated *2n* for two sets of chromosomes. The products of meiosis, haploid gametes, are designated *1n.* Diploidy is reestablished when opposite gametes combine at fertilization.

As in mitosis, meiosis is preceded by an interphase period when DNA replicates (table 10.1). The cell in which meiosis begins has **homologous pairs** of chromosomes, or homologs for short. Homologs look alike and carry the genes for the same traits in the same sequence. One homolog comes from the person's mother, and its mate comes from the father. When meiosis begins, the DNA of each homolog replicates, forming two chromatids, each carrying the same sequence of genes, joined by a centromere (fig. 10.6). Although the cell has doubled its genetic material, the chromosomes are not yet condensed enough to be visible.

Interphase is followed by **prophase I** (so called because it is the prophase of meiosis I). Early in prophase I, replicated chromosomes condense and become visible (fig. 10.7). Toward the middle of prophase I, the homologs line up next to one another, gene by gene, a

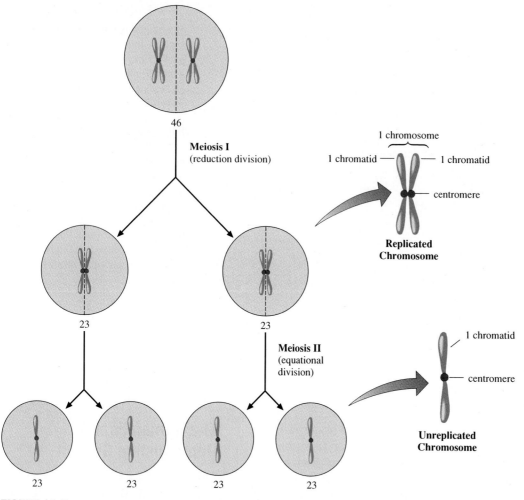

FIGURE 10.6

Meiosis is a special form of cell division in which certain cells are set aside to give rise to haploid germ cells. In humans, the first meiotic division reduces the number of chromosomes to 23, all in the replicated form (inset). In the second meiotic division, each cell from the first division essentially undergoes mitosis. The result of the two divisions of meiosis is four haploid cells.

the **endometrium.** Over the next nine months, the fertilized ovum develops into a new human being. If the oocyte is not fertilized, both endometrium and oocyte are expelled as the **menstrual flow.**

The lower end of the uterus narrows to form the **cervix,** which opens into the tubelike **vagina** that exits from the body. The vaginal opening is protected on the outside by two pairs of fleshy folds: the **labia majora** (major lips), and the thinner, underlying flaps of tissue they protect, called the **labia minora** (minor lips). At the upper juncture of both pairs of labia is a 2-centimeter-long structure called the **clitoris,** which is anatomically similar to the penis. Rubbing the clitoris stimulates females to experience orgasm.

Hormones control the cycle of egg maturation and the preparation of the uterus to nurture it.

KEY CONCEPTS

The female reproductive system also includes gonads containing reproductive cells, transport tubes, and glands that secrete hormones. Oocytes develop in the female gonads, the ovaries. Each month, one oocyte is released from an ovary and is captured by fingerlike projections from a fallopian tube. Each fallopian tube leads to the uterus. If the oocyte is fertilized by a sperm, it nestles into the endometrium to develop. Otherwise, it exits the body during the monthly menstrual flow, passing through the cervix and vagina. Hormones control the cycle of oocyte development.

Table 10.1   Comparison of Mitosis and Meiosis

Mitosis	Meiosis
One division	Two divisions
Two daughter cells per cycle	Four daughter cells per cycle
Daughter cells genetically identical	Daughter cells genetically different
Chromosome number in daughter cells same as in parent cell (2n)	Chromosome number in daughter cells half that in parent cell (n)
Occurs in somatic cells	Occurs in germ-line cells
Occurs throughout life cycle	In humans, completed only after sexual maturity
Used for growth, repair, and asexual reproduction	Used for sexual reproduction, in which new gene combinations arise

phenomenon called **synapsis.** The paired chromosomes are held together by a conglomeration of RNA and protein. At this point, all of the genes on a homolog came from one parent.

Toward the end of prophase I, the synapsed chromosomes separate but remain attached at a few points along their lengths (fig. 10.8). Here, the homologs exchange parts in a process called **crossing over.** After crossing over, each homolog contains some genes from each parent. (Recall that a gene is a sequence of DNA, part of a chromosome, that encodes a protein or controls the expression of another gene.) New gene combinations arise when the

2*n*

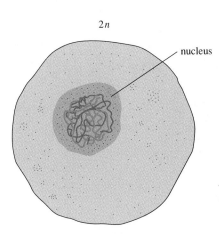

a. prophase I (early)

2*n*

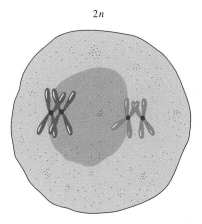

b. prophase I (late)

2*n*

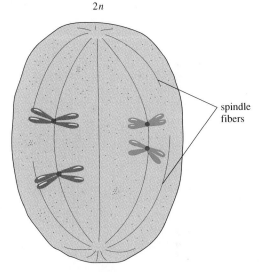

c. metaphase I

2*n*

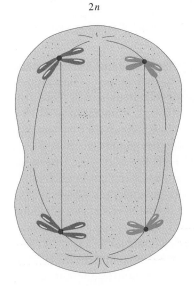

d. anaphase I

1*n* or 2*n*

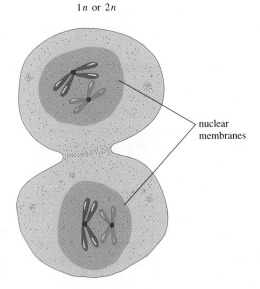

e. telophase I

FIGURE 10.7

a. In early prophase I, replicated chromosomes condense and become visible as a tangle within the nucleus. *b.* By late prophase I, the pairs are aligned and the homologs cross over. *c.* In metaphase I, spindle fibers align the homologs, and *d.* in anaphase I, the homologs move to opposite poles. *e.* In telophase I, the genetic material is partitioned into two daughter nuclei, each containing only one homolog from each pair. In most species, cytokinesis occurs between the meiotic divisions, forming two cells after telophase I.

199

parents carry different forms of the same gene, called **alleles.** Crossing over is one reason why siblings (except for identical twins) are never exactly alike.

Consider a simplified example of how crossing over mixes trait combinations. Suppose that one homolog carries genes for hair color, eye color, and finger length. One of the chromosomes in the homolog pair—perhaps the one that came from the person's father—carries alleles for blond hair, blue eyes, and short fingers. Its mate, the homolog from the mother, carries alleles for black hair, brown eyes, and long fingers. After crossing over, one of the chromosomes might bear alleles for blond hair, brown eyes, and long fingers, and the other bear alleles for black hair, blue eyes, and short fingers. Either combination might appear later in the person's offspring.

Meiosis continues in **metaphase I,** when the homologs align down the center of the cell (fig. 10.7c). A spindle forms, and each chromosome attaches to a spindle fiber stretching to the opposite pole. The chromosomes' alignment during metaphase I is important in generating genetic diversity. Within each homolog pair, the maternally derived and paternally derived members attach to each pole at random. The greater the number of chromosomes, the greater the genetic diversity; different combinations of maternal and paternal homologs move to each pole.

For two pairs of homologs, four (2^2) different metaphase configurations are possible (fig. 10.9). For three pairs of homologs, eight (2^3) configurations can occur. Our 23 chromosome pairs can thus line up in 8,388,608 (2^{23}) different ways. The random arrangement of homologs in the metaphase cell is called

FIGURE 10.8

Crossing over recombines genes. The capital and lowercase forms of the same letter represent different forms (alleles) of the same gene.

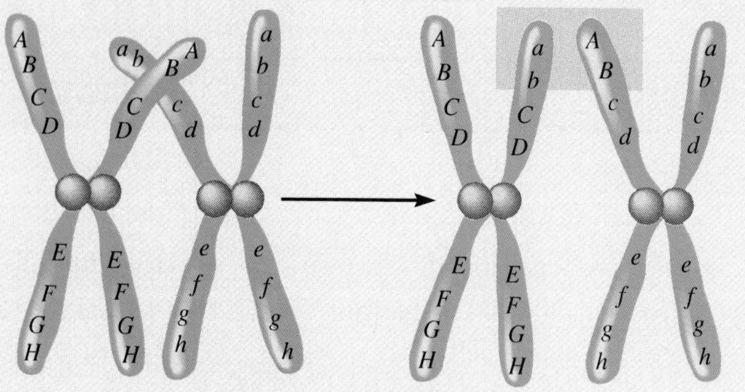

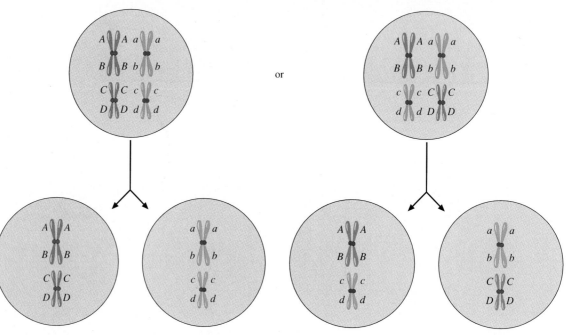

FIGURE 10.9

Independent assortment. The pattern in which homologs align during metaphase I determines the combination of chromosomes in the daughter cells. This illustration follows two chromosome pairs with different alleles of the same gene indicated by capital and lowercase forms of the same letter. Two pairs of chromosomes can align in two different ways to produce four different possibilities in the daughter cells. The potential variability generated by meiosis skyrockets when one considers all 23 chromosome pairs and the effects of crossing over.

independent assortment. It accounts for a basic law of inheritance, which is discussed in the next unit.

Homologs separate in **anaphase I** (fig. 10.7d), and they complete their movement to opposite poles in **telophase I** (fig. 10.7e). During this second interphase, the chromosomes unfold into very thin threads. Proteins are manufactured, but the genetic material is not replicated a second time. (The single DNA replication, followed by the double division of meiosis, halves the chromosome number.) In most species, the cell divides in two after telophase I.

Prophase II marks the start of the second meiotic division. The chromosomes again condense and become visible (fig. 10.10a). In **metaphase II** (fig. 10.10b), the replicated chromosomes align down the center of the cell. In **anaphase II** (fig. 10.10c), the centromeres split and the chromatids move to opposite poles. In **telophase II** (fig. 10.10d), nuclear envelopes form around the four nuclei containing the separated sets of chromosomes.

Cytokinesis completes the process of meiosis by separating the four nuclei into individual cells (fig. 10.10e). In

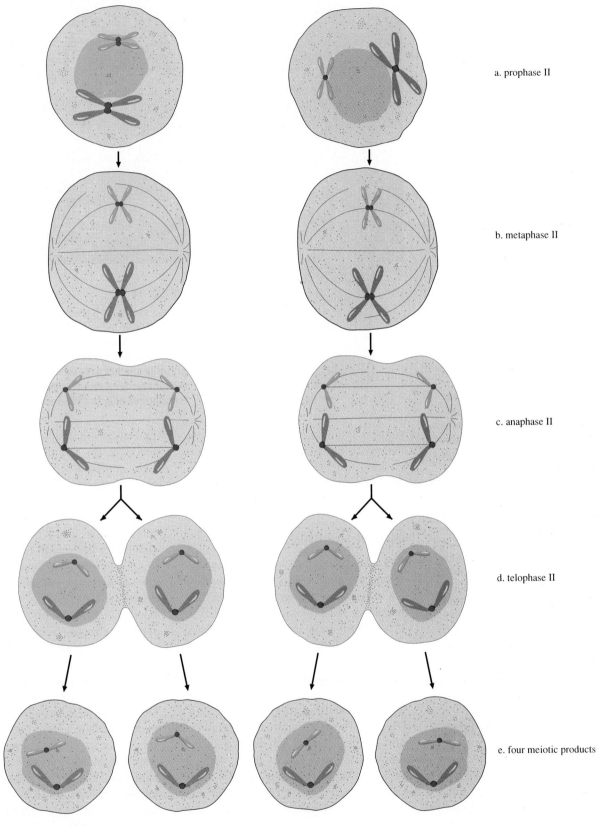

a. prophase II

b. metaphase II

c. anaphase II

d. telophase II

e. four meiotic products

FIGURE 10.10

Meiosis II. The second meiotic division is very similar to mitosis. *a.* In prophase II, the chromosomes are visible. *b.* In metaphase II, the spindle apparatus aligns the chromosomes. *c.* In anaphase II, the centromeres split, and each chromatid pair divides into two chromosomes, which are pulled toward opposite poles. *d.* In telophase II in some species, each of the two separated sets of chromosomes is enclosed in a nuclear membrane, and *e.* in cytokinesis, they are partitioned into individual haploid cells. The net yield from the entire process of meiosis: four haploid daughter cells.

most species, cytokinesis occurs between the first and second meiotic divisions, forming first two cells, and then four. In some species, cytokinesis occurs only following telophase II, separating the four nuclei enclosed in one large cell into four smaller haploid cells.

Meiosis and Genetic Variability

Meiosis generates astounding genetic variety. Any one of a person's more than 8 million possible chromosome combinations can combine with any one of the more than 8 million combinations of his or her mate, raising potential variability to more than 70 trillion ($8,388,608^2$) genetically unique individuals! Crossing over contributes even more genetic variability.

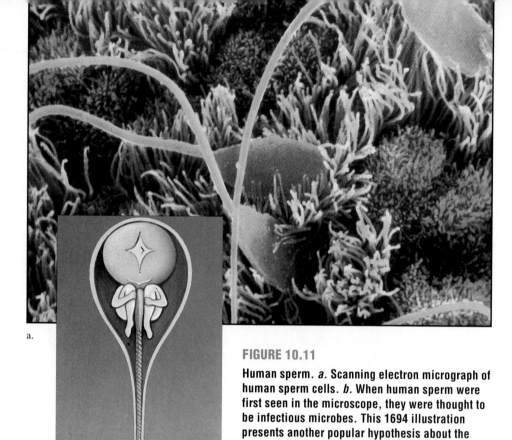

a.

b.

Gamete Maturation

Meiosis occurs in both sexes, but the sperm looks nothing like the ovum. Although each type of gamete has a haploid set of chromosomes, different distributions of other cell components create differences between them. The product of male meiosis is the motile, lightweight sperm, which must travel

FIGURE 10.11

Human sperm. *a.* Scanning electron micrograph of human sperm cells. *b.* When human sperm were first seen in the microscope, they were thought to be infectious microbes. This 1694 illustration presents another popular hypothesis about the role of sperm—some thought they were carriers of a preformed human called a homunculus.

to meet its mate; the result of female meiosis is the relatively large ovum, packed with nutrients and organelles. We will look first at gamete maturation in the male.

Sperm Development

Biologists have long been fascinated by sperm cells, perhaps in part because they are so intriguing to watch under a microscope (fig. 10.11a). Sperm cells look a little like microorganisms; indeed, when Anton van Leeuwenhoek first viewed sperm in 1678, he thought they were parasites in the semen. By 1685, he modified his view, suggesting that each sperm is a seed containing a preformed being (fig. 10.11b) and requiring a period of nurturing in the female to develop into a new life.

In the 1770s, Italian biologist Lazzaro Spallanzani experimented by placing toad semen onto filter paper. He observed that the material that passed through the tiny holes in the filter paper—seminal fluid minus the sperm—could not fertilize eggs. Even though he

had shown that sperm are required for fertilization, Spallanzani, like van Leeuwenhoek before him, concluded that the sperm cells were contaminants. Several researchers in the nineteenth century finally showed that sperm were not microbial invaders, but human cells that play a key role in reproduction.

The specialization of sperm cells is called **spermatogenesis** (fig. 10.12). A diploid cell that divides, yielding daughter cells that become sperm cells, is a **spermatogonium.** Several spermatogonia are attached to each other by bridges of cytoplasm, and they undergo meiosis together. First, the spermatogonia accumulate cytoplasm and replicate their DNA. They are now called **primary spermatocytes.** Next, during reduction division (meiosis I), each primary spermatocyte divides, forming two equal-sized haploid cells called **secondary spermatocytes.** In meiosis II, each secondary spermatocyte divides, yielding two equal-sized **spermatids.** Each spermatid then specializes, developing the characteristic sperm tail, or flagellum. The base of the tail has many mitochondria and ATP

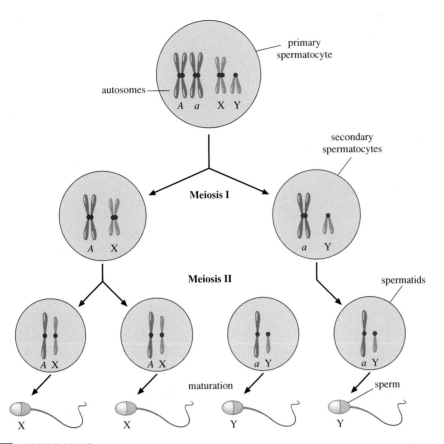

FIGURE 10.12

Sperm formation. In humans, primary spermatocytes have the normal diploid number of 23 chromosome pairs. The pair of blue chromosomes represents a pair of autosomes, and the green pair represents the sex chromosomes.

molecules, forming an energy system that enables the sperm to swim inside the female reproductive tract. (A sperm, which is a mere 0.0023 inch long, must travel about 7 inches to reach the ovum!) After the spermatids form, some of the cytoplasm connecting the cells falls away, leaving mature, tadpole-shaped **spermatozoa,** or sperm (fig. 10.13).

The mature sperm has three functional parts: a haploid nucleus, a locomotion system, and enzymes to penetrate an egg. Each sperm cell consists of a tail, body or midpiece, and head region. A small bump on the front end, the **acrosome,** contains enzymes that help the cell penetrate the ovum's outer membrane. Within the bulbous sperm head, DNA is wrapped around proteins called **protamines.** The sperm's DNA at this time is genetically inactive. A male

manufactures trillions of sperm in his lifetime. Only a very few will come near an ovum.

Male meiosis begins in the seminiferous tubules. Spermatogonia reside at the side of the tubule farthest from the lumen (the central cavity). When a spermatogonium divides, one daughter cell moves towards the lumen and accumulates cytoplasm, becoming a primary spermatocyte. The other daughter cell remains in the tubule wall. There, it acts as a stem cell—it remains unspecialized, but continually gives rise to cells that specialize.

The developing sperm cells travel towards the lumen of the seminiferous tubule, and by the time they reach it, they are spermatids. The spermatids are stored in the epididymis, where they complete their differentiation into sper-

matozoa. The entire process, from spermatogonium to spermatocyte, takes about two months.

When the epididymis contracts during ejaculation, the sperm cells are propelled into the vas deferens. The sperm, along with the secretions from the accessory glands, form semen, which exits the body through the penis. If the sperm fertilizes an ovum, a new organism begins to develop.

Male meiosis has some built-in protections against birth defects. Spermatogonia exposed to toxins tend to be so damaged that they never differentiate into mature sperm. More mature sperm cells exposed to toxins are often so damaged they cannot swim. However, some toxic drugs are carried in the seminal fluid without affecting the sperm. They can affect a developing organism by harming the uterus, or they can enter a woman's circulatory system and move to the **placenta,** the organ connecting the woman to the **fetus.** Cocaine can affect a fetus by another route—it attaches to thousands of binding sites on the sperm, without harming the cells or impeding their movements. Therefore, sperm can ferry cocaine to an ovum.

KEY CONCEPTS

Sperm development begins with spermatogonia, located deep within the walls of the seminiferous tubules. A diploid spermatogonium divides mitotically, yielding one stem cell and one cell that accumulates cytoplasm and becomes a primary spermatocyte as it moves toward the lumen of the tubule. In meiosis I, each spermatogonium halves its genetic material to form two secondary spermatocytes. In meiosis II, each secondary spermatocyte divides to yield two equal-sized spermatids. In the epididymis, the spermatids specialize further, developing their characteristic mitochondria-laden flagella. These developing sperm cells are attached to each other by bridges of cytoplasm. When the spermatids mature, the cells separate and shed the excess cytoplasm. The mature spermatocyte has a tail, body, and head region. A bump on the head, the acrosome, contains enzymes that enable the sperm to penetrate an ovum.

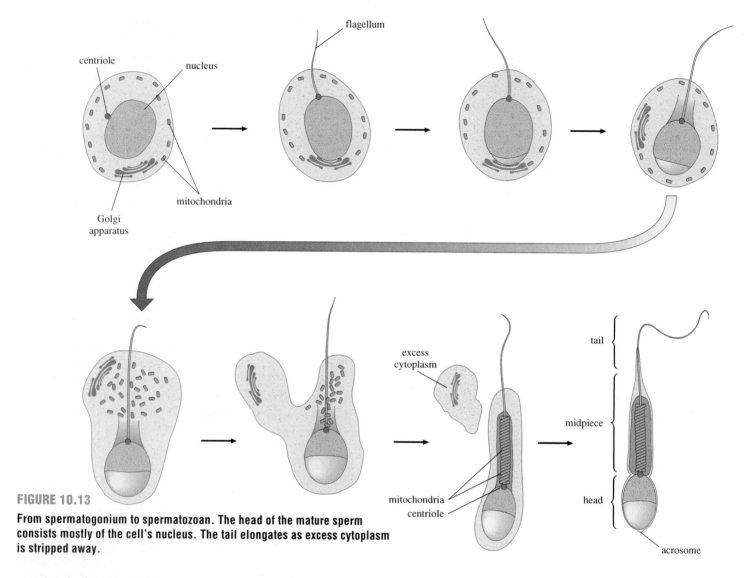

FIGURE 10.13

From spermatogonium to spermatozoan. The head of the mature sperm consists mostly of the cell's nucleus. The tail elongates as excess cytoplasm is stripped away.

Ovum Development

Meiosis in the female is called **oogenesis** (egg-making). It begins, like spermatogenesis, with a diploid cell, an **oogonium.** Unlike the male spermatogonia, oogonia are not attached to each other, but are surrounded by a layer of follicle cells. Each oogonium grows, accumulates cytoplasm and replicates its DNA to become a **primary oocyte.** The ensuing meiotic division in oogenesis, unlike that in spermatogenesis, produces cells of different sizes.

In meiosis I, the primary oocyte divides into a small cell with very little cytoplasm, called a **polar body,** and a much larger cell called a **secondary oocyte** (fig. 10.14). Each is haploid. In meiosis II, the tiny polar body may divide

to yield two polar bodies of equal size or may decompose. The secondary oocyte, however, divides unequally in meiosis II to produce a small polar body and the mature egg cell or ovum, which contains a large amount of cytoplasm. Therefore, each cell undergoing meiosis in the female can potentially divide to yield a maximum of four cells, only one of which will become the ovum; and most of the cytoplasm among the meiotic products is concentrated in it. Polar bodies are absorbed by the woman's body and normally play no further role in development. However, about 1 in 100 miscarriages is caused by a fertilized polar body. It develops for a few weeks into a mass of tissue that does not resemble an embryo, and then the woman's body rejects it.

Geneticists are trying to diagnose some inherited conditions by probing polar bodies, removing the oocyte with its clinging polar bodies from a woman's body. By identifying a disease-causing gene in the polar body, researchers can infer that the oocyte contains the healthy variant of the gene. They can then fertilize it in the laboratory and return it to the woman, where it continues to develop (fig. 10.15).

The ovum is packed with biochemicals that the new organism will use until its own genes begin to function. These biochemicals include proteins, RNA, ribosomes, and molecules that influence cell specialization in the early embryo. An ovum can be an impressive storehouse. Some amphibian oocytes contain 10^{12} ribosomes during meiosis!

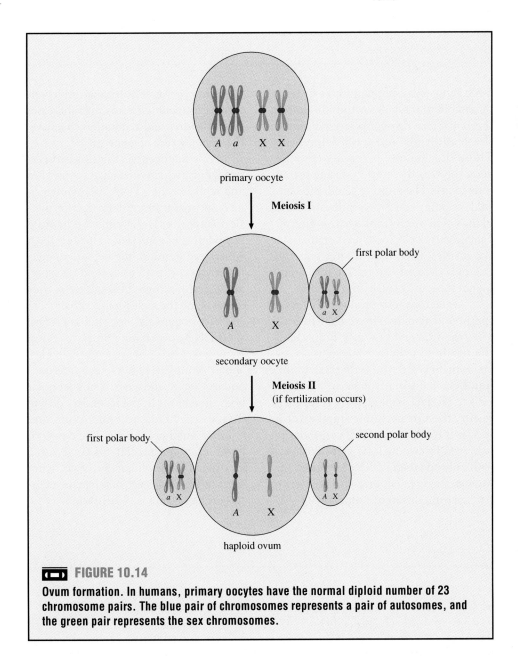

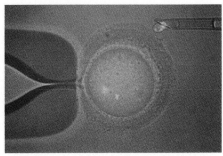

FIGURE 10.15

The fact that an oocyte shares the mother's divided genetic material with a much smaller companion, the polar body, is the basis for a new technique to diagnose genetic disorders before an oocyte is fertilized by a sperm. If a woman is known to be a carrier of a genetic disorder and the polar body contains the disease-causing gene variant, it can be inferred that the oocyte has received the "healthy" version of the gene. This diagnostic technique, called a polar body biopsy, is possible because the polar body remains attached to the oocyte. The oocyte is held in place by a large pipette, and the attached polar body is captured by drawing it into a smaller pipette. Genetic tests may then be performed on the polar body.

FIGURE 10.14

Ovum formation. In humans, primary oocytes have the normal diploid number of 23 chromosome pairs. The blue pair of chromosomes represents a pair of autosomes, and the green pair represents the sex chromosomes.

A human female will only ovulate about 400 oocytes between puberty and menopause, and only a very few of these will likely contact a sperm cell. Only one in three oocytes that do meet and merge with a sperm cell will continue to grow, divide, and specialize to form a new human life. The next chapter explores the journey of human development.

KEY CONCEPTS

Ovum development begins in the ovary in an oogonium. An oogonium accumulates cytoplasm and replicates its chromosomes, becoming a primary oocyte. In meiosis I, the primary oocyte divides to form a small polar body and a large, haploid secondary oocyte. In meiosis II, the secondary oocyte divides to yield another small polar body and a mature ovum. The million oocytes that a female is born with are arrested at prophase I. After puberty, one oocyte a month continues meiosis as it is ovulated, and meiosis is completed only when a secondary oocyte is fertilized.

The timetable for ovum formation differs greatly from that of sperm formation. Six months before a human female is born, her ovaries contain 2 million or more primary oocytes. From then on, the number of primary oocytes declines. A million are present by birth, and about 400,000 remain by the time of puberty. At birth, the oocytes are arrested in prophase I. After puberty, just one or a few oocytes complete meiosis I each month. These oocytes stop meiosis again, this time at metaphase II. Specific hormonal cues each month prompt an ovary to ovulate, or release a secondary oocyte. If a sperm penetrates the oocyte membrane, meiosis is completed, and a fertilized ovum forms. Therefore, female meiosis does not even finish unless the process of fertilization has already begun. If the secondary oocyte is not fertilized, it degenerates and leaves the body in the menstrual flow.

The stage of meiosis occurring when a sperm fertilizes an oocyte varies with species. In many worms, fertilization occurs just as the primary oocyte forms, whereas in foxes and dogs, it occurs in an older primary oocyte. In mollusks (including shellfish, octopuses, and squids), fertilization takes place during metaphase I, and in the sea urchin, it follows the completion of meiosis.

SUMMARY

Reproduction is essential to species survival. Asexual reproduction, such as by binary fission, can be successful in an unchanging environment. Sexual reproduction mixes traits and therefore provides diversity that could increase the likelihood of a species's survival in a changing environment. Sexual reproduction combines genes from different individuals and increases the number of organisms. Some species reproduce both asexually and sexually.

Conjugation, a form of gene transfer some microorganisms use, is similar to sexual reproduction because one individual transfers genetic material to another; however, no additional individual forms. *Chlamydomonas* undergoes cell fusion, which may have been a forerunner of sexual reproduction.

The cells that combine to form a new human are produced in the male and female reproductive systems. These systems include paired structures called gonads in which sperm and ova are manufactured, networks of tubes, and glands.

Male germ cells (spermatozoa) originate in seminiferous tubules within the paired testes and pass through a series of tubes, the epididymis and vasa deferentia. There they mature before exiting the body through the urethra during sexual intercourse. The prostate gland, the seminal vesicles, and the bulbourethral glands add secretions to the sperm. Female germ cells (oocytes) originate in the ovaries. Each month after puberty, one ovary releases an oocyte into a fallopian tube and travels to the uterus.

Gametogenesis produces the male sperm and female oocyte that fuse to form the first cell of a new individual, the fertilized oocyte. Meiosis is a special form of cell division that halves the number of chromosomes in somatic cells to produce haploid germ cells. As a result, the chromosome number of a species stays the same. This constancy is achieved by two meiotic cell divisions, with only one round of DNA replication. Meiosis ensures genetic variability by partitioning different combinations of genes into germ cells through independent assortment. Crossing over increases the variability, and maturation completes the manufacture of gametes.

Spermatogenesis begins with spermatogonia, which accumulate cytoplasm and replicate their DNA, becoming primary spermatocytes. After reduction division (meiosis I), the cells are haploid secondary spermatocytes. In meiosis II, an equational division, the secondary spermatocytes divide to each yield two spermatids, which then differentiate as they travel along the male reproductive tract.

In oogenesis, oogonia replicate their DNA, becoming primary oocytes. In meiosis I, the primary oocyte divides, apportioning cytoplasm to one large secondary oocyte and a much smaller polar body. In meiosis II, the secondary oocyte divides to yield the large ovum and another small polar body. The development of a sperm cell takes about 2 months. The development of an ovum takes at least 13 years, and female meiosis completes only at fertilization.

KEY TERMS

acrosome 203
allele 200
anaphase I 200
anaphase II 200
asexual reproduction 192
binary fission 192
bulbourethral glands 197
cervix 198
clitoris 198
conjugation 193
crossing over 199
diploid cell 193
ejaculation 197
endometrium 198
epididymis 197
equational division 198
fallopian tube 197
fertilized ovum 197

fetus 203
follicle cell 197
gametes 196
gametogenesis 198
germ cells 196
germ-line cells 198
gonads 196
haploid cell 193
homologous pairs
    (homologs) 198
independent assortment 200
labia majora 198
labia minora 198
meiosis 195
menstrual flow 198
metaphase I 200
metaphase II 200
oocytes 197

oogenesis 204
oogonium 204
orgasm 197
ovary 197
ovum 196
parthenogenesis 194
penis 197
placenta 203
polar body 204
primary oocyte 204
primary spermatocyte 202
prophase I 198
prophase II 200
protamine 203
reduction division 198
scrotum 196
secondary oocyte 204
secondary spermatocyte 202

seminal fluid 197
seminal vesicle 197
seminiferous tubule 196
sex pilus 193
sexual reproduction 193
sperm 196
spermatids 202
spermatogenesis 202
spermatogonium 202
spermatozoa 203
synapsis 199
telophase I 200
telophase II 200
testis 196
urethra 197
uterus 197
vagina 198
vas deferens 197

REVIEW QUESTIONS

1. A dog has 39 pairs of chromosomes. Considering only independent assortment (the random lining up of maternally and paternally derived chromosomes), how many genetically different puppies are possible from the mating of two dogs? Is this number an underestimate or an overestimate? Why?

2. Men who have the genetic disorder cystic fibrosis often lack vasa deferentia. Why are they unable to father children?

3. Define the following terms.
   a. crossing over   d. homolog
   b. gamete   e. synapsis
   c. haploid   f. diploid

4. How are the male and female reproductive tracts similar?

5. How many sets of chromosomes are present in each of the following cell types?
   a. an oogonium
   b. a primary spermatocyte
   c. a spermatid
   d. a cell from either sex during anaphase of meiosis I
   e. a cell from either sex during anaphase of meiosis II
   f. a secondary oocyte
   g. a polar body derived from a primary oocyte

6. Do the following examples illustrate asexual or sexual reproduction?
   a. A fundatrix aphid produces dozens of daughters and granddaughters without a mate.
   b. An amoeba divides, producing two from one.
   c. A somatic cell is taken from a carrot plant and nurtured in a laboratory dish. The cell divides many times, forming a whitish lump, from which a tiny new carrot plant sprouts.
   d. A piece of a wild cucumber plant is replanted in a garden, where a new plant grows.
   e. A brown guinea pig with a white tuft of hair on his head mates with a white guinea pig with a brown splotch on her back. Two of the offspring resemble the father. The third does not look like either parent, but has several brown splotches in a white background.

7. How does a human oocyte rid itself of excess genetic material? How does a human sperm shed excess cytoplasm?

TO THINK ABOUT

1. Why is the halving of the chromosome number that is accomplished in meiosis necessary for sexual reproduction?

2. Why is it extremely unlikely that a human child will be genetically identical to a parent?

3. Many veterans of the Vietnam War who were exposed to the herbicide Agent Orange claim that their children—born years after the exposure—have birth defects caused by a contaminant in the herbicide called dioxin. What types of cells would the chemical have to affect in these men to cause birth defects years later? Explain your answer.

4. A woman who is about four weeks pregnant suddenly begins to bleed and pass tissue through her vagina. After a physician examines the material, he explains to her that a sperm fertilized a polar body instead of an ovum, so that an embryo could not develop. Why do you think a polar body cannot support the development of an embryo, while an ovum, which is genetically identical to it, can?

5. How do the structures of the male and female human germ cells aid them in performing their functions?

SUGGESTED READINGS

Alberts, B., et al. 1992. *The molecular biology of the cell.* New York: Garland. Chapter 15 presents an in-depth look at meiosis and germ cell formation.

Anderson, Alun. July 17, 1992. The evolution of sexes. *Science,* vol. 257. Did sexual reproduction evolve to temper competition between organelles?

Bell, Graham. February 1992. Dividing they stand. *Natural History.* Paramecia reproduce without sex and have sex without reproduction.

Gould, Stephen Jay. 1980. Dr. Down's syndrome. In *The Panda's Thumb,* New York: W. W. Norton. An entertaining essay on the importance of meiosis, as evidenced by what can happen when it goes awry.

Howard, R. Stephen, and Curtis M. Lively. February 10, 1994. Parasitism, mutation, accumulation, and the maintenance of sex. *Science,* vol. 367. Sexual reproduction provides the variety a species often needs to survive such threats as mutations and parasites.

Keller, Sylvia Fox. 1983. *A feeling for the organism: The life and work of Barbara McClintock.* San Francisco: W. H. Freeman. Chapter 3 clearly explains meiosis.

Levine, Richard J., et al. July 5, 1990. Differences in the quality of semen in outdoor workers during summer and winter. *The New England Journal of Medicine.* Lower birth rates in the spring may reflect sperm sensitivity to the heat of the previous summer.

Lewis, Ricki. March 1993. Choosing a perfect child. *The World and I.* The diagnosis of a genetic disease is possible before conception by probing the genes of polar bodies.

Moran, Nancy A. April 1992. Quantum leapers. *Natural History.* Sumac gall aphids shift from a parthenogenic to a sexual lifestyle.

CHAPTER 11

Animal Development

CHAPTER OUTLINE

What child hasn't delighted in watching tadpoles develop? These tree frogs are attempting to mate.

LEARNING OBJECTIVES

By the chapter's end you should be able to answer these questions:

1. How did early embryologists account for development before birth?

2. What types of experiments did embryologists conduct in the eighteenth and nineteenth centuries?

3. How does an unspecialized cell specialize?

4. What roles do model organisms play in the study of developmental biology?

5. What basic stages of early development are common to many species?

6. What structures form during the three major stages of human prenatal development? What are their functions?

7. What structures support the embryo and fetus? How do they do so?

8. What happens during labor and birth?

9. What evidences of aging become noticeable at different stages in a human lifetime?

10. In what ways is aging a passive process? How can it also be an active process?

11. How has technology affected the human life span, longevity, and common causes of death?

A Descriptive Science Becomes Experimental

When kindergarten children gather daily around the incubator to watch a clutch of chicken eggs hatch in early spring, they echo the observations made by the Greek philosopher Aristotle in the fourth century B.C. The children follow chick development by looking at photographs (fig. 11.1); Aristotle actually broke open an egg each day and made intricate drawings of what he saw. Both

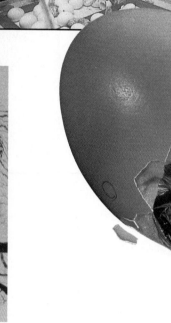

a.

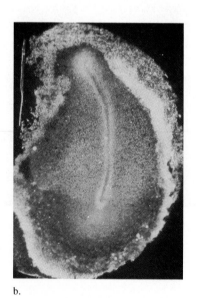

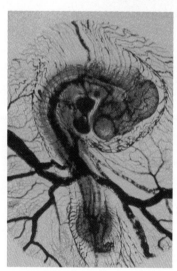

FIGURE 11.1

a. Using photographs, these children can follow the day-by-day development of chicks—echoing the observations of Aristotle and other biologists. *b.* A developing chick first appears to be a streak of tissue. Layers and then distinct structures rapidly form. By the twenty-first day, a chick is ready to peck its way out of the shell.

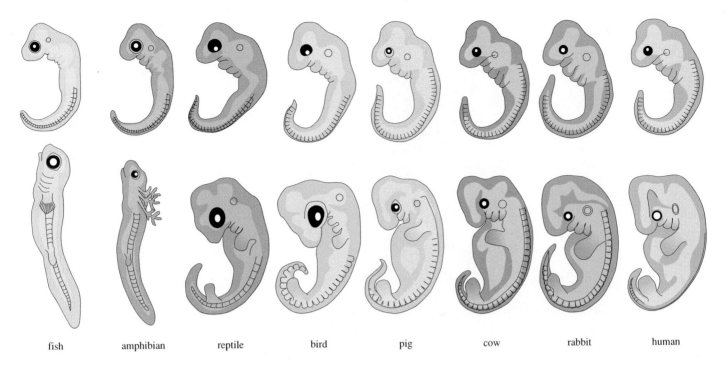

fish   amphibian   reptile   bird   pig   cow   rabbit   human

FIGURE 11.2

In 1828, embryologist Karl Ernst von Baer neglected to label two embryos in an experiment. The embryos were so similar in appearance that without the labels, he couldn't tell whether they were embryos of mammals, reptiles, or birds. Notice how similar the early embryos in the top row are—each has gill slits and a tail reminiscent of fish. Later in development, as depicted in the bottom row, the embryos become more distinctive.

Source: Gilbert, *Developmental Biology*, 1991, Sinauer Associates, Inc., Publishers, Sunderland, MA.

the ancient philosopher and biologists of all ages today are mesmerized by the rapid formation of a baby chick from what appears to be a thin strip of tissue.

Preformation versus Epigenesis

Until the mid-eighteenth century, most biologists believed in **preformation,** the idea that a fertilized egg (or even a gamete) contains a tiny, preformed organism, or "germ" that simply increases in size (see the homunculus in figure 10.11b). In 1759, German physiologist Kaspar Friedrich Wolff published a revolutionary pamphlet. In it, Wolff stated that a fertilized ovum contains no preformed germ, but unspecialized tissue that gradually becomes specialized. Other scientists considered Wolff's hypothesis of **epigenesis** so bizarre that they did not take it seriously for the next 50 years.

Through the nineteenth century, biologists continued to observe and sketch embryos. A German scientist, Karl Ernst von Baer, studied dog, chick, frog, and salamander embryos; he eventually came to be known as a founder of embryology. Baer watched, fascinated, as a fertilized egg divided into two cells, then four, then eight, gradually forming a solid ball that then hollowed out. The hollow balls would indent, then form two layers, then build up what Baer thought were four layers (later, scientists corrected this to three). Baer's observation—that specialized cell layers arise from nonlayered, unspecialized cells—confirmed Wolff's theory of epigenesis.

When Baer compared embryos of several vertebrate species, he noted that they were incredibly similar early on but then gradually developed species-distinct characteristics (fig. 11.2). For example, an early-appearing bump looked similar on embryos from four species, but one developed as a paw, one an arm, another

a wing, and another a flipper. Charles Darwin interpreted Baer's observations as evidence that different species descended from a common ancestor. Baer disagreed vehemently with Darwin on this matter for the rest of his days.

Later, German naturalist Ernst Haeckel extended Darwin's interpretation. Haeckel suggested that vertebrate embryos proceed through a series of stages that represent adult forms of the species that preceded them in evolutionary time. Today, most biologists reject the biogenetic law, as Haeckel's view is called. Structures in early embryos probably resemble each other simply because they have not yet specialized.

Early Embryological Experiments

During the twentieth century, research took a new direction. Embryologists began intervening in development, disrupting

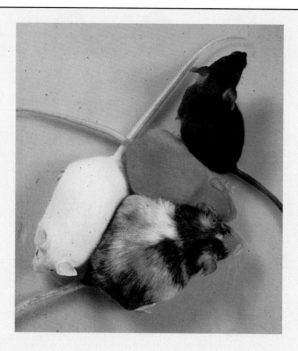

FIGURE 11.3

The tricolored mouse has six parents—a pair for each of the three original embryos combined to form the one individual. The tricolored mouse was bred to a white mouse, and she gave birth to the three mice shown above her in the photograph. The fact that the tricolored mouse gave birth to solid-colored mice of the original three colors indicates that some of her germ cells descended from each of the original three donor embryos.

embryonic structures to see what happened. A student of Haeckel's, Wilhelm Roux, was one of the first to alter embryos. He shook them, suspended them upside down, destroyed certain cells with heat or cold, and observed fairly consistent results—disturbed embryos often cease developing.

In the 1920s, German zoologist Hans Spemann experimented on newt and frog embryos. He teased apart the cells of a very early embryo and watched them develop, separately, into distinct but identical individuals. Apparently each cell of the original embryo retained the potential to give rise to a complete individual, a capacity called **pluripotency.** Scientists repeated Spemann's experiment a few years ago but introduced a variation: they separated the cells of three different mice embryos and reassembled the cells to create one new embryo—which then developed into a mouse with six parents (fig. 11.3). The

mouse experiment only works at early stages of development; by the time an embryo contains 1,000 cells, the cells have shut off the genes that would allow them to generate a complete organism.

Along with his pluripotency research, Spemann also pioneered surgery on embryos. First he removed part of a region called the dorsal lip from an embryo at the stage when the hollow ball of cells begins to indent. Then he transplanted the dorsal lip to the corresponding region of an embryo at the same stage but of a different species. A new embryo grew at the site where the dorsal lip tissue was grafted onto the second embryo. This region of the dorsal lip, named Spemann's organizer, is apparently a vital trigger to embryonic development.

Numerous transplant experiments involving many species helped researchers decipher the fates and functions of embryonic structures. These experimenters sought to determine when

a cell becomes committed to follow a specific developmental pathway, or the series of events that culminate in the formation of a particular tissue such as nerve or muscle. If transplanted cells develop normally, as if they had not moved, we know their developmental pathway has been determined before the time of transplantation. If, on the other hand, a transplanted cell develops as other cells do in that location, the cell was still pluripotent (or unspecialized) when it was transplanted.

With the discovery of the nature of genetic material in 1953, embryologists—now known by the broader term developmental biologists—could incorporate the ability to detect gene activity into their experiments. The discovery of DNA's role as the molecule of heredity provided a vital piece of information about early development because differential gene expression underlies development. In other words, all cells contain all of the organism's genes, but as cells become specialized, or **differentiated,** they express only certain genes, whose products endow the cell with its distinctive characteristics. DNA underlies the transformation from a generic clump of cells to a complex set of specialized tissues.

Differentiation and Selective Gene Activation

Elegant experiments conducted on the African clawed frog, *Xenopus laevis,* in the early 1960s demonstrated that differentiated cells contain a complete genetic package but only use some of the information in it. British developmental biologist J. B. Gurdon designed a way to "turn on" the genes that a differentiated cell normally "turns off." He removed the nucleus from a specialized cell lining the small intestine of a tadpole and injected it into an egg of the same species whose nucleus had been destroyed.

Can you deduce what happened? Some of the altered eggs developed into tadpoles, a few actually developing into perfectly normal adult frogs (fig. 11.4).

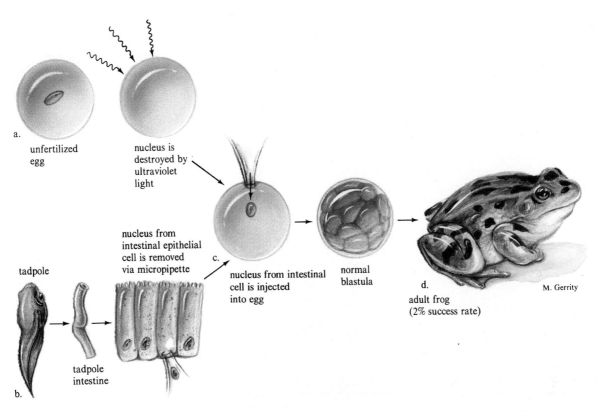

a. unfertilized egg

nucleus is destroyed by ultraviolet light

tadpole

nucleus from intestinal epithelial cell is removed via micropipette

tadpole intestine

b.

c. nucleus from intestinal cell is injected into egg

normal blastula

d. adult frog (2% success rate)

M. Gerrity

FIGURE 11.4

Nuclear transplantation shows that the nucleus of a differentiated cell can support complete development. *a.* In the first step of the procedure, a frog egg nucleus is inactivated with radiation. *b.* A nucleus is removed from a tadpole's differentiated cell. *c.* Next, the nucleus from the differentiated cell is transferred to the enucleated egg. *d.* The egg, controlled by its transplanted nucleus, develops into a tadpole and then a frog.

(The experiment did not work every time partly because it is physically difficult to manipulate single cells and nuclei without damaging them.) The fact that a nucleus from a differentiated cell (the intestinal cell) could support normal development from the single-cell stage proved that such a nucleus contains the complete set of genetic instructions for the species. Genes must therefore be inactivated—not lost, as some investigators once thought—as the life of an individual multicellular organism proceeds. Today, we know that genes are selectively turned off by proteins that physically block their activation. Gurdon's work has been repeated in other species, using different types of specialized cells.

Experimental developmental biology has become increasingly molecular as developmental biologists seek to decipher the precise natures and distributions of gene products that guide pluripotent cells to grow into highly specialized structures. Consider muscle formation. Researchers have identified a "master control gene" that turns on dozens of genes, causing the cell to produce muscle-specific proteins. When the master control gene is placed in chick connective tissue cells growing in culture, the cells elongate, and then pulsate—they differentiate into muscle! Figure 11.5 shows a mouse embryo actively producing muscle protein.

Studying human development before birth is problematical for a variety of legal, ethical, and practical reasons. But because many developmental processes are strikingly similar in differ-

ent species, researchers often work with model organisms to learn more about human development. The remainder of this chapter takes a short look at observations and experiments that have revealed much about development in two model organisms, and then a longer look at our own development. The next chapter explores technologies that manipulate human development.

KEY CONCEPTS

As biologists observed embryos, they replaced the theory of preformation with the idea of epigenesis: specialized tissue forms from unspecialized tissue. Early vertebrate embryos look similar, supporting epigenetic theory. Today, biologists alter embryos to learn exactly when a particular cell's fate is established, signaling the differential gene activity that causes the cell to specialize.

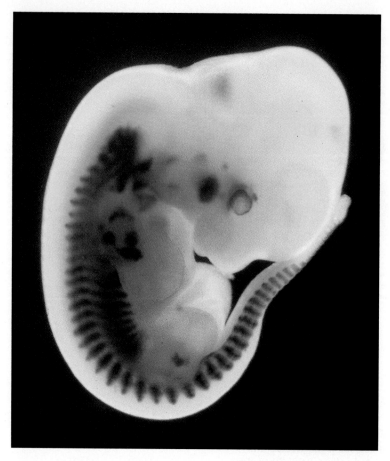

FIGURE 11.5

Highlighting embryonic muscle development. This 11.5-day-old mouse embryo has been genetically engineered so that a gene encoding a key muscle protein is attached to a gene that produces a blue color. The color appearing in the embryo indicates where the muscle protein is actively being produced.

Two Beginnings: Early Development in a Worm and a Fly

A millimeter-long worm, a fruit fly, and a human may not appear to have much in common. However, each organism meets the same challenges in embryonic and later development: it must develop from a single cell to an adult with a variety of specialized cell types.

A Worm— *Caenorhabditis elegans*

Like other multicellular organisms, the nematode worm *Caenorhabditis elegans* begins as a fertilized egg. Unlike other multicellular organisms, this tiny, transparent, soil-dwelling worm is so simple—with fewer than 1,000 cells comprising the adult—that very patient researchers have been able to trace the number and timing of cell divisions that occur over the worm's three-day lifetime (fig. 11.6).

From 1975 to 1980, in the tradition of the early embryologists, Einhard Schierenberg of West Germany and John Sulston of England stared for hours at worms developing under a light microscope, cataloging the fates of each and every cell. Thanks to the efforts of these and other "worm people," as *C. elegans* researchers call themselves, we now have a complete picture of the development of one organism—a complex process, even in a simple animal.

For the first few cell divisions, one cell in the worm looks much like another. By the fourth mitosis for some cells, and later for others, certain cells "commit" to follow certain developmental pathways—that is, to become part of the outer covering (cuticle), the intestine, muscle, nerve, pharynx, vulva, egg, or gonad. These specialized structures do not actually appear until a certain number of cell divisions later. Sometime during the initial hours of rapid cell division, the transparent bag of identical, seemingly separate cells becomes a moving, coordinated multicellular organism—a vivid example of an emergent principle, or the idea that new qualities emerge at more complex levels of organization. A half day after the first cell division, a larva of 558 cells hatches from an egg laid by the adult. After more rounds of cell division, and the death of precisely 113 cells, the 959-celled adult worm is mature.

A Fly—*Drosophila melanogaster*

Bizarre-looking mutants of the well-studied fruit fly *Drosophila melanogaster* offer clues to the fly's normal development. Consider a mutant fruit fly embryo with two rear ends. The fact that this mutation can happen indicates that cell position is an important part of specialization. Biochemicals must provide some type of positional information to the developing organism.

In this mutant, the headless embryo does lack a protein called bicoid, which means "two tails." Bicoid protein is normally present in the fruit fly egg before it is fertilized. The greater con-

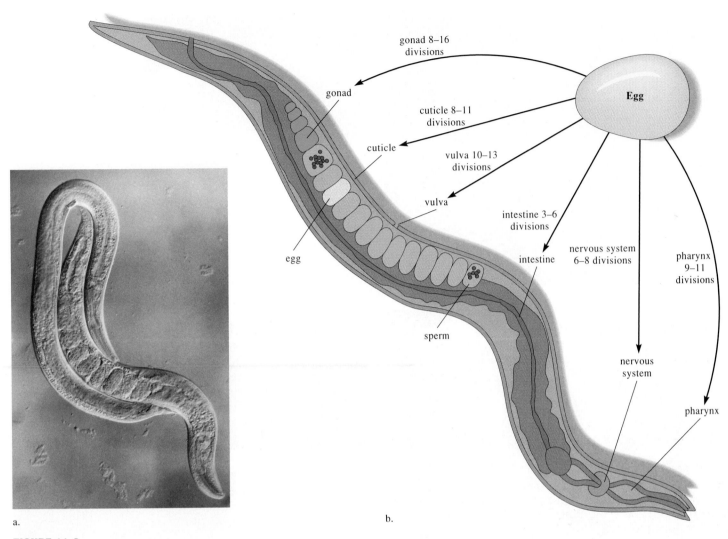

FIGURE 11.6

Cell fates are followed in the simple roundworm *C. elegans. a.* In this photo, the long, thin worm on the outside, a male, is attempting to mate with a shorter worm, a mutant called "dumpy." Magnification, × 100. *C. elegans* can also fertilize itself. *b.* "Worm people" have meticulously built intricate fate maps that diagram the relationships of all cells in the development of the larva and adult *C. elegans.*

centration of bicoid in the anterior (head) end of the fly causes it to differentiate the tissues characteristic of the head. Similarly, at the back end of the animal, higher concentrations of a protein called nanos signal the abdominal tissues to differentiate.

When the concentration of a substance increases or decreases over an area, it constitutes a gradient. Gradients of gene-controlled biochemicals act as guides to development that the first embryologists could not see—although

they could watch their effects, or disrupt them. Proteins whose gradients profoundly influence development are called **morphogens** (**morphology** means "form").

Mutant flies with legs where their antennae should be provide clues to later development (fig. 11.7a). These flies, called **homeotic** mutants, have mixed-up body parts because cells in the embryo receive incorrect signals that alter the distribution of morphogens. The flies develop normal structures, but in abnormal places.

A chain reaction of sorts begins as gradients of morphogens wash through the embryo (fig. 11.7b). The varying concentrations of different morphogens distinguish parts of the embryo, sending signals that stimulate cells to produce yet other proteins, which ultimately regulate the formation of the specific components of a differentiated structure. A cell destined to be part of the adult fly's antenna, for example, has different morphogen concentrations than a cell

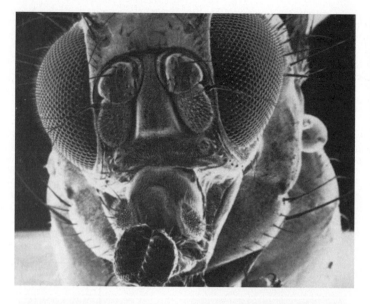

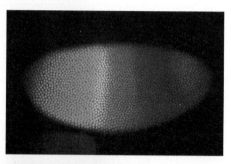

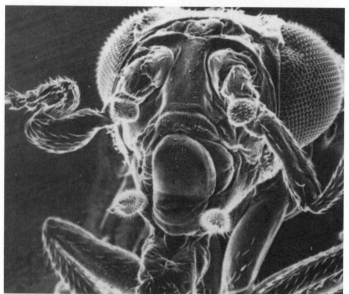

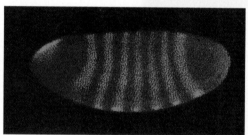

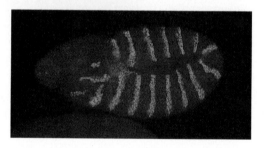

a.

b.

FIGURE 11.7

Morphogen gradients underlie development. *a.* A homeotic mutation sends the wrong morphogen signals to the antenna of a fruit fly, *Drosophila melanogaster,* making the tissue differentiate as leg. Top photo is a normal fly and bottom photo is a homeotic fly. *b.* Very early in the development of *D. melanogaster,* a gradient of bicoid protein, present from the mother in the fertilized ovum, distinguishes one end of the embryo from the other. In the first panel of this computer-enhanced image, different colors represent different concentrations of bicoid. In the second panel, bicoid protein directs the synthesis of two other proteins, one depicted in red and one in green, adding complexity to the protein gradient. A half hour later, as shown in the third panel, another gene directs the production of a protein that divides the embryo into seven stripes. In the final panel, yet another gene oversees dividing each existing section of the embryo in two. Overall, genes produce protein gradients that set up distinct biochemical environments in different parts of the embryo, and these biochemicals further influence differentiation.

whose descendants will become part of the eye. When a homeotic gene mutates, or is altered, the signals go off in the wrong place.

The most astounding discovery about homeotic genes is that they are not peculiar to a few types of insects, as biologists once thought, but appear among an incredible diversity of species. Cows, earthworms, humans, frogs, lampreys, corn, mice, flour beetles, locusts, bacteria, yeast, chickens, roundworms, and mosquitoes are only some of the organisms that carry homeotic genes.

Even though humans hardly sprout legs from their heads, they do develop diseases caused by homeotic genes—for example, some forms of leukemia, a cancer in which white blood cells follow the "wrong" developmental pathway. The fact that homeotic genes are found in so many very different living things indicates that this gene type is both ancient and very important.

The Stages of Human Prenatal Development

During the 38 weeks of human prenatal development, a most amazing biological process unfolds. Through an enormous number of cell divisions and precise cell specializations and interactions, a single cell gradually gives rise to a newborn baby. The newborn consists of trillions of cells and can move, breathe, digest, excrete, think, sense, and respond to the environment.

Developmental biologists recognize three stages of prenatal development. The first 2 weeks, which occur before the three layers that define an embryo appear, comprise the **preembryonic stage.** This stage includes fertilization; division of the fertilized ovum, or **zygote,** as it travels through the woman's fallopian tube towards the uterus; implantation of the ovum into the uterine wall; and initial folding into layers.

· The **embryonic stage** lasts from the start of the third week until the end of the eighth week. During this vital period, the cells of the three layers grow, specialize, and interact to form all the body's organs. Structures that support the embryo—the **placenta, umbilical cord,** and **extraembryonic membranes**—also develop during this stage.

The third stage of prenatal development is the **fetal period,** from the beginning of the ninth week through the full 38 weeks of development. During this stage, organs begin to function and coordinate to form organ systems. Growth is

very rapid. Prenatal development ends with **labor** and **parturition,** which is the birth of a baby. Table 11.1 lists the major stages in the prenatal development of humans and some other animals.

The Preembryonic Stage

Fertilization

The first step in prenatal development is the initial contact between the male's sperm cell and the female's secondary oocyte. (Table 11.2 outlines this step and those that follow.) Recall that the female cell arrests in the metaphase of the sec-

ond meiotic division. It is encased in a thin, clear layer of protein and sugars called the **zona pellucida,** which is further surrounded by a layer of cells called the **corona radiata.** The sperm must penetrate these layers to fertilize the ovum (fig. 11.8a).

About 100 million sperm cells are deposited in the vagina during sexual intercourse. A sperm cell can survive in the woman's body for up to 3 days, but it can only fertilize the oocyte in the 12 to 24 hours after ovulation.

Once inside the woman's body, sperm are chemically activated by a

Table 11.1 Stages of Animal Development

Stage	Events	Timing in Humans
Gametogenesis	Manufacture of sperm and oocytes	Ongoing during much of adulthood
Fertilization	Sperm and oocyte meet	Within 24 hours after ovulation
Cleavage	Rapid cell division	Days 1–3
Gastrulation	Germ layers form	Week 2
Organogenesis	Cells specialize, forming organs	Weeks 3 through 8
Further development	Growth and tissue specialization	Week 9 throughout life

Table 11.2 Stages and Principal Events during Early Human Development

Stage	Time Period	Principal Events
Fertilized ovum	12–24 hours following ovulation	Egg is fertilized; zygote has 23 pairs of chromosomes (diploid) from haploid sperm and haploid egg; genetically unique
Cleavage	30 hours to third day	Mitotic divisions increase number of cells
Morula	Third to fourth day	Solid ball of cells forms
Blastocyst	Fifth day to end of second week	Ball-like structure hollows to single layer of thickness; inner cell mass and trophoblast form; implantation occurs; embryonic disc forms, followed by primary germ layers

From Kent M. Van De Graaff, *Human Anatomy,* 3d ed. Copyright © 1992 Wm. C. Brown Communications, Inc., Dubuque, Iowa. All rights reserved. Reprinted by permission.

217

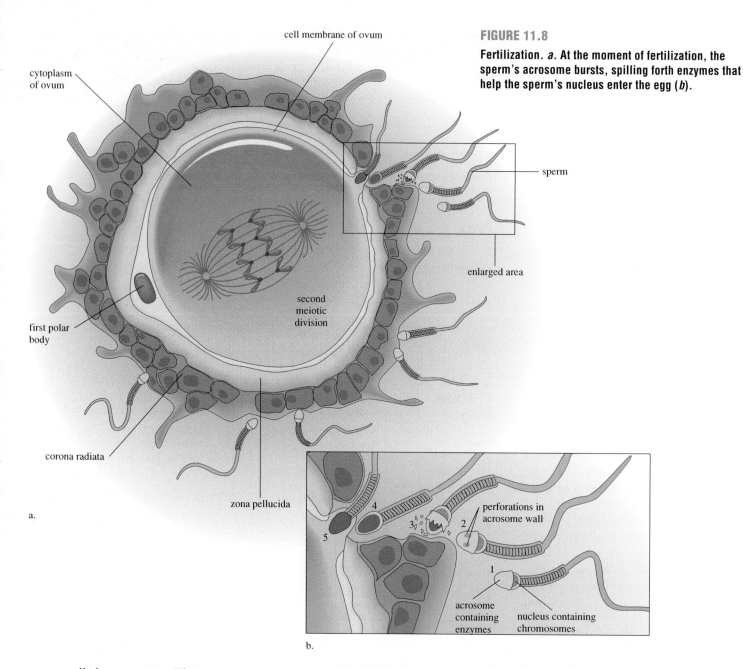

FIGURE 11.8

Fertilization. *a.* At the moment of fertilization, the sperm's acrosome bursts, spilling forth enzymes that help the sperm's nucleus enter the egg (*b*).

cell membrane of ovum

cytoplasm of ovum

sperm

enlarged area

first polar body

second meiotic division

corona radiata

zona pellucida

a.

perforations in acrosome wall

4

3

2

5

1

acrosome containing enzymes

nucleus containing chromosomes

b.

process called **capacitation.** This causes a chemical change in their cell surfaces that enables them to enter an oocyte. In addition, researchers have found that in some species, the oocyte secretes a chemical that attracts the sperm. This may be true in humans, although no attractant has yet been identified. The female's muscle contractions, the movement of the sperm tails, the surrounding mucus, and the waving, hairlike fringes on the cells of the female reproductive tract combine to propel the sperm toward the ovum. Despite this help, only 200 or so sperm near the ovum, and only one will

penetrate it (fig. 11.9). Moreover, even if a sperm contacts an ovum, fertilization may not occur.

When a particular sperm contacts a secondary oocyte, its acrosome bursts, spilling enzymes that digest the zona pellucida and corona radiata protecting the oocyte (fig. 11.8b). Fertilization, or conception, begins when the outer membranes of the sperm and secondary oocyte meet (fig. 11.10). The meeting is dramatic. A wave of electricity spreads physical and chemical changes across the entire oocyte surface to keep other sperm out; accordingly, these changes are called a "block to polyspermy." If more than

one sperm does enter a single oocyte, the resulting cell has too much genetic material to develop normally. Very rarely, such individuals survive until birth, but they have defects in many organs and die within days. When two sperm fertilize two oocytes, fraternal twins result.

As the sperm enters the secondary oocyte, the female cell completes meiosis and becomes a fertilized ovum. Fertilization is not complete, however, until the genetic packages, or **pronuclei,** of the sperm and ovum meet. Within 12 hours of the sperm's penetration, the nuclear membrane of the ovum disappears, and the two new sets of chromo-

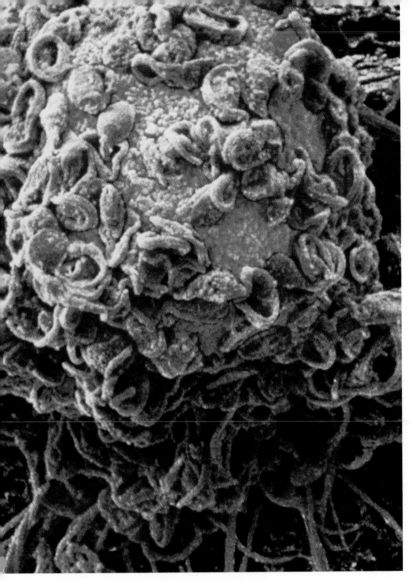

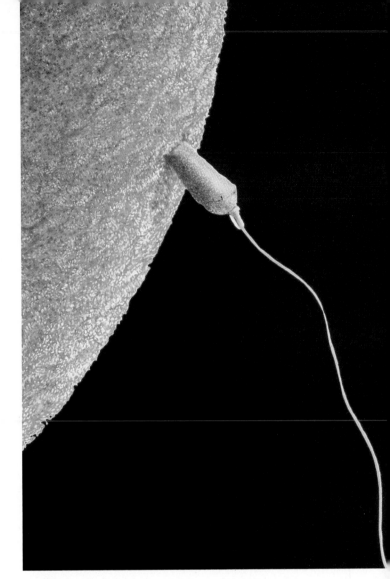

FIGURE 11.9

This electron micrograph shows many sperm near the egg's surface. Only one will penetrate.

FIGURE 11.10

The moment of fertilization.

somes mingle (fig. 11.11). The first cell of the new individual thus forms. It has 23 pairs of chromosomes—one chromosome from each parent in each pair. This is the zygote (Greek for "yolked together"). At this point, the cell is still within a fallopian tube.

Cleavage

About 30 hours after fertilization, the zygote undergoes its first mitotic division. This starts a period of rapid cell division called **cleavage,** resulting in cells called **blastomeres.** Sixty hours after fertilization, another division occurs. The zygote now consists of 4 cells. Cleavage continues until a solid ball of 16 or more cells forms. This ball is called a **morula,** Latin for "mulberry," which it resembles (fig. 11.12).

Three days have passed since fertilization, and the new organism is still within the fallopian tube, although it is drawing closer to the uterus. It is still about the same size as the fertilized ovum, because the initial cleavage divisions produce daughter cells about half the size of the parent cell. After the first few cleavage divisions, the cells have about as much cytoplasm as any somatic cell. During cleavage, organelles and molecules from the secondary oocyte's cytoplasm still control most cellular activity, but some of the developing organism's genes are becoming active.

The morula travels down the fallopian tube and arrives in the uterus sometime between days 3 and 6. The ball of cells hollows out, and its center fills with

fluid that seeps in from the uterus. The organism is now called a **blastocyst,** Greek for "germ bag" (fig. 11.13). The blastomeres are now located either in an outer layer or inside the blastocyst. Some blastomeres form the outer layer of single cells called the **trophoblast** (Greek for "nourishment of germ"). Certain trophoblast cells will develop into a membrane called the **chorion,** which eventually forms the fetal portion of the placenta, the organ that brings oxygen and nutrients to the fetus and remove wastes from it.

A clump of blastomeres inside the blastocyst forms the **inner cell mass.** These cells will develop into the embryo plus its supportive extraembryonic membranes. The fluid-filled center of the ball of cells is called the **blastocyst cavity.**

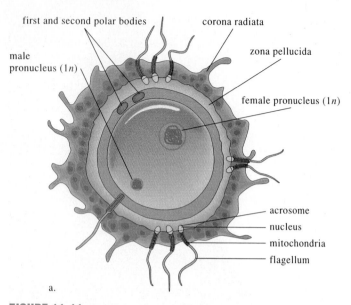

first and second polar bodies

corona radiata

male pronucleus (1*n*)

zona pellucida

female pronucleus (1*n*)

acrosome

nucleus

mitochondria

flagellum

a.

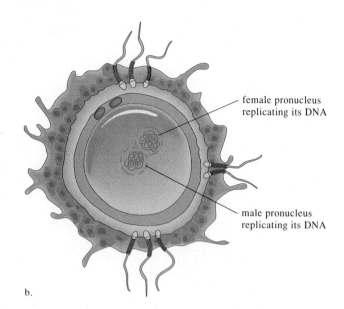

female pronucleus replicating its DNA

male pronucleus replicating its DNA

b.

FIGURE 11.11

a. Once the sperm cell enters the egg, (*b*) its pronucleus approaches the egg's pronucleus, and (*c*) finally brings the 23 sets of chromosomes together. *d.* The first cell of the new organism now exists.

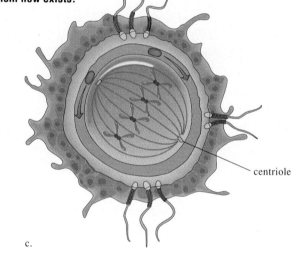

centriole

c.

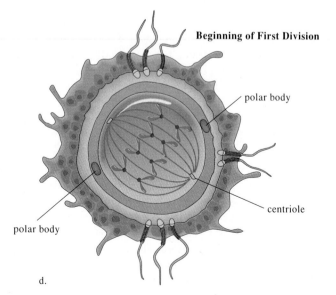

Beginning of First Division

polar body

centriole

polar body

d.

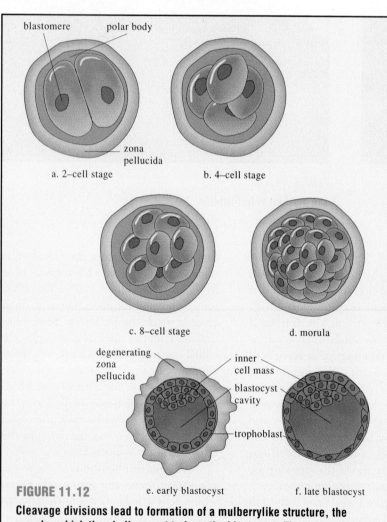

blastomere

polar body

zona pellucida

a. 2–cell stage

b. 4–cell stage

c. 8–cell stage

d. morula

degenerating zona pellucida

inner cell mass

blastocyst cavity

trophoblast

e. early blastocyst

f. late blastocyst

FIGURE 11.12

Cleavage divisions lead to formation of a mulberrylike structure, the morula, which then hollows out to form the blastocyst. The outer cell layer is the trophoblast, and the clump of cells on one side of the interior is the inner cell mass.

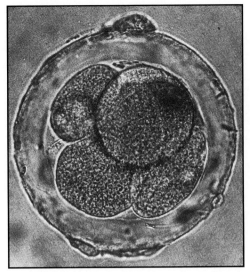

a.

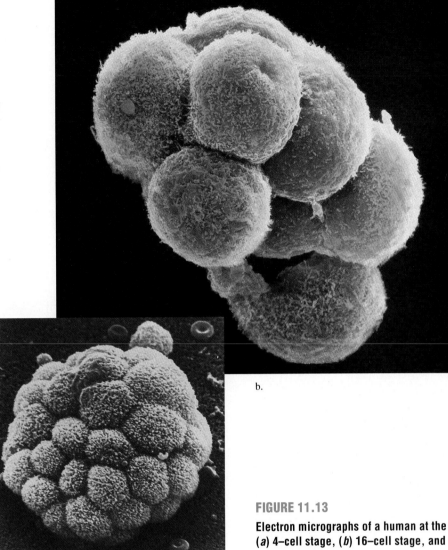

b.

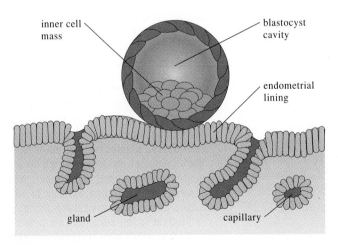

Implantation

Sometime between days 5 and 7, the blastocyst attaches to the uterine lining or **endometrium** (fig. 11.14). The inner cell mass within the blastocyst settles against the endometrium. Digestive enzymes produced by the trophoblast eat through the outer epithelial layer of the uterine lining, and ruptured blood vessels surround the blastocyst, bathing it in nutrient-rich blood. This nestling of the blastocyst into the uterine lining is called **implantation.** The trophoblast layer directly beneath the inner cell mass thickens and sends out fingerlike projections into the endometrium at the site of implantation (fig. 11.15). These projections develop into the chorion.

c.

FIGURE 11.13

Electron micrographs of a human at the (*a*) 4–cell stage, (*b*) 16–cell stage, and (*c*) morula stage.

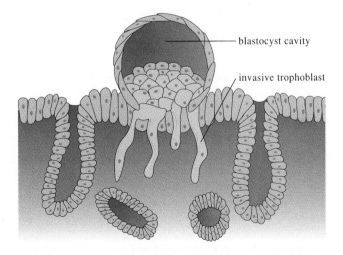

FIGURE 11.14

Implantation. On about day 6, the blastocyst implants in the uterine lining, the endometrium.

FIGURE 11.15

Invasion of the endometrium. Part of the trophoblast adjacent to the endometrium sends out projections, which will develop into the chorion.

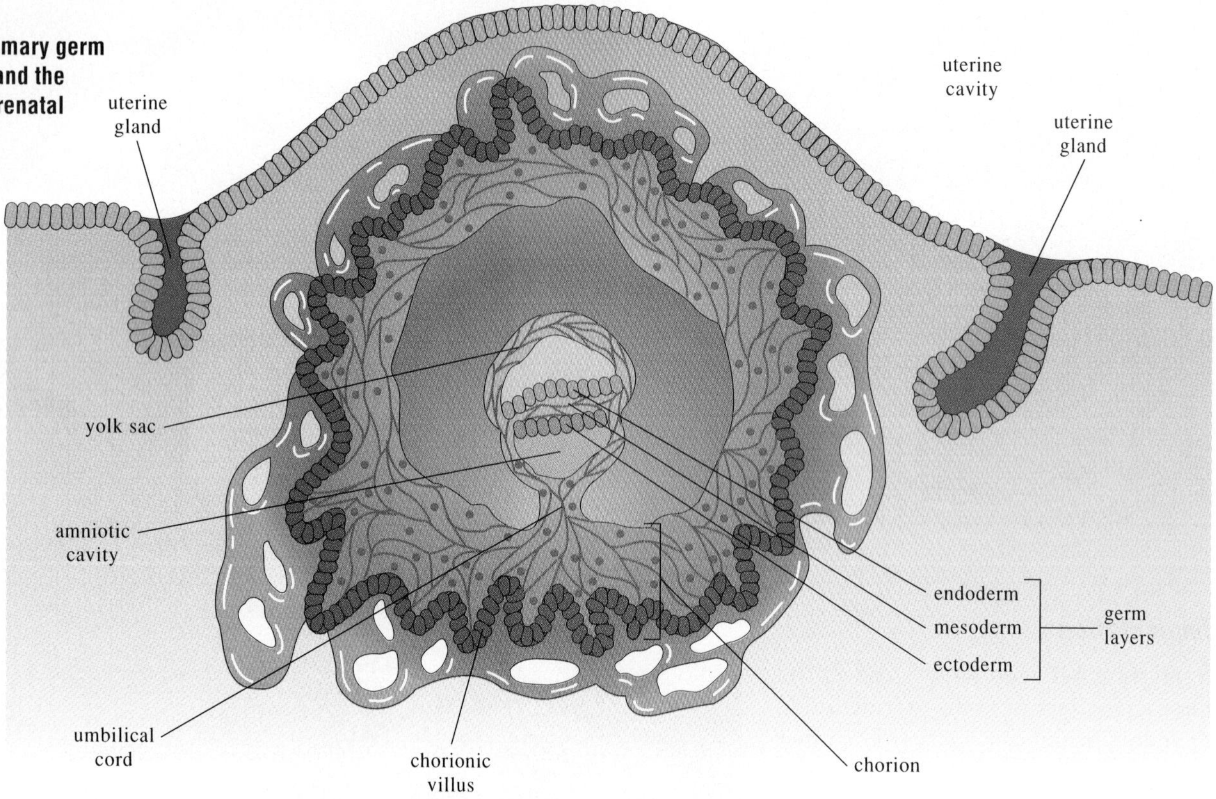

FIGURE 11.16

After 2 weeks, the primary germ layers have formed, and the embryonic stage of prenatal development begins.

The trophoblast cells now secrete a hormone, **human chorionic gonadotropin** (hCG), which prevents menstruation. In this way, the blastocyst helps to ensure its own survival, for if menstruation occurs, the new organism would be eliminated from the woman's body along with the tissue in the uterus. The cells continue to produce hCG for about 10 weeks.

Gastrulation

During the second week of prenatal development, the blastocyst completes implantation, and the inner cell mass changes. A space called the **amniotic cavity** forms between the inner cell mass and the portion of trophoblast that "invades" the endometrium. The inner cell mass then flattens and is called the **embryonic disc.**

The embryonic disc at first consists of two layers. The outer layer, nearest the amniotic cavity, is called the **ectoderm** (Greek for "outside skin"). The inner layer, closer to the blastocyst cavity, is the **endoderm** (Greek for "inside skin"). Shortly after, a third and middle layer called the **mesoderm** (Greek for "middle skin") forms. This three-layered structure

is the **primordial embryo** or **gastrula.** The process of forming the primordial embryo is called gastrulation, and the layers are called **germ layers** (fig. 11.16).

Gastrulation is an important process in prenatal development because a cell's fate is determined by which layer it is in (fig. 11.17). Ectoderm cells develop into the nervous system, sense organs, the outer skin layer (epidermis), hair, nails, and skin glands. Mesoderm cells develop into bone, muscles, blood, the inner skin layer (dermis), reproductive organs, and connective tissue. Endoderm cells eventually form the organs and the linings of the digestive, respiratory, and urinary systems. Gastrulation marks the start of **morphogenesis,** the series of events that mold the newborn. Diagrams depicting the correspondence between cell regions in a prenatal organism and adult body structures are called **fate maps.** Developmental biologists have used many experimental techniques to derive fate maps (Biology in Action 11.1).

As the preembryonic stage ends, it is the end of the second week of prenatal development. Although the woman has not yet "missed" her menstrual period, she might suspect something is happen-

ing: her breasts may be swollen and tender, and she may be unusually tired. She may even be carrying twins (fig. 11.18). By now her urine contains enough hCG to respond positively to an at-home pregnancy test. Highly sensitive blood tests can detect hCG as early as three days after conception.

KEY CONCEPTS

The preembryonic stage encompasses fertilization, cleavage, implantation, and gastrulation. During fertilization, a single sperm's acrosome bursts, releasing enzymes that enable the sperm to enter the oocyte; then the oocyte's surface changes to prevent other sperm from entering. The chromosomes from the male and female cells meet to form the zygote. During cleavage, the single-celled zygote divides repeatedly, eventually forming the two-layered blastocyst. The outer layer, the trophoblast, eventually forms the placenta, while the inner layer or cell mass forms the embryo and extraembryonic membranes. Enzymes help the blastocyst to implant in the endometrium between days 5 and 7. There, trophoblast cells secrete hCG. During gastrulation, the three germ layers of the primordial embryo form. Cells in each germ layer later become part of particular organ systems.

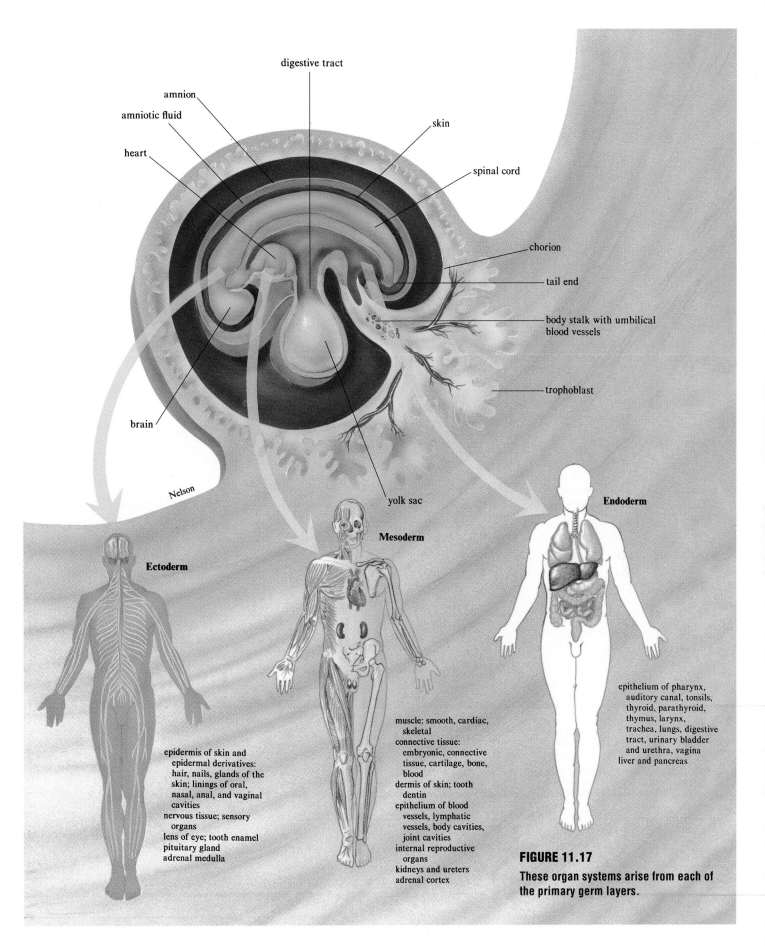

digestive tract

amnion

amniotic fluid

skin

heart

spinal cord

chorion

tail end

body stalk with umbilical blood vessels

trophoblast

brain

Nelson

yolk sac

Endoderm

Mesoderm

Ectoderm

epidermis of skin and
  epidermal derivatives:
  hair, nails, glands of the
  skin; linings of oral,
  nasal, anal, and vaginal
  cavities
nervous tissue; sensory
  organs
lens of eye; tooth enamel
pituitary gland
adrenal medulla

muscle: smooth, cardiac,
  skeletal
connective tissue:
  embryonic, connective
  tissue, cartilage, bone,
  blood
dermis of skin; tooth
  dentin
epithelium of blood
  vessels, lymphatic
  vessels, body cavities,
  joint cavities
internal reproductive
  organs
kidneys and ureters
adrenal cortex

epithelium of pharynx,
  auditory canal, tonsils,
  thyroid, parathyroid,
  thymus, larynx,
  trachea, lungs, digestive
  tract, urinary bladder
  and urethra, vagina
liver and pancreas

FIGURE 11.17

These organ systems arise from each of the primary germ layers.

Biology in ACTION

A Zebrafish Fate Map

As in many other organisms, the early embryo of the zebrafish *Brachydenio rerio* consists of three layers: the endoderm, mesoderm, and ectoderm. At what point do cells in each layer commit to develop into a specific adult structure? Robert Ho from Princeton University and Charles B. Kimmel of the University of Oregon tackled this question by transplanting embryonic cells from different stages into a recipient embryo, watching to see what structures the cells would become. By labeling the cells and enlisting the help of time-lapse videography, Ho and Kimmel obtained a view of early developmental "decision making" (fig. 1).

Five hours after fertilization, the zebrafish embryo is at the early gastrula stage, resembling a hollow ball that has just been punched in, yielding the three germ layers. Ho and Kimmel took cells from the newly formed mesoderm that they knew were destined to develop into muscle. They transplanted the cells to another 5-hour embryo, placing them with cells that would normally differentiate into either eye or nervous tissue. The transplanted cells, apparently taking cues from the cells in their new home, developed as eye or neuron.

When they repeated the experiment, transplanting mesoderm cells from a 6.5-hour embryo into a 5-hour recipient, the researchers obtained a different result—a third of the transplanted cells retained the specialization associated with their original position. When an 8-hour embryo served as donor, nearly all of the transplanted cells followed their original fate and developed as muscle tissue. However, another unexpected result cropped up—the transplanted cells differentiated as muscle, but they also migrated. The cells had been transplanted to the brain of the second embryo, but they moved to a more appropriate location. Not only do late-gastrula-stage cells seem to "know" what tissues they will become, they also "know" where to develop.

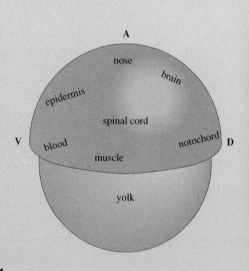
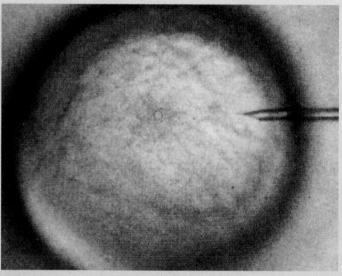

FIGURE 1

A fate map of the zebrafish gastrula indicates the structures that cells in certain regions will become. The photograph of the gastrula looks down on the point labeled "A" on the fate map. The instrumen on the right is placing a transplanted cell atop the recipient embryo.

From Robert K. Ho and Charles B. Kimmel, "Commitment of Cell Fate in the Early Zebrafish Embryo," in *Science* 261:109, July 2, 1993. Copyright © 1993 by the American Association for the Advancement of Science, Washington, D.C. Reprinted by permission of the publisher and authors.

The Embryonic Stage

The embryonic stage extends from the third to eighth weeks of prenatal development. During this stage, organs begin to develop and structures form that will nurture and protect the developing organism. The embryo begins to organize around a central axis. Towards the end of this period, the embryo develops rapidly in a precisely regulated sequence. The embryonic stage includes the development of the embryo and formation of the supportive structures. These include the placenta, the umbilical cord, and certain extraembryonic membranes. The **yolk sac,** and **allantois** are extraembryonic membranes that form during the embryo period. Two others, the chorion and **amnion,** begin to form during the preembryonic period.

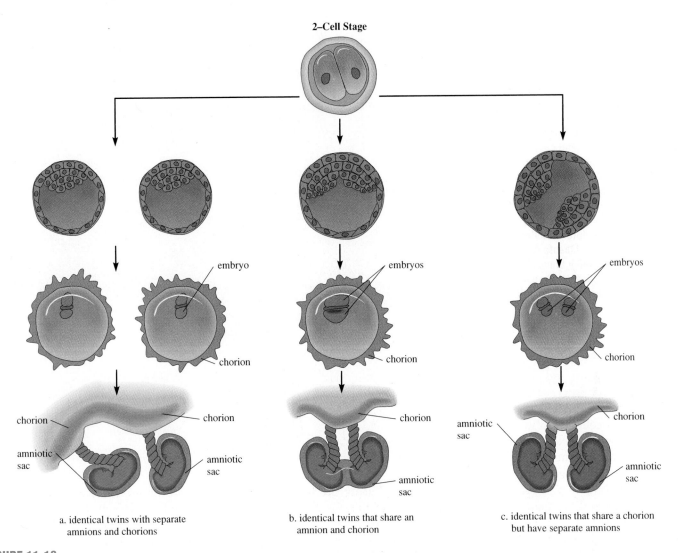

a. identical twins with separate
amnions and chorions

b. identical twins that share an
amnion and chorion

c. identical twins that share a chorion
but have separate amnions

FIGURE 11.18

Facts about twins. One in 81 births in the United States produces twins. Identical twins originate at three points in development. *a.* In about one-third of identical twins, separation of cells into two groups occurs before the trophoblast forms on day 5. These twins have separate chorions and amnions. *b.* About 1% of identical twins share a single amnion and chorion, because the tissue splits into two groups after these structures have already formed. *c.* In about two-thirds of identical twins, the split occurs after day 5 but before day 9. These twins share a chorion but have separate amnions. Fraternal twins result from two sperm fertilizing two secondary oocytes. The twins develop their own amniotic sacs, yolk sacs, allantois, placentae, and umbilical cords. Fraternal twins can have two different fathers, if the mother ovulated twice and had intercourse with two men within a short period of time. Fraternal twins run in families. In about 55% of all twin conceptions, one twin dies before birth. Many times the parents do not even know that the surviving baby was a twin!

Supportive Structures

By about the third week after conception, as the embryo is folding into its three-layered form, the fingerlike projections from the chorion, called **chorionic villi,** extend further into the uterine lining. These embryonic outgrowths, reaching towards the mother's blood supply, are the beginnings of the placenta. The placenta is a unique organ because one side of it—the chorion tissue—comes from the new organism, and the other side consists of blood pools that come from the pregnant woman's circulatory system (fig.11.19).

The blood systems of mother and embryo are separate, but they lie side by side. The chorionic villi extend between the pools of maternal blood. This proximity makes it possible for nutrients and oxygen to diffuse from the mother's circulatory system across the chorionic villi cells to the embryo, and for wastes to leave the embryo's circulation and enter the mother's, which eventually excretes them.

The placenta is completely developed at 10 weeks, establishing a vital link between mother and fetus that will last throughout pregnancy. When fully formed, the placenta is a reddish-brown disc that weighs about two pounds. In addition to providing a lifeline to the embryo and fetus, the placenta secretes hormones that maintain the pregnancy. Placental hormones also alter the mother's glucose metabolism so that the fetus receives much of the energy-rich sugar.

A generation ago, physiologists thought that the placenta filtered out substances in the mother's circulation that could harm the embryo. Today we know that although the placenta can detoxify some substances, many chemicals pass readily from mother to unborn

FIGURE 11.19

The circulations of mother and unborn child come into close contact at the site of the chorionic villi, but they do not actually mix. Branches of the mother's arteries in the wall of her uterus open into pools near the chorionic villi. The nutrients and oxygen from the mother's blood cross the villi cell membranes and enter fetal capillaries (tiny blood vessels) on the other side of the villi cells. The fetal capillaries lead into the umbilical vein, which is enclosed in the umbilical cord. From here, the fresh blood circulates through the fetus's body. Blood that the fetus has depleted of nutrients and oxygen returns to the placenta in the umbilical arteries, which branch into capillaries. Waste products then diffuse to the maternal side from these capillaries.

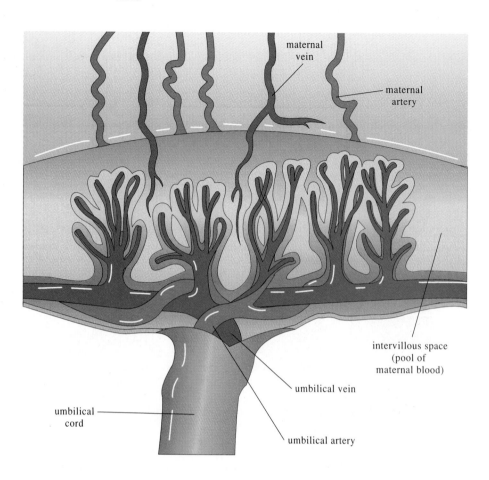

child. Nicotine from cigarette smoke crosses the placenta and stunts fetal growth. Addictive drugs such as heroin and cocaine reach the fetus, addicting the newborn. Alcohol can also cross the placenta and harm the fetus. Viruses are small enough to cross the placenta; they may cause devastating effects in the fetus, even when they produce only minor symptoms in the pregnant woman. The human immunodeficiency virus that causes AIDS crosses the placenta in about a third of infected pregnant women. Chapter 12 discusses these and other environmental factors that may cause birth defects.

During the early embryonic period, the extraembryonic membranes also form. The yolk sac begins to appear beneath the embryonic disc at the end of the second week. It manufactures blood cells until about the sixth week, when the liver takes over, and then it starts to shrink. Parts of it eventually develop into the intestines and germ cells. Despite its name, the human yolk sac does not actually contain yolk. Similar structures in other animals do contain yolk, which provides nutrients to the developing embryo.

By the third week, an outpouching of the yolk sac forms the allantois, another extraembryonic membrane. It, too, manufactures blood cells, and it also gives rise to the fetal umbilical arteries

and vein. During the second month, most of the allantois degenerates, but part of it persists in the adult as a ligament supporting the urinary bladder.

As the yolk sac and allantois develop, the amniotic cavity, the space within the amnion, swells with amniotic fluid. This "bag of waters" cushions the embryo, maintains a constant temperature and pressure, protects the embryo if the mother falls, and allows development to proceed unhampered by the forces of gravity. The amniotic fluid is derived from the mother's blood. It also contains fetal urine and cells from the amniotic sac, the placenta, and the fetus.

The "amniote egg," in which the embryo is enclosed in a watery sac, is the legacy of an evolutionary giant step that may have occurred about 30 million years ago. Such eggs appeared first in reptiles; today, the embryo is immersed in fluid in reptiles, birds, and mammals. More primitive vertebrates such as amphibians must return to water to lay their eggs. In contrast, an amniote egg has built-in water and food supplies.

Toward the end of the embryonic period, as the yolk sac shrinks and the amniotic sac swells, another structure forms: the umbilical cord. The cord is 2 feet (0.6 meter) long and 1 inch (2.5 centimeters) in diameter. It houses two umbilical arteries, which transport oxygen-depleted blood from the embryo to the placenta, and one umbilical vein, which brings oxygen-rich blood to the embryo. The umbilical cord attaches to the center of the placenta. It twists because the umbilical vein is slightly longer than the umbilical arteries.

Extraembryonic structures are used in medical procedures. For example, fetal cells from amniotic fluid or chorionic villi cells provide information on the genetic health of a fetus. Umbilical cord blood is a rich source of stem cells for gene therapy.

The Embryo

As the days and weeks proceed, different rates of cell division in different parts of the embryo cause its tissues to fold in intricate patterns. In a little

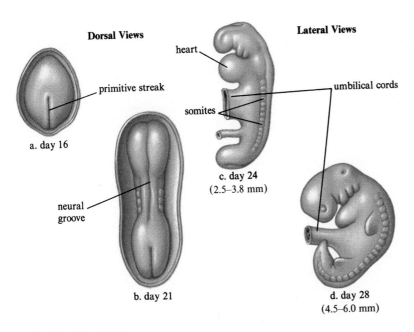

Dorsal Views — Lateral Views

heart

primitive streak

umbilical cords

somites

a. day 16

neural groove

c. day 24 (2.5–3.8 mm)

b. day 21

d. day 28 (4.5–6.0 mm)

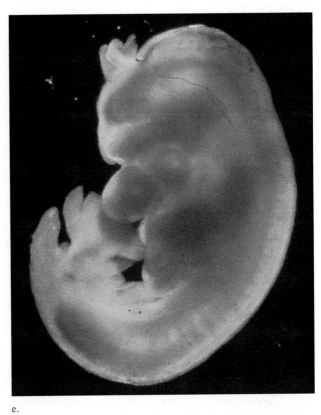

e.

FIGURE 11.20

Early embryos. It takes about a month for the embryo to look like a "typical" embryo. At first all that can be distinguished is the primitive streak (*a*), but soon the central nervous system begins to form (*b*). By the 24th day, the heart becomes prominent as a bulge (*c*), and by the 28th day, the organism is beginning to look human (*d* and *e*).

understood process called **embryonic induction,** the specialization of one group of cells causes adjacent groups of cells to specialize. Gradually, these changes mold the three primary germ layers into organs and organ systems. **Organogenesis** (making organs) is the term that describes the transformation of the structurally simple, three-layered embryo into an individual with distinct organs. During the weeks of organogenesis, the organism is particularly sensitive to environmental factors such as chemicals and viruses.

During the third week of prenatal development, a band called the **primitive streak** appears along the back of the embryonic disc. It gradually elongates to form an axis, an anatomical reference point which other structures organize around as they develop. The primitive streak eventually gives rise to connective tissue precursor cells and the **notochord,** a structure that forms the basic framework of the skeleton. The notochord induces overlying ectoderm to specialize into a hollow **neural tube,** which goes on to develop into the brain and spinal cord (central nervous system).

Formation of the neural tube, or **neurulation,** is a key event in early development because it marks the beginning of organ formation. Soon after neurulation begins, a reddish bulge containing the heart appears. It begins to beat around day 18. Soon the central nervous system starts to form. Figure 11.20 shows some early embryos undergoing these changes.

The fourth week of embryonic existence is one of spectacularly rapid growth differentiation. Blood cells begin to form and to fill primitive blood vessels. Immature lungs and kidneys appear. If the neural tube does not close normally at about day 28, a deformity called a neural tube defect results, so that an area of the spine is open and parts of the brain or spinal cord protrude. If this happens, a fetal liver biochemical called alpha fetoprotein leaks at an abnormally rapid rate into the pregnant woman's circulation. If a maternal blood test at the fifteenth week of pregnancy detects high levels of alpha fetoprotein, the fetus may have a neural tube defect.

Also during the fourth week, small buds appear where the arms and legs will develop. In the early 1960s,

European women who took the mild tranquilizer thalidomide between days 28 and 42 gave birth to babies whose limbs developed abnormally during the embryonic period. About ten thousand children were born with flipperlike stumps where their arms and legs should have been. Several "thalidomide babies" were born in 1994 to women in several South American nations where they were not warned about the drug's effects.

The 4-week embryo has a distinct head and jaw, and the eyes, ears, and nose are beginning to take shape. The inklings of a digestive system appear as a long, hollow tube that will become the intestines. A woman carrying this embryo, which is now only 1/4 inch (0.6 centimeter) long, may suspect that she is pregnant, because her menstrual period is about 2 weeks late.

By the fifth week, the embryo's head appears far too large for its body. Limbs extend from the body, ending in platelike structures. Tiny ridges run down the plates, and by week 6 the ridges deepen, molding fingers and toes. The eyes are open, but they do not yet have eyelids or irises. The head still seems large, and cells in the brain are rapidly differentiating. The embryo is now about 1/2 inch (1.3 centimeters) from head to buttocks.

During the seventh and eighth weeks, a cartilage skeleton appears. The placenta is now almost fully formed and functional, secreting hormones that maintain the blood-rich uterine lining. The embryo is about the size and weight of a paper clip. The eyes are sealed shut and will stay that way until the seventh month. The nostrils are also closed. A neck appears as the head begins to comprise proportionately less of the embryo, and the abdomen flattens somewhat. The 7- or 8-week-old organism, on the border between embryo and fetus, already looks quite human (fig. 11.21).

KEY CONCEPTS

The embryonic period of prenatal development extends from week 3 through week 8. During the third week, the chorionic villi reach through the uterine lining toward the maternal circulation. The blood supplies of mother and embryo lie adjacent to one another. Nutrients and oxygen enter the embryo's circulation, and wastes are passed from the embryo into the mother's circulation. Many substances can pass through the placenta to the embryo. The yolk sac forms early in the embryonic period and manufactures blood cells until the liver takes over at week 6. The allantois forms from part of the yolk sac; it produces blood cells and eventually forms the umbilical arteries and vein. The amniotic cavity expands with fluid as the yolk sac and allantois shrink. The umbilical cord forms late in the embryonic period. During the third week the primitive streak appears, and then, rapidly, other rudiments of organs form, including the central nervous system, notochord, neural tube, heart, limbs, fingers, toes, eyes, ears, and nose.

The Fetal Period

The body proportions of the fetus appear more like those of a newborn. The ears lie low, and the eyes are widely spaced. Bone begins to form and will eventually replace most of the softer cartilage. Soon, as the nerves and muscles coordinate, the fetus will move its arms and legs.

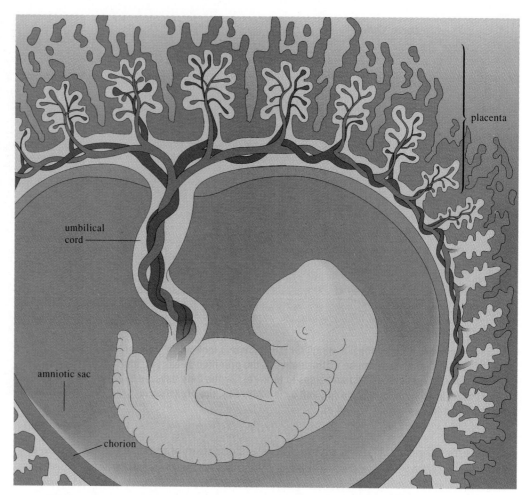

FIGURE 11.21

At seven weeks of development, the chorionic villi and the tissue between them form the placenta. Blood from the embryo flows to and from the placenta in the umbilical vein and arteries.

Sex is determined at conception, when a sperm bearing an X or Y chromosome meets an oocyte, which always has an X chromosome. An individual with two X chromosomes is a female, and one with an X and a Y is a male. A gene on the Y chromosome, called SRY (for "sex-determining region of the Y") determines maleness. Differences between the sexes do not appear until week 6, after the SRY gene activates. Male hormones then stimulate male reproductive organs and glands to differentiate from existing structures that have the capacity to develop as either sex. In a female, the indifferent structures of the early embryo develop as female organs and glands. The fetus is obviously male or female by the twelfth week.

Also by the twelfth week, the fetus sucks its thumb, kicks, makes fists and faces, and begins to form baby teeth. It breathes the amniotic fluid and urinates and defecates into it. The first trimester (three months) of pregnancy is over.

During the second trimester, the body proportions of the fetus become even more like those of a baby. By the fourth month, it has hair, eyebrows, lashes, nipples, and nails. Some fetuses even scratch themselves before birth. Bone continues to replace the cartilage skeleton. The fetus's muscle movements become stronger, and the woman may begin to feel slight flutterings. By the end of the fourth month, the fetus is about 8 inches (20 centimeters) long and weighs about 6 ounces (170 grams).

During the fifth month, the fetus becomes covered with an oily substance called vernix caseosa, which looks like cottage cheese. It protects the skin, which is just beginning to grow. The vernix is held in place by white, downy hair called lanugo, which may persist for a few weeks after birth. By 18 weeks, the vocal cords form, but the fetus makes no sounds because it does not breathe air. By the end of the fifth month, the fetus is curled in the classic head-to-knees position. It weighs about 1 pound (454 grams) and is 9 inches (30.5 centimeters) long.

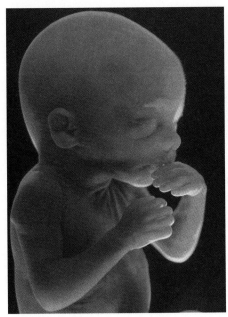

FIGURE 11.22
A fetus at 20 weeks.

During the sixth month (fig. 11.22), the skin appears wrinkled because there isn't much fat beneath it. The skin turns pink as blood-filled capillaries extend into it. By the end of the second trimester, the woman feels distinct kicks and jabs and may even detect a fetal hiccup. The fetus is now about 12 inches long.

In the final trimester, fetal brain cells rapidly form networks, and organs elaborate and grow. A layer of fat develops beneath the skin. The digestive and respiratory systems mature last, which is why infants born prematurely often have difficulty digesting milk and breathing.

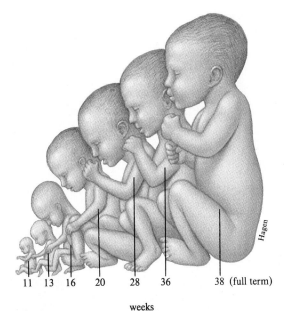

11 13 16 20 28 36 38 (full term)

weeks

Hagen

FIGURE 11.23

Note the changing proportions of the head to the rest of the body in a prenatal human.

Figure 11.23 shows human fetuses at various stages. Approximately 266 days after a single sperm burrowed its way into an oocyte, a baby is ready to be born.

KEY CONCEPTS

During the fetal period, from week 9 until birth, structures grow and increase in detail and begin to interact with one another. Bone replaces cartilage in the skeleton, the body grows in proportion to the head, and sex organs become distinct. The fetus is active now. During the fifth month, the fetus becomes coated with vernix caseosa, and downy hair called lanugo appears. In the final trimester, fetal movements become more pronounced, fat fills out the skin, and growth is rapid. After 38 weeks, the fetus is ready for birth.

Labor and Birth

Strenuous work by the pregnant woman, appropriately called labor, immediately precedes parturition, or birth. Labor may begin with an abrupt leaking of amniotic fluid as the fetus presses down and ruptures the sac. This exposes the fetus to the environment; if birth doesn't occur within 24 hours, the baby may be born with an infection. Labor may also begin with a slight discharge of blood and mucus from the vagina. Or a woman may feel mild contractions in her lower abdomen about every 20 minutes.

As labor proceeds, the hormone-prompted uterine contractions gradually increase in frequency and intensity. During the first stage of labor, the baby presses against the cervix with each contraction. The cervix dilates (opens) a little more each time. At the start of labor, the cervix is a thick, closed band of tissue. By the end of the first stage of labor, the cervix stretches open to about 10 centimeters. The cervix sometimes takes several days to open, with mild labor beginning well before a woman realizes it has started (fig. 11.24).

The second stage of labor is the delivery of the baby (figs. 11.24 and 11.25). It begins when the cervix is completely dilated to 10 centimeters. At this point, the woman feels a tremendous urge to push. Anywhere from a few minutes to 2 hours later, the baby descends through the cervix and vagina and is born. In the third and last stage of labor, uterine contractions expel the placenta and extraembryonic membranes from the woman's body.

Sometimes the fetus is born through a surgical procedure called a Caesarian section. In this procedure, the surgeon removes the baby through an incision made in the mother's abdomen. For a variety of reasons, about one in five births in the United States is performed by Caesarian section. The baby may be too large to fit through the mother's pelvis; it may be positioned feet or buttocks down rather than head down (breech), so that the head may become caught if the baby is delivered vaginally; or the fetus may be side-to-side and refuse to budge. If the umbilical cord wraps around the fetus's neck as it moves into the birth canal, a Caesarian section can save the baby's life.

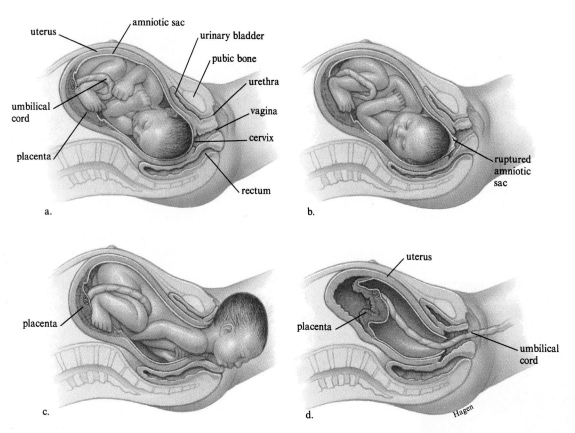

uterus
amniotic sac
urinary bladder
pubic bone
urethra
vagina
cervix
umbilical cord
placenta
rectum

a.

ruptured amniotic sac

b.

placenta

c.

uterus
placenta
umbilical cord

d.

Hagen

FIGURE 11.24

About two weeks before birth, the baby "drops" in the woman's pelvis and the cervix may begin to dilate *(a)*. At the onset of labor, the amniotic sac may break *(b)*. The baby *(c)* and then the placenta *(d)* and extraembryonic membranes are pushed out of the birth canal.

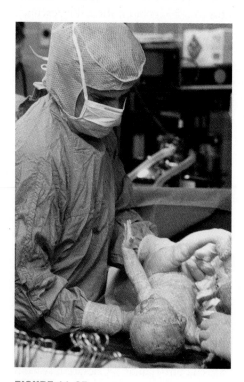

FIGURE 11.25

The moment of birth.

The birth of a live, healthy baby is against the odds, considering human development from the beginning. Of every 100 secondary oocytes exposed to sperm, 84 are fertilized. Of these, 69 implant in the uterus, 42 survive 1 week or longer, 37 survive 6 weeks or longer, and only 31 are born alive. Of those that do not survive, about half have chromosomal abnormalities too severe to maintain life.

KEY CONCEPTS

Any of several events may signal that labor and parturition are about to begin—the amniotic sac rupturing, a bloody mucus discharge, or mild labor contractions. During labor, uterine contractions push the baby down against the cervix, which gradually dilates. After the baby is born, the uterus expels the placenta and extraembryonic membranes. In some situations that threaten the mother or child, a Caesarian section is performed.

Maturation and Aging

Development does not culminate when the fetus enters the world outside the uterus, or even when a child passes through adolescence into adulthood. The structural and functional changes in the body that constitute aging occur throughout life, although they may not be particularly obvious until adulthood (fig.11.26). These changes occur at different rates in different individuals, but many people notice some general bodily changes at certain ages.

Aging Over a Lifetime

Aging actually begins even before fertilization, because the primary oocyte has been dormant in the ovary since the female reached puberty. The fact that children born to women over age 35 are more likely to inherit an abnormal number of chromosomes may be related to the

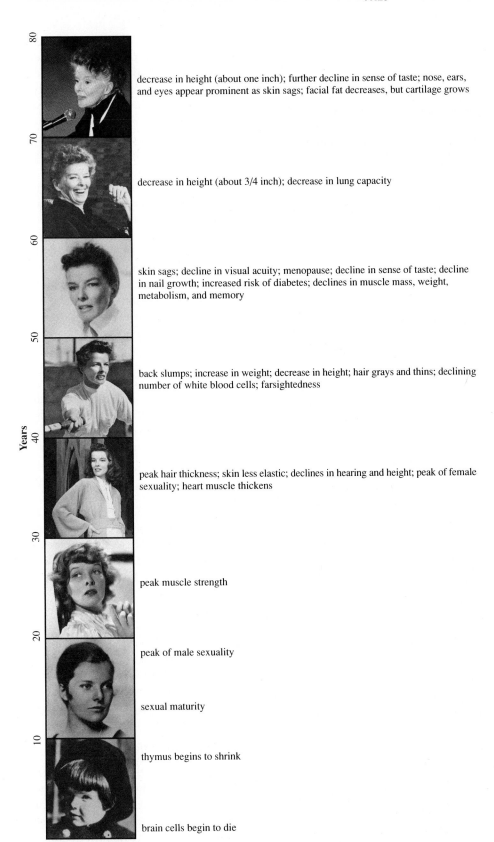

decrease in height (about one inch); further decline in sense of taste; nose, ears, and eyes appear prominent as skin sags; facial fat decreases, but cartilage grows

decrease in height (about 3/4 inch); decrease in lung capacity

skin sags; decline in visual acuity; menopause; decline in sense of taste; decline in nail growth; increased risk of diabetes; declines in muscle mass, weight, metabolism, and memory

back slumps; increase in weight; decrease in height; hair grays and thins; declining number of white blood cells; farsightedness

peak hair thickness; skin less elastic; declines in hearing and height; peak of female sexuality; heart muscle thickens

peak muscle strength

peak of male sexuality

sexual maturity

thymus begins to shrink

brain cells begin to die

FIGURE 11.26

Although many biological changes ensue as we grow older, photographs of actress Katharine Hepburn at various stages of life indicate that we can age with grace and beauty.

greater age of the female germ cell at conception. "Older" oocytes have had more time to accumulate genetic damage from long-term exposure to chemicals, viruses, and radiation, which may disrupt meiosis so that oocytes form with abnormal numbers of chromosomes. An abnormal chromosome number usually results in miscarriage or severe birth defects. .

Human cells begin to die even before an individual is born—for example, fingers and toes are gradually carved from weblike extremities by cells in the webbing that die. Cells die and cells form throughout human life. The life spans of cells are reflected as biological structures and functions wax and wane, and as different abilities peak and then decline at characteristic rates.

By the age of 2 years, a child has as many brain cells as he or she will ever have. The brain grows at the same rate over the first two years of life as it does in the last 6 months in the uterus. At age 10, a person's hearing is the best it will ever be. Shortly before an individual reaches sexual maturity (around age 12 for a girl, 14 for a boy), a small gland in the chest called the thymus reaches its greatest size (about the size of a walnut). It then begins to shrink until it is almost microscopic by about the seventh decade of life. The thymus is part of the body's immune system. Its declining activity may play a key role in a person's increasing susceptibility to certain illnesses as he or she ages.

Muscle strength peaks in both sexes in the 20s. Hair is at its fullest in the 30s, and each hair is as thick as it will ever be. By the end of the third decade of life, the first obvious signs of aging may appear as the facial skin becomes less elastic, creating small wrinkles around the mouth and eyes. Height is already starting to decrease, but not yet at a detectable level.

Age 30 seems to be a developmental turning point. Hearing often becomes less acute. Heart muscle begins to thicken. Elasticity of ligaments between the small bones in the back lessens, setting the stage for the slumping posture characteristic of the later years. Some researchers estimate that beginning roughly at age 30, the human body becomes functionally less efficient by about 0.8% every year. Yet growing older has its benefits—many people feel the strongest desire for sex during their 30s.

During their 40s, many people weigh 10 to 20 pounds (4.5 to 9 kilograms) more than they did at age 20, thanks to slowing metabolism and decreased activity levels. They may be 1/8 inch (0.3 centimeter) shorter. Hair may be graying or thinning, and the person may become farsighted or nearsighted. Some of the immune system's white blood cells lose efficiency, making the body more susceptible to infection and cancer.

The early 50s bring further declines in the functioning of the human body. Nail growth slows, taste buds die, and the skin continues to lose elasticity. For most people, the ability to see close objects becomes impaired, but for the nearsighted, vision improves. Women stop menstruating (menopause), although this does not necessarily mean that they lose interest in sex. Decreased activity of the pancreas may lead to diabetes. By the decade's end, muscle mass and weight begin to decrease. A male produces less semen, though he may still be sexually active. His voice may become higher as his vocal cords degenerate. A man has half the strength in his arm muscles and half the lung function he did at age 25.

The 60-year-old may experience minor memory losses. He or she may be about 3/4 inch (2 centimeters) shorter. A few million of the person's trillion or so brain cells have died over his or her lifetime, but for the most part, intellect remains sharp.

By age 70, height decreases a full inch (2.5 centimeters). Sagging skin and loss of connective tissue, combined with continued cartilage growth, make the nose, ears, and eyes more prominent. For some people, life ceases when they are in their 70s.

KEY CONCEPTS

Aging begins before fertilization and continues throughout the human lifetime. Certain cells die at particular times. Although organs such as the brain and thymus cease cell division early in life, aging usually becomes apparent after age 30. Physical aging becomes more noticeable as an individual enters the 40s and 50s, and mental symptoms such as memory loss often begin in the 60s and 70s.

What Triggers Aging?

The aging process is difficult to analyze because of the intricate interactions of the body's organ systems. One structure's breakdown ultimately affects the way others function. The field of gerontology examines the biological changes of aging at the molecular, cellular, organismal, and population levels. Aging has both passive and active components—existing structures break down, and new substances and structures form.

Aging as a Passive Process

Aging as a passive process entails the breakdown of structures and slowing of functions. At the molecular level, passive aging is seen in the degeneration of the connective tissue's elastin and collagen proteins. As these proteins progressively break down, skin loses its elasticity and begins to sag, and muscle tissue slackens.

During a long lifetime, biochemical abnormalities accumulate. For example, throughout life, mistakes occur when DNA replicates in dividing cells. Usually "repair" enzymes correct this damage immediately. But over many years, exposure to chemicals, viruses, and radiation disrupts DNA repair, so that the burden of fixing the errors becomes too great. The cell may die as a result of faulty genetic instructions.

Another sign of passive aging at the biochemical level is lipid breakdown. As aging cell and organelle membranes leak due to lipid degeneration, a fatty, brown pigment called lipofuscin accumulates. Lipofuscin does not cause aging, but it is characteristic of old cells. Mitochondria also begin to break down in older cells, decreasing the supply of chemical energy available to power the cell's functions.

The cellular degradation associated with aging may be set into action by highly reactive metabolic by-products called free radicals. A free radical molecule has an unpaired electron in its outermost valence shell. This unusual situation causes the molecule to grab electrons from other molecules. As free radicals steal electrons, they make the molecules they steal from less stable, and a chain reaction of chemical instability begins that could kill the cell. Free radicals orginate when nearly any kind of stable molecule is exposed to radiation or toxic chemicals (such as those introduced into the body by cigarette smoke). Enzymes that usually inactivate free radicals before they damage cellular constituents diminish in number and activity in the later years. One such enzyme, superoxide dismutase (SOD), is promoted as an antiaging remedy at some health food stores. Even though this enzyme is a natural free radical fighter, no evidence exists that it, or other natural antioxidants such as vitamins C and E and beta carotene, stalls aging on a whole-body level.

Passive aging is also apparent at the organ system level. The thymus gland produces molecules called thymosins, which seem to control the longevity of the immune system. By age 70, the thymus is 1/10 the size it was at age 10. Declining efficiency of the immune system increases a person's susceptibility to infection and cancer.

Aging as an Active Process

Aging can entail the active initiation of new activities or new substances, as well as the passive breakdown of structure and function. One such active "aging substance" may be the lipofuscin granules that build up in aging muscle and nerve cells. Accumulation of lipofuscin illustrates both passive and active aging. Lipofuscin actively builds up with age, but it does so because of the passive breakdown of lipids. Another example of active aging is autoimmunity, in which the immune system turns against the body, attacking its cells as if they were invading organisms. Rheumatoid arthritis is an autoimmune disorder.

Active aging begins before birth, as certain cells die as part of the developmental program encoded in the genes. This "programmed cell death," also called **apoptosis,** occurs regularly in the embryo, degrading certain structures to pave the way for new ones. In the adult, programmed cell death occurs in the brain cells that die as we grow older and in blood and skin cells that die and are replaced at a continuously high rate. Programmed cell death occurs dramatically in various species that undergo metamorphosis, as adult structures replace larval forms. Cell death, then, is not restricted to the aged. It is a normal part of life.

Accelerated Aging Disorders: Providing Clues to Normal Aging

Just as biologists decipher developmental stages by studying abnormal development in model organisms such as roundworms and fruit flies, we can learn about aging in humans by examining inherited diseases that accelerate the aging timetable.

The most severe aging disorders are the **progerias** (fig. 11.27). In Hutchinson-Gilford syndrome, a child appears normal at birth but, by the first birthday, displays obviously retarded growth. Within just a few years, the child ages with astounding rapidity, developing wrinkles, baldness, and the prominent facial features characteristic of advanced age. The body is aging on the inside as well, as arteries clog with fatty deposits. The child usually dies of a heart attack or stroke by age 12, although some patients live into their 20s. Medical literature reports only a few dozen cases of this syn-

FIGURE 11.27

Progeria causes rapid aging. Eight-year-old Fransie Geringer of South Africa thought he was the only person with progeria in the world. But in 1981, he met nine-year-old Mickey Hays of Texas. The two children enjoyed a day at Disneyland, where they were met by yet a third progeria patient. People from the hometowns of the three elderly-looking children raised money to grant them their wish of visiting Disneyland.

drome. In an "adult" form of progeria called Werner syndrome, which appears before the 20th birthday, death from old age usually occurs when the individual is in his or her 40s.

Not surprisingly, the cells of progeria patients show aging-related changes. Recall that normal cells in culture divide only 50 times before dying. Cells from progeria patients die in culture after only 10 to 30 divisions, as if they were programmed to die prematurely. Certain structures seen in normal cultured cells as they near the 50-division limit (glycogen particles, lipofuscin granules, many lysosomes and vacuoles, and a few ribosomes) appear in cells of progeria patients early on. Understanding the mechanisms that cause these diseased cells to race through the aging process may help us not only to treat the disease but also to understand the biological aspects of normal aging. Biology in Action 11.2 discusses another disorder associated with aging—Alzheimer disease.

KEY CONCEPTS

Aging is both an active and a passive process. In passive aging, structures composed of collagen, elastin, muscles, and membranes begin to break down, and DNA repair becomes less efficient. Free radicals threaten cell stability, the thymus shrinks, and cholesterol deposits hamper circulation. Active aging involves the accumulation of lipofuscin pigments in cells and programmed cell death. The progerias and Alzheimer disease demonstrate the role genes play in aging.

The Human Life Span

In the age-old quest to prolong life, people have sampled everything from turtle soup to owl meat to human blood. A Russian-French microbiologist, Ilya Mechnikov, believed that a human could attain a life span of 150 years with the help of a steady diet of milk cultured with bacteria. He thought that the

Biology in ACTION

Unlocking the Secrets of Alzheimer Disease

In many families, Alzheimer disease robs older members of two of their most precious possessions—their memories and intellects. German neurologist Alois Alzheimer first identified the condition in 1907 in people in their 40s and 50s, and it initially became known as "presenile dementia." Alzheimer disaese affects an estimated 5% of U.S. citizens over age 65 and 25% of those over age 85. The symptoms of the disease, including memory loss and inability to reason, are so common that for a long time they were regarded by the medical community as normal aging. Today, Alzheimer disease is considered a disorder. It strikes 2 million Americans annually, killing 100,000. It may be inherited

Alzheimer disease begins gradually. Mental function declines for 3 to 10 years after the first symptoms appear. Confused and forgetful Alzheimer patients often wander away and become lost. As the disease reaches its final stages, the patient cannot perform basic functions such as speaking or eating and usually must be cared for in a hospital or nursing home. Death is often due to infection, a common killer of bedridden people.

The brain of a person with Alzheimer disease accumulates a short, abnormal form of a protein called beta-amyloid (fig. 1). On autopsy, the brains of Alzheimer patients contain gummy plaques of beta-amyloid in the learning and memory centers. Alzheimer-affected brains also contain structures called **neurofibrillary tangles,** which consist of a protein called **tau.** Tau binds to and disrupts microtubules in nerve cell branches, destroying the shape of the cells.

Do the plaques and tangles that strangle the brain of an individual with Alzheimer disease cause the symptoms, or result from them? Researchers are attacking this question by genetically engineering mice, inserting the human gene for Alzheimer disease and genetic instructions to overproduce the gummy beta-amyloid protein. The mice develop plaques in the same parts of their brains as Alzheimer patients. The next step is to see if the mice grow forgetful as they age. Using an animal model provides researchers with the knowledge they need to understand and treat Alzheimer disease and other human disorders.

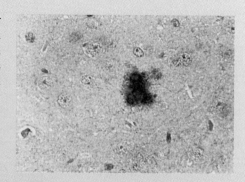

FIGURE 1

The dark area in this slice of mouse brain is a beta-amyloid plaque, produced under instructions from a human gene engineered into each of the mouse's cells. If this mouse model truly mirrors Alzheimer disease in humans, it would provide a means to test new treatments before conducting human clinical trials.

---

bacteria would take up residence in the large intestine and somehow increase longevity. (He died at 71.)

The human *life span*—theoretically, the longest length of time a human can live—is approximately 120 years. Of course, most people succumb to disease or injury long before they reach such an age. Life expectancy is a more realistic projection of how long an individual will live, based on epidemiological information. In the United States, life expectancy is 79 years for males and 83 years for females.

Causes of Death

Life expectancy draws closer to life span as technology conquers diseases and alters the effects of the most prevalent killers. However, the recent uprising of infectious diseases warns us not to become complacent. The development of antibiotic drugs removed infectious diseases such as pneumonia and influenza from the top of the "leading causes of death" list early in the century, but the evolution of antibiotic-resistant bacteria has placed some bacterial illnesses near the top of the list (table 11.3).

Is Longevity Inherited?

Aging reflects genetic activity plus a lifetime of environmental influences. Families with many very aged members usually have a fortuitous collection of genes, plus shared environmental influences, such as sound nutrition, excellent health care, devoted relatives, and a myriad of other nurturing advantages.

Teasing apart the inborn from the environmental influences on life span is difficult. One approach compares the health of adopted individuals to that of their biological and adoptive parents. One study reported that Danish adoptees with one biological parent who died of natural causes before age 50 were more than twice as likely to also die before age 50 as adoptees whose biological parents lived beyond this age. This suggests an inherited component to longevity. Interestingly, adoptees whose natural parents died early due to an infectious disease were more than 5 times as likely to also die early of infection, perhaps because of inherited immune system deficiencies. The adoptive parents' age at death showed no correlation with the longevity of their adopted children, suggesting that heredity may play a more important role than environment in an individual's life expectancy.

As we learn more about aging, we can improve not only the length but the quality of life.

KEY CONCEPTS

The maximum possible human life span is estimated at 120 years, but death usually occurs much sooner from disease or injury. Technological advances conquer some major causes of death but new causes present new challenges. Longevity depends on genetics and the environment. Studying both factors may help us increase both the length and the quality of life.

Table 11.3	**Five Leading Causes of Death: 1900, 1986, and 1993**	
1900 Rank	**1986 Rank**	**1993 Rank**
1. Pneumonia and influenza	1. Heart disease	1. Heart and blood vessel diseases
2. Tuberculosis	2. Cancer	2. Diarrheal diseases
3. Diarrhea and enteritis	3. Stroke	3. Cancer
4. Heart disease	4. Chronic respiratory disease	4. Pneumonia
5. Stroke	5. Injuries	5. Tuberculosis

Source: Data from Centers for Disease Control and Prevention, Atlanta, GA.

SUMMARY

In the mid-eighteenth century, the theory of preformation, or the idea that a preformed adult occupies fertilized ova or gametes, gave way to the theory of epigenesis as biologists observed embryos becoming more specialized over time. Baer noted that early vertebrate embryos resemble each other, but then gradually become more distinctive; he also identified primary germ layers. Haeckel proposed that embryonic stages repeat the adult structures of more primitive species. Modern scientists believe that vertebrate embryos simply share similar early stages.

The cells of the early embryo are pluripotent. Gradually, morphogen gradients establish biochemically distinct regions of the embryo that differentiate down specific developmental pathways. A cell commits to its future fate at a certain point in development. Before this time, a transplanted cell develops according to its new surroundings; after this point, it retains the specialization other cells exhibit in its original location. Differential gene expression underlies cell specialization.

Developmental biology has evolved from making observations to disrupting development to tracing the developmental effects of certain mole-cules. Biologists have developed a complete fate map for the worm C. *elegans*. When homeotic mutations occur, an abnormal morphogen pattern causes a normal structure to develop in an abnormal place.

Human prenatal development begins at fertilization. A capacitated sperm cell reaches a secondary oocyte, burrows through the zona pellucida and corona radiata with acrosomal enzymes, and joins its pronucleus with the oocyte's pronucleus. Only one sperm enters. Cleavage ensues, and a 16-celled morula forms. Between days 3 and 6, the morula arrives at the uterus and hollows out to form the blastocyst, made up of individual cells called blastomeres. The trophoblast layer and inner cell mass form. Around day 6 or 7, the blastocyst implants in the endometrium, and trophoblast cells secrete human chorionic gonadotropin, which prevents menstruation.

During the second week, the amniotic cavity forms as the inner cell mass flattens, forming the embryonic disc. Ectoderm and endoderm form, and then mesoderm appears, establishing the germ layers of this primordial embryo or gastrula. Cells in a particular germ layer develop into parts of specific organ systems.

During the third week, the embryonic period begins; the chorion begins to develop into the placenta, and the yolk sac, allantois, and umbilical cord begin to form as the amniotic sac swells with fluid. Organogenesis occurs throughout the embryonic period. Gradually structures appear, including the primitive streak, the notochord, the neural tube, arm and leg buds, the heart, facial structures and skin specializations, and a skeleton.

The fetal period begins in the third month. The organs that begin developing in the embryonic period grow and specialize as the organism develops. Gradually the fetus's body proportions more closely resemble those of a baby. The fetus becomes active. During the fifth month, vernix caseosa and lanugo cover the fetus's developing skin. In the last trimester, the brain develops rapidly and fat fills out the skin. The digestive and respiratory systems mature. At about 38 weeks, strong uterine contractions dilate the cervix and push the baby out.

Development does not end at birth. Even before conception, aging begins as cells form and die at certain times. The brain cells are formed by age 2, and by the time an individual reaches sexual maturity, the thymus gland has attained its peak size. Noticeable physical signs of aging appear in one's 30s and accelerate in the 40s and 50s. Memory loss, a decrease in height, and cartilage growth mark the 60s and 70s.

Aging is both passive and active. In passive aging, structures such as collagen, elastin, muscles, and membranes break down, and DNA repair becomes less efficient. Free radicals build up, threatening cell stability, the thymus shrinks, and blood vessels narrow as cholesterol deposits build up. In active aging, lipofuscin accumulates in cells and cells undergo apoptosis, or programmed cell death.

The theoretical maximum human life span is 120 years. The most common causes of death shift as technologies conquer various diseases. Both heredity and the environment affect longevity. Studying both may increase not only the length but the quality of human life.

KEY TERMS

allantois 224
amnion 224
amniotic cavity 222
apoptosis 233
blastocyst 219
blastocyst cavity 219
blastomere 219
capacitation 218
chorion 219
chorionic villi 225
cleavage 219
corona radiata 217
differentiate 212
ectoderm 222

embryonic disc 222
embryonic induction 227
embryonic stage 217
endoderm 222
endometrium 221
epigenesis 211
extraembryonic membrane 217
fate map 222
fetal period 217
gastrula 222
germ layers 222
homeotic 215
human chorionic gonadotropin
    (hCG) 222

implantation 221
inner cell mass 219
labor 217
mesoderm 222
morphogenesis 222
morphogens 215
morphology 215
morula 219
neural tube 227
neurulation 227
notochord 227
organogenesis 227
parturition 217
placenta 217

pluripotency 212
preembryonic stage 217
preformation 211
primitive streak 227
primordial embryo 222
progerias 233
pronucleus 218
trophoblast 219
umbilical cord 217
yolk sac 224
zona pellucida 217
zygote 217

REVIEW QUESTIONS

1. Why would the scientific method support the hypothesis of epigenesis better than the theory of preformation?

2. Why can very different specialized cells have identical genes?

3. How do morphogen gradients cause cells to differentiate?

4. Describe experiments in the mouse, zebrafish, and frog that illustrate how cells lose pluripotency as they develop.

5. Why did developmental biologists choose to develop a complete fate map for *C. elegans*? Why didn't they construct such a map for humans?

6. Why is the discovery that homeotic genes exist in many diverse species significant?

7. Arrange these prenatal humans into order from youngest to oldest: morula, gastrula zygote, fetus, blastocyst, and embryo.

8. What events must take place for fertilization to occur?

9. How do the developmental fates of a trophoblast cell and an inner cell mass cell differ? How do fates differ for cells in the ectoderm, endoderm, and mesoderm?

10. Toxins usually cause more severe medical problems if exposure occurs during the first 8 weeks of pregnancy rather than during the later weeks. Why?

11. Why can't a fetus born in the fourth month survive?

12. What evidence demonstrates programmed cell death in *C. elegans*? in humans?

13. What is the difference between life span and life expectancy?

14. What types of studies distinguish between inherited and environmental influences on longevity?

TO THINK ABOUT

1. In the nineteenth century, numerous people observed chick embryos developing. According to evolutionary biologist and geologist Stephen Jay Gould, preformationists concluded, "Yes, it looks like a simple streak of tissue developing into a complex organism, but the streak is actually a preformed chick too small to be seen." Those who believed in epigenesis interpreted what they saw literally—they argued that the prenatal chick increases in complexity as specialized tissue arises from unspecialized tissue.

How can the scientific method resolve differing interpretations of the same observations or data?

2. Is the presence of a complete set of genetic instructions in a fertilized ovum consistent with the hypothesis of preformation or epigenesis?

3. When an isolated sperm is mechanically injected into an oocyte that is not located in a female's reproductive tract, and the treated egg is then placed in the woman's body, no embryo develops. Based on what you know about fertilization, why doesn't development ensue? (Hint: What key event is missing?)

4. A woman has been pregnant for 41 weeks. Her doctor performs a Caesarian section to deliver the child because biochemical tests indicate that the cells of the placenta are dying. Why is it vital to the child's survival to perform the Caesarian section?

5. Rats raised in an "enriched environment" (provided with toys and lots of attention from their keepers) live 33% longer than normal, and their brains show cellular patterns similar to those of much younger rats. Other animal studies have correlated lower than normal body temperature and lower caloric intake with longevity. How might these findings be applied to the study of human aging?

6. When does aging begin? What structures exhibit aging long before a person appears to be elderly?

7. What factors do you think contribute to longevity?

8. Why might lists of leading causes of death differ in different parts of the world, or even in different communities within a nation?

9. Explain how cancer might be caused by a defect in programmed cell death.

10. In the year 2050, 1 in every 20 persons in the United States is projected to be over the age of 85. What provisions would you like to see made for the elderly of the future (especially if you will be one of them)? What can you do now to increase the probability that your final years will be healthy and enjoyable?

11. What problems would a scientist likely encounter in studying the inheritance of longevity?

SUGGESTED READINGS

Alper, Joseph. January 1994. Two firms tackle cord-blood transplants. *Bio/Technology*, vol 12. Special blood cells taken from a newborn's umbilical cord can be used to treat people with certain blood disorders, replacing bone marrow transplants.

Gould, Stephen Jay. 1977. *Ontogeny and phylogeny*. New York: W. W. Norton. An in-depth but readable analysis of the biogenetic law.

———— 1977. *Ever since Darwin.* New York: W. W. Norton. The essay "On Heroes and Fools in Science" argues that preformation made sense in its era.

Hart, Stephen. September 1992. MyoD: master gene or cog? *BioScience*, vol. 42, no.8. MyoD is a gene that triggers the cascade of gene action that prompts a cell to specialize as muscle.

Henig, Robin Marantz. 1985. *How a woman ages* and *How a man ages*. New York: Ballantine Books. A step-by-step look at the various changes associated with aging in each sex.

Ho, Robert K., and Charles B. Kimmel. July 2, 1993. Commitment of cell fate in the early zebrafish embryo. *Science*, vol. 261. An elegant experiment shows how developmental choices narrow over time.

Kline, Douglas. February 1991. Activation of the egg by the sperm. *Bioscience*, vol. 41, no. 2. Fertilization requires much more than egg meeting sperm.

Kosik, Kenneth S. May 8, 1992. Alzheimer's disease: A cell biological perspective. *Science*, vol. 256. We are beginning to understand Alzheimer disease at the cellular level, with the help of families that inherit it.

Lewis, Ricki. January 1990. Prenatal peeks. *Biology Digest*. Technology has lifted much of the mystery of prenatal existence.

———— March 1993. Choosing a perfect child. *The World and I*. Preimplantation genetic diagnosis depends upon the fact that the cells of an 8-celled preembryo retain the potential to develop into a complete newborn.

Marx, Jean. September 4, 1992. Alzheimer's debate boils over. *Science*, vol. 257. Excess beta-amyloid is part of Alzheimer disease, but its precise role remains elusive.

Pollen, Daniel A. 1993. *Hannah's heirs*. New York: Oxford University Press. A fascinating historical discussion of Alzheimer disease, first identified in 1844 in a woman named Hannah.

EXPLORATIONS   **Interactive Software**

Life Span and Lifestyle

This interactive exercise allows students to explore how the lifestyle decisions they make influence how long they can expect to live. The exercise presents a diagram of a typical human life span, shown as an age distribution of U.S. life expectancies (how long a U.S. citizen of a certain age can expect to live). The student explores the change in probability of survival when certain activities are initiated. The student may initiate or stop smoking at any age (cigarettes induce cancer as well as lung and heart disease), eat or stop eating a diet high in animal fat (which induces heart disease and stroke), eat or stop eating a high protein diet (which seems to induce colon cancer), and initiate or cease regular exercise (which counteracts cardiovascular disease). By varying patterns of behavior, the student soon learns that survival reflects lifestyle.

1. Can exercise counteract the lowered life expectancy due to smoking?

2. Is it more beneficial to give up smoking or a red-meat diet?

3. Is smoking more harmful at one age than another?

4. Is there any age where diet doesn't matter?

5. What lifestyle is associated with the longest life expectancy? The shortest?

CHAPTER 12
....
Human Reproductive and Developmental Problems

CHAPTER OUTLINE

Awaiting motherhood.

LEARNING OBJECTIVES

By the chapter's end, you should be able to answer these questions:

1. What are some of the physical causes of human infertility, both male and female? How can infertility be treated?

2. How can the reproductive technologies of artificial insemination, in vitro fertilization, and embryo transfer help infertile couples have children?

3. What are some possible causes of spontaneous abortion?

4. What are some of the medical problems faced by a baby born prematurely?

5. How do certain toxic chemicals interfere with prenatal development?

6. What are some commonly used and experimental birth control methods, and how do they work?

7. What are some sexually transmitted diseases, and how do they affect health?

Two Louises: Research and Reproductive Technology

They are both named Louise, and each, in her special way, has contributed much to our knowledge of prenatal development. Louise Joy Brown was born in England in 1978, a normal little girl in every way except the way she was conceived—in a piece of laboratory glassware. "Adorable Louise" was born in 1985, a normal donkey in every way except the place where she developed prenatally—in the uterus of a horse.

Louise Brown was the first human conceived outside the body and transferred as a preembryo to her mother's uterus, where she continued development. Scientists had previously perfected the techniques that led to Louise Brown's birth in nonhuman animals, and Louise's historic beginning has since guided the technology enabling the births of more than a thousand "test tube" babies.

Adorable Louise was conceived in a female donkey; but eight days after fertilization, veterinary researchers at Cornell University transferred her embryo to the uterus of a horse. Normally, the immune system of a female horse would abort the "foreign" donkey embryo. The Cornell researchers wanted to prevent rejection by exposing the horse to white blood cells from Louise's natural donkey parents before the embryo transfer. It worked. The horse's

a.   b.

FIGURE 12.1

Two Louises. *a.* Louise Joy Brown was conceived when her father's sperm cell met her mother's egg cell in the laboratory. Here, Louise—at age 10, the oldest "test tube baby"— holds one of the youngest, Andrew Macheta. The children were among 600 test tube babies attending a 10-year anniversary party at the clinic near London where they were conceived. *b.* Adorable Louise looks like a normal donkey, but a horse gave birth to her following embryo transfer. The techniques used to allow Louise to gestate in a horse led to treatments that can prevent miscarriage in humans.

immune system, "tricked" into recognizing donkey cells as its own, accepted the donkey embryo. A year later, the horse gave birth to Adorable Louise (fig. 12.1).

Infertility specialists can now offer a similar technology to some women who suffer repeated early pregnancy losses. The mothers' immune systems apparently do not "ignore" the develop-

ing embryos, as they normally would; instead, they attack the embryos as if they were foreign substances. When doctors inject the women with their husbands' white blood cells before they attempt pregnancy, their immune systems later accept the embryos.

The two Louises illustrate a trend in reproductive biology—not only to learn from such studies in their own right, but to transfer basic research and agricultural work to human health care. As medical scientists learn more about the complex biological interactions that form a newborn human from a fertilized egg, they discover more ways to enable or hinder conception, to prevent or treat birth defects, and to treat the fetus or newborn.

Infertility

For one out of every six couples in the United States, trying to conceive a child is not a joyous prospect, but instead is a period of growing anxiety and frustration as the couple realizes that they may be infertile. **Infertility** is the inability of the sperm nucleus to merge with the egg nucleus. Couples are considered infertile if they are unable to conceive after one year of frequent intercourse without the use of contraceptives. Physicians can identify a physical cause for infertility in 90% of all cases and can treat or correct many of these problems. But infertility seems to be on the rise in the United States.

Among the cases involving an identifiable physical cause, 30% entail physical difficulties primarily in the male and 60% entail problems primarily in the female. The statistics are somewhat unclear, because in 20% of the cases in which specialists discover a problem, both partners have a medical condition that could contribute to infertility. A common combination, for example, might be a woman with an irregular menstrual cycle and a man with a low sperm count.

Figure 12.2 summarizes the frequency of different causes of infertility.

Male Infertility

Male infertility is easier to detect but sometimes harder to treat than female infertility. Some men have difficulty fathering a child because they produce many fewer than the 100 million sperm found in a normal ejaculate; if a hormonal imbalance is lowering sperm count, administering the appropriate hormones may boost sperm output.

MALES		
Problems	**Possible Causes**	**Treatment**
Low sperm count	Hormone imbalance, varicose vein in scrotum, excessive exposure to heat by wearing tight pants or taking hot showers	Hormone therapy, surgery, avoiding excessive heat
Immobile sperm	Abnormal sperm shape	None
	Infection	Antibiotics
	Malfunctioning prostate	Hormones
Antibodies against sperm	Problem in immune system	Drugs

FEMALES		
Problems	**Possible Causes**	**Treatment**
Ovulation problems	Pituitary or ovarian tumor	Surgery
	Underactive thyroid	Drugs
Antisperm secretions	Unknown	Acid or alkaline douche, estrogen therapy
Blocked fallopian tubes	Infection caused by IUD or abortion or by sexually transmitted disease	Laparotomy, egg removed from ovary and placed in uterus
Endometriosis	Delayed parenthood until the 30's	Hormones, laparotomy

FIGURE 12.2
Causes of infertility.

Sometimes a man's immune system produces antibodies that cover the sperm and prevent them from binding to oocytes. Male infertility can also be caused by a varicose vein in the scrotum. This enlarged vein overheats the developing sperm so that they cannot mature. Surgeons can remove a scrotal varicose vein and improve fertility.

For many men with low sperm counts, fatherhood is just a matter of time. If an ejaculate contains at least 60 million sperm cells, fertilization is likely to eventually occur. To speed conception, a man with a low sperm count can donate several semen samples over a period of weeks at a fertility clinic. The clinic keeps the samples in cold storage, then pools them. They withdraw some of the seminal fluid to leave a sperm cell "concentrate," which they then inject into the woman's reproductive tract. This procedure isn't very romantic, but it is highly effective at achieving pregnancy.

Sperm quality is even more important than quantity. Sperm cells that are unable to move—a common problem—cannot reach an oocyte. If lack of motility is due to a physical defect, as when sperm tails are misshapen or missing, no treatment is presently available. If the cause is hormonal, however, hormone therapy can sometimes restore sperm motility. Computers can track sperm shape, as well as the speed and direction of their movements (Biology in Action 12.1).

KEY CONCEPTS

Infertility is the inability of the sperm nucleus to merge with the egg nucleus. A couple is considered infertile if they are unable to conceive a child after a year of unprotected intercourse. For the 90% of cases in which medical specialists can determine a physical cause, the male is primarily implicated in 30%, while the female is primarily involved in 60%. Both partners may contribute to infertility. Male infertility can be caused by low sperm count, an immune reaction against sperm, a varicose vein in the scrotum, or malformed sperm.

Biology in ACTION

Scrutinizing Sperm

Both quality and quantity of sperm are essential to a man's ability to father a child. If a sperm head is misshapen, if the sperm cannot swim properly, or if the man simply produces too few sperm cells, the sperm may be unable to complete the arduous journey and entry into the well-protected oocyte.

Until recently, sperm analysis was largely subjective; a urologist observed the sperm cells under a microscope. Now, computer-aided sperm analysis—CASA, for short—is standardizing and expanding the criteria for normalcy in human seminal fluid and sperm cells.

To obtain a sperm analysis, a man abstains from intercourse for two to three days, then provides a sperm sample, which must be analyzed within 30 minutes. The man must also provide information about his reproductive history and possible toxic exposures. The sperm sample is placed on a slide under a microscope, and then technology intervenes. A video camera sends an image to a VCR, which projects a live or digitized image, and to a computer, which traces out sperm trajectories and displays them on a monitor or prints them out. Figure 1 shows a CASA of normal sperm, depicting their changing swimming patterns along their journey.

CASA systems are also able to use sperm as "biomarkers" of toxic exposure. In one study, researchers compared sperm from two groups of men. One group worked in the dry cleaning industry and were exposed to the solvent perchloroethylene (believed to damage sperm). The other men worked in the laundry industry and were exposed to many of the same substances, but not to this one chemical. CASA showed a difference in sperm motility directly related to level of perchloroethylene exposure. This supported the reproductive evidence—although the men in both groups had the same numbers of children, it took far longer for the dry cleaners and their wives to achieve pregnancy than it did for the launderers and their wives.

FIGURE 1

A computer tracks sperm movements. In semen, sperm swim in a straight line (*a*), but as they are activated by biochemicals normally found in the woman's body, their trajectories widen (*b*). The sperm in (*c*) are in the mucus of a woman's cervix, and the sperm in (*d*) are attempting to penetrate the structures surrounding an oocyte.

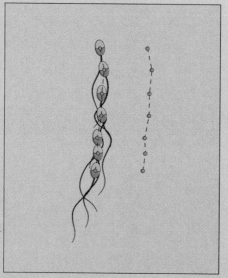

a.

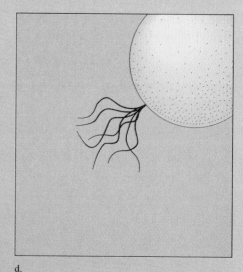

b.

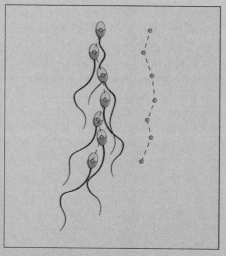

c.

d.

FIGURE 12.3

Sites of reproductive problems in the human female.

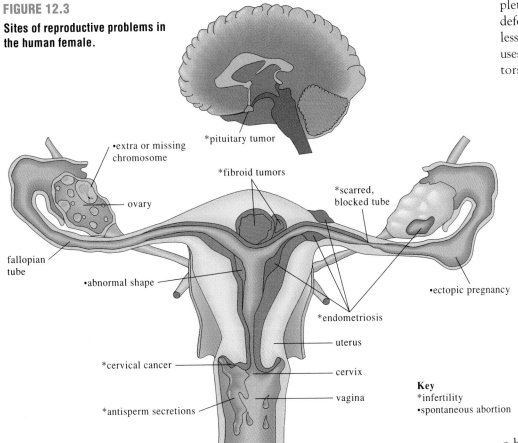

- •extra or missing chromosome
- *pituitary tumor
- *fibroid tumors
- *scarred, blocked tube
- ovary
- fallopian tube
- •abnormal shape
- •ectopic pregnancy
- *endometriosis
- uterus
- *cervical cancer
- cervix
- vagina
- *antisperm secretions

Key
*infertility
•spontaneous abortion

Female Infertility

Female infertility can be linked to abnormalities in any part of the reproductive system (fig 12.3). Many infertile women have irregular menstrual cycles, which makes it difficult to pinpoint when they are most likely to conceive. In an average 28-day menstrual cycle, ovulation usually occurs around the fourteenth day after menstruation begins, and this is when a woman is most likely to conceive.

The average woman under 30 years old and not using birth control is likely to become pregnant within three or four months. A woman with irregular menstrual periods may have much more difficulty. She can improve her chances by using an ovulation predictor test, which detects oncoming ovulation by measuring rising levels of a certain hormone, or by recording her body temperature each morning using a special thermometer

with very fine subdivisions. A slight rise in this "basal body temperature" occurs about a day before ovulation.

Many conditions can cause the hormonal imbalance that usually underlies irregular ovulation—a tumor in the ovary or in the brain's pituitary gland, which hormonally controls the reproductive system; an underactive thyroid gland; or steroid-based drugs such as cortisone. Sometimes a woman produces too much prolactin, the hormone that normally promotes milk production and suppresses ovulation in new mothers. If prolactin is abundant in a nonpregnant woman, she will not ovulate and therefore cannot conceive.

Fertility drugs can stimulate ovulation, but they may also cause women to "superovulate," producing more than one egg each month. One such drug, clomiphene, raises the chance of bearing twins from the normal 1–2% to 4–6%. If a woman's ovaries are com-

pletely inactive or absent (due to a birth defect or surgery), fertility drugs are useless; she can become pregnant only if she uses a donor oocyte. Although some doctors have tried ovary transplants, they have only succeeded when donor and recipient are identical twins.

Blocked fallopian tubes are a common cause of female infertility. Blockage can prevent sperm from reaching the oocyte, or keep the fertilized ovum from descending into the uterus, resulting in an **ectopic** ("out of place") **pregnancy** in the tube. Fallopian tubes can be blocked due to a birth defect or, more likely, because of an infection such as pelvic inflammatory disease, a complication of some sexually transmitted diseases. A woman may not know she has blocked fallopian tubes until she has difficulty conceiving and medical tests discover the problem. X rays or an ultrasound exam may reveal the blockage, or more invasive procedures may be needed. In a hysterosalpingogram, dye is injected into the fallopian tubes; in laparoscopy, a telescopelike instrument is inserted through a small incision made near the naval, enabling a doctor to view the tubes. Blocked tubes can sometimes be opened and cleared surgically.

Excess tissue growing in the uterine lining can cause female infertility. Benign **fibroid** tumors or **endometriosis,** a condition in which tissue builds up in and sometimes outside of the uterus, can make the uterine lining inhospitable to an embryo. In endometriosis, not only the lining but the excess tissue bleeds during menstruation, causing painful cramps. Endometriosis can make conception difficult, but curiously, once a woman with endometriosis has been pregnant, the painful cramps and bleeding usually subside. Other uterine conditions may impair fertility. A woman whose mother took a drug called diethylstilbestrol (DES) while pregnant can have an abnormally shaped uterus that cannot maintain pregnancy.

Table 12.1 **Semen Analysis**	
Characteristic	**Normal Value**
Volume	1.5–5.0 milliliters/ejaculate
Sperm density	20 million cells/milliliter
Percent motile	>40%
Motile sperm density	>8 million/milliliter
Average velocity	>20 microns/second
Motility	>8 microns/second
Percent abnormal morphology	>40%
White blood cells	>5 million/milliliter

Table 12.2 **Female Infertility Tests**	
Test	**What It Checks**
Hormone levels	Whether ovulation occurs
Ultrasound	Placement and appearance of reproductive organs and structures
Postcoital test	Whether cervical mucus (collected soon after unprotected intercourse) is thin enough to allow sperm to reach cervix
Endometrial biopsy	Whether uterine lining (sampled and viewed under microscope) can support an embryo
Hysterosalpingogram	Whether fallopian tubes are clear or blocked (determined by scanning movement of injected dye)
Laparoscopy	Whether fallopian tubes are blocked by scar tissue (determined by inserting an optical device through small incision near navel)

Even if a woman's uterus is normal, secretions in the vagina and cervix may be hostile to sperm. If the mucus produced here is unusually thick or sticky (as can happen during infection), sperm quickly become trapped and cannot move far enough to encounter an oocyte. Acidic or alkaline vaginal secretions can weaken or kill sperm cells. Infertility specialists treat some mucus secretion problems with low doses of the hormone estrogen. They may also instruct the woman to douche daily with either an acidic solution, such as acetic acid (vinegar), or an alkaline solution, such as bicarbonate (baking soda), to alter the pH of the vagina and make it more receptive to sperm cells.

While most of the foregoing causes are not age-related, we know that the likelihood of female infertility does increase with age. Older women are more likely to produce oocytes with abnormal numbers of chromosomes. The effects of a chromosomal imbalance are often so severe that the pregnancy ends before the woman detects it. The tendency to lose very early embryos may be mistaken for infertility because the bleeding caused by the aborted embryo appears to be the normal menstrual flow. Older women also experience a higher incidence of meiotic error. This may be because their oocytes have had longer or more exposures to harmful chemicals, viruses, and radiation. Because so many conditions cause infertility, detecting a particular couple's problems may be a difficult and expensive process.

Infertility Tests

A number of medical tests seek to identify the cause or causes of a couple's infertility. Most specialists check the man first, because it is easier, less costly, and certainly less painful to obtain sperm than oocytes. Sperm are checked for number (sperm count), motility, and morphology (shape). An ejaculate may contain abnormal forms numbering up to 40% and still be considered normal; more than 40% can interfere with fertility. Table 12.1 lists the normal test values.

If sperm tests do not reveal male infertility, the next step is for the woman to visit a gynecologist, who may use some of the techniques described in table 12.2 to see that the many components of the woman's reproductive system are present and functioning. However, some cases of infertility seem to have no explanation.

> **KEY CONCEPTS**
> Infertility may involve the male or female partner, or both. Male infertility can be due to a low sperm count or sperm that cannot swim or are abnormal in structure. Female infertility can be caused by an irregular menstrual cycle or blocked fallopian tubes. Fibroid tumors, endometriosis, or a misshapen uterus may prevent a fertilized ovum from implanting, secretions in the vagina and cervix may inactivate or immobilize sperm, or oocytes may fail to release a sperm-attracting biochemical. Early pregnancy losses due to abnormal numbers of chromosomes, more common among older women, may be mistaken for infertility. While the causes are many and complex, a variety of medical tests can usually pinpoint the cause of infertility.

Table 12.3   Causes and Treatments of Human Infertility

Males

Problem	Possible Causes	Diagnostic Test	Treatment
Low sperm count	Hormone imbalance, scrotal varicose vein, excessive exposure to heat from wearing tight pants or taking hot showers	Examine sperm under microscope	Hormone therapy, surgery, avoiding excessive heat
Immobile sperm	Abnormal sperm shape	Examine sperm under microscope	None
	Infection	Examine semen for infective agents	Antibiotics
	Malfunctioning prostate	Test semen for chemical abnormalities	Hormones
Sperm-attacking antibodies	Immune system abnormality	Test semen for antibodies	Drugs

Females

Problem	Possible Causes	Diagnostic Test	Treatment
Ovulation problems	Pituitary or ovarian tumor	Measure hormone levels, take daily temperature readings, X rays	Surgery
	Underactive thyroid	Same as above	Drugs
Antisperm secretions	Unknown	Postcoital test	Acid or alkaline douche, estrogen therapy
Blocked fallopian tubes	Infection caused by IUD or abortion or by sexually transmitted disease	X rays, ultrasound examination, hysterosalpingogram, laparoscopy	Laparotomy to remove scar tissue; egg removed from ovary and placed in uterus
Endometriosis	Delaying parenthood until age 30 or over	Laparoscopy	Hormones, laparotomy to remove excess tissue

Treating Infertility

Several standard infertility treatments prove successful for many couples (table 12.3). More drastic reproductive technologies are used less frequently as standard treatments. Many of these approaches were perfected in nonhuman animals long before they became available to treat human infertility (table 12.4).

Donated Sperm—Artificial Insemination

The oldest reproductive alternative is **artificial insemination,** in which a doctor introduces donated sperm into a woman whose partner is infertile or who desires a child but not a mate. More than 250,000 babies have been born worldwide as a result of this procedure.

The first human artificial inseminations were done in the 1890s. For many years, physicians donated sperm, and this became a way for male medical students to earn a few extra dollars. By 1953, sperm could be frozen and stored for later use. Today, donated sperm is frozen and stored in sperm banks, which provide the cells to obstetricians who perform inseminations.

Today, a woman or couple who chooses artificial insemination can select sperm from a catalog that lists the personal characteristics of the donors, including race, blood type, hair and eye color, build, and even educational level and interests. A blond-haired, blue-eyed infertile man who loves athletics and music might scan the catalog for a sperm donor with these same traits. This can, of course, border on silliness, because not all of these traits are inherited.

Problems can arise in artificial insemination, as they can in any pregnancy. One woman used the same sperm donor twice, and twice conceived and gave birth to a child with the same devastating inborn metabolic inherited illness. Although the disease was too rare for the sperm bank to screen for it, the bank apparently made no effort to find

Table 12.4 **Landmarks in Reproductive Technology**

	In Animals	**In Humans**
1782	Use of artificial insemination in dogs	
1799		Pregnancy reported from artificial insemination
1890s	Birth from embryo transplantation in rabbits	Artificial insemination by donor
1949	Use of cryoprotectant to successfully freeze and thaw animal sperm	
1951	First calf born after embryo transplantation	
1952	Live calf born after insemination with frozen sperm	
1953		First reported pregnancy after insemination with frozen sperm
1959	Live rabbit offspring produced from in vitro ("test tube") fertilization (IVF)	
1972	Live offspring from frozen mouse embryos	
1976		First reported commercial surrogate motherhood arrangement in the United States
1978	Transplantation of ovaries from one cow to another	Baby born after IVF in United Kingdom
1980		Baby born after IVF in Australia
1981	Calf born after IVF	Baby born after IVF in United States
1982	Sexing of embryos in rabbits	
	Cattle embryos split to produce genetically identical twins	
1983		Embryo transfer after uterine lavage
1984		Baby born in Australia from frozen and thawed embryo
1985		Baby born after gamete intrafallopian transfer (GIFT)
		First reported gestational surrogacy arrangement in the United States
1986		Baby born in the United States from frozen and thawed embryo
1989		First preimplantation genetic diagnosis

Source: Data from Office of Technology Assessment, *Infertility: Medical and Social Choices*, U.S. Congress, Government Printing Office, Washington, DC, May 1988.

a different donor after the woman had her first child. In a more notorious case, a physician named Cecil Jacobson artificially inseminated 15 women with his own sperm, but told them it was from anonymous donors.

A Donated Uterus—Surrogate Motherhood

A male's role in reproductive technologies is simple compared to a woman's role. This is because a man can be a genetic parent, providing half of his genetic self in his sperm, but a woman can be both a genetic parent and a gestational parent, providing oocyte, uterus, or both.

If a man produces healthy sperm but his mate's uterus is absent or cannot maintain a pregnancy, a **surrogate mother** may be a possibility. A surrogate mother is artificially inseminated with the man's donated sperm. When the child is born, the surrogate gives the baby to the couple. In this situation, the surrogate is both the genetic and the gestational mother.

Attorneys usually arrange surrogate mother relationships. The surrogate mother signs a statement signifying her intent to give up the baby, and she is paid $10,000 or more for her nine-month job. The United States is the only country where surrogacy is legal. More than 1,000 babies have been born by this method since 1988.

The dilemma with surrogate motherhood is that a woman may not be able to predict her responses to pregnancy and childbirth as she signs paperwork in a lawyer's office. When a surrogate mother changes her mind, the results are wrenching for all. A prominent case involved a young woman named Mary Beth Whitehead, who carried the child of a married man for a fee, and then changed her mind when the baby was born. Whitehead's ties to Sarah, called "baby M" while her three parents battled over custody, were both genetic and gestational. Today, Sarah spends time with both families.

Another type of surrogate mother lends only her uterus but not her oocyte. She receives a fertilized egg from a woman who has healthy ovaries but lacks a functional uterus. The gestational-only surrogate mother turns the child over to the donors of the genetic material. This arrangement works as long as all parties agree, as was the case for Arlette Schweitzer, who gladly lent her uterus to house her grandchildren-to-be (see Biology in Action 12.2). However, gestational surrogacy can also go drastically awry. Consider what happened to Anna Johnson, Mark and Crispina Calvert, and the child conceived by the Calverts and carried by Johnson.

Crispina Calvert's uterus had been removed, but her ovaries were intact, healthy, and producing viable oocytes. The Calverts contracted Anna Johnson, a young, single nurse, to be a gestational

surrogate and carry their fertilized ovum to term for $10,000. Near the end of the pregnancy, Johnson had misgivings. She went to court, asking to be declared the natural mother and to have visitation rights with the child.

The question in Johnson v. Calvert and Calvert boiled down to the very essence of motherhood—was Crispina Calvert baby Christopher's mother, or was Anna Johnson? Christopher wouldn't have been conceived without Calvert's oocyte, but he wouldn't have developed from a fertilized ovum to a baby without the nurturing uterus Johnson provided.

The California court ruled in favor of the genetic ties. Judge Richard N. Parslow, Jr., told the courtroom:

> *Anna Johnson is the gestational carrier of the child, a host in a sense . . . she and the child are genetic hereditary strangers . . . Anna's relationship to the child is analogous to that of a foster parent providing care, protection, and nurture during the period of time that the natural mother, Crispina Calvert, was unable to care for the child.*

The judge based his ruling on several factors: scientific evidence indicating the importance of genes in establishing one's characteristics; the validity of the contract; and the best interests of the child. A court of appeals unanimously affirmed the ruling.

Not everyone agreed with the court's decision. Many a woman who has felt the first rumblings of prenatal life within, felt fetal movements, and then experienced childbirth, would argue that this role qualifies a woman for motherhood more than donating a cell. The American College of Obstetricians and Gynecologists states that gestation, not genetics, defines motherhood. The American Academy of Pediatrics defines a surrogate mother as "a woman who carries a pregnancy for another woman," avoiding the genetic-gestation distinction. It also recommends that

> *surrogate parenting arrangements be considered a tentative, preconception adoption agreement in which the surrogate mother is the sole decision*

> *maker until after she gives birth to the infant. After birth, applicable local adoption rules and practices should be followed.*

With such divergent views, we will certainly hear more about surrogate motherhood.

In Vitro Fertilization

In **in vitro fertilization,** or IVF, which means "fertilization in glass," sperm meets oocyte outside the woman's body. The fertilized ovum undergoes two or three cleavage divisions and is then introduced into the oocyte donor's (or another woman's) uterus. If all goes well, the preembryo will implant into the lining.

A woman might undergo IVF if her ovaries and uterus work but her fallopian tubes are blocked. To begin, the woman takes a superovulation drug that causes her ovaries to release more than one mature oocyte at a time. Using a laparoscope to view the ovaries and fallopian tubes, a physician removes several of the largest oocytes and transfers them to a laboratory dish. Chemicals that mimic those in the female reproductive tract are added, and sperm donated by the woman's mate are applied to the oocytes. If the sperm cannot readily penetrate the oocyte, they may be sucked up into a tiny syringe and microinjected into the female cell. Alternatively, in a procedure called *zona* drilling, the outer layer of the oocyte, the *zona pellucida,* is gently pierced before the oocyte is exposed to sperm.

A day or so after sperm wash over the oocytes in the laboratory dish, a physician transfers some of the preembryos—balls of 8 or 16 cells—to the woman's uterus. If the pregnancy hormone human chorionic gonadotropin appears in her blood a few days later, and if its level rises, the procedure has worked—the woman is pregnant.

IVF costs from $5,000 to $10,000 per attempt. The success rate is only about 14% in most facilities. In contrast, over two-thirds of naturally conceived and developed preembryos successfully implant in the uterus.

New Ways to Make Babies

Grandmother and Mother to the Same Child

In October 1991, 42-year-old Arlette Schweitzer gave birth to twins, Chelsea and Chad, at St. Luke's Midland Hospital in Aberdeen, South Dakota. Arlette's twins were rather unusual—they were her genetic grandchildren, but her gestational children.

Two decades earlier, Arlette had given birth to daughter Christa. When Christa reached her late teens and still didn't menstruate, she went to a doctor. An ultrasound scan revealed that although she had healthy ovaries and oocytes, Christa lacked a uterus. She could not have children.

A few years later, Christa was happily married and desperately wanted children. Her mother stepped in to help. Christa's oocytes and her husband Kevin's sperm were mixed in a laboratory dish, where they joined to form fertilized ova. After a few cell divisions, preembryos were implanted into Arlette's uterus. (Arlette's hormonal cycle had been manipulated with drugs to make the uterine lining receptive to pregnancy.) Christa and Kevin's unusual route to parenthood was a resounding success because the technology worked, and because the people involved shared the same goal.

Midlife Mothers

At age 53, Mary Shearing gave birth to twins Amy Leigh and Kelly Ann. Already in the last stages of menopause, Mary had three grown children from a previous marriage, as well as two grandchildren, when she and her 32-year-old husband, Don, decided to try to have a child. Don's sperm was used to fertilize donated oocytes in a laboratory dish, and the fertilized ova were implanted in Mary's uterus. The Shearings' success showed that the age of the oocyte, not the condition of the uterine lining in an older woman, is the most important factor in conceiving, carrying, and delivering a healthy baby.

Most women who undergo the treatment Shearing underwent are under 45 years of age and have lost ovaries to disease. Sometimes the egg donor is the recipient's sister, as was the case for Vicki and Larry Miceli. Vicki's sister, Bonny De Irueste,

Table 1	Counting the Costs of Conception
It took Pamela and Jonathon Loew five years to conceive Alexandra. Here is a breakdown of the costs involved:	
Two GIFTS	$18,000
Eight artificial inseminations	$8,000
One frozen embryo transfer	$1,000
Other tests	$3,000
	$30,000

donated the oocyte that, when fertilized, eventually became baby Anthony (figure 1).

A Five-Year Wait

The experience of Pamela and Jonathon Loew is more typical than that of Arlette Schweitzer or the Shearings. For five years, the Loews worked their way through a long list of techniques, trying to conceive (table 1). Happily, they eventually did, and Alexandra was born in 1990.

The Loews' quest began by ruling out various causes of infertility. Pamela underwent hormone therapy, then had a physician place Jonathon's sperm near her cervix eight times through artificial insemination. One time, this resulted in pregnancy, but it was **ectopic;** the preembryo lodged in one of the fallopian tubes. Doctors had to remove the embryo to save Pamela's life.

Next, the Loews tried sperm washing, a procedure that removed protein antigens from the surfaces of Jonathon's sperm. This would prevent the antigen molecules from eliciting a response from Pamela's immune system. Doctors placed the washed sperm in her fallopian tube, close to the uterus, along with oocytes they had removed from Pamela's ovary. If all went well, the gametes would meet and merge. The first time the Loews used this procedure, called *gamete intrafallopian transfer* (GIFT), it didn't work. The second time, it produced Alexandra.

Whatever happened to "boy meets girl" and "baby makes three"? Whenever possible, parents still conceive and carry babies

FIGURE 1

Little Anthony Miceli was born from Vicki Miceli's uterus, but he was conceived when father Larry Miceli's sperm fertilized an oocyte donated by Bonny De Irueste, Vicki's sister. Said Bonny, "I'm his aunt, that's it." Not all instances where new reproductive technologies are used are so joyous.

the natural way, but when this is not possible, new reproductive technologies can help. At the same time, these approaches reveal more about human prenatal development. Some people argue that in a world plagued by overpopulation, and where many children are not adequately cared for, these high-tech routes to reproduction are not really necessary—perhaps even frivolous. To people like the Shearings, the Micelis, and the Loews, however, these procedures offer the hope of conceiving a new life.

FIGURE 12.4

In 1988, Michele and Ray L'Esperance underwent IVF because Michele's diseased fallopian tubes had been removed. Five fertilized ova were implanted in her uterus, in hopes that one or two would complete development. All five did; now they are Erica, Alexandria, Veronica, Danielle, and Raymond. Today, only two or three fertilized ova or preembryos are usually implanted.

in the ovary and that may provide needed growth factors.

Developing tests to determine which preembryos are most likely to successfully implant in the uterus. Such tests might assess metabolism or numbers of chromosomes. One developmental biologist who has performed hundreds of IVF procedures claims that he can predict which preembryos will "take" just by their appearance!

Figure 12.5 compares IVF, surrogacy, and artificial insemination.

Gamete Intrafallopian Transfer

As the world marveled at the miracle of Louise Joy Brown and other test tube babies, disillusioned couples were learning that IVF is costly, time consuming, aggravating, and rarely works. The reason why it rarely works may be the artificial fertilization environment. In the mid-1980s, Ricardo Asch, at the University of Texas in San Antonio, developed a procedure called **GIFT,** which stands for **gamete intrafallopian transfer.** GIFT assists fertilization, but in the woman's body rather than in glassware. Certain religions approve of GIFT, but not IVF, because GIFT preserves more natural reproductive function.

In GIFT, a woman takes a superovulation drug for a week, and then has several of her largest oocytes removed. The man submits a sperm sample, and the most active cells are separated from it. The collected oocytes and sperm are deposited together in the woman's fallopian tube, at a site past any obstruction so that implantation can occur. GIFT is about 40% successful.

A variation of GIFT is **ZIFT,** or **zygote intrafallopian transfer.** In this procedure, an in vitro fertilized ovum is introduced into the woman's fallopian tube, rather than into her uterus, as in IVF. Allowing the fertilized ovum to make its own way to the uterus seems to increase the chance that it will implant.

To increase the odds that IVF will lead to a birth, several preembryos are usually transferred. This can result in multiple births, as Michele and Ray L'Esperance learned in 1988 when their attempt to have one baby via IVF resulted in five (fig. 12.4)! Today, as IVF is increasingly successful, fewer preembryos are implanted. When too many preembryos implant, some doctors remove some of them so that two or three have room to develop.

Even when more than one preembryo is implanted, several more preembryos have usually formed in the lab. What should be done with the "extras"? Often, extra preembryos are donated to an infertile couple or frozen for the genetic parents' later use. Freezing apparently causes some damage, because two out of three frozen and thawed preembryos fail to develop.

The ability to freeze embryos can create strange situations. One man sued his wife for possession of their frozen embryos as part of the divorce settlement. In another case, identical twins were born 18 months apart because one was frozen as a preembryo. In 1985, a plane crash killed the wealthy parents of two preembryos stored at −320° F in a hospital in Melbourne, Australia. Adult children of the couple found themselves in a curious position—possibly having to share their estate with two eight-celled microscopic relatives.

Researchers are taking several measures to improve the chance that IVF will result in a birth. These measures include:

Blocking certain of a woman's hormones during the superovulation stage to produce more mature oocytes, which are more likely to be fertilized and develop.

Transferring preembryos slightly later in development.

Culturing fertilized ova and early preembryos with other "helper" cells that normally surround the oocyte

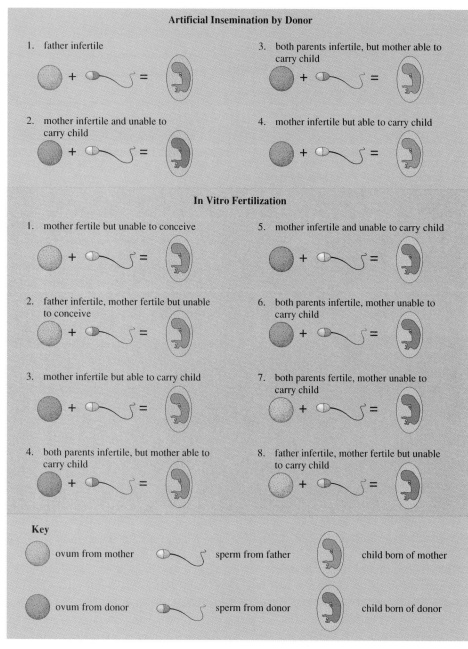

Artificial Insemination by Donor

1. father infertile

2. mother infertile and unable to carry child

3. both parents infertile, but mother able to carry child

4. mother infertile but able to carry child

In Vitro Fertilization

1. mother fertile but unable to conceive

2. father infertile, mother fertile but unable to conceive

3. mother infertile but able to carry child

4. both parents infertile, but mother able to carry child

5. mother infertile and unable to carry child

6. both parents infertile, mother unable to carry child

7. both parents fertile, mother unable to carry child

8. father infertile, mother fertile but unable to carry child

Key

ovum from mother   sperm from father   child born of mother

ovum from donor   sperm from donor   child born of donor

FIGURE 12.5

Reproductive alternatives. Artificial insemination, in vitro fertilization, surrogate mothering, and embryo freezing and transfer have provided many ways to produce a baby other than the "natural" way. This figure shows possible parental combinations with artificial insemination by donor and in vitro fertilization.

Oocyte Banking and Donation

Can female gametes be frozen and stored, as sperm are? If so, a woman wishing to become a parent later in life, when fertility declines, could set aside oocytes while she is young. The ability to store healthy oocytes would also benefit women who must undergo a medical treatment that could damage oocytes, such as cancer chemotherapy, or who work with toxins. Although Korean researchers have successfully frozen and thawed oocytes, nurtured them in vitro, and fertilized them, the procedure presents problems.

Oocytes are frozen in liquid nitrogen at –86 to –104° F (–30 to –40° C), while they are at metaphase of the second meiotic division. The chromosomes are held in suspended animation along the spindle—and the spindle is sensitive to temperature extremes. If the spindle comes apart as the cell freezes, the oocyte may lose a chromosome. This would cause a developmental disaster. Another problem that occurs when oocytes are frozen is that they sometimes retain a polar body, resulting in a diploid oocyte. Can you envision how this would hamper development?

Researchers are developing ways to nurture oocyte-packed, one-millimeter-square sections of ovaries in the laboratory. So far, culturing oocytes in the lab appears to require a high level of oxygen and a complex combination of biochemicals. If researchers can successfully culture oocytes, oocyte banking may become more feasible.

Another technology allows young women to donate "fresh" oocytes to older women. A medical journal article entitled "A Preliminary Report on Oocyte Donation Extending Reproductive Potential to Women Over 40," set off a flurry of headlines as people envisioned 80-year-old women cradling their newborns. So far, a few women in their 60s have given birth. The American Fertility Society considers the ordinary childbearing age span to extend from age 15 to age 44. But oocyte donation can benefit younger women as well—women who have entered menopause prematurely, or who have an inherited disease they do not wish to risk passing to their offspring.

Oocyte donation's success rate ranges from 20 to 50%, and it generally costs about $9,000. Oocytes are usually donated from women who have "leftover" oocytes from IVF, or from a female relative of the infertile woman. But donating oocytes isn't easy—the donor must receive several daily shots of a superovulation drug, have her blood checked each morning for three weeks, then have laparoscopic surgery to collect the oocytes. The prospective father's sperm and the donor's oocytes are then placed in the recipient's uterus.

Embryo Adoption

Embryo adoption is a variation on oocyte donation. In this procedure, a woman who has malfunctioning ovaries but a healthy uterus carries an "adopted" embryo. The process begins when her mate supplies sperm to artificially inseminate a woman who produces healthy oocytes. If the woman conceives, a physician gently "lavages" the preembryo, or flushes it out of her uterus a week after the insemination. The preembryo is then inserted through the cervix and into the uterus of the infertile woman. The embryo is the genetic child of the woman's male partner and the woman who carries it for the first week, but the baby is born from the woman who cannot produce healthy oocytes.

Preimplantation Genetic Diagnosis

Medical specialists can scrutinize technology-aided pregnancies using the same prenatal diagnostic tests they use to monitor natural pregnancies—tests such as chorionic villus sampling and amniocentesis. However, reproductive technologies also make possible earlier detection of genetic and chromosomal abnormalities, even before the preembryo implants in the uterus on the sixth day. Intervention at this very early stage is called **preimplantation genetic diagnosis.**

In this technique, a doctor removes a cell from an 8-cell preembryo and examines its genetic material (fig. 12.6). If the cell's DNA indicates that this particular preembryo has not inherited the condition in question, then the remaining 7-cell preembryo is implanted in the woman, where it completes development. The 7-cell preembryo is apparently unharmed by the loss of 1 cell.

One of the first beneficiaries of preimplantation genetic diagnosis was Chloe O'Brien. Chloe's brother has cys-

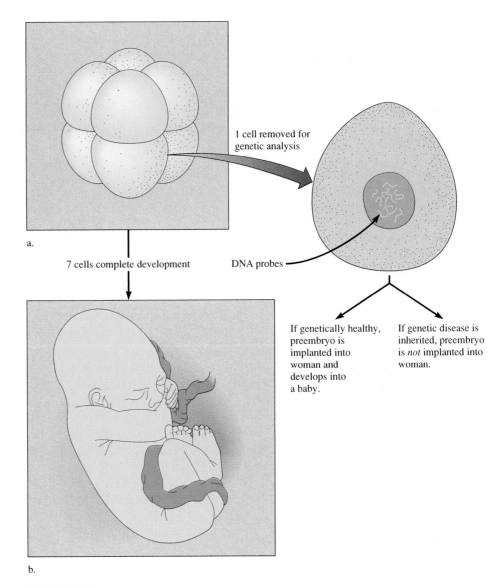

a.

1 cell removed for genetic analysis

7 cells complete development

DNA probes

If genetically healthy, preembryo is implanted into woman and develops into a baby.

If genetic disease is inherited, preembryo is *not* implanted into woman.

b.

FIGURE 12.6

Preimplantation genetic diagnosis probes disease-causing genes in an 8-cell preembryo.

tic fibrosis. Chloe was checked, as an 8-cell preembryo, to see if she had also inherited the disease. She hadn't; she was born, a healthy little girl, in March 1992.

Although preimplantation genetic diagnosis offers an alternative to terminating a pregnancy in much later stages, the technology may not become widespread because it is expensive (at least $10,000 for the IVF procedure, DNA amplification, and use of genetic probes) and technically difficult.

Prenatal development is so complex, with so many steps and intricately coordinated activities, that the birth of a healthy baby seems a miracle. Indeed,

in some pregnancies, the intricate developmental pattern does go awry. Following are some of the more common problems encountered as a new human life takes form.

Spontaneous Abortion

In a **spontaneous abortion,** a pregnancy ends naturally before the fetus is developed enough to survive in the outside world. Sometimes the embryo or fetus is deformed. Sometimes it is normal, but some problem in the woman's body prevents her from carrying the fetus to full term. Nearly one in five women who know they are pregnant have spontaneous abortions. The actual number of spontaneous abortions is estimated at two of every three pregnancies, but many occur so early that the woman is unaware that she is pregnant and doesn't realize she has aborted.

A spontaneous abortion is similar to an unusually heavy menstrual period, with cramping and copious bleeding. The more advanced the pregnancy, the more severe the symptoms. If the pregnancy is past the first trimester, the spontaneous abortion is called a miscarriage. If a woman in the second trimester notices that the fetus has not moved in 24 hours, it may be a sign that a miscarriage is imminent. A doctor will usually advise immediate bed rest, but most miscarriages cannot be prevented. In a pregnancy that is 6 months or more along, the woman must endure labor to give birth to a child that she knows will be born dead, or stillborn.

Why don't all embryos and fetuses complete development? Many embryos cannot survive for more than a few weeks because they have an abnormal number of chromosomes. Sometimes the preembryo is normal, but the trophoblast cells that surround it do not produce enough hCG to suppress menstruation, preventing implantation. A fetus may literally fall out of a woman's body if the uterus is malformed or the cervix too loose. An "incompetent" cervix can be closed with sutures, which are removed when labor begins.

Fibroid tumors in the uterus may crowd the developing embryo or fetus, stunting its growth and possibly interfering with its nutrient supply. If the preembryo implants in a fallopian tube rather than in the uterus in a dangerous and painful ectopic pregnancy, spontaneous abortion may occur, but the woman usually requires an immediate therapeutic abortion. Otherwise, the growing embryo will eventually burst the tube, and the woman can lose a great

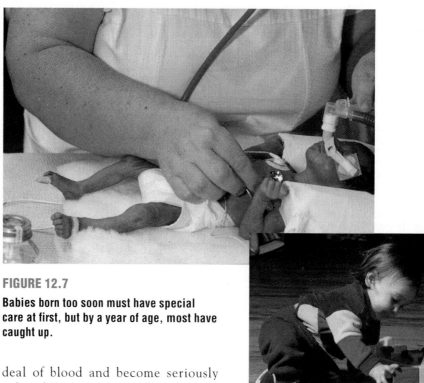

FIGURE 12.7

Babies born too soon must have special care at first, but by a year of age, most have caught up.

deal of blood and become seriously infected. Ectopic pregnancies are on the rise because of an increase in infections that damage the fallopian tubes and, curiously, because of the ability to treat them. (Technology might also contribute to the reported increase as doctors detect more ectopic pregnancies.) Before doctors could diagnose pelvic infections and treat them promptly with antibiotics, the tubes would completely close with scar tissue, so pregnancy could not occur. Today, antibiotic treatment halts infection so that the tubes are only partially obstructed by scar tissue. Later, this can entrap fertilized ova, resulting in an ectopic pregnancy.

A couple experiencing spontaneous abortion often grieves for their expected child, wondering, "Can this happen again?" or "Was it my fault?" For most couples, the chance that a chromosomal abnormality will recur is only 5 to 20%. If the spontaneous abortion results from a genetic defect inherited from two carrier parents, the recurrence risk is 25%. If a hormonal imbalance or a physical malformation causes a spontaneous abortion, hormone therapy or surgery may prevent a recurrence.

However, some women spontaneously abort, for unknown reasons, at about the same point in several successive pregnancies. These women are advised to stay in bed during this critical time in future pregnancies.

KEY CONCEPTS

Spontaneous abortion may be caused by chromosomal defects, hormonal imbalances, fibroid tumors, or a malformed uterus or cervix. In an ectopic pregnancy, the preembryo implants in a fallopian tube and must be surgically removed.

Born Too Soon— Prematurity

Human prenatal development normally takes about 38 weeks. Babies born more than 4 weeks early, or weighing less than 5 pounds (2.3 kilograms), are **premature**

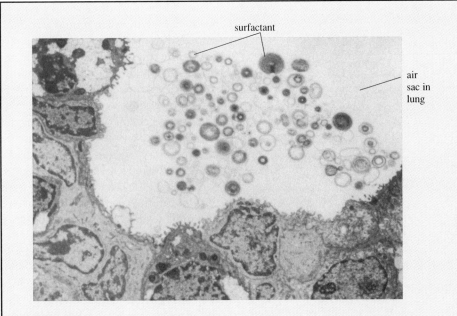

FIGURE 12.8

Towards the end of prenatal development, the fetus's lungs produce surfactant, or "surface-active agent." This foamy liquid enables the microscopic air sacs in the lungs to inflate by reducing surface tension between inhaled, dry air and moist lung surfaces. Premature babies often suffer from respiratory distress syndrome because they do not yet produce enough surfactant. Fortunately, the missing biochemical can be provided.

control mechanisms combine to make them lose heat rapidly.

Sometimes a fetus is carried full term but still weighs less than 5 pounds (2.3 kilograms) at birth. A "small for gestational age" infant usually has few problems beyond rapid heat loss due to low birth weight. Unlike the infant born too soon, this tiny tot has mature breathing and sucking reflexes. Women who are malnourished or who smoke during pregnancy often have full-term babies of low weight. Abnormal development of the placenta may also account for small-for-gestational-age infants.

KEY CONCEPTS

Premature babies, born more than 4 weeks early or weighing less than 5 pounds (2.3 kilograms), often suffer health problems because of their small sizes and immature reflexes. These infants may have difficulty sucking, breathing, and staying warm. Some suffer from respiratory distress syndrome, but this can often be treated with synthetic surfactant.

(fig. 12.7). The longer they remain in the uterus and the more they weigh, the greater their chances of survival. Babies born more than 12 weeks prematurely often live only a few days, but rapidly developing technology is greatly increasing the chances that even the tiniest, most immature infants will survive. Each year about 10,000, or 8%, of the babies born in the United States are premature. Twins and triplets are more likely to be born prematurely because of their cramped prenatal environment. Teens and malnourished women also are more likely to deliver prematurely. However, a "preemie" can be born to even a well-nourished, healthy woman, for unknown reasons.

The most serious problem premature infants have is immature breathing and sucking reflexes. The baby born too soon is often not yet able to breathe properly, because normally the placenta would still be providing oxygen and car-

rying away carbon dioxide. **Respiratory distress syndrome** develops when the baby's lungs are not sufficiently covered with a soapy substance called **surfactant** (fig. 12.8). The small air sacs in the lungs cannot inflate without it. Straining to draw dry air into deflated lungs damages the delicate lung tissue.

Fortunately, doctors can drip a synthetic surfactant into a baby's windpipe to assist breathing, then place the infant on a respirator. Within minutes, a rosy blush spreads through the baby's skin, indicating a functioning respiratory system.

Often preemies cannot suck and must be fed either intravenously or through a stomach tube inserted through the nose. They are also more vulnerable to birth injuries and more prone to infection. Preemies must be kept warm in heated units called isolettes because their small body size, low body fat, and immature temperature-

Birth Defects

When genetic abnormalities or toxic exposures affect an embryo, developmental errors occur, resulting in birth defects. Although many things can go wrong during the complex period of prenatal development, it is reassuring to know that most babies—about 97%—are apparently normal at birth.

The Critical Period

The specific nature of a birth defect usually depends on which structures are developing when the damage occurs. The time when genetic abnormalities, toxic substances, or viruses can alter a specific structure is its **critical period** (fig. 12.9). Some body parts, such as fingers and toes, are sensitive for short periods of time. In contrast, brain development can be disrupted throughout prenatal development, as well as during the first two years of life. Because of the

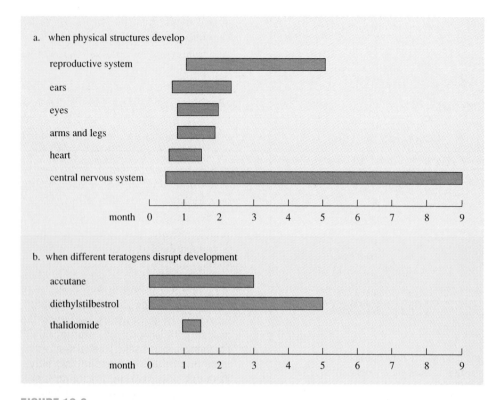

a. when physical structures develop

reproductive system
ears
eyes
arms and legs
heart
central nervous system

month   0   1   2   3   4   5   6   7   8   9

b. when different teratogens disrupt development

accutane
diethylstilbestrol
thalidomide

month   0   1   2   3   4   5   6   7   8   9

FIGURE 12.9

By comparing the times when a drug acts and when certain structures in the embryo or fetus are sensitive, one can predict the nature of a birth defect resulting from exposure to the drug.

brain's extensive critical period, many birth defects involve mental retardation. The continuing sensitivity of the brain after birth explains why toddlers who eat lead-based paint demonstrate impaired learning.

About two-thirds of all birth defects stem from a disruption during the embryonic period. More subtle defects that become noticeable only after infancy, such as learning disabilities, are often caused by disruptions during the fetal period. For example, chemicals that disrupt brain development in the first trimester are likely to produce a mentally retarded newborn. A substance that affects brain function in the seventh month of pregnancy, however, might cause a more subtle difficulty in learning to read.

Some birth defects can be attributed to an abnormal gene that exerts its influence at a specific point in prenatal development. In a rare inherited condition, phocomelia (seal limbs), an abnormal gene halts limb development from the third to fifth week of the embryonic

period, causing the infant to be born with "flippers" where arms and legs should be. Geneticists can predict the chances that a genetically caused birth defect will recur in the family.

Many birth defects, however, are caused by toxic substances the expectant mother ingests. These environmentally caused problems cannot be passed on to future generations and will not recur unless the exposure occurs again. Chemicals or other agents that cause birth defects are called **teratogens** (Greek for "monster causing"). Table 12.5 lists some commonly used teratogens.

Teratogens

Thalidomide

The idea that the placenta protects the embryo and fetus from harmful substances was tragically disproven between 1957 and 1961, when 10,000 children born in Europe displayed what seemed, at first, to be phocomelia. Because doctors realized that this genetic disorder is very rare, they began to look for another

cause. They soon discovered that the mothers had all taken a mild tranquilizer, **thalidomide,** early in pregnancy, during the time an embryo's limbs form. The "thalidomide babies" were born with stumps or nothing at all where their legs and arms should have been (fig. 12.10). The United States was spared from the thalidomide disaster because an astute government physician noted the drug's adverse effects on laboratory monkeys. Still, several "thalidomide babies" were born in South America in 1994, where pregnant women were given the drug without warnings.

Rubella

Ironically, at about the same time doctors recognized the severity of the thalidomide crisis, another teratogen, this one a virus, was sweeping the United States. Australian physicians first noted the teratogenic effects of the **rubella** virus that causes German measles in 1941, but in the United States, public attention did not focus on rubella until the early 1960s, when a rubella epidemic caused 20,000 birth defects and 30,000 stillbirths. Women who contract the virus during the first trimester of pregnancy run a high risk of bearing children with cataracts, deafness, and heart defects. Rubella's effects on fetuses exposed during the second or third trimesters of pregnancy include learning disabilities, speech and hearing problems, and juvenile-onset diabetes.

Alcohol

Alcohol is a teratogen. A pregnant woman who has just one or two drinks a day, or perhaps a large amount at a single crucial time in prenatal development, risks **fetal alcohol syndrome** in her unborn child. Because scientists have not yet determined the effects of small amounts of alcohol at different stages of pregnancy, and because each woman metabolizes alcohol slightly differently, it is best to entirely avoid alcohol when pregnant, or when trying to become pregnant.

A child with fetal alcohol syndrome (FAS) has a characteristic small head, misshapen eyes, and a flat face and nose (fig. 12.11). He or she grows slowly

Table 12.5 **Teratogenic Drugs**

Drug	Medical Use	Risk to Fetus
Alkylating agents	Cancer chemotherapy	Growth retardation
Aminopterin, methotrexate	Cancer chemotherapy	Skeletal and brain malformations
Coumadin derivatives	Seizure disorders	Tiny nose Hearing loss Bone defects Blindness
Danazol	Endometriosis	Masculinization of female structures
Diphenylhydantoin (Dilantin)	Seizures	Cleft lip, palate Heart defects Small head
Diethylstilbestrol (DES)	Repeat miscarriage	Vaginal cancer, vaginal adenosis Small penis
Isotretinoin (Accutane)	Severe acne	Cleft palate Heart defects Abnormal thymus Eye defects Brain malformation
Lithium	Manic-Depression	Heart and blood vessel defects
Penicillamine	Rheumatoid arthritis	Connective tissue abnormalities
Progesterones in birth control pills	Contraception	Heart and blood vessel defects Masculinization of female structures
Pseudoephedrine	Nasal decongestant	Stomach defects
Tetracycline	Antibiotic	Stained teeth
Thalidomide	Morning sickness	Limb defects

FIGURE 12.10

This child is 1 of about 10,000 children born between 1957 and 1961 in Europe to mothers who had taken the tranquilizer thalidomide early in pregnancy. The drug acted as a teratogen on developing limbs, causing infants to be born with stumps in place of arms and/or legs. It is important that the public be made aware of thalidomide's danger, because the drug is used to treat leprosy and possibly AIDS.

before and after birth. The child's intellect is impaired, ranging from minor learning disabilities to mental retardation. One study reported that if a mother-to-be consumed three mixed drinks a day, her child lost five IQ points. Aristotle noticed problems in children of alcoholic mothers more than 23 centuries ago. In the United States today, fetal alcohol syndrome is the third most common cause of mental retardation in newborns, and 1 to 3 in every 1,000 infants has the syndrome—more than 40,000 born each year.

The effects of FAS continue beyond childhood. Teens and young adults with the syndrome are short and have small heads. More than 80% of them retain facial characteristics of a young child with FAS, including abnormal lips, misaligned or malformed teeth, and a wide space between the upper lip and the nose. These facial traits make people with FAS look similar, but not abnormal.

The long-term mental effects of prenatal alcohol exposure are more severe than the physical vestiges. Many adult individuals with FAS function at early grade school level. They often lack

social and communication skills and find it difficult to understand the consequences of actions, form friendships, take initiative, and interpret social cues.

Cocaine

Cocaine is very dangerous to the unborn. It can cause spontaneous abortion by inducing a stroke in the fetus, and cocaine-exposed infants who do survive seem far more distracted and unable to concentrate on their surroundings than normal infants. Other health and behavioral problems arise as these children grow. A problem in evaluating the effects of cocaine is that affected children are often exposed to other substances and situations that could also account for their symptoms.

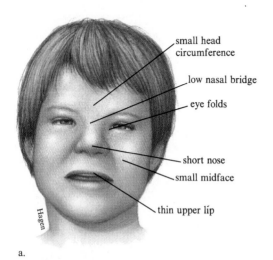

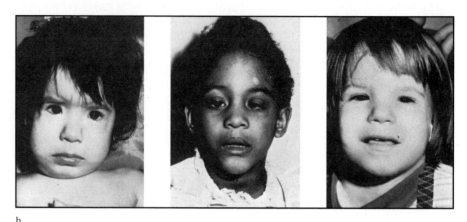

a.

b.

FIGURE 12.11

Fetal alcohol syndrome. Some children whose mothers drank heavily during pregnancy have characteristic flat faces (*a*) that are strikingly similar in children of different races (*b*). Alcoholic mothers-to-be have a 30 to 45% chance of giving birth to a child who is affected to some degree by prenatal exposure to alcohol. Early research indicates that damage begins to occur when the mother consumes two mixed drinks per day, but researchers have not yet confirmed this.

Cigarettes

Chemicals in cigarette smoke stress a fetus. Carbon monoxide crosses the placenta and plugs up sites on the fetus's hemoglobin molecules that would normally bind oxygen. Other chemicals in smoke prevent nutrients from reaching the fetus. Studies comparing the placentas of smokers and nonsmokers show that smoke-exposed placentas lack important growth factors, causing poor growth before and after birth. Cigarette smoking during pregnancy is linked to spontaneous abortion, stillbirth, prematurity, and low birth weight.

Nutrients

Certain nutrients ingested in large amounts, particularly vitamins, act as drugs in the human body. The acne medicine isotretinoin (Accutane) is a vitamin A derivative that causes spontaneous abortions and defects of the heart, nervous system, and face. The tragic effects of this drug were first noted exactly nine months after dermatologists began prescribing it to young women in the early 1980s. Today, the drug package bears prominent warnings, and it is never prescribed to pregnant women. A vitamin A-based drug used to treat psoriasis, as well as excesses of vitamin A itself, also cause birth defects. Some forms of vitamin A are stored in body fat for up to three years after ingestion.

Another nutrient that can harm a fetus when the pregnant woman takes it in excess is vitamin C. The fetus becomes accustomed to the large amounts the woman takes. After birth, when the baby is taking in far less of the vitamin, symptoms of vitamin C deficiency (scurvy) set in. The baby bruises easily and is prone to infection.

Malnutrition in a pregnant woman also threatens the fetus. A woman needs to consume extra calories while she is pregnant or breastfeeding. Obstetrical records of pregnant women before, during, and after World War II link inadequate nutrition in early pregnancy to an increase in the incidence of spontaneous abortion. The aborted fetuses had very little brain tissue. Poor nutrition later in pregnancy affects the development of the placenta, which can cause low birth weight, short stature, tooth decay, delayed sexual development, learning disabilities, and possibly mental retardation.

Occupational Hazards

Some women (and men) encounter teratogens in the workplace. Researchers note increased rates of spontaneous abortion and children born with birth defects among women who work with textile dyes, lead, certain photographic chemicals, semiconductor materials, mercury, and cadmium. We do not know much about the role the male plays in transmitting environmentally caused birth defects. The male does have one built-in protection—if a sperm cell is severely damaged, it will not be able to move. However, men whose jobs expose them to sustained heat, such as smelter workers, glass manufacturers, and bakers, may produce sperm that can move, fertilize an oocyte, and possibly cause spontaneous abortion or a birth defect. A virus or a toxic chemical carried in semen may also cause a birth defect.

The Baby Doe Dilemma

Thirty years ago, if a woman went into labor after only 24 weeks of pregnancy, the result was almost always a spontaneous abortion or an infant that died within a few hours after birth. Today, nearly half of such pregnancies produce extremely small live infants. Thanks to the modern medical technology of **neonatology** (study of the newborn), many of these infants survive. Some of them, however, are born with severe and sometimes multiple medical problems. Often new parents and physicians must make difficult decisions over whether it is kinder to subject a severely ill newborn to corrective surgery and drug treatments or to leave the child alone, letting nature take its course.

Consider the case of Andrew. Because the placenta detached early, Andrew was born only 24 weeks after he was conceived. The doctors pronounced the 1-pound, 12-ounce (794 grams) baby "marginally viable," giving him less than a 5% chance of survival. After several months of medical intervention, Andrew died. In his brief life, Andrew suffered from dehydration, broken bones, collapsed lungs, seizures, and diseases of the blood, eyes, urinary tract, liver, and heart. Before his death, his mother summed up her confusion: "I'm afraid my baby is going to die. I'm afraid my baby is going to live."

The anguish parents of severely ill newborns feel is called the "Baby Doe dilemma," after a baby whose situation was widely reported but who was publicly known only as Baby Doe. Baby Doe was born with a blocked esophagus and Down syndrome in Bloomington, Indiana, in 1982. Unlike Andrew's parents, who elected to intervene medically, Baby Doe's parents chose to withhold food, water, and medical help, because even though doctors could surgically unblock the esophagus, the baby would still have Down syndrome. The baby died at a few months of age. In the years that followed, other Baby Does came to national attention, but the troubling questions of how—or whether—to treat them are still handled on a case-by-case basis. A presidential commission established in 1979 to develop ethics guidelines concerning the denial of medical treatment concluded that the effects associated with Down syndrome are *not* severe enough to justify withholding medical treatment.

The Fetus as a Patient

A partial solution to the Baby Doe dilemma is to treat the fetus before the baby is born. Some prenatal medical problems can be treated by administering drugs to the mother or by altering her diet. An abnormally small fetus can receive a nutritional boost if the mother eats a high-protein diet. A fetus that cannot produce adequate amounts of a specific vitamin can sometimes overcome the deficiency if the mother takes large doses of the vitamin. It is also possible to treat some prenatal medical problems directly: A tube inserted into the uterus can drain the dangerously swollen bladder of a fetus with a blocked urinary tract, providing relief until the problem can be surgically corrected at birth. A similar procedure can remove excess fluid from the brain of a hydrocephalic (fluid trapped in the brain) fetus. Drugs can reach the fetus, bypassing the mother, through a tube placed in the umbilical cord.

Little Blake Schultz made medical history when he underwent major surgery seven weeks before his birth. Ultrasound had revealed that his stomach, spleen, and intestines protruded through a hole in his diaphragm, the muscle sheet separating the abdomen from the chest. This defect would have suffocated him shortly after birth were it not for pioneering surgery by Michael Harrison at the University of California at San Francisco. Harrison's surgical team made an incision in the mother, exposed Blake's left side, gently tucked his organs into place, and patched the hole with a synthetic material used in clothing.

Sex and Health

The scope of reproductive health is broader than protecting the health of pregnant woman and fetus. The reproductive system is unique among the body's organ systems in that it does not become active until sexual maturity. When sexual activity begins, it becomes important to prevent pregnancy and sexually transmitted diseases (STDs).

Birth Control

Contraception is the use of devices or practices that work "against conception." Birth control methods work in two basic ways: either they physically block the meeting of sperm and oocyte, or they make the female system's environment hostile to sperm or to a preembryo's implantation (table 12.6).

In 1960, the first contraceptive pill became available. The "combination pill" contains synthetic versions of the hormones estrogen and progesterone, which hinder conception in three ways: they suppress ovulation, alter the

Table 12.6 Birth Control Methods

	Method	Mechanism	Advantages	Disadvantages	Pregnancies per Year per 100 Women*
	None				80
Barrier and Spermicidal	Condom	Worn over penis, keeps sperm out of vagina	Protection against sexually transmitted diseases	Disrupts spontaneity, can break, reduces sensation in male	2–12
	Condom and spermicide	Worn over penis (male) or inserted into vagina (female), keeps sperm out of vagina, and kills sperm that escape	Protection against sexually transmitted diseases	Disrupts spontaneity, reduces sensation in male	2–5
	Diaphragm and spermicide	Kills sperm and blocks uterus	Inexpensive	Disrupts spontaneity, messy, must be fitted by doctor	3–17
	Cervical cap and spermicide	Kills sperm and blocks uterus	Inexpensive, can be left in 24 hours	May slip out of place, messy, must be fitted by doctor	5–20
	Spermicidal foam or jelly	Kills sperm and blocks vagina	Inexpensive	Messy	5–22
	Spermicidal suppository	Kills sperm and blocks vagina	Easy to use and carry	Irritates 25% of users, male and female	3–15
	Contraceptive sponge	Kills, blocks, and absorbs sperm in vagina	Can be left in for 24 hours	Expensive	6–28
Hormonal	Combination birth control pill	Prevents ovulation and implantation, thickens cervical mucus	Does not interrupt spontaneity, lowers risk of some cancers, decreases menstrual flow	Raises risk of cardiovascular disease in some women, causes weight gain and breast tenderness	0–10
	Minipill	Blocks implantation, deactivates sperm, thickens cervical mucus	Fewer side effects	Weight gain	1–10
	Depo-Provera	Prevents ovulation, alters uterine lining	Easy to use, lasts 3 months	Menstrual changes, weight gain, doctor must give injection	1
	Norplant	Prevents ovulation, thickens cervical mucus	Easy to use, lasts 5 years	Menstrual changes, doctor must implant	0.2
Behavioral	Rhythm method	No intercourse during fertile times	No cost	Difficult to do, hard to predict timing	13–21
	Withdrawal	Removal of penis from vagina before ejaculation	No cost	Difficult to do	9–25
Surgical	Vasectomy	Sperm cells never reach penis	Permanent, does not interrupt spontaneity	Requires minor surgery, difficult or impossible to reverse	0.15
	Tubal ligation	Egg cells never reach uterus	Permanent, does not interrupt spontaneity	Requires surgery, some risk of infection, difficult to reverse	0.4
Other	Intrauterine device	Prevents implantation	Does not interrupt spontaneity	Severe menstrual cramps, increased risk of infection	1–5

*The lower figures apply when the contraceptive device or technique is used correctly. The higher figures take into account human error.

Table 12.7 The Events of Early Development and New Contraceptive Approaches

Event	Block
	Vaccine prompts woman to produce antibodies that bind to glycoproteins on zona, preventing sperm
	Vaccine consists of antibodies that cover sperm surface molecules, preventing them from binding to zona
	Premature induction of "blocks to polyspermy" prevents any sperm from entering zona or ovum
	Pill blocks progesterone receptors on uterine lining; vaccine blocks human chorionic gonadotropin, a hormone produced by a fertilized ovum that maintains uterine lining; both prevent implantation

uterine lining so that a preembryo cannot implant, and thicken the mucus in the cervix so that sperm cells cannot get through.

Two newer contraceptive methods available to women in the United States include a synthetic version of the hormone progesterone, called Depo-Provera, and Norplant, an implant version of the hormone used in birth control pills. Depo-Provera is given by injection every three months. Norplant consists of six matchstick-shaped implants placed beneath the skin on the woman's upper arm. It remains effective for five years.

A promising candidate for male birth control, a substance called gossypol, is derived from the seeds, stems, and roots of the cotton plant. Its dampening effects on sperm count were discovered in a rural community in China in the 1950s, when researchers traced male infertility to the use of cottonseed oil in cooking. Gossypol decreases levels of lactate dehydrogenase in the testicles, where it is needed for sperm production. Although thousands of Chinese men have successfully used gossypol as a birth control pill, its side effects include appetite loss, weakness, heart rhythm changes, lowered sex drive, and occasional permanent infertility. Gossypol is not available in the United States.

The idea of using the immune system to develop a vaccine against pregnancy stems from studies of women whose vaginal secretions contain substances that inactivate sperm cells. One experimental vaccine prompts a female's immune system to produce antibody molecules that attach to the zona pellucida that surrounds each oocyte (table 12.7). The bound antibodies block a sperm's entry. Another experimental female birth control vaccine is a piece of a molecule normally found on the surfaces of sperm cells. When this fragment is injected into the woman's bloodstream, it programs her immune system to recognize the molecule—and the sperm cells that carry it—as foreign. The sperm cells are then destroyed by the woman's immune system. A single injection seems to be effective for a few months.

In the early 1970s, 45 million women in the United States used **intrauterine devices** (IUDs). These devices, which a physician must place in the uterus, disrupt the uterine lining, preventing implantation. However, one brand of IUDs caused uterine infections so severe that in 1974, 20 women died from them. This prompted other manufacturers to cease producing the devices. An IUD was removed by a string hanging from the vagina; this string provided a very effective conduit for bacteria. IUD users frequently experienced heavy and painful menstrual periods. Pregnancies sometimes occurred despite the device, and some babies were born accompanied by a dislodged IUD! Today, only one brand of IUD is available, and it is designed so that it is unlikely to cause infection. IUDs are a commonly used form of contraception in many nations.

Contraceptives that provide a physical and/or chemical barrier between sperm and oocyte are very popular because they provide some protection against sexually transmitted diseases. A male **condom** is a sheath worn over the erect penis. A female condom is a baglike device inserted in the vagina. Either type blocks sperm from ascending the female reproductive tract. Spermicidal (sperm-killing) jellies, foams, creams, and suppositories are inserted into the vagina, and they can be used in conjunction with devices that block the cervix, such as the **diaphragm, contraceptive sponge,** or **cervical cap.** Most products now on the market use the spermicide nonoxynol-9.

The sponge and cervical cap are ancient devices. A contraceptive sponge was mentioned in the Talmud, a religious document written in 800 A.D. that reflected practices over the preceding thousand years. Centuries ago, Europeans used sponges soaked in lemon juice to kill sperm. Cervical caps, which adhere to the cervix by suction, have been constructed of materials such as beeswax, opium poppy fibers, lemon halves, and paper. Camel drivers still insert stones in the reproductive tracts of female camels to prevent pregnancies on long trips.

"Natural" birth control or "fertility awareness" methods rely on restricting sexual activity during times when conception is likely. In the **rhythm method,** the woman charts her menstrual cycle. If a woman menstruates every 28 days, she is most likely to conceive on the 12th to 16th days following the onset of bleeding. In a more precise version, a woman takes her temperature every morning with a specially calibrated thermometer. Body temperature rises slightly just before ovulation, indicating that the couple should avoid intercourse.

The consistency and appearance of vaginal mucus also provides clues to when a woman is likely to conceive. The secretion is thin, elastic, and clear during fertile periods, and cloudy, sticky, or absent at other times. A few women can feel differences in the position and texture of the cervix when they are ovulating.

Sterilization provides permanent birth control. A **vasectomy** involves cutting or burning the vas deferens so that sperm cannot leave the testicles. The germ cells degenerate in the man's body. Females can have their fallopian tubes tied off in a **tubal ligation.** Vasectomies and tubal ligations can sometimes be reversed. Figure 12.12 summarizes birth control devices.

Terminating a Pregnancy

An unwanted pregnancy can be terminated by preventing the preembryo's implantation in the uterus, or, later on, by removing an embryo or fetus from the uterus. First-trimester **abortions** use a sucking or scraping device to remove the embryo or fetus, while later abortions utilize more intense scraping, injection of a salt solution into the amniotic sac, or prostaglandin supposi-tories, which trigger uterine contractions and expulsion of the fetus within 24 hours. A drug developed in France, **RU 486,** can induce abortion if it is taken during the first seven weeks of pregnancy. Because it is taken orally, it is safer than a surgical abortion. RU 486 blocks the hormone progesterone, which is necessary for implantation and early development (fig. 12.13).

Many people feel that abortion differs from devices and methods that actually prevent conception, such as the diaphragm, birth control pill, or condom. In the United States, abortion has been a controversial subject since it was first legalized in the early 1970s. In the former Soviet Union, however, two-thirds of all women have had at least one abortion because of the inavailability of contraceptives. Abortion is often chosen when a fetus is diagnosed with a devastating or untreatable medical condition.

Sexually Transmitted Diseases

The 20 recognized sexually transmitted diseases (STDs) are often called "silent infections" because in the early stages they may not produce symptoms, especially in women (table 12.8). By the time symptoms appear, it is often too late to prevent complications or the spread of the infection to sexual partners. Because many STDs have similar symptoms, and some of the symptoms may also arise from diseases or allergies that are unrelated to sexual activity, most physicians suggest that a consultation is in order if one or a combination of these symptoms appears:

1. Burning sensation during urination
2. Pain in the lower abdomen
3. Fever or swollen glands in the neck
4. Abnormal discharge from the vagina or penis
5. Pain, itch, or inflammation in the genital or anal area
6. Pain during intercourse

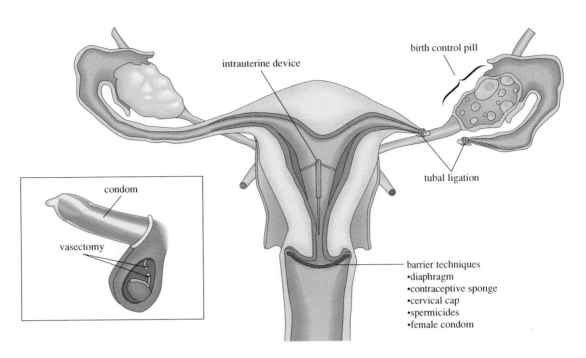

FIGURE 12.12

Sites of birth control action. Birth control pills prevent the release of egg cells from the ovary. In a tubal ligation, the fallopian tubes are tied so that egg cells cannot pass through them to the uterus. An intrauterine device embeds in the uterine wall, preventing implantation of a zygote. The diaphragm, contraceptive sponge, female condom, and cervical cap block the cervix so that sperm cells cannot enter the uterus. Spermicidal jellies, creams, and suppositories enhance the effectiveness of barrier contraceptives. A man can prevent impregnating a woman by wearing a condom or by undergoing a vasectomy, which can now be done without surgery.

7. Sores, blisters, bumps, or a rash anywhere on the body, particularly the mouth or genitals
8. Itchy, runny eyes

One possible complication of two STDs, gonorrhea and chlamydia, is **pelvic inflammatory disease.** In this disease, bacteria enter the vagina and spread throughout the reproductive organs. The first symptoms are intermittent cramps, followed by sudden fever, chills, weakness, and more severe cramps. Hospitalization and intravenous antibiotics can stop the infection. The uterus and fallopian tubes are often scarred by this disease, resulting in infertility and increased risk of ectopic pregnancy.

In **acquired immune deficiency syndrome** (AIDS), the immune system steadily deteriorates. The body becomes overrun by infection and often by cancer, diseases that immune system cells and biochemicals normally attack and conquer. The virus thought to cause AIDS is passed from one person to another in body fluids such as semen, blood, and vaginal secretions. It is most frequently transmitted during anal intercourse or by using a needle containing contaminated blood.

The human reproductive system is unique among organ systems because it does not begin to function until several years after birth. It is also unique because an individual can survive without it. On the population level, however, the reproductive system is essential to the survival of the species. A healthy reproductive system can make life very fulfilling, both in providing the pleasures of sexuality and in producing new life.

KEY CONCEPTS

Maintaining reproductive health means obtaining prenatal care, controlling fertility, and preventing the spread of sexually transmitted diseases. Contraceptive methods include the birth control pill and newer hormonal implants and injections; the IUD; barrier methods such as the condom, diaphragm, cervical cap, and contraceptive sponge; the spermicide nonoxynol-9; vasectomy and tubal ligation; and monitoring natural changes that indicate when a woman is most fertile. Experimental contraceptives include gossypol and vaccines. Symptoms of the 20 known sexually transmitted diseases overlap. Two of the more serious are pelvic inflammatory disease and AIDS.

How RU 486 Works

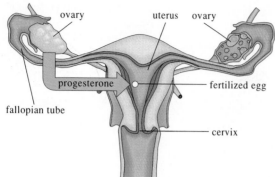

Before taking RU 486: progesterone suppresses menstruation, maintains pregnancy.

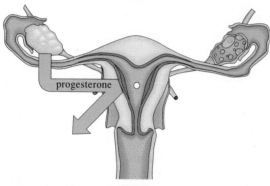

RU 486 taken early in pregnancy: RU 486 blocks progesterone.

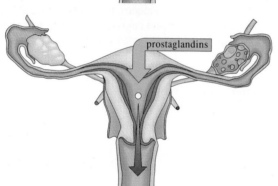

Prostaglandins contract uterus. Preembryo is expelled.

FIGURE 12.13

RU 486 terminates pregnancy by blocking progesterone.

SUMMARY

Human developmental and reproductive problems are becoming more amenable to treatment as we learn more about reproductive health and functioning. A major area of concern is infertility, the inability of a couple to conceive a child after a year of trying. Causes of infertility in the male include a low sperm count, a malfunctioning immune system, a varicose vein in the scrotum, heat exposure, structural sperm defects, and abnormal hormone levels. Female infertility can be due to absent or irregular ovulation, blocked fallopian tubes, an inhospitable uterus, or antisperm secretions. Older women generally take longer to conceive.

Infertile couples are increasingly turning to reproductive alternatives such as artificial insemination, in vitro fertilization, embryo transfer, frozen embryos, and surrogate mothers. While relatively new to humans, these techniques are standard practices in animal breeding. These reproductive technologies have spawned many ethical and legal questions. Other techniques, such as preimplantation genetic diagnosis, allow early detection of fetal abnormalities and increase a couple's chances of giving birth to a healthy child.

An embryo or fetus spontaneously aborts in about one out of every five recognized pregnancies, usually because the fetus has an extra or missing chromosome or because of a hormonal or structural abnormality in the woman's

Table 12.8 Some Sexually Transmitted Diseases

Disease	Cause	Symptoms	Number of Reported Cases (U.S.)	Effects on Fetus	Treatment	Complications
Acquired immune deficiency syndrome (AIDS)	Human immunodeficiency virus (HIV)	Fever, weakness, infections, cancer	14 million+ (infected)	Exposure to AIDS virus and other infections	Drugs to treat or delay symptoms; no cure	Dementia and death
Chlamydia	*Chlamydia* bacteria	Painful urination and intercourse, mucus discharge from penis or vagina	3–10 million	Premature birth, blindness, pneumonia	Antibiotics	Pelvic inflammatory disease, infertility, arthritis, ectopic pregnancy
Genital herpes	Herpes simplex virus type I or II	Genital sores, fever	20 million	Brain damage, stillbirth	Antiviral drug (acyclovir)	Increased risk of cervical cancer
Genital warts	Human papilloma virus	Warts on genitals	1 million	None known	Chemical or surgical removal	Increased risk of cervical cancer
Gonorrhea	*Neisseria gonorrheae* bacteria	In women, usually none; in men, painful urination	2 million	Blindness, stillbirth	Antibiotics	Arthritis, rash, infertility, pelvic inflammatory disease
Syphilis	*Treponema pallidum* bacteria	Initial chancre sore usually on genitals or mouth; rash 6 months later; several years with no symptoms as infection spreads; finally damage to heart, liver, nerves, brain	90,000	Miscarriage, prematurity, birth defects, stillbirth	Antibiotics	Death
Hepatitis B	Hepatitis B virus	Fatigue, persistent low-grade fever, jaundice (yellowing skin), rash, abdominal pain	300,000 per year	Low birth weight	Rest, alpha interferon	Cirrhosis, liver cancer

reproductive tract. Children born a month or more prematurely often experience a range of health problems, the most serious of which are immature lungs and breathing reflexes. Birth defects can be traced to genetic anomalies or to teratogens the woman has been exposed to at a critical period in prenatal development. The field of neonatology deals with the medical problems of newborns.

Contraceptives prevent pregnancy by blocking the meeting of sperm cell and oocyte, or by making the uterus inhospitable to the implantation of the zygote. Birth control devices include hormone-containing pills, injections, or implants, barrier contraceptives such as condoms or devices that block the cervix, spermicides, and intrauterine devices. Vasectomy and tubal ligation are permanent means of birth control. Contraception in the future, for both sexes, may rely on vaccines developed by manipulating the immune response.

Once sexual activity begins, sexually transmitted diseases can become a health threat. Prevalent STDs include chlamydia, gonorrhea, syphilis, herpes simplex infection, and acquired immune deficiency syndrome. Preventing STDs, preventing or achieving pregnancy, and producing healthy offspring are important aspects of human reproductive health.

KEY TERMS

abortion 260
acquired immune deficiency
   syndrome (AIDS) 261
artificial insemination 245
cervical cap 260
condom 260
contraception 257
contraceptive sponge 260
critical period 253

diaphragm 260
ectopic pregnancy 243
embryo adoption 251
endometriosis 243
fetal alcohol syndrome 254
fibroid 243
gamete intrafallopian transfer
   (GIFT) 249
infertility 241
intrauterine devices 260

in vitro fertilization 247
neonatology 257
pelvic inflammatory disease 261
preimplantation genetic
   diagnosis 251
premature 252
respiratory distress syndrome 253
rhythm method 260
RU 486 260
rubella 254

spontaneous abortion 251
surfactant 253
surrogate mother 247
teratogen 254
thalidomide 254
tubal ligation 260
vasectomy 260
zygote intrafallopian transfer
   (ZIFT) 249

REVIEW QUESTIONS

1. Knowledge about infertility sometimes leads to the development of new birth control methods. Match each cause of infertility with the contraceptive method based on it.

Cause of infertility	*Contraceptive device*
a. Failure to ovulate due to hormonal imbalance	Intrauterine device
b. Fibroid tumor on the uterine lining	Tubal ligation
c. Fallopian tubes blocked by endometriosis tissue	Birth control pill
d. Low sperm count	Birth control vaccine
e. Immune infertility	Gossypol

2. A couple is having trouble conceiving a child. After several visits to fertility specialists, they find their problem is twofold. The male has a sperm count of 50 million sperm per ejaculate, and the woman ovulates only three times a year. Suggest a way to help the couple to conceive.

3. Big Tom is a bull with valuable genetic traits. His sperm are used to conceive 1,000 calves in many different cows.

   Mist, a dairy cow with exceptional milk output, is given a superovulation drug, and many oocytes are removed from her ovaries. Once fertilized, the oocytes are implanted into other cows. With their help, Mist becomes the genetic mother of 100 calves—far more than she could give birth to naturally.

   Which human technologies do the experiences of Big Tom and Mist mirror?

4. How can "fertility awareness" methods help a couple to conceive a child? How can they be used as contraceptive measures?

5. Ultrasound detects twin embryos in a woman who is six weeks pregnant. Seven months later, only one baby is born. What happened?

6. A few years ago, a doll manufacturer sold a cute, pudgy "preemie" model. Is this an accurate representation of an infant born too soon? Why or why not?

7. In Aldous Huxley's book *Brave New World,* egg cells are fertilized in a laboratory and allowed to develop assembly-line style. To render some of the embryos less intelligent than others, lab workers give alcohol to some of them. What human medical condition does this scenario invoke?

8. Contraception literally means "against conception." According to this definition, is an intrauterine device a contraceptive? Why or why not?

TO THINK ABOUT

1. State who the genetic parents are and who the gestational mother is in each of the following scenarios:

   a. A 26-year-old woman has her uterus removed because of cancer. However, her ovaries are intact and her oocytes are healthy. She takes drugs to superovulate and has oocytes removed, and they are fertilized in vitro with her husband's sperm. Two fertilized ova are implanted into the uterus of the woman's best friend.

   b. Max and Tina had a child by IVF in 1986. At that time, they had three extra preembryos frozen. Two are thawed in 1994 and implanted into the uterus of Tina's sister, Karen. Karen's uterus is healthy, but she has ovarian cysts that often prevent her from ovulating.

   c. Forty-year-old Christensen von Wormer wanted children, but not a mate. He donated sperm and hired an Indiana woman to undergo artificial insemination. On September 5, 1990, von Wormer held his newly born daughter, Kelsey, for the first time.

   d. Two men who live together want to raise a child. They go to a fertility clinic, have their sperm collected and mixed, and use it to artificially inseminate a female friend, who turns the baby over to them.

2. The intervention of lawyers and government regulatory agencies in biological matters is mentioned several times in this chapter. What position do you think the law should take in the following procedures or situations?

   a. Treatment of multiple physically and mentally challenged newborns

   b. Treatment of "extra" fertilized ova or embryos growing in vitro

   c. Malpractice liability if an embryo or fetus conceived by a reproductive alternative is spontaneously aborted or is born with a birth defect

   d. Custody of a child if a surrogate mother decides, after giving birth, that she wants to keep the child rather than surrendering it to its genetic parents

   e. Under what circumstances a woman can obtain an abortion

   f. Whether a minor's parents should be informed that he or she has sought birth control

   g. Whether a new birth control vaccine should be approved

   h. Which items of personal information a sperm donor should provide

3. Studies have shown that many people with the mental disorder schizophrenia were exposed to influenza virus when their mothers contracted the infection during the second or third trimester of pregnancy. What information might be important in investigating whether or not the flu virus causes schizophrenia in people exposed to it before birth?

4. Conjoined (Siamese) twins were born in Indiana in August 1993. Seven weeks later, surgery separated the twins, killing Amy but enabling Angela to survive. Angela was doing well, but died at 10 months of age. The medical bill topped $1 millon, and because of this, some people claimed that the twins should have been left untreated. Do you think they should have been treated? Cite a reason for your answer.

5. A study showed that male college students who regularly smoked marijuana had lower-than-average sperm counts. However, the study did not take into account that some of these students drank alcohol, took other drugs, and ejaculated frequently. Does the study support the conclusion that marijuana smokers have lower sperm counts? How could such a study be conducted in a more thorough manner?

6. Do you think that an AIDS test that a person can perform at home is a good idea? Why or why not?

7. Adoption is one reproductive alternative for a couple with a fertility problem. Traditionally, identity of the biological parents has often been kept confidential. Do you think this is a wise policy? Why or why not? Suggest another reproductive alternative in which confidentiality is an issue.

8. Our society treats some fetuses with medical problems in the uterus and undertakes "heroic measures" to prolong the life of premature and possibly very ill infants. It also allows the abortion of healthy fetuses. What inconsistencies are at work? How, in your opinion, can they be resolved?

SUGGESTED READINGS

Ames, Katrine. September 30, 1991. And donor makes three. *Newsweek.* Oocyte donors give children to couples who otherwise could not have them.

Annas, George J. February 6, 1992. Using genes to define motherhood—the California solution. *The New England Journal of Medicine,* vol. 326, no. 6. Not everyone agrees that genes alone define parenthood.

Chattingius, Sven, M.D., et al. August 19, 1992. Delayed childbearing and risk of adverse perinatal outcome. *The Journal of the American Medical Association,* vol. 268, no. 7. Older women face increased risk of pregnancy complications.

Chervenak, Frank A., and Laurence B. McCullough. October 1992. Fetus as patient: an ethical concept. *Contemporary Ob/Gyn.* Physicians are finding that the fetus, in some situations, has ethical rights.

Cohn, Jeffrey P. October 1991. Reproductive biotechnology. *Bioscience.* Many human reproductive procedures were perfected on domestic animals.

Edwards, Robert, and Patrick Steptoe. 1980. *A matter of life.* New York: William Morrow. The story of Louise Joy Brown—the first "test tube baby."

Elmer-Dewitt, Philip. September 30, 1991. Making babies. *Time.* A journey through prenatal development, pinpointing what can go wrong and how it can be corrected.

Farley, Dixie. January–February 1993. Endometriosis: Painful, but treatable. *FDA Consumer.* Teenage girls can suffer agonizing pain from this menstrual condition.

Fleming, Anne Taylor. March 29, 1987. Our fascination with Baby M. *The New York Times Magazine.* A detailed look at a famous surrogate mother case.

Grifo, Jamie A., et al. August 12, 1992. Pregnancy after embryo biopsy and coamplification of DNA from X and Y chromosomes. *The Journal of the American Medical Association,* vol. 268, no. 6. A photo essay vividly depicts the steps of preimplantation genetic diagnosis.

Handyside, Alan H., et al. September 24, 1992. Birth of a normal girl after in vitro fertilization and preimplantation diagnostic testing for cystic fibrosis. *The New England Journal of Medicine,* vol. 327, no. 13. Chloe O'Brien was born with two normal CFTR alleles after being checked genetically at the 8-cell stage.

Jones, J. W.,Jr., and J.P. Toner. December 2, 1993. Current concepts: the infertile couple. *The New England Journal of Medicine,* vol. 329. Most infertile couples eventually are helped to have children.

Lewis, Ricki. March 1993. Choosing a perfect child. *The World and I.* Will preimplantation genetic diagnosis and the human genome project team up to create a Brave New World, where every fertilized ovum is carefully screened and selected?

Segal, Marian. May 1991. Norplant: Birth control at arm's reach. *FDA Consumer.* Approved in 1991, Norplant is an unusual form of a familiar contraceptive.

Schwartz, R. M. et al. May 26, 1994. Effect of surfactant on morbidity, mortality, and resource use in newborn infants weighing 500 to 1500 g. *The New England Journal of Medicine,* vol. 330. Giving very low birth weight infants lung surfactant saves many lives.

Smolowe, Jill. June 14, 1993. New, improved and ready for battle. *Time.* RU 486 will make abortions far safer, because a pill can replace surgery. But will they make abortion too easy?

Stehlin, Dori. March 1993. Depo-Provera: The quarterly contraceptive. *FDA Consumer.* For some women, a shot of contraceptive every three months is preferable to remembering to take a daily pill, or enduring the mess of barrier contraceptives.

Willis, Judith Levine. June 1993. Preventing STDs. *FDA Consumer.* A primer for young adults on how to prevent sexually transmitted diseases.

Winston, R. M. C., and A. H. Handyside. May 14, 1993. New challenges in human in vitro fertilization. *Science,* vol. 260. Researchers are exploring several methods to improve the success rate of IVF.

UNIT

4

Genetics

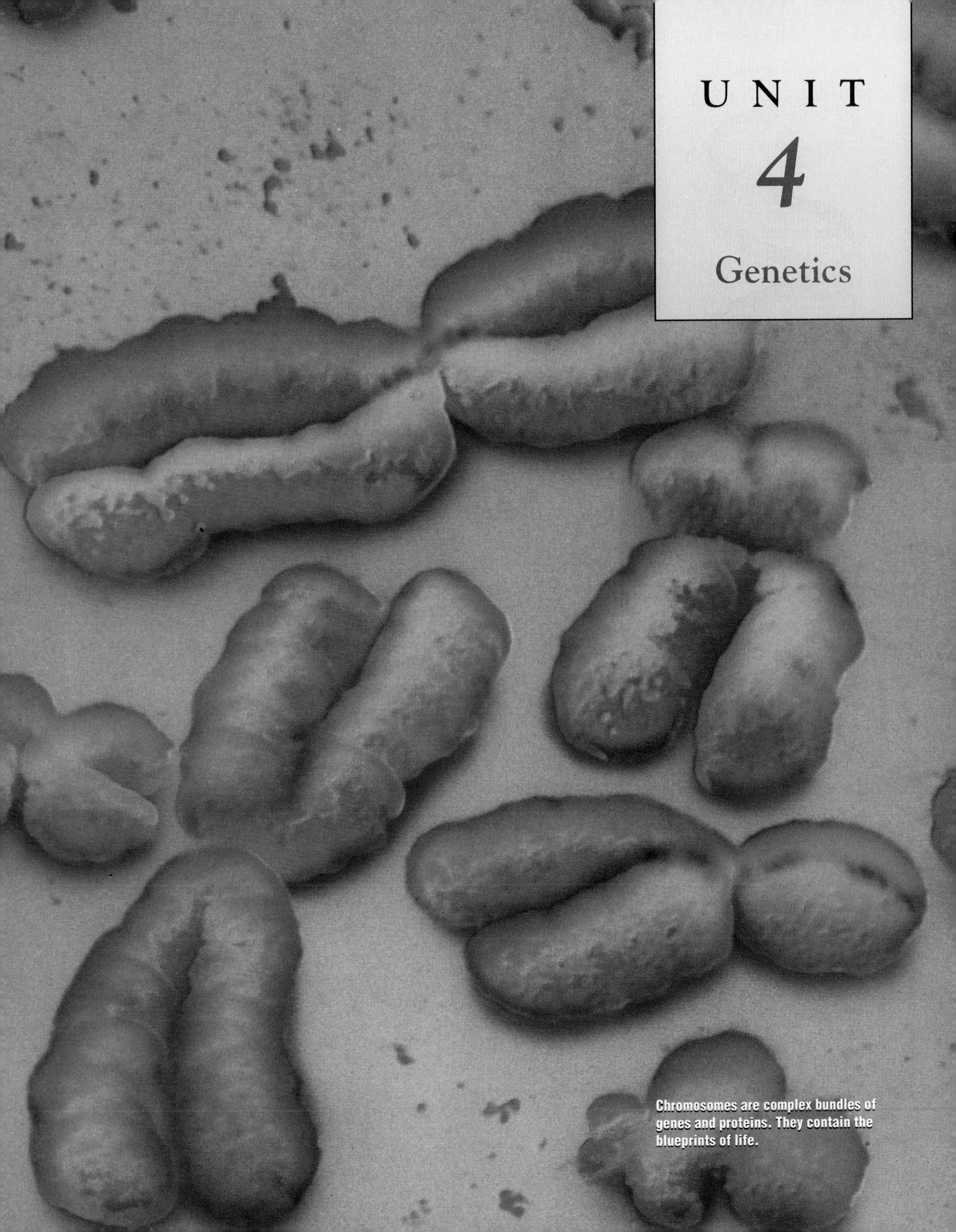

Chromosomes are complex bundles of genes and proteins. They contain the blueprints of life.

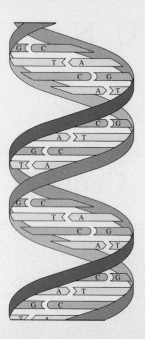

CONNECTIONS

Triple-Headed Purple Monsters

The triple-headed purple monster first appeared in the late 1970s, dubbed by a genetic researcher attempting to describe the public's fear of the fledgling field of recombinant DNA technology. This technology, better known as genetic engineering, allows a gene from one species to be expressed in another.

Public alarm at the prospect of recombinant DNA experiments was understandable, because at the time, researchers themselves expressed concerns that an altered bacterium might somehow escape the confines of the lab. Biologists exploring the new field met to design clever controls and safeguards for their planned work. Within a few years, the first drug produced by the new technology—human insulin—became available to people with diabetes. For many of them, the new drug was an improved treatment over the pig or cow insulin they had been using.

Despite the success of human insulin and then a handful of other genetically engineered drugs—particularly when the AIDS epidemic made some traditional drug sources unsafe—the triple-headed purple monster still loomed. Genetic research continued to alarm the public, possibly because very few people understand exactly what a gene is and what it does. Sometimes, the fear escalated into violence.

In 1987, a young researcher released genetically engineered bacteria onto strawberry plants growing in a field in Tulelake, California. Nearby residents, who had held near-hysterical town meetings to block the experiment, ripped out the experimental plants, even after the researcher replanted them. Despite being given reams of explanations by the Environmental Protection Agency, people failed to grasp that the bacteria sprayed on the strawberries were abnormal because they *lacked* a gene, not because they had been endowed with some mysterious, threatening new trait. So again, the monster—fear caused by lack of knowledge—reared its triple purple head.

In the 1980s and early 1990s, new drugs produced by recombinant DNA technology continued to be approved by the Food and Drug Administration. Gene therapy experiments began, offering hope to sufferers of cancer and inherited disease. But the triple-headed purple monster emerged in a new guise, applying genetic engineering to agriculture. A new tomato with an extended shelf life debuting on supermarket shelves was rejected outright by consumers, who demanded that labels indicate which tomatoes were the products of genetic engineering. Again, the headlines and news shows glossed over the fact that the new tomato did not have a scary new trait, but *lacked* a ripening enzyme.

By 1994, with gene therapy experiments actually working, and new DNA-based drugs and diagnostic tests on the market, the monster arose again. This time the problem was genetically-engineered bovine somatotropin, a protein growth hormone that stimulates dairy cows to produce more milk. People had many objections to this application, mostly that in a country with a milk surplus and struggling small farmers, it simply wasn't necessary. But many consumers and newspaper columnists roared against the genetic engineering part of the issue—failing to understand that it is the excess hormone, and not its source, that is responsible for extra milk production and health problems in the cows that might result from it. As with the slow-to-ripen tomato, consumers demanded special labeling on milk from cows given the genetically-engineered hormone—although it is chemically identical to the natural protein.

If the people of Tulelake or consumers facing genetically-altered tomatoes or milk understood just what had been done, their fear, if not dissipated, would at least be based on knowledge, not emotion. But even today, if you were to ask several people, "What is a gene? What does a gene do?," only a small number would know the answers.

However, you will. As the chapters in this unit point out, genetics is a field that will impact on all of us in the coming years. To judge the coming technologies wisely, arm yourself with knowledge. Fight that triple-headed purple monster!

Transmission Genetics

CHAPTER OUTLINE

This mosquito, which carries the parasite that causes malaria in humans, has mutant white eyes.

LEARNING OBJECTIVES

By the chapter's end, you should be able to answer these questions:

1. How and where did the modern science of genetics originate?

2. How did Gregor Mendel follow the inheritance of a single gene, and of two genes, by observing the offspring of crosses of pea plants?

3. How did Mendel's observations reflect meiotic events?

4. How can one predict the outcomes of genetic crosses?

5. What terms are commonly used in genetics?

6. What phenomena can disrupt Mendelian ratios? Which is the only one that is a true exception to Mendelian inheritance?

7. How do interactions between genes and the environment mold phenotypes?

The History of Genetics

When radio listeners heard the gravelly, slightly whiny voice of the new rock singer, some felt as if they had traveled back in time. The voice was virtually identical to that of 1960s folksinger Bob Dylan, though the song and style were different. In fact, the new voice *was* a Dylan's—but it was Bob's son, Jakob's.

Heredity can be startlingly obvious, as in the Dylans' voices or in the striking resemblance between father and son actors Kirk and Michael Douglas (fig. 13.1). Although growing up in a musical or theatrical environment can shape a talent, heredity plays a strong role, too. As figure 13.2 indicates, interest in heredity must be as old as humankind itself, as people throughout time have wondered at their similarities—and fought over their differences.

Farmers in Mexico applied genetic principles six thousand years ago when they chose seed from the hardiest wild grass plants to sow the next season's crop. In this way, over many plant generations, they bred domesticated corn. Four thousand years earlier, humans had domesticated wheat. But early knowledge of genetics was not confined to agriculture. The Talmud, an ancient book of Jewish law, reveals a remarkable comprehension of human heredity. It tells the story of a woman whose three sisters all had sons who died after circumcision. When she asked a rabbi about circumcising her newborn son, he advised her not to. This

FIGURE 13.1

Kirk and Michael Douglas—obviously father and son.

demonstrates an ancient recognition of hemophilia's transmission through a female carrier.

Nineteenth century biologists thought that particular parts of the body controlled trait transmission. Different biologists gave the units of inheritance, today called **genes**, a variety of colorful names, including pangens, idioblasts, bioblasts, plastidules, nuclein, plasomes, stirps, gemmules, or just characters. But an investigator who used the term elementen made the most lasting impression on what would become the science of genetics. His name was Gregor Mendel.

As a child in what is now Czechoslovakia near the Polish border, Mendel learned farming from his family, tending fruit trees for the lord of a manor. He overcame extreme poverty to obtain a university education in science, mathematics, and teaching, and eventually entered a monastery. There he learned more about plant breeding from the abbot, who had built a greenhouse for scientific research.

Although peas had been popular with plant breeders since the 1820s, Mendel was the first to use them to study the mechanisms of heredity. From 1857 to 1865, Mendel carefully set up pea plant crosses. He observed that consistent ratios of traits in the offspring seemed to indicate that the plants transmitted distinct "elementen." Mendel knew of the cell theory, but no one had yet visualized chromosomes. Although he did not understand how organisms

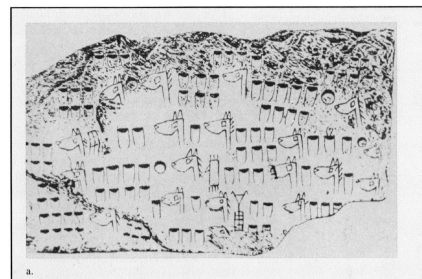

FIGURE 13.2

Ancient cultures recognized inheritance patterns. *a.* A horse breeder in Asia 4,000 years ago etched a record of his animal's physical characteristics in stone. *b.* In the ninth century B.C., Assyrian priests hand-pollinated date palms.

physically transmit hereditary information, his two general laws of inheritance, segregation and independent assortment, eventually proved to apply to any diploid species, including humans.

Following the Inheritance of One Gene—Segregation

Mendel worked with many pea plant variants, but he was especially interested in seven easily distinguishable characteristics (fig. 13.3). He noted that short plants bred to other short plants were "true-breeding," consistently yielding seeds that gave rise only to more short plants. Tall plants were not always true-breeding. Some tall plants, when crossed with a short plant or another tall plant, produced only tall plants in the next generation. But when certain other tall plants were crossed with each other, about one-quarter of the plants in the next generation were short, suggesting that tallness could mask the potential to transmit shortness. Of the remaining plants, one-third (one-quarter of the total number of offspring) proved through further crosses with short plants to be "true-breeding tall." The other two-thirds produced some short plants in the next generation (fig. 13.4).

Mendel suggested that the gametes distribute elementen, because these cells physically link generations. The elementen would separate from each other as gametes form. When gametes join at fertilization, the elementen would group into new combinations. Mendel reasoned that each elementen was packaged in a separate gamete, and if opposite-sex gametes combine at random, then he could mathematically explain the different ratios of traits produced from his pea plant crossings. Mendel's idea that elementen separate in the gametes would later be called the **law of segregation.**

Few people realized Mendel was demonstrating principles that applied far beyond peas. Even after he published "Experiments with Plant Hybrids" in a natural history journal in 1866, few scientists were impressed enough to attempt to repeat his tests. By 1869, a frustrated Mendel turned his energies to monastery administration.

It wasn't until 1900 that botanists realized that Mendel's conclusions explained data gathered in crosses of many plant species. Also in the early 1900s, several researchers noted that chromosomes behave much like Mendel's elementen. Both paired elementen and paired chromosomes separate at each generation and pass—one from each parent—to the offspring. Both elementen and chromosomes are inherited in random combinations. Chromosomes provide a physical mechanism for Mendel's hypotheses.

In 1909, Wilhelm Johannsen renamed Mendel's elementen: the units of inheritance became **genes** (Greek for "give birth to"). Soon thereafter, English biologist William Bateson coined the term genetics to describe the new science of heredity.

For the next few years, the gene remained a black box; many biologists accepted its existence, but few ventured any suggestions about its physical nature. By the 1940s, scientists began to turn their attention to discovering the gene's chemical basis. Although James Watson and Francis Crick deciphered the structure of the hereditary molecule DNA in

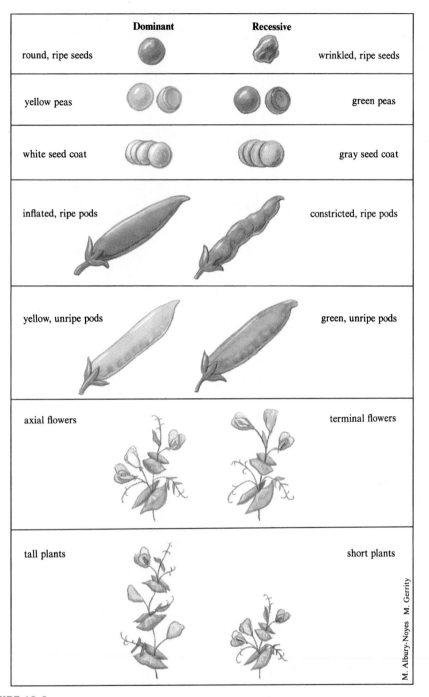

M. Albury-Noyes M. Gerrity

FIGURE 13.3

Gregor Mendel studied the transmission of seven traits in the pea plant. Each trait had two expressions, easily distinguished from each other. Seeds were either round or wrinkled; peas were either yellow or green; seed coats were either white or gray; ripe pods were either inflated or constricted; unripe pods were either yellow or green; flowers were either axial (arising sideways from the stems) or terminal (emanating only from the top of the plant); and plants were either tall (6 to 7 feet) or short (3/4 to 1 1/2 feet).

1953, we are continually learning new details about how genes control each other. The field of genetics is still young.

How Biologists Describe Single Genes

Mendel observed two different expressions of a trait—short and tall. Traits are expressed in different ways because a gene can exist in alternate forms, or **alleles.** An individual who has two identical alleles for a gene is **homozygous** for that gene. An individual with two different alleles is **heterozygous.**

Mendel noted that for some genes, one variant could mask the expression of another. The allele that masks the other is completely **dominant,** and the masked allele is **recessive.** When Mendel crossed a true-breeding tall plant to a short plant, the tall allele was completely dominant to the short allele. The plants of the next generation were all tall.

When a gene has two alleles, it is common to symbolize the dominant allele with a capital letter, and the recessive with the corresponding small letter. If both alleles are recessive, the individual is **homozygous recessive.** Two small letters, such as *tt* for short plants, symbolize this. An individual with two dominant alleles is **homozygous dominant.** Two capital letters, such as *TT* for tall pea plants, represent homozygous dominance. Another possible allele combination is one dominant and one recessive allele—*Tt* for non true-breeding tall pea plants. An individual with these alleles is a **heterozygote.**

An organism's appearance does not always reveal its alleles. Both a *TT* and a *Tt* pea plant are tall, but the first is a homozygote, the second a heterozygote. The **genotype** describes the organism's alleles, while the **phenotype** describes the outward expression of an allele combination. A pea plant with a tall phenotype may have genotype *TT* or *Tt*. A

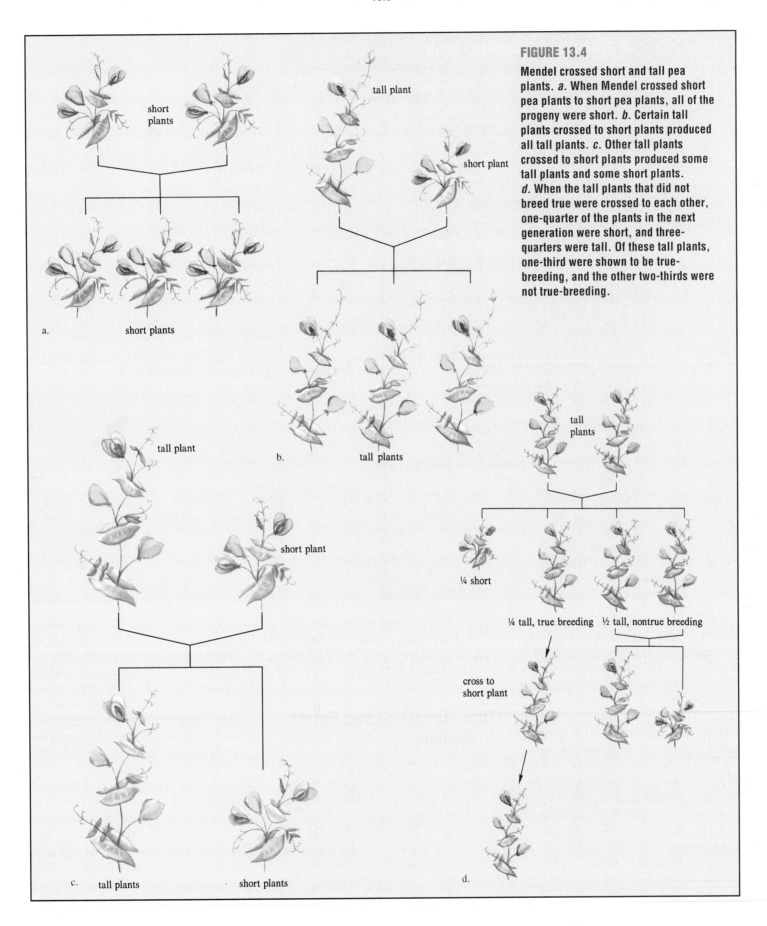

FIGURE 13.4

Mendel crossed short and tall pea plants. *a.* When Mendel crossed short pea plants to short pea plants, all of the progeny were short. *b.* Certain tall plants crossed to short plants produced all tall plants. *c.* Other tall plants crossed to short plants produced some tall plants and some short plants. *d.* When the tall plants that did not breed true were crossed to each other, one-quarter of the plants in the next generation were short, and three-quarters were tall. Of these tall plants, one-third were shown to be true-breeding, and the other two-thirds were not true-breeding.

Table 13.1 A Glossary of Genetic Terms

Term	Definition	Term	Definition
Allele	One of two or more alternate forms of any given gene.	Meiosis	A form of cell division that produces gametes (sperm or egg cells).
Autosome	Any chromosome other than a sex chromosome.	Monohybrid	An individual heterozygous for one particular gene.
Chromosome	A rod-shaped structure in a cell's nucleus composed of DNA and protein.	Mutant	A phenotype or allele that differs from the most common for a certain gene in a population.
Dihybrid	An individual heterozygous for two particular genes.	P_1	The parental generation.
Dominant	An allele that masks the expression of another allele.	Phenotype	The observable expression of a specific genetic constitution in a specific environment.
F_1	The first filial generation; offspring.	Recessive	An allele whose expression is masked by the activity of another allele.
F_2	The second filial generation; offspring of offspring.	Segregation	Mendel's first law. States that alleles of a gene separate into equal numbers of gametes during meiosis.
Gene	A segment of a chromosome, composed of DNA, that directs the cell's synthesis of a particular protein or controls the activity of another gene.	Sex chromosome	A chromosome that determines sex.
Genotype	The genetic constitution of an individual.		
Heterozygous	Possessing different alleles of the same gene.	Sex linked	A gene located on the X chromosome or a trait that results from the activity of a gene on the X chromosome.
Homozygous	Possessing identical copies of an allele of the same gene.		
Independent assortment	Mendel's second law states that a gene on one chromosome does not influence the inheritance of a gene on a different (nonhomologous) chromosome because chromosomes are packaged randomly into gametes during meiosis.	Wild type	The most common phenotype or allele for a certain gene in a population.

wild type phenotype is the most common expression of a particular gene in a population. An allele that arises as a change, or **mutation,** from wild type is a **mutant.** Although wild type is often taken to mean standard or normal, a mutant is a variant, not necessarily harmful and sometimes advantageous.

When biologists analyze genetic crosses, they call the first generation the **parental generation,** or P_1; the second generation is the **first filial generation,** or F_1; the next generation is the **second filial generation,** or F_2, and so on. If you considered your grandparents the P_1 generation, your parents would be the F_1 generation, and you and your siblings the F_2 generation. Table 13.1 summarizes these and other commonly used genetic terms.

How Mendel's Laws Relate to Meiosis

Mendel's observations on inheritance of a single gene reflect the events of meiosis. When a gamete is produced, the two copies of a particular gene separate, as the homologs that carry them do. In a plant of genotype Tt, for example, gametes carrying either T or t form in equal numbers during anaphase I. When gametes meet to start the next generation, they combine at random. That is, a t-bearing oocyte is neither more nor less attractive to a t-bearing sperm than a T-bearing sperm. These two factors—equal allele distribution and random combinations of gametes—underlie Mendel's law. Figure 13.5 shows the role they play in segregation.

Mendel crossed short plants (tt) with true-breeding tall plants (TT). The resulting seeds grew into F_1 plants of the same phenotype: tall (genotype Tt). Next, he crossed the F_1 plants with each other in a **monohybrid** cross, so called because one trait is followed by crossing

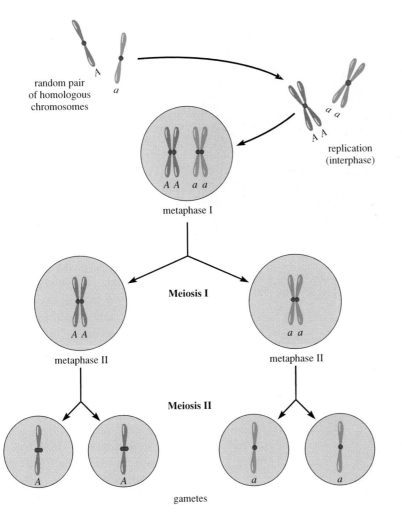

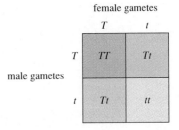

FIGURE 13.6

A Punnett square, a diagram of gametes and how they can combine in a cross between two particular individuals, is helpful in following the transmission of traits. The different types of female gametes are listed along the top of the square; male gametes are listed on the left-hand side. Each compartment within the square contains the genotypes that result when gametes that correspond to that compartment join. The Punnett square here describes Mendel's monohybrid cross of two tall pea plants. In the progeny, tall plants outnumber short plants 3:1. Can you determine the genotypic ratio?

FIGURE 13.5

Mendel's first law—gene segregation. During meiosis, homologous pairs of chromosomes (and the genes that comprise them) separate from one another and are packaged into separate gametes. At fertilization, gametes combine at random to form the individuals of a new generation.

two heterozygous, or hybrid, individuals. The three possible outcomes of such a cross are *TT*, *tt*, and *Tt*. A *TT* individual results when a *T* sperm fertilizes a *T* oocyte; a *tt* plant results when a *t* oocyte meets a *t* sperm; and a *Tt* individual results when either a *t* sperm fertilizes a *T* oocyte, or a *T* sperm fertilizes a *t* oocyte.

Because two of the four possible sperm-egg combinations produce a heterozygote, and each of the others produces one of the homozygotes, the **genotypic ratio** expected of this monohybrid cross is 1 *TT*: 2 *Tt*: 1 *tt*. The corresponding **phenotypic ratio** is three tall plants to one short plant, a 3:1

ratio. These ratios can be determined by constructing a diagram called a **Punnett square** (fig. 13.6). Experiments yield numbers of offspring that approximate these ratios. Mendel observed 787 tall plants and 277 short plants when he crossed *Tt* and other *Tt* pea plants—close to the predicted 3 tall to 1 short phenotypic ratio.

Mendel distinguished the two genotypes resulting in tall progeny—*TT* from *Tt*—by carrying out additional crosses. He bred tall plants of unknown genotype to short (*tt*) plants. If a tall plant crossed to a *tt* plant produced both tall and short progeny, Mendel knew it

was genotype *Tt*. But if it produced only tall plants, he knew it must be *TT*. The technique of crossing an individual of unknown genotype to an individual with a homozygous recessive genotype is called a **test cross.** The homozygous recessive is the only genotype that can be identified by observing phenotype— that is, a short plant can be positively identified as genotype *tt*. The homozygous recessive serves as a "known" that can reveal the unknown genotype of another individual when the two are crossed. Figures 13.7 and 13.8 show the results of a monohybrid cross in two other familiar species.

Single Mendelian Traits in Humans

Mendel's laws apply to all diploid species. Figure 13.9 depicts a monohybrid cross in humans. The ratios Mendel's law predicts apply individually to each child conceived, just as a coin has a 50% chance of coming up heads on each toss, no matter how many heads have already been thrown. Some people believe that if

FIGURE 13.7

Kernels on an ear of corn represent progeny of a single cross. When a plant with a dominant allele for purple kernels and a recessive allele for yellow kernels is self-crossed, the resulting ear has approximately three purple kernels for every yellow kernel.

FIGURE 13.8

a. Albinism can result from a monohybrid cross in a variety of organisms. A heterozygote, or "carrier," for albinism has one allele that directs the synthesis of an enzyme needed to manufacture the skin pigment melanin and one allele that fails to make the enzyme. Each child of two carriers has a one in four chance of inheriting the two deficient alleles and being unable to manufacture melanin. Such a child has albinism. *b.* An albino mouse with a normally pigmented sibling.

they have one child with a recessive disorder, the next three are guaranteed to be healthy. Can you see why this is a false assumption?

The nature of the phenotype is important when evaluating transmission of Mendelian traits. For example, each adult sibling of a person who is a known carrier of Tay-Sachs disease has a two-thirds chance of also being a carrier. This is because there are only three genotypic possibilities in an adult—homozygous for the wild type allele (no Tay-Sachs disease and no possibility of transmitting it to offspring), heterozygous with one wild type allele from the mother and one mutant allele from the father (a carrier), or heterozygous with one wild type allele from the father and one mutant allele from the mother (also a carrier). A homozygous recessive individual would never survive childhood because Tay-Sachs so severely damages the central nervous system that it is fatal early in life. Biology in Action 13.1 discusses several other genetic diseases.

In humans, disorders or traits caused by a single gene are called **Mendelian traits.** Even the most prevalent Mendelian disorders, such as cystic fibrosis and Duchenne muscular dystrophy, are very rare, usually affecting 1 in 10,000 or fewer births. About twenty-five hundred Mendelian disorders are known, and some twenty-five hundred other conditions are suspected to be Mendelian, based on their recurrence patterns in large families.

Geneticists who study human traits and illnesses cannot set up test crosses, as Mendel did, but they can pool information from families whose members have the same illness. Consider a simplified example of 50 couples in which both people carry the allele that causes sickle cell disease, plus one normal allele. Suppose that they give birth to 100 children. About 25 of the children would be

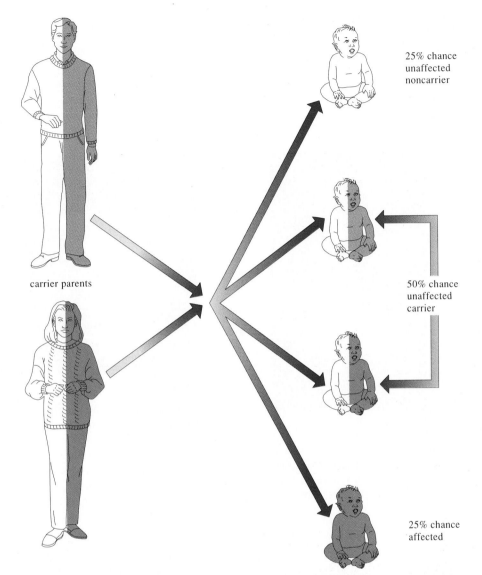

carrier parents

25% chance
unaffected
noncarrier

50% chance
unaffected
carrier

25% chance
affected

FIGURE 13.9

A 1:2:1 genotypic ratio results from a monohybrid cross, whether in peas or people. When parents are carriers for the same autosomal recessive trait or disorder, such as cystic fibrosis, each child faces a 25% risk of inheriting the condition (a sperm carrying a CF allele fertilizing an oocyte carrying a CF allele); a 50% chance of being a carrier like the parents (a CF-carrying sperm with a wild type oocyte, and vice versa); and a 25% chance of inheriting two wild type CF alleles.

expected to inherit sickle cell disease and suffer joint pain, frequent infection, and other symptoms caused by blocked circulation. Of the remaining 75 children, about 50 would be carriers like their parents, and the remaining 25 would have two nonmutant alleles. Table 13.2 lists some less harmful single-gene traits in humans.

Pedigree Analysis

Today, pedigrees are the human geneticist's primary tool, and the bigger the family, the better. The more children in a generation, the easier it is to discern patterns or modes of inheritance. People have drawn diagrams depicting family relationships since the fifteenth century. Such a chart is called a **pedigree** (from the French *pre de grue,* meaning "crane's

---

Table 13.2 **A Catalog of Single-Gene Human Traits**

Freckles, widow's peak, cleft chin, tongue rolling, and ear wiggling are familiar, single-gene inherited traits in humans. Here are a few stranger ones, taken from thousands that appear in Victor McKusick's compendium, *Mendelian Inheritance in Man***:**

Uncombable hair	Extra, missing, protruding, fused, snowcapped or shovel-shaped teeth
Misshapen toes or teeth	Grooved tongue
Pigmented tongue tip	Duckbill-shaped lips
Inability to smell freesia flowers, musk, or skunk	Flared ears
Lack of teeth, eyebrows, nasal bones, or thumbnails	Egg-shaped pupils
White forelock	Three rows of eyelashes
Whorl in the eyebrow	Spotted nails
Tone deafness	Double toenails
Hairs that are triangular in cross-section or that have multiple hues	Magenta urine after eating beets
Hairy nosetips, knuckles, palms, soles, or elbows	Sneezing fits in bright sunlight

Living with Genetic Disease

Thousands of inherited disorders affect humans, yet all of them are quite rare. As we identify the genetic instructions that go awry in genetic disease, we not only better understand basic biological phenomena, but can also develop new ways to detect and treat symptoms. Following are glimpses of a few of the more common inherited autosomal (nonsex chromosome) disorders.

Autosomal Recessive Disorders

Osteogenesis Imperfecta

The young mother was distraught when she brought her 10-month-old son to the emergency room. His arm had begun dangling at an abnormal angle after he had hauled himself up to a standing position for the first time.

Two months earlier, the child had inexplicably broken two fingers, also following normal activity. Concerned, the child's pediatrician consulted the medical records for the boy's older sister. He discovered that she, too, had suffered an unusually large number of broken bones. The physician decided that before he accused the parents of child abuse, he would have to question them very carefully and run some tests.

The pediatrician made an astute decision, for it wasn't the parents' behavior that caused the children to break bones—it was the parents' genes. Both children had inherited a rare form of *osteogenesis imperfecta* from healthy parents who were carriers. As a result, the children lacked a type of collagen (a bone protein). Their unusually fragile bones may have even broken before birth (fig. 1). Although these parents were relieved that they were no longer suspected of child abuse, they knew that protecting their childrens' delicate bones wouldn't be easy.

Cystic Fibrosis

Cystic fibrosis is the most prevalent genetic disease among white Americans, affecting 1 in 2,000 newborns each year. A child with cystic fibrosis produces abnormally large amounts of thick mucus, which block the lungs and pancreas, causing respiratory dis-

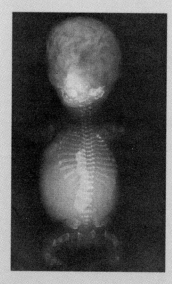

FIGURE 1

Osteogenesis imperfecta causes easy bone breakage. Numerous skeletal injuries are already present at birth in some infants who have inherited osteogenesis imperfecta.

tress and preventing the intestines from absorbing fats. The defective gene specifies an abnormal membrane protein, that entraps chloride ions inside cells, drawing water in and thickening mucus in affected tissues.

The child with severe cystic fibrosis fails to gain weight and has frequent respiratory and ear infections. Sometimes a doctor initially diagnoses the child with "failure to thrive," confirming cystic fibrosis only after detecting a characteristic high salt content in the child's sweat and locating the responsible mutant gene. Although researchers did not discover the gene until 1989, the disease has been recognized for centuries. A seventeenth century English saying, evidently refers to cystic fibrosis: "A child that is salty to taste will die shortly after birth." We now know that there are many mild cases too.

Before cystic fibrosis was clinically described in 1938, affected children usually died in infancy. Today, thanks to antibiotics, physical therapy performed by the parents, new drugs, and even gene therapy, many affected individuals are living well into adulthood. Alex Deford was not so

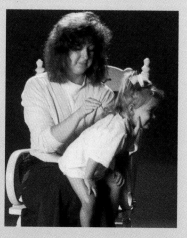

FIGURE 2

A physical therapist administers chest therapy with postural drainage to a young cystic fibrosis patient.

lucky. She died at the age of 8 years; her father, Frank Deford, describes her battle in *Alex, the Life of a Child*. Many people with cystic fibrosis still follow a daily regimen similar to Alex's, including postural drainage (fig. 2):

. . . Alex's day would start with an inhalation treatment that took several minutes. This was a powerful decongestant mist that she drew in from an inhaler to loosen the mucus that had settled in her lungs. Then, for a half hour or more, we would give her postural drainage treatment to accomplish the same ends physically. It is quite primitive, really, but all we had, the most effective weapon against the disease. Alex had to endure 11 different positions, each corresponding to a section of the lung, and Carol or I would pound away at her, thumping her chest, her back, her sides, palms cupped to better "catch" the mucus. Then, after each position, we would press hard about the lungs with our fingers, rolling them as we pushed on her in ways that were often more uncomfortable than the pounding.

Some positions Alex could do sitting up, others laying flat on our laps. But a full 4 of the 11 she had to endure nearly

FIGURE 3

Folksinger Woody Guthrie lingered for more than a decade with the mental and physical degeneration of Huntington disease. The left photo was taken before symptoms arose; the right photo shows how this condition makes one prematurely age.

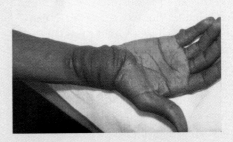

FIGURE 4

A patient with neurofibromatosis type I may have brown cafe-au-lait spots or tumors. In severe cases, tumors cover the entire body.

upside down, the blood rushing to her head, as I banged away on her little chest, pounding her, rattling her, trying somehow to shake loose that vile mucus that was trying to take her life away. One of her first full sentences was, "No, not the down ones now, Daddy." . . .

. . . Only slowly did the recognition come that she was singled out for these things. Then she began to grope for the implications. One spring day when she was four, Alex came into my office and said she had a question. Just one was all she would bother me with. All right, I asked, what was it. And Alex said, "I won't have to do therapy when I'm a lady, will I?"

It was a leading question; she knew exactly where she was taking me.

As directly as I could I said. " No, Alex"—not because I would lie outright about it, but because I knew the score by then. I knew that she would not grow up to be a lady unless a cure was found. . . .

Autosomal Dominant Disorders

Huntington Disease

At age 37, Suzanne began to experience peculiar mood changes, lashing out at her family for no apparent reason. She became clumsy and often lost her balance, walking as if she was intoxicated. When she fell in the street one day and some children taunted her for being drunk, she went to her physician.

At the doctor's office, the nurse took Suzanne's family history. Suzanne men-tioned that her father had also become clumsy before he died in a car accident at age 42, and that her father's mother was presently in a mental institution. On the basis of Suzanne's symptoms, her family history, and a DNA test that detected a certain "expanded" gene at one tip of the fourth largest chromosome, the doctor concluded that Suzanne was showing the early symptoms of *Huntington disease*, an autosomal dominant condition that she had inherited from her father, and he from his mother. (Folksinger Woody Guthrie had Huntington disease—see figure 3.) Suzanne would continue to decline mentally and physically until she needed constant care. Her children would need to be tested to see if they, too, would develop symptoms later in life.

Suzanne is still in the early stages of the illness. She hopes that the 1993 discovery of the gene responsible for Huntington disease will soon lead to a treatment.

Neurofibromatosis Type I

Rhonda was not even aware that she had an inherited disorder until her infant son Patrick was diagnosed with *neurofibromatosis type I* (NF1). His otherwise pale skin was dotted with dozens of light brown "cafe-au-lait" spots. More troublesome were the abundant tumors beneath his skin (fig. 4). These growths, which originated from nervous tissue, resembled cauliflower. Fortunately, they were benign, but Patrick's NF1 put him at high risk of developing a cancerous tumor.

NF1 is inherited as an autosomal domi-nant trait, meaning that one of Patrick's parents also has the disorder. When Patrick was diagnosed, Rhonda realized that the 9 brown spots on her body, normally hidden by clothing, were the only signs of her mild case of NF1. She also realized that her father, paternal grandmother, and one of her sisters had mild cases too.

The earliest known depiction of NF1 is a thirteenth-century illustration of a man covered with the characteristic skin growths. In 1768, an English physician implicitly recognized the hereditary nature of NF1 when he reported the case of a severely affected man whose father also had the condition. The doctor had removed the younger man's lesions, only to see them return in more severe form—an occurrence still seen today. Cases accumulated, and in 1882, a medical student named Friedrich Daniel von Recklinghausen put all of the information together, describing what we now call NF1, or *von Recklinghausen neurofibromatosis*.

NF1 affects 1 in 3,000 people. In 1990, when researchers identified the gene that causes NF1, they also shed some light on how some of the symptoms arise. Normally, a protein called neurofibromin suppresses tumors by blocking a growth factor from acting at the cellular level. When an individual has NF1, the gene that produces neurofibromin is defective. Missing or abnormal neurofibromin fails to block the growth factor; the growth factor, in turn, activates a protein that hikes the cell division rate. If the rate increases enough, tumors form.

In the past decade, researchers have identified the genes responsible for many genetic diseases. While this knowledge does not always lead to an immediate cure, it does offer hope—and sometimes improved treatment—to thousands of people who suffer from genetic disorders.

foot," because the lines connecting parents to multiple children resemble a bird's claw). Prior to the twentieth century, pedigrees were used primarily in genealogy, showing family relationships. With the birth of the science of genetics in the twentieth century, pedigrees were adapted to track traits through families—of humans, pedigreed cats and dogs, and thoroughbred racehorses.

A pedigree consists of shapes connected by lines. Vertical lines represent generations; horizontal lines connect parents; siblings are connected to their parents by vertical lines and joined by a horizontal line. Squares indicate males,

circles females, and diamonds individuals of unknown sex. Colored shapes indicate individuals who express a particular trait, and half-filled shapes are known carriers—people who carry the allele along with a dominant allele that masks it. Figure 13.10 shows commonly used pedigree symbols.

Pedigrees can reveal modes of inheritance. An autosomal trait is carried on an autosome, or a chromosome that does not determine sex. An **autosomal recessive** trait can affect both sexes and can skip generations, because carriers show no symptoms (fig. 13.11). If a condition is known to be autosomal recessive, individuals who have affected (homozygous recessive) children must both be carriers. An **autosomal dominant** trait can affect both sexes. It affects every generation (fig. 13.12) unless a generation of individuals does not inherit the causative gene, in which case transmission permanently stops. Two

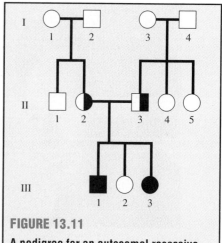

FIGURE 13.11

A pedigree for an autosomal recessive trait. Albinism (absence of pigment in the skin and hair) does not show up in this family until the third generation. Individuals II–2 and II–3 must each be carriers.

Symbols

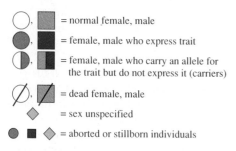

○, ■ = normal female, male

●, ■ = female, male who express trait

◐, ◧ = female, male who carry an allele for the trait but do not express it (carriers)

⊘, ⊘ = dead female, male

◇ = sex unspecified

●, ■, ◆ = aborted or stillborn individuals

Lines

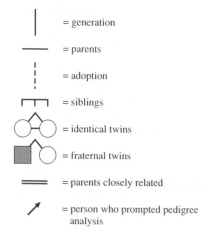

| = generation

— = parents

┆ = adoption

┌┬┐ = siblings

= identical twins

= fraternal twins

= parents closely related

↗ = person who prompted pedigree analysis

Numbers

Roman numerals = generations

Arabic numerals = individuals

FIGURE 13.10

Symbols used in pedigree construction are connected to form a pedigree chart, which displays the inheritance patterns of particular traits.

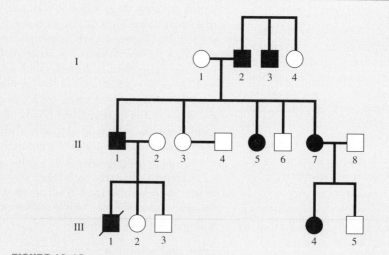

FIGURE 13.12

A pedigree for an autosomal dominant trait. When individual III-1 died suddenly of heart failure while playing basketball at age 19, his family and doctors were perplexed; he had seemed so healthy. But the family history and microscopic examination of his heart muscle at autopsy led to a diagnosis of familial hypertrophic cardiomyopathy. It is inherited as an autosomal dominant, it affects both sexes, and it occurs in every generation. The boy's father had died at a young age of heart failure, as had two of the father's sisters. After the diagnosis was made, blood cells from the young man's two cousins, whose mother also had the condition, were tested for the presence of the mutant gene. One of them was diagnosed, although she was still healthy. Restricting her exercise to swimming may extend her life. The young man's paternal grandfather and a paternal great-uncle also had the disorder.

alopecia as an autosomal recessive trait

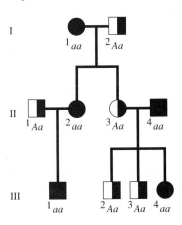

if *A* = normal hair
    a = alopecia

alopecia as an autosomal dominant trait

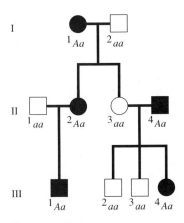

if *a* = normal hair
    A = alopecia

FIGURE 13.13

An inherited trait that does not impair health enough to prevent reproduction can result in an inconclusive pedigree. This is the case for alopecia, in which some hair on the head falls out, grows back, and falls out again. Alopecia can be inherited as an autosomal recessive or as an autosomal dominant trait. In the family depicted in this pedigree, the mode of inheritance cannot be discerned—either autosomal recessive or dominant may fit the data.

other modes of inheritance, involving the sex chromosomes, are discussed in the next chapter. Although pedigrees are very useful, they are often inconclusive (fig. 13.13).

Pedigrees can be difficult to construct and interpret for several reasons. People sometimes hesitate to supply needed information because they are embarrassed by symptoms affecting behavior or mental stability. Tracing family relationships can be complicated by adoption, children born out of wedlock, serial marriages and the resulting blended families, and reproductive alternatives, such as surrogate mothers and artificial insemination (chapter 12). In the United States, many people cannot trace their families back more than three or four generations, which may not provide sufficient evidence to reach a conclusion about a mode of inheritance.

KEY CONCEPTS

The study of genetics sprang from the observations of Gregor Mendel. From crossing pea plants with different traits, Mendel deduced the law of segregation, which states that the two characters or genes for plant height segregate from one another during meiosis, then combine at random with the character from the opposite gamete at fertilization. Every individual thus has two alleles for each gene; a homozygote has two identical alleles, and a heterozygote has two different alleles for a gene. In a heterozygote, the expressed allele is dominant and the unexpressed allele is recessive. The alleles constitute the genotype, and their expressions constitute the phenotype. The most common phenotype for any trait is called the wild type. An allele that has undergone a change is mutant. A monohybrid cross yields a genotypic ratio of 1:2:1, and a phenotypic ratio of 3:1. A test cross uses a homozygous recessive individual to reveal the unknown genotype of another individual. Mendel's laws operate in humans and all other diploid species. Geneticists use pedigrees to track Mendelian traits in families.

Following the Inheritance of Two Genes— Independent Assortment

The law of segregation follows the inheritance pattern of two alleles of a single gene. In a second set of experiments with pea plants, Mendel examined the inheritance of two different traits, each of which has two different alleles: he looked at seed shape, which may be either round or wrinkled (determined by the *R* gene), and seed color, which may be either yellow or green (determined by the *Y* gene). When Mendel crossed plants with round yellow seeds to plants with wrinkled green seeds, all the progeny had round yellow seeds. Therefore, round is completely dominant to wrinkled, and yellow is completely dominant to green.

Next, Mendel took F_1 plants (genotype *RrYy*) and crossed them to each other. This is a **dihybrid cross,** because the individuals are heterozygous for two genes. Mendel found four types of seeds in the F_2 generation: round yellow (315 plants), round green (108 plants), wrinkled yellow (101 plants), and wrinkled green (32 plants). This is an approximate ratio of 9:3:3:1 (fig. 13.14).

Mendel took each plant from the F_2 generation and crossed it to wrinkled green (*rryy*) plants. These test crosses established whether each F_2 plant was true-breeding for both genes (*RRYY* or *rryy*), true-breeding for one but heterozygous for the other (*RRYy, RrYY, rrYy,* or *Rryy*), or heterozygous for both genes (*RrYy*). Based upon the results of the dihybrid cross, Mendel proposed what is now known as the law of **independent assortment,** which states that a gene for one trait does not influence the transmission of a gene for another trait. This second law is true only for genes on different chromosomes. The seed shape and seed color genes that Mendel worked with meet this criterion.

Mendel had again inferred a principle of inheritance that has its physical basis in meiosis. Independent assortment occurs because chromosomes from each parent combine in a random fashion (fig. 13.15). In Mendel's dihybrid cross, each parent produces equal numbers of gametes of four different types: *RY, Ry, rY,* and *ry.* (Note that each of these combinations has one gene for each trait.) A Punnett square for this

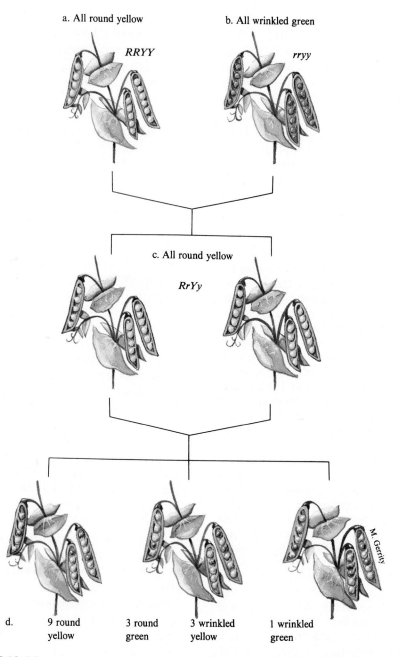

a. All round yellow   *RRYY*

b. All wrinkled green   *rryy*

c. All round yellow   *RrYy*

d.   9 round yellow   3 round green   3 wrinkled yellow   1 wrinkled green

M. Gerrity

FIGURE 13.14

Mendel's crosses involving two genes. To study the inheritance pattern of two genes, Mendel crossed *a.* a true-breeding plant with round yellow seeds (*RRYY*) to *b.* a plant with wrinkled green seeds (*rryy*). *c.* The peas of the F₁ generation were all round and yellow (*RrYy*). *d.* When the F₁ dihybrid pea plants were crossed to each other, the F₂ generation exhibited the phenotypes round yellow (*RrYr*, *RRYy*, *RrYY*, and *RRYY*), round green (*Rryy* and *RRyy*), wrinkled yellow (*rrYY* and *rrYy*), and wrinkled green (*rryy*) in a 9:3:3:1 ratio.

ier way to predict genotypes and phenotypes is to use the mathematical laws of probability Punnett squares are based on. Probability predicts the likelihood of an event. The **product rule** states that the chance that two independent events will both occur—such as two alleles both being inherited—equals the product of the individual chances that each event will occur.

The product rule can be used to predict the chance of obtaining a wrinkled green (*rryy*) plant from dihybrid (*RrYy*) parents. Consider the dihybrid one gene at a time. A Punnett square for *Rr* crossed to *Rr* shows that the probability that two *Rr* plants will produce *rr* progeny is 25%, or 1/4. Similarly, the chance of two *Yy* plants producing a *yy* individual is 1/4. According to the product rule, the chance of dihybrid parents (*RrYy*) producing homozygous recessive (*rryy*) offspring is 1/4 multiplied by 1/4, or 1/16. Now consult the 16-box Punnett square for Mendel's dihybrid cross in figure 13.16. Only 1 of the 16 boxes is *rryy*.

KEY CONCEPTS

Mendel's second law, independent assortment, states that one gene does not influence the transmission of any other gene on another chromosome. Mendel arrived at this conclusion by crossing dihybrid heterozygotes for seed color and shape. He obtained a phenotypic ratio of 9:3:3:1, which supported the idea of independent assortment. Segregation and independent assortment are explained by meiotic events. Either Punnett squares or probability rules can predict and follow the results of independent assortment.

When Gene Expression Appears to Alter Mendelian Ratios

Mendel's crosses yielded easily distinguishable offspring. A pea is either yellow or green, round or wrinkled; a plant is either tall or short. At times, however, offspring traits do not occur in the

cross (fig. 13.16) predicts that the four types of seeds—round yellow (*RRYY*, *RrYY*, *RRYy*, and *RrYy*); round green (*RRyy*, *Rryy*); wrinkled yellow (*rrYY*, *rrYy*); and wrinkled green (*rryy*)—will occur in the ratio 9:3:3:1, just as Mendel found.

Using Probability to Analyze More than Two Genes

Punnett squares become cumbersome when analyzing more than two genes—a Punnett square for three genes has 64 boxes; for four genes, 256 boxes. An eas-

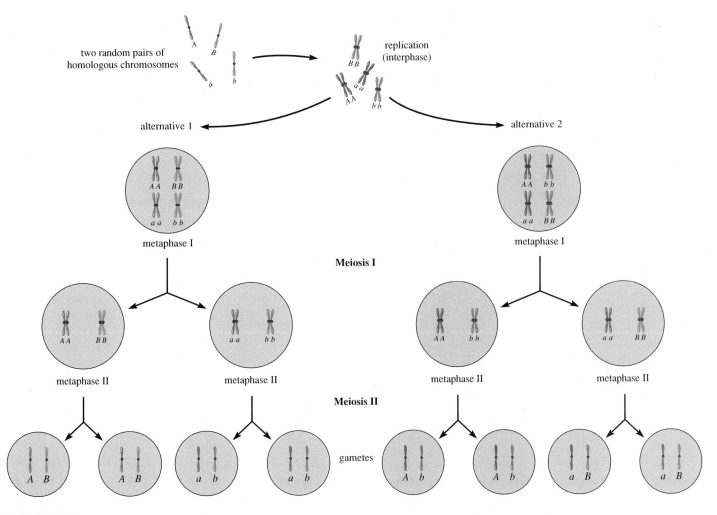

FIGURE 13.15

The independent assortment of genes carried on different chromosomes results from the random alignment of chromosome pairs during metaphase of meiosis I. An individual of genotype *AaBb,* for example, manufactures four types of gametes, containing the dominant alleles of both genes (*AB*), the recessive alleles of both genes (*ab*), and a dominant allele of one with a recessive allele of the other (*Ab* and *aB*). The allele combination depends upon which chromosomes are packaged together into the same gamete—and this happens at random.

proportions predicted by Punnett squares or probabilities. In such cases, it appears that Mendel's laws do not apply—but they do. The underlying genotypic ratios remain unchanged, but either the nature of the phenotype or influences from other genes or the environment alter phenotypic ratios.

Lethal Allele Combinations

Genes begin to function soon after fertilization. Some allele combinations cause problems so early in prenatal development that the individual is never born, and therefore never counted as a phenotypic class of offspring. An allele that causes such early death is a **lethal allele.**

Plants of certain genotypes die at fertilization, during seed development, or as seedlings. In humans, lethal alleles can cause spontaneous abortion (miscarriage). When a man and woman each carry a recessive lethal allele for the same gene, each pregnancy has a 25% chance of spontaneously aborting—a figure that represents the homozygous recessive class.

Multiple Alleles

Although a diploid organism always has two alleles for a particular gene on an autosome, a gene can exist in more than two allelic forms. This is because a change, or mutation, in any of the hundreds or thousands of DNA bases of a gene can lead to a different allele. The change is detectable if the protein product the new allele encodes changes in a way that affects the phenotype.

Many disease-causing genes have multiple alleles. Consider cystic fibrosis (CF), the autosomal recessive disorder in which certain cells cannot export salt. People with CF have thickened secretions that impair breathing and digestion. Within a year of identifying the most common allele that causes cystic fibrosis, researchers found more than 100 others. Sorting out the phenotypes associated with the thousands of possible genotypes for the cystic fibrosis gene may prove more daunting than the original gene discovery!

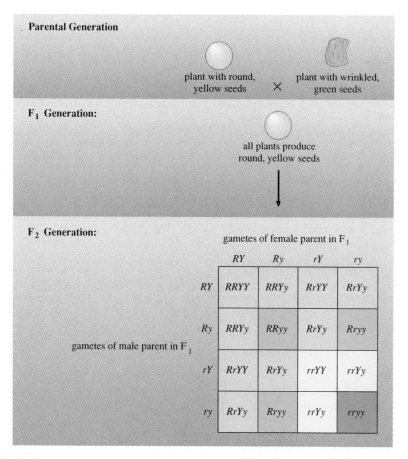

FIGURE 13.16

A Punnett square can be used to represent the random combinations of gametes produced by dihybrid individuals.

Different Dominance Relationships

Some genes show **incomplete dominance,** with the heterozygous phenotype intermediate between the two homozygotes. In familial hypercholesterolemia (FH), a person with two disease-causing alleles completely lacks receptors on liver cells that take up cholesterol from the bloodstream. A person with one disease-causing allele has half the normal number of cholesterol receptors and someone with two wild type alleles has the normal number of receptors. The phenotypes parallel the number of receptors—those with two mutant alleles die as children of heart attacks, those with one mutant allele die in young adulthood, and those with two wild type alleles have genetically healthy hearts.

A classic example of incomplete dominance is the snapdragon plant (fig. 13.17). A red-flowered plant of genotype *RR* crossed to a white-flowered *rr* plant gives rise to a *Rr* plant—which has pink flowers. This intermediate color is presumably due to an intermediate amount of a pigment.

Different alleles that are both expressed in a heterozygote are **codominant.** Two alleles of the *I* gene, which determines human ABO blood type, are codominant. People of blood type A have a molecule called antigen A on the surfaces of their red blood cells. People of blood type B have red blood cells with antigen B. A person with type AB has red blood cells with both the A and B antigens, and the red cells of a person with type O blood carry neither antigen.

The *I* gene has three alleles that encode the enzymes that place the A and B antigens on red blood cell surfaces. They are I^A, I^B, and *i*. People with type A blood may be either genotype $I^A I^A$ or $I^A i$; type B corresponds to $I^B I^B$ or $I^B i$; type AB to $I^A I^B$; and type O to *ii*. Even though the I^A and I^B alleles are codominant, they segregate between generations (fig. 13.18 and table 13.3).

Epistasis—Gene Masking at *General Hospital*

If one gene masks the effect of another, a phenomenon called **epistasis,** it may appear that Mendel's laws are not operating. (Do not confuse this with the situation in which a dominant allele masks a recessive allele of the same gene.) Epistasis was part of a plot in the television soap opera *General Hospital* (fig. 13.19). It concerned the Bombay phenotype, a result of interacting *I* and *H* genes. The relationship of these two genes affects the expression of the ABO blood type.

The *H* allele produces an enzyme that inserts a sugar molecule onto a particular glycoprotein on the red blood cell surface. The A and B antigens then attach to this sugar molecule. The recessive *h* allele produces an inactive form of the enzyme, which cannot insert the sugar. As long as at least one *H* allele is present, the ABO genotype dictates the person's blood type. However, in a person with genotype *hh*, the A and B antigens cannot adhere to the red blood cells. The person tests as blood type O, but may actually have any ABO genotype.

Penetrance and Expressivity

The same allele combination can produce different degrees of the phenotype in different individuals, even siblings, because of outside influences such as nutrition, toxic exposures, illnesses, and other genes. Most disease-causing allele combinations are **completely penetrant,** which means that all individuals that

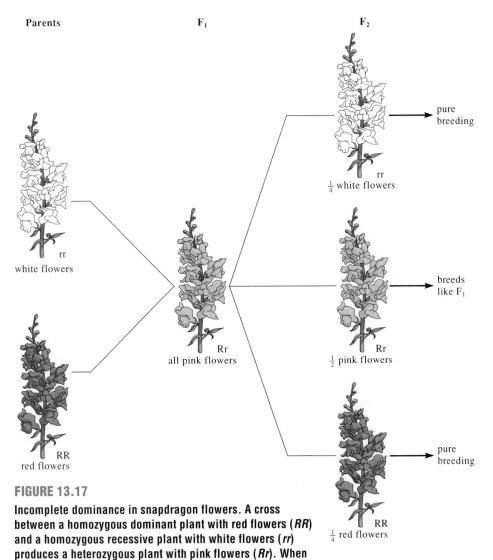

FIGURE 13.17

Incomplete dominance in snapdragon flowers. A cross between a homozygous dominant plant with red flowers (*RR*) and a homozygous recessive plant with white flowers (*rr*) produces a heterozygous plant with pink flowers (*Rr*). When *Rr* pollen fertilizes *Rr* egg cells, one-quarter of the progeny are red-flowered (*RR*), one-half are pink-flowered (*Rr*), and one-quarter are white-flowered (*rr*). The phenotypic ratio of this monohybrid cross is 1:2:1 (instead of the 3:1 seen in cases of complete dominance) because the heterozygous class has a phenotype different from that of the homozygous dominant class.

Parents — **F₁** — **F₂**

rr white flowers

RR red flowers

Rr all pink flowers

rr ¼ white flowers → pure breeding

Rr ½ pink flowers → breeds like F₁

RR ¼ red flowers → pure breeding

FIGURE 13.18

Even though the I^A and I^B alleles of the I gene are codominant, they still follow Mendel's law of segregation. These Punnett squares follow the genotypes that could result by crossing a person with type A blood with a person with type B blood. Is it possible for parents with type A and type B blood to have a child who is type O?

	I^A	I^A
I^B	I^{A}I^B	I^{A}I^B
I^B	I^{A}I^B	I^{A}I^B

	I^A	i
I^B	I^{A}I^B	I^{B}i
I^B	I^{A}I^B	I^{B}i

	I^A	I^A
I^B	I^{A}I^B	I^{A}I^B
i	I^{A}i	I^{A}i

	I^A	i
I^B	I^{A}I^B	I^{B}i
i	I^{A}i	ii

Table 13.3 Codominance in Human ABO Blood Types

Phenotype (Blood Type)	Genotype
A	I^{A}I^A or I^{A}i
B	I^{B}I^B or I^{B}i
AB	I^{A}I^B
O	ii

inherit the allele combination have some symptoms. A genotype is **incompletely penetrant** if some individuals do not express the phenotype. Polydactyly, having extra fingers or toes, is incompletely penetrant (fig. 13.20). Cats who inherit the dominant polydactyly allele have more than 4 toes on at least one paw, yet others who are known to have the allele (because they have an affected parent and offspring) have the normal number of toes. The penetrance of a gene is described numerically. If 80 of 100 cats who inherit the dominant polydactyly allele have extra digits, the allele is 80% penetrant.

A phenotype is **variably expressive** if the symptoms vary in intensity in different individuals. One cat with polydactyly might have an extra digit on two paws; another might have two extra digits on all four paws; a third cat might have just one extra toe. Penetrance refers to the all-or-none expression of a genotype; expressivity refers to the severity of a phenotype. Biology in Action 13.2

Biology in *ACTION*

Tracing a Mendelian Trait—Distal Symphalangism

Student James Poush saw Mendel's in action when he explored his own family for a science fair project. His work was the first full report on an autosomal dominant condition called *distal symphalangism*.

James had never thought his stiff fingers and toes with their tiny nails were particularly odd. Others in his family had them, too. When he studied genetics, James realized that this might be a Mendelian trait. After much detective work, James identified 27 affected individuals among 156 relatives. He even went on to analyze a second family with the disorder.

James concluded that the trait is autosomal dominant. Of 63 relatives with an affected parent, 27 (43%) expressed the phenotype, close to the 50% expected when an allele is autosomal dominant. His figure was lower, James reasoned, because of underreporting and a few instances of nonpenetrance. Some relatives did not realize they had inherited the condition until James pointed it out to them!

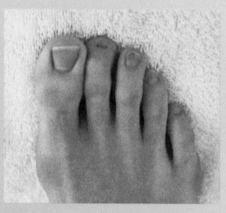

FIGURE 1

Rudimentary nails, shown here on the second and fifth toes, are characteristic of distal symphalangism.

The dominant allele responsible for distal symphalangism is variably expressive (different digits are affected, and to different extents), incompletely penetrant, (some

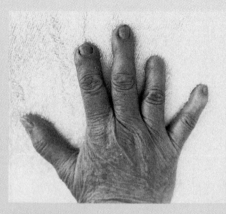

FIGURE 2

Another symptom of distal symphalangism is brachydactyly, or short fingers.

people who inherit the allele have normal fingers and toes) and pleiotropic (nails, bones, joints, and skin are affected).

FIGURE 13.19

Gene masking at *General Hospital*. Monica has just had a baby. But who is the father—her husband, Alan, or friend and fellow doctor, Rick? Monica's blood type is A, Alan's is AB, and Rick's is O. The child's blood type is O. At first glance, it looks as if Rick is the father. *a*. Whether Monica's genotype is *I*^{A}*i* or *I*^{A}*I*^{A}, she cannot have a type O child with Alan, whose genotype is *I*^{A}*I*^{B}. *b*. If Monica's genotype is *I*^{A}*i*, she could have a type O child with Rick, whose genotype is *ii*. Unless . . . nosy nurse-in-training Amy has just learned about the Bombay phenotype and suggests that the baby have a blood test to see if he manufactures the *H* protein.

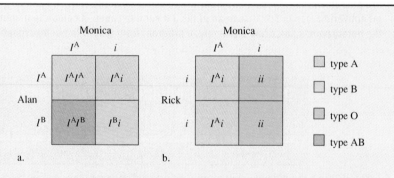

She also advises the adults to look into their family histories to see if any other relatives' blood types are not predicted by their parents' blood types. Sure enough, the baby is of genotype *hh*, and Alan's family has had past incidences in which the blood types of a child did not match what was predicted by parents' blood types. On further testing, both Monica and Alan are found to be *Hh*, but Rick is *HH*. Alan Jr. is *hh*. He has a type O phenotype, but his ABO genotype can be either *I*^{A}*I*^{A}, *I*^{A}*i*, *I*^{B}*i*, or *I*^{A}*I*^{B}. What he cannot be is Rick's son.

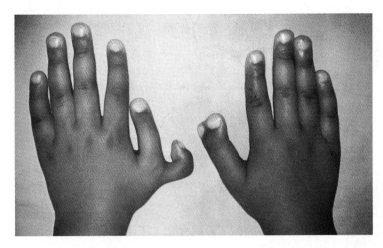

FIGURE 13.20

The dominant allele that causes polydactyly is incompletely penetrant and variably expressive. Some people who inherit the allele have the normal number of fingers and toes. Those in whom the allele is penetrant express the phenotype to differing degrees.

describes another interesting inherited condition that affects the digits and is incompletely penetrant and variably expressive. It was discovered by a high school student!

KEY CONCEPTS

Several situations can appear to disrupt Mendelian ratios. A homozygous lethal allele will not show up in progeny. A gene may also have many alleles, although any single diploid individual only has one or two of them, and some alleles may be expressed to different degrees in different individuals. In incomplete dominance, the heterozygote phenotype is intermediate between those of the homozygotes and in epistasis, genes interact, altering Mendelian phenotypic ratios. Genotypes vary in penetrance (the percent of individuals affected) and expressivity (intensity of symptoms). Though all these circumstances appear to change Mendelian ratios, the underlying genotypes affirm Mendel's laws.

Environmental Influences

Some gene expressions are exquisitely sensitive to the environment. Temperature influences gene expression in some familiar animals—Siamese cats and Himalayan rabbits have dark ears, noses, feet, and tails because these anatomical extremities tend to be colder than the animals' abdomens. Heat destroys pigment molecules that provide coat color. Temperature-sensitive alleles also lead to striking phenotypes in the fruit fly (fig. 13.21).

Pleiotropy

A Mendelian disorder with many symptoms is **pleiotropic.** Such conditions can be difficult to trace through families, because individuals with different subsets of symptoms may appear to have different disorders. A child with sickle cell disease, for example, may suffer from frequent infections and have an enlarged liver and spleen; his brother, though he has the same disease, might experience mostly severe joint pain.

Pleiotropy occurs in genetic diseases that affect a single protein found in different parts of the body. This is the case for Marfan syndrome, an autosomal dominant defect in an elastic connective tissue protein called fibrillin. Fibrillin is abundant in the lens of the eye, in the aorta (the largest artery in the body, leading from the heart), and in the bones of the limbs, fingers, and ribs. Marfan syndrome symptoms include lens dislocation, long limbs, spindly fingers, and a caved-in chest. The most serious symptom is a life-threatening weakening in the aortic wall that sometimes causes the vessel to suddenly burst. However, if the weakening is detected early, surgeons can graft in a synthetic artery wall to replace the weakened section.

Phenocopies—When It's Not Really in the Genes

A **phenocopy** is an environmentally caused trait that appears to be inherited. It can resemble a known Mendelian disorder or mimic inheritance by occurring in certain relatives. The limb birth defect caused by the drug thalidomide, discussed in chapter 12, is a phenocopy of the inherited illness phocomelia. An infection can also be mistaken as a Mendelian disorder when it affects several family members.

KEY CONCEPTS

Environmental influences may appear to disrupt Mendelian laws. Temperature can affect gene expression. A pleiotropic gene may produce a variety of symptoms. A phenocopy is a trait caused by the environment but resembling a known genetic trait.

Uniparental Disomy—An Exception to Mendel's Law of Segregation

According to the law of segregation, parents contribute equally, gene by gene, to offspring. However, occasionally an offspring inherits two copies of a gene (or genes) from one parent, and none from the other.

In 1988, Arthur Beaudet of the Baylor College of Medicine saw a very unusual cystic fibrosis patient. Beaudet was comparing CF alleles in the patient, her father, and, because her mother was dead, her maternal aunt. Oddly, only the mother (inferred from the aunt) was a CF carrier—the father had two wild type alleles. Didn't both parents have

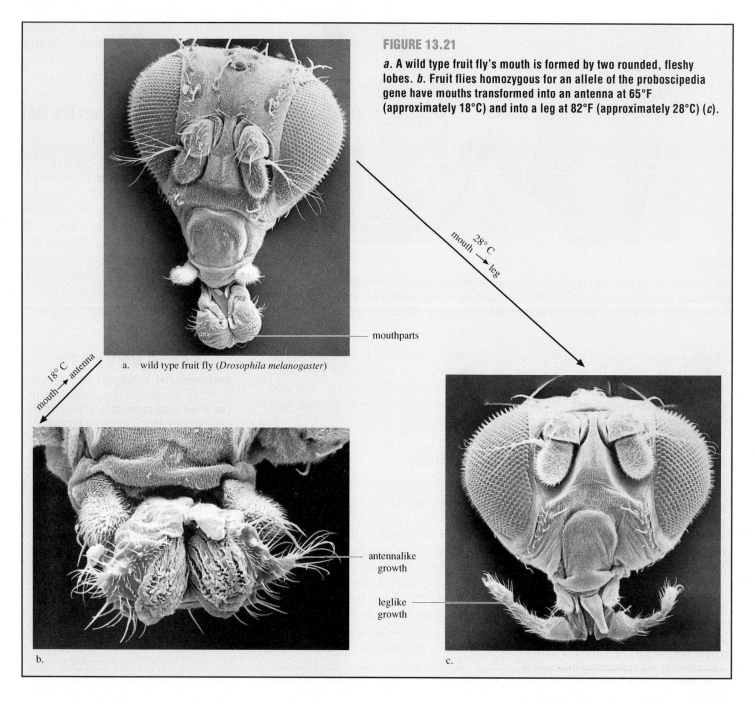

FIGURE 13.21

a. A wild type fruit fly's mouth is formed by two rounded, fleshy lobes. *b.* Fruit flies homozygous for an allele of the proboscipedia gene have mouths transformed into an antenna at 65°F (approximately 18°C) and into a leg at 82°F (approximately 28°C) (*c*).

mouth → leg 28° C

mouthparts

a.   wild type fruit fly (*Drosophila melanogaster*)

mouth → antenna 18° C

antennalike growth

leglike growth

b.

c.

to carry CF for a child to inherit the disorder? Also, neither of the patient's seventh largest chromosomes, the home of the CF gene, matched her father's—both came from her mother. How did this happen?

Apparently an error occurred in the patient's mother—the chromatids for each chromosome 7 should have separated in meiosis II but didn't. This caused an oocyte bearing two identical chromosome 7s, instead of one, to form. The unlucky patient had inherited two copies of her mother's CF-bearing chromosome

7, and, somehow, none from her father. Inheriting two of the same chromosome from one parent shatters the protection offered by the genetic combinations of sexual reproduction. This very rare inheritance of a double dose of genetic material from one parent, but none from the other, is called **uniparental disomy,** which literally means "two bodies from one parent."

The woman with CF had the chromosome pair she inherited from her mother in all of her cells, indicating that her condition was present from concep-

tion. Uniparental disomy can also occur in somatic tissue, affecting only parts of a person's body.

Uniparental disomy may explain two syndromes that are associated with a particular region of the fifteenth largest chromosome. Researchers originally thought they were different illnesses, but they may actually be different expressions of the same gene, depending upon which parent transmits it. Differing phenotypes based on the sex of the parent who passes it on is called **genomic imprinting.**

The two disorders that seem to exhibit genomic imprinting caused by uniparental disomy are Prader-Willi syndrome and Angelman syndrome. They are both extremely rare, cause mental retardation, and are sometimes associated with missing genetic material on chromosome 15. A child with Prader-Willi syndrome is obese, has small hands and feet, has an uncontrollable appetite, and does not develop signs of puberty. Doctors first described Angelman syndrome in 1967 as "happy puppet syndrome" because of the child's appearance. A child with this disorder laughs excessively and uncontrollably, and has an extended tongue, floppy muscle coordination, a large jaw, and peculiar convulsions that make the arms flap.

In 1989, geneticists closely examined the chromosome region affected in Prader-Willi syndrome in several children. Surprisingly, the regions on each homolog matched the mothers' chromosomes. Yet the chromosomes lacked the corresponding DNA from the father, which contradicted the law of segregation. For Angelman syndrome, the situation was reversed. Children had a double dose of their father's DNA, but no maternal contribution, for the same chromosomal region implicated in Prader-Willi syndrome. Could Prader-Willi and Angelman syndromes be different, gender-directed manifestations of the same genetic variant? We do not yet know. The story of this fascinating, only known true exception to Mendel's laws continues to unfold.

KEY CONCEPTS

In uniparental disomy, an offspring receives two genes from one parent, and none from the other. This may occur in gametes, fertilized ova, or somatic tissue. Uniparental disomy may play a role in genomic imprinting, and may account for the differences between Prader-Willi and Angelman syndromes. These syndromes appear to be different expressions of the same genetic abnormality. Whether an affected person has one or the other is determined by which parent's DNA is present in duplicate.

Multifactorial Traits—"It Runs in the Family"

A woman who is a prolific writer has a daughter who becomes a successful novelist. An overweight man and woman have obese children. A man whose father was an alcoholic is himself an alcoholic. Are these characteristics—writing talent, obesity, and alcoholism—inherited or imitated?

The medical conditions mentioned so far are single-gene disorders inherited according to Mendel's laws. Geneticists can predict the probability that a certain family member will inherit the condition. Some traits and diseases, though, seem to "run in families," appearing in a few relatives with no apparent pattern. Sometimes a single-gene problem may appear to follow an unpredictable pattern because of variable expressivity and nonpenetrance. In other cases, however, deviations from a Mendelian ratio can reflect the influence of more than one gene, of the environment—or both.

Inherited and Environmental Causes of Multifactorial Traits

Traits determined by more than one gene are **polygenic**—meaning "many genes"—or quantitative traits. Usually, several genes each contribute to the overall phenotype in equal, small degrees. The combined actions of many genes produce a continuum, or **continuously varying** expression, of the trait (table 13.4). The genes do follow Mendel's laws (unless they are linked on the same chromosome), but they don't produce typical ratios because they contribute equally to the phenotype, and are neither dominant nor recessive with respect to each other. Height and skin color are familiar examples of polygenic traits in humans (fig. 13.22). In plants, quantitative traits include flower color, petal length, blade length, and stomata density on the leaves.

Traits molded by one or more genes plus the environment are **multifactorial traits.** Consider favism, a sex-linked recessive deficiency in glucose-6-phosphate dehydrogenase (G6PD), an enzyme important in red blood cell energy metabolism. Different allele combinations cause varying degrees of deficiency of the enzyme. However, the fatigue of anemia occurs only when the person with favism takes certain drugs, contracts certain infections, or eats fava beans. If these environmental factors are absent, the phenotype is normal. Other possible multifactorial traits in humans include schizophrenia, homosexuality, and intelligence.

Measuring Multifactorial Traits

Empiric Risk

The inherited component of multifactorial traits can be assessed using information from population studies and from family relationships. It is possible to predict the risk that a single-gene or Mendelian disorder or trait will recur when one knows the mechanism of gene transmission, such as autosomal dominant or recessive. To predict the risk that a multifactorial trait will recur, it is necessary to determine the **empiric risk,** which depends on the trait's incidence in specific populations. In general, empiric risk increases with the severity of the disorder, the number of affected family members, and how closely related the person is to affected individuals.

To illustrate the medical role of empiric risk, consider the task of predicting the likelihood that a neural tube defect will recur. This defect is an opening or lesion in the brain or spinal cord that occurs on about the 28th day of prenatal development. In the United States, the overall population's risk of carrying a fetus with an NTD is about 1 in 1,000. However, if a sibling has an NTD, the risk of recurrence increases to 3%, and if two siblings are affected, the risk to a third child is even greater. Still, only about 10% of children with this common birth defect have an affected older

Table 13.4 Polygenic Model of Skin Color Inheritance

Multilocus Genotypes	Number of Pigment Genes
AABBCC	6
AaBBCC, AABbCC, AABBCc	5
aaBBCC, AAbbCC, AABBcc, AaBbCC, AaBBCc, AABbCc	4
AaBbCc, aaBbCC, AAbbCc, AabbCC, AABbcc, aaBBCc, AaBBcc	3
AaBbcc, AabbCc, AAbbcc, aaBBcc, aabbCC	2
Aabbcc, aaBbcc, aabbCc	1
aabbcc	0

Studies that classify skin color by measuring the amount of light reflected from the skin surface suggest that three or four or more different genes, probably with several alleles each, interact to produce pigment in the skin. The greater the number of pigment-specifying genes, the darker the skin. The inheritance of skin color is most likely even more complicated than this three-gene model indicates.

Number of individuals	1	0	0	1	5	7	7	22	25	26	27	17	11	17	4	4	1
Height in inches	58	59	60	61	62	63	64	65	66	67	68	69	70	71	72	73	74

FIGURE 13.22

Height is a continuously varying trait. When 175 soldiers were asked to line up according to their heights, they formed a characteristic bell-shaped, continuous distribution.

sibling. Medical researchers believe that NTDs are caused by a combination of genes and a folic acid deficiency. The environmental cause of NTDs may be greatly reduced by supplementing the diet with folic acid, especially in early pregnancy.

Heritability—The Genetic Contribution to a Trait

A large biology class has a few students as short as 5′1″, and several members of the basketball team who tower over the others at 6′5″. In between are students of many other heights. Certainly hered-

ity influences height, but nutrient intake, an environmental factor, also affects height by providing the raw material to build muscle and bone.

Heritability estimates the proportion of phenotypic variation in a group that can be attributed to genes. It is calculated as double the difference of the variance between two groups of individuals. Approximately 80% of height variance is due to heredity. Table 13.5 lists the heritabilities of this and other common traits.

Heritability describes a particular group and is not a property of any gene.

This means it can vary in different environments. For example, the influence nutrition has on egg production in chickens is greater in a starving population than in a well-fed one.

Coefficient of Relationship

The closer the relationship between two individuals, the more genes they have in common, and the greater the probability that they will share a trait. For example, the risk that cleft lip will recur is 40% for the identical twin of an affected individual, but only 4% for a sibling, and less than 1% for a niece, nephew, or first

Table 13.5 Heritabilities for Some Human Traits

Trait	Heritability
Clubfoot	0.8
Height	0.8
Weight	0.7
Blood pressure	0.6
Menstrual pain	0.4
Verbal aptitude	0.7
Mathematical aptitude	0.3
Spelling aptitude	0.5
Total fingerprint ridge count	0.9
Intelligence	0.5–0.8
Total serum cholesterol	0.6

Table 13.6 Coefficient of Relationship and Shared Genes

Relationship	Degree of Relationship	Percent Shared Genes
Sibling to sibling	1°	50% (1/2)
Parent to child	1°	50% (1/2)
Uncle/aunt to niece/nephew	2°	25% (1/4)
First cousin to first cousin	3°	12 1/2% (1/8)

age than a similar person in the general population. This may be because inherited variants in immune system genes increase susceptibility to certain infectious agents. In support of this hypothesis, the risk that adoptees would die young from infection did not correlate with adoptive parents' death from infection before age 50.

Although researchers concluded from this investigation that length of life is mostly determined by heredity, they did find evidence of environmental influences. For example, if adoptive parents died before age 50 of cardiovascular disease, their adopted children were three times as likely to die of heart and blood vessel disease as a person in the general population. Can you think of a factor that might account for this correlation?

Twins

Since 1979, more than 100 sets of twins and triplets who were separated at birth have visited the laboratories of Thomas Bouchard at the University of Minnesota. There, for a week or more, each set of twins undergoes a battery of physical and psychological tests. These "Minnesota twins" have helped unravel how genes and the environment influence an astonishingly wide variety of traits.

Twins separated at birth provide natural experiments for distinguishing nature from nurture. Much of what they have in common can be attributed to genetics, especially if their environments have been very different (fig. 13.23). Their differences reflect differences in their upbringing. Since the study of genetics blossomed in the early twentieth century, genetic researchers have recognized the value twins offer in unearthing the roots of complex traits.

Twins occur in 1 out of 80 births. Identical or **monozygotic** (MZ) twins result when a single fertilized egg splits. Therefore, identical twins are always of the same sex and always have identical genes. Fraternal or **dizygotic** (DZ) twins arise from two fertilized eggs. Therefore, fraternal twins are no more similar genetically than any other two siblings, although they share the same prenatal environment.

cousin. The **coefficient of relationship** depends on how closely related individuals are (table 13.6).

Multifactorial inheritance analysis does not easily lend itself to the scientific method, which ideally entails the study of one variable. However, when studying humans, geneticists can turn to two types of people to tease apart the genetic and environmental components of complex traits—twins and adopted individuals.

Adoptees and Their Families

A person adopted by nonrelatives shares environmental influences, but not genes, with his or her adoptive family. Conversely, adoptees share genes, but not the exact environment, with their biological parents. Therefore, biologists

assume that similarities between adoptees and adoptive parents reflect environmental influences, whereas similarities between adoptees and their biological parents mostly reflect genetic influences. Information on both sets of parents can reveal to what degree heredity and the environment each contribute to the development of a trait.

Many adoption studies use the Danish Adoption Register, a compendium of all adopted Danish children and their families from 1924 to 1947. One study examined correlations between causes of death among biological and adoptive parents and among adoptees. The study found that if a biological parent died of infection before age 50, the adopted child was five times more likely to die of infection at a young

Separated at birth, the Mallifert twins meet accidentally.

FIGURE 13.23

Drawing by Chas. Addams; © 1981 The New Yorker Magazine, Inc.

If a trait tends to occur more frequently in both members of identical twin pairs than in both members of fraternal twin pairs, it is at least partly controlled by heredity. Geneticists calculate the **concordance** of a trait, or the degree to which it is inherited, by calculating the percentage of pairs in which both twins express the trait.

Diseases caused by single genes, whether dominant or recessive, are always 100% concordant in MZ twins—that is, if one twin has it, so does the other. However, among DZ twins, concordance is 50% for a dominant single-gene trait and 25% for a recessive trait. These are the same Mendelian values that apply to any two siblings. For a trait determined by several genes, concordance values for MZ twins are significantly greater than for DZ twins. Finally, a trait molded mostly by the environment exhibits similar concordance values for both types of twins.

Twin studies have been useful in assessing complex behavioral characteristics such as personality traits. In 1976, John C. Loehlin, a psychologist at the University of Texas in Austin, administered a battery of personality tests to 514 MZ twins and 336 DZ twins selected from a national sample of high school seniors. Personality concordance was 50% for MZ twins and 28% for

DZ twins. The heritability for these measures was calculated to be 44% [(50-28) ×2], which means that heredity accounts for slightly less than half of the personality similarities between these twins—leaving quite a lot of room for environmental influences.

Comparing twin types assumes, perhaps incorrectly, that both types of twins share similar experiences. In fact, identical twins are often closer than fraternal twins, particularly those of opposite sex. This discrepancy led to some misleading results in twin studies conducted in the 1940s. One study concluded that tuberculosis is inherited because concordance among identical twins was higher than among fraternal twins. Actually, the infectious disease was more readily passed between identical twins because their parents kept them in close physical contact.

Today, the Minnesota Study goes well beyond twin studies of the past by tracking a huge number of traits. Each pair of twins undergoes a 6-day battery of tests that measures both physical and behavioral traits, including 24 different blood types, handedness, direction of hair growth, fingerprint pattern, height, weight, functioning of all organ systems, intelligence, allergies, and dental patterns. Researchers videotape the twins' facial expressions and body movements in different circumstances and probe their fears, vocational interests, and superstitions.

Researchers have found that identical twins separated at birth and reunited later are remarkably similar, even when they grow up in very different adoptive families. Idiosyncrasies are particularly striking. A pair of identical male twins raised in different countries practicing different religions were astounded, when they were reunited as adults, to find that they both laugh when someone sneezes and flush the toilet before using it. Twins who met for the first time in their 30s responded identically to questions; each paused for 30 seconds, rotated a gold necklace she was wearing three times, and then answered the question. Coincidence, or genetics?

The "twins reared apart" approach is not a perfectly controlled way to separate nature from nurture. Identical twins share an environment in the uterus and possibly in early infancy that may affect later development. Siblings, whether adoptive or biological, do not always share identical home environments. Differences in age, sex, general health, school and peer experiences, temperament, and personality affect each individual's perception of such environmental influences as parental affection and discipline.

Adoption studies, likewise, are not perfect experiments. Adoptions agencies often search for adoptive families who have socioeconomic or religious backgrounds similar to those of the biological parents. Thus, even when different families adopt separated twins, their environments may not be as different as they might be for two unrelated adoptees. However, at the present time, twins reared apart are providing intriguing insights into the number of body movements, psychological quirks, interests, and other personality traits that seem to be rooted in our genes.

KEY CONCEPTS

Multifactorial traits are molded by a combination of genes and the environment and do not express Mendelian ratios. Polygenic traits vary continuously.

Although geneticists predict single-gene outcomes through Mendelian ratios, they must calculate empiric risk to predict the likelihood that a complex multifactorial trait will recur. Risk increases with severity of the trait, number of affected relatives, and increasing relatedness to an affected individual.

Heritability estimates the proportion of variation in a multifactorial trait's expression in a population that is attributable to genetics. The coefficient of relationship indicates the proportion of genes different types of relatives share.

Researchers can compare adoptees and their adoptive and biological parents to assess whether adoptees' traits stem from heredity or the environment. MZ twins separated at birth provide perhaps the best distinction between nature and nurture in molding a complex characteristic.

SUMMARY

For centuries people have noted the inheritance of certain traits. Between 1857 and 1865, an Austrian monk, Gregor Mendel, became the first to systematically investigate the nature of inheritance. He set up crosses between certain pea plants and followed the inheritance of one or two traits at a time. The genes for the traits he studied were carried on different chromosomes, and each trait had two easily distinguished forms. From Mendel's data, geneticists a generation later, who knew more about chromosomes, deduced the two basic laws of inheritance.

Mendel's first law, segregation, states that inherited "elementen" separate in meiosis, so each new individual receives one copy of each elementen from each parent. Mendelian traits and illnesses are caused by a single gene. Geneticists use pedigrees to trace traits.

Mendel's second law, independent assortment, follows the transmission of two or more characters located on different chromosomes. Because maternally and paternally derived chromosomes (and the genes they carry) sort randomly in meiosis, different gametes have different combinations of genes.

An allele is an alternate form of a gene. An allele whose expression masks another is dominant; an allele whose expression is masked by a dominant allele is recessive. A heterozygote has two different alleles of a particular gene; a homozygous recessive individual has two recessive alleles; a homozygous dominant individual has two dominant alleles. The combination of alleles constitutes a genotype, and the expression of a particular genotype is its phenotype. Punnett squares, which use the principles of mathematical probability, can predict the outcomes of genetic crosses.

In some crosses, the proportions of phenotypes do not seem to follow Mendel's laws. Individuals who have homozygous recessive lethal alleles cease developing before birth and are never detected as a particular phenotype. A gene can have multiple alleles because its DNA sequence may be altered in many ways. In epistasis, one gene masks the effect of another. Different types of dominance relationships between alleles also influence the phenotypic ratios of offspring classes. Heterozygotes of incompletely dominant alleles have phenotypes intermediate between those of the two homozygotes. Codominant alleles are both expressed. Any of these types of alleles may appear to disrupt the expected Mendelian ratios.

An incompletely penetrant genotype, one that is not expressed in all who inherit it, creates another situation that appears to contradict Mendel's laws. Phenotypes that vary in intensity are variably expressive, and the environment can also influence gene expression. Pleiotropic genes have several expressions. Sometimes a characteristic that appears to be inherited is not genetic at all, but is caused by an environmental factor. Uniparental disomy is the one true exception to Mendelian inheritance; one parent contributes a double dose of genetic material and the other does not contribute the corresponding gene. Multifactorial traits are influenced by the environment and genes. A polygenic trait is determined by more than one gene and varies continuously. Multifactorial traits do not follow Mendel's laws, but they do tend to recur in families. Empiric risk measures the prevalence of a multifactorial trait and its likelihood of recurring. The risk rises with genetic closeness to an affected person, severity of the phenotype, and the number of affected relatives.

Heritability estimates what proportion of variation in a multifactorial trait is attributable to genetics. It equals twice the difference of the variance between two groups of individuals. Heritability is not a gene characteristic; it varies from population to population. The coefficient of relationship describes the proportion of genes shared by different types of relatives. Concordance measures the expression of a trait in MZ or DZ twins. The more a trait is determined by genes, the higher the concordance value, and the more likely both twins will have the trait.

Characteristics shared by adoptees and their biological parents are mostly inherited, whereas similarities between adoptees and adoptive parents reflect environmental influences. Similarities between identical twins raised apart also indicate genetically influenced characteristics. Twin studies offer the best distinction between nature and nurture in molding individual human traits.

KEY TERMS

allele 272
autosomal dominant 280
autosomal recessive 280
codominant alleles 284
coefficient of relationship 291
complete penetrance 284
concordance 292
continuous variation 289
dihybrid cross 281
dizygotic (DZ) twins 291
dominant 272

empiric risk 289
epistasis 284
first filial generation (F$_1$) 274
genes 270
genomic imprinting 288
genotype 272
genotypic ratio 275
heterozygote 272
heterozygous 272
heritability 290
homozygous 272
homozygous dominant 272
homozygous recessive 272

incomplete dominance 284
incomplete penetrance 285
independent assortment 281
law of segregation 271
lethal allele 283
Mendelian trait 276
monohybrid 274
monozygotic (MZ) twins 291
multifactorial trait 289
mutant 274
mutation 274
parental generation (P$_1$) 274
pedigree 277

phenocopy 287
phenotype 272
phenotypic ratio 275
pleiotropic 287
polygenic 289
product rule 282
Punnett square 275
recessive 272
second filial generation (F$_2$) 274
test cross 275
uniparental disomy 288
variably expressive 285
wild type 274

REVIEW QUESTIONS

1. Bob and Wanita know from a blood test that they are each heterozygous for the autosomal recessive gene that causes beta thalassemia, a type of anemia. If their first three children are healthy, what is the probability that their fourth child will have the disease?

2. Mendel crossed pea plants heterozygous for the height gene (*Tt*) and obtained the monohybrid phenotypic ratio of 3:1 and the genotypic ratio of 1:2:1. Calculate the genotype and phenotype ratios for the following crosses:

   a. Homozygous dominant crossed to homozygous recessive

   b. Homozygous dominant crossed to heterozygote

   c. Homozygous recessive crossed to heterozygote

3. A man and a woman each have dark eyes, dark hair, and freckles. The genes for these traits assort independently. The woman is heterozygous for each of these traits, but the man is homozygous. The dominance relationships of the alleles are as follows:

   B = dark eyes   *b* = blue eyes
   H = dark hair   *h* = blond hair
   F = freckles   *f* = no freckles

   a. What is the probability that their child will share their phenotype?

   b. What is the probability that their child will share the same genotype as the mother? as the father?

Use probability or a Punnett square to obtain your answers. Which method is easier?

4. Which genetic principle does each of the following examples illustrate?

   a. Genes on two different chromosomes affect hearing in humans. The *D* gene controls development of the auditory nerve, with allele *d* impairing development. The *F* gene controls development of the cochlea, which is part of the inner ear, with allele *f* blocking that development. If a person is homozygous recessive for either gene (*dd* or *ff*), he or she is deaf, irrespective of the genotype at the other locus.

   b. A woman develops dark patches on her face. Her family physician suspects that she may have alkaptonuria, an inherited deficiency of the enzyme homogentisic acid oxidase that causes darkened skin, a stiff spine, darkened ear tips, and urine that turns black when it contacts air. However, a dermatologist the woman sees discovers that she has been using a facial cream containing hydroquinone, which is known to cause darker patches in dark-skinned people.

5. A dominant allele causes severe mental retardation and heart defects in a heterozygote. Geneticists never see homozygotes of the dominant allele. Why not?

6. Explain how each of the following appear to disrupt Mendelian ratios. Which is a true exception to Mendel's law of segregation?

   a. pleiotropy
   b. variable expressivity
   c. incomplete penetrance
   d. lethal alleles
   e. uniparental disomy

7. A man with type AB blood has children with a woman who has type O blood. What are the chances that a child they conceive will have type A blood? type B? AB? O?

8. A couple who have a child with cystic fibrosis have oocytes removed from the woman and fertilized with the husband's sperm in a laboratory dish. If 16 of their preembryos are screened for the cystic fibrosis gene, how many of them, based on Mendel's first law, would be expected to have two copies of the CF allele? How many would theoretically have one CF allele and one wild type allele? How many would have only wild type alleles and lack an allele that could cause CF?

9. Two mice have a litter of 12. Three of the babies cannot move their hind legs because of a defective nerve in their thighs. The parents are healthy.

    a. What is the mode of inheritance for this trait?
    b. Draw a pedigree of this family.
    c. What is the probability that a healthy F_1 individual is a carrier of this condition?

10. Enamel hypoplasia is a hereditary defect that causes holes and cracks to appear around the crowns of the baby teeth. It is inherited as an autosomal dominant trait, but with incomplete penetrance and variable expressivity. Below is a partial pedigree for a family with this trait.

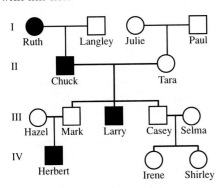

    a. Which individual appears to display incomplete penetrance for this trait?
    b. How might the variable expressivity appear among the affected members of the family?
    c. What term describes an environmental health problem that resembles a known inherited disorder, such as tooth decay caused by sleeping with a juice bottle in the mouth ?
    d. What is the probability that Herbert's child will inherit enamel hypoplasia?
    e. The family cleans their attic and finds old baby photos that shed some light on the family tooth anomaly. Ruth's mother, Lucy, had a toddler's grin full of oddly shaped teeth. Her father, Jerry, had normal teeth. Ruth also discovers that her older brother Fred and older sister Anna had abnormal teeth, but she remembers that her younger sister Lulubelle had beautiful, healthy teeth. Add this information to the pedigree.

11. Domesticated hens with white feathers and large, single combs mate with roosters that have dark feathers and small combs. The offspring all resemble their mothers for these two traits.

    a. Which alleles of each trait are dominant, and which are recessive?
    b. If the F_1 are crossed to each other, what percentage of the F_2 would be expected to have dark feathers and large, single combs?

12. Two genes carried on different chromosomes impart pigment to potato skin. The *D* gene provides red color, and the *P* gene provides blue. However, if the dominant allele of the *D* gene is present, the potatoes are red, no matter which *P* gene allele is present. Which phenomenon that can alter Mendelian ratios does this illustrate?

TO THINK ABOUT

1. Do you think Mendel's observations would have differed if the traits he chose to study were part of the same chromosome?

2. If you wanted to confirm Mendel's law of independent assortment by recording the phenotypes and genotypes of the offspring of a dihybrid cross, what sorts of traits would you choose to examine, and in what organism?

3. A white woman with fair skin, blond hair, and blue eyes and a black man with dark brown skin, hair, and eyes have fraternal twins. One twin has blond hair, brown eyes, and light skin, and the other has dark hair, brown eyes, and dark skin. What Mendelian principle does this real-life case illustrate?

4. How can parents carry recessive lethal genes and not know it? Why is there no physical evidence for the existence of dominant lethal mutations?

5. A man and woman want to have children, but each is a known carrier of Tay-Sachs disease. To avoid the possibility that they will conceive a child who will inherit this untreatable, fatal disorder, the woman is artificially inseminated with sperm from a man who does not carry the Tay-Sachs gene. Why is this procedure (discussed in chapter 12) a solution to the couple's problem?

6. Many plants have more than two sets of chromosomes. How would having four (rather than two) copies of a chromosome more effectively mask expression of a recessive allele?

7. In an attempt to breed winter barley that is resistant to barley mild mosaic virus, agricultural researchers cross a susceptible domesticated strain to a resistant wild strain. The F_1 plants are all susceptible, but when the F_1 are crossed to each other, some of the F_2 individuals are resistant. Is the viral resistance gene recessive or dominant? How do you know?

SUGGESTED READINGS

Bouchard, T. J., Jr., June 17,1994. Genes, environment, and personality. *Science,* vol. 264. Identical twins separated at birth offer clues to hereditary and environmental influences on behavior.

Carmelli, Dorit, Ph. D., et. al. Sept. 17, 1992. Genetic influence on smoking—a study of male twins. *The New England Journal of Medicine,* vol. 327. Smoking behavior among thousands of twin pairs suggests a hereditary influence.

Elmer-Dewitt, Philip. October 5, 1992. Catching a bad gene in the tiniest of embryos. *Time.* Screening preembryos for autosomal recessive, disease-causing alleles validates Mendel's first law.

Feinberg, Andrew P. June 1993. Genomic imprinting and gene activation in cancer. *Nature Genetics,* vol. 4. Some childhood cancers may be caused by uniparental disomy.

Grebe, Theresa A., et al. March 1994. Genetic analysis of hispanic individuals with cystic fibrosis. *The American Journal of Human Genetics,* vol. 54. Different ethnic groups have different allele frequencies for the cystic fibrosis gene.

Heim, Werner G. February 1991. What is a recessive allele? *The American Biology Teacher.* We now understand the pea plant traits Mendel studied at the molecular level.

King, Richard A., Jerome I. Rotter, and Arno G. Motulsky, eds. 1992. *The genetic basis of common diseases.* New York: Oxford University Press. Many common disorders scientists thought were not inherited in a Mendelian fashion nevertheless have hereditary components.

McKusick, Victor A. July 25, 1991. The defect in Marfan syndrome. *Nature,* vol. 352. Marfan syndrome is pleiotropic because of the widespread distribution of fibrillin in the body.

Mascari, Maria, Wayne Gottlieb, M. S., Peter K. Rogan, Ph.D., et. al. June 11, 1992. The frequency of uniparental disomy in Prader-Willi syndrome. *The New England Journal of Medicine,* vol. 326, no. 24. Prader-Willi and Angelman syndromes illustrate genomic imprinting in humans.

Morell, Virginia. June 18, 1993. Evidence found for a possible 'aggression gene.' *Science,* vol. 260. A mutation in the gene encoding the enzyme monoamine oxidase A may account for the extreme aggression of many males in a large family.

Platt, Orah S. March 17, 1994. Easing the suffering caused by sickle cell disease. *The New England Journal of Medicine,* vol. 330. Understanding the basis of blood vessel blockage in sickle cell disease may help physicians to relieve the terrible pain.

Plomin, R., M.J. Owen, P. McGuffin. June 17, 1994. The genetic basis of complex human behaviors. *Science,* vol. 264. Separating nature from nurture can be very challenging.

Reed, Martha L., and Stuart E. Leff. February 1994. Maternal imprinting of human SNRPN, a gene deleted in Prader-Willi syndrome. *Nature Genetics,* vol. 6. Researchers are closing in on the specific genes implicated in Prader-Willi syndrome.

Resta, Robert G. December 1993. The crane's foot: The rise of the pedigree in human genetics. *Journal of Genetic Counseling,* vol. 2. People have diagrammed family relationships using pedigree charts since the fifteenth century.

Rudolph, Jeffrey, Sharon J. Spier, Glen Byrns, et. al. October 1992. Periodic paralysis in quarter horses: A sodium channel mutation disseminated by selective breeding. *Nature Genetics,* vol. 2. Skeletal muscle paralysis in quarter horses is caused by an autosomal dominant gene. This paper contains some interesting pedigrees.

Walker, R. April 6, 1991. New evidence supports genomic imprinting. *Science News.* Can inheriting double doses of the same gene from one's mother or father produce two different diseases?

CHAPTER 14

Chromosomes

CHAPTER OUTLINE

Early Theories about Chromosomes

Linked Genes—An Exception to Mendel's Laws

Linkage Maps

Sex Linkage and Sex Chromosomes
Sex Determination in Humans
Sex-linked Recessive Inheritance
Sex-linked Dominant Inheritance
Y Linkage

X Inactivation—Equalizing the Sexes

Gender Effects on Phenotype
Sex-limited Traits
Sex-influenced Traits
Genomic Imprinting

Genetic Heterogeneity

Awaiting the Amino Result

Chromosome Structure

Abnormal Chromosome Numbers
Extra Chromosome Sets—Polyploidy
Extra and Missing Chromosomes—Aneuploidy
   Down Syndrome—A Closer Look
   Sex Chromosome Aneuploids

Chromosome Rearrangements
Translocations
Inversions

Other Chromosomal Abnormalities

Why Study Chromosomes?

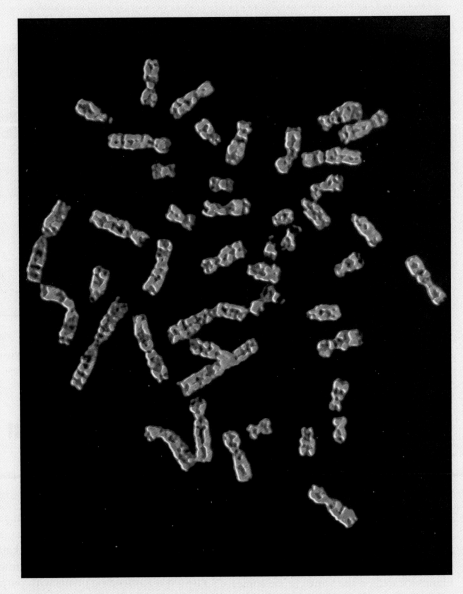

Replicated chromosomes.

LEARNING OBJECTIVES

By the chapter's end, you should be able to answer these questions:

1. How are genes on the same chromosome inherited differently than genes located on different chromosomes?

2. How do geneticists use crossover frequency between pairs of genes to construct linkage maps?

3. How are genes located on the sex chromosomes inherited differently in males and females?

4. How is sex chromosome constitution rendered equal in the two sexes in mammals?

5. How does gender influence gene expression?

6. What is the structure of a chromosome?

7. How do changes in chromosome number and chromosome structure occur?

8. In what ways can chromosomal abnormalities affect the phenotype?

9. What types of information can scientists and physicians obtain from chromosome studies?

Early Theories about Chromosomes

When biologists discovered at the beginning of the twentieth century that genes are actually part of chromosomes, they realized that the number of observable inherited traits in an organism far exceeds the number of chromosomes. Fruit flies, for example, have four pairs of chromosomes, but dozens of different bristle patterns, body colors, eye colors, wing shapes, and other characteristics. This was confusing; scientists had only recently rediscovered Mendel's laws and renamed his "characters" genes, and now another genetic unit—the chromosome—came along. How might a few chromosomes control so many traits?

The first hypothesis was that during meiosis, chromosomes somehow break into gene-sized pieces. This idea accounted for the discrepancy between the number of inherited traits and the number of chromosomes and also explained why the genes Mendel studied behaved as if they were physically separate when they assorted into gametes. However, a different explanation emerged as researchers found that some traits did not assort independently: each chromosome carries many genes. Genes located on the same chromosome are **linked**, or inherited together, and do not assort independently. The seven traits Mendel followed in his pea plants were specified by different chromosomes. Had the same chromosome carried these

Table 14.1	Observed and Expected Phenotypes for a Dihybrid Pea Plant Cross	
Phenotype P=purple flower p=red flower L=long pollen grain l=round pollen grain	Number of Observed Plants	Number of Expected Plants (9:3:3:1 ratio)
Purple, long (P_L_)	284	215
Purple, round (P_ll)	21	71
Red, long (ppL_)	21	71
Red, round (ppll)	55	24
	381	381

genes near each other, Mendel would have generated markedly different results in his dihybrid crosses.

Linked Genes—An Exception to Mendel's Laws

In the early 1900s, William Bateson and R. C. Punnett first observed the peculiar offspring ratios indicating gene linkage, again in pea plants. Bateson and Punnett crossed true-breeding plants with purple flowers and long pollen grains (genotype PPLL) to true-breeding plants with red flowers and round pollen grains (genotype ppll). Then they crossed the F1 plants, of genotype PpLl, to each other.

Surprisingly, the F2 generation did not show the expected 9:3:3:1 ratio for an independently assorting dihybrid cross (table 14.1).

Bateson and Punnett noticed that two types of F2 peas—those with the same phenotypes as the parents, P_L_ and ppll—were more abundant than predicted, while the other two progeny classes—ppL_ and P_ll—were less common. They hypothesized that the prevalent parental allele combinations reflected genes transmitted on the same chromosome, which therefore did not separate during meiosis. The other two offspring classes could also be explained by a meiotic event—crossing over. Recall that this is an exchange between homologs which mixes up maternal and paternal gene combinations in the

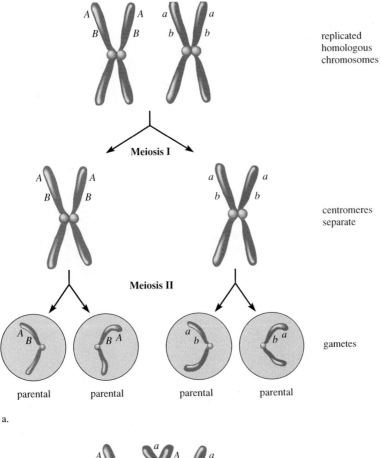

replicated
homologous
chromosomes

Meiosis I

centromeres
separate

Meiosis II

gametes

parental   parental   parental   parental

a.

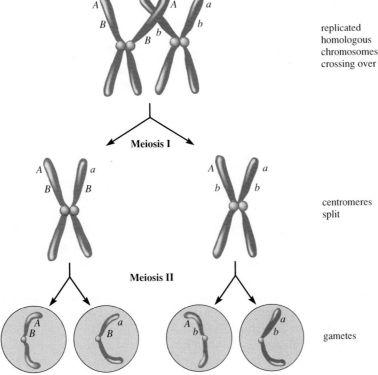

replicated
homologous
chromosomes
crossing over

Meiosis I

centromeres
split

Meiosis II

gametes

parental   recombinant   recombinant   parental

b.

FIGURE 14.1

a. Genes that are linked closely to one another on the same chromosome are usually inherited together when that chromosome becomes packaged into a gamete. *b.* Linkage between two genes can be interrupted if the chromosome they are located on crosses over with its homolog at a point between the two genes. Such crossing over packages recombinant arrangements of the genes into gametes. (The centromere is exaggerated here; it is actually a constriction.)

gametes, but retains the sequence of genes on the chromosomes (figs. 10.8 and 14.1).

Recombinant progeny represent the mixing of maternal and paternal alleles into new combinations in the gamete. **Parental** gametes retain the gene combinations passed from the organism's mother and father. Parental and recombinant are relative terms, depending on the parents' allele combinations. Had the parents in Bateson and Punnett's crosses been of genotypes *ppL_* and *P_ll* (phenotypes red flowers, long pollen grains and purple flowers, round pollen grains), then *P_L_* and *ppll* would be the recombinant rather than the parental classes.

Two other terms describe the arrangement of linked genes in heterozygotes. Consider a pea plant with genotype *PpLl*. These alleles can be arranged on the chromosomes in two ways. If the two dominant alleles travel on one chromosome and the two recessive alleles are transmitted on the other, the genes are in coupling. In the opposite configuration, when one dominant allele is located near one recessive allele on a chromosome, the two genes are in repulsion (fig. 14.2).

Alleles in coupling tend to be transmitted together, because they reside on the same homolog. However, alleles on different homologs separate at meiosis and may not be passed on together.

As Bateson and Punnett were deciphering linkage in peas, geneticist Thomas Hunt Morgan and his coworkers at Columbia University were setting up crosses of the fruit fly *Drosophila*

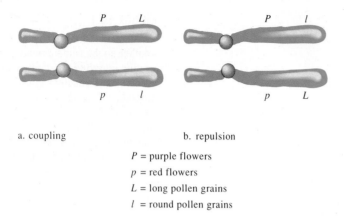

a. coupling   b. repulsion

P = purple flowers
p = red flowers
L = long pollen grains
l = round pollen grains

FIGURE 14.2

Parental type progeny can be distinguished from recombinant progeny only if one knows the allele configuration of the two genes—they are either in coupling (*a*) or in repulsion (*b*).

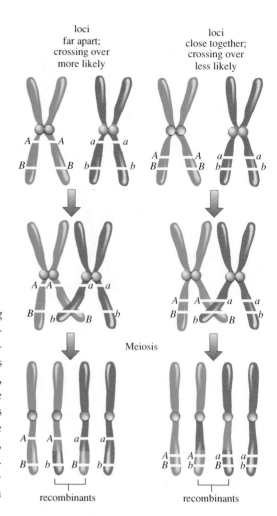

FIGURE 14.3

Crossing over is more likely to occur between the widely spaced linked genes on the left than between the more closely spaced linked genes on the right.

melanogaster to assess whether various combinations of two traits were linked. As their data accumulated, the researchers realized that pairs of traits fell into four groups. Within each group, crossed dihybrids did not produce the proportions of offspring predicted by Mendel's second law. This indicated four sets of linked genes. Not coincidentally, the number of these linkage groups is exactly the number of chromosome pairs in the organism. The traits fell into four groups because the genes causing traits that are inherited together travel on the same chromosome.

Morgan wondered why the size of the recombinant classes varied among the crosses. Might the differences reflect different physical relationships between the genes on the chromosome? Alfred Sturtevant, Morgan's undergraduate assistant, explored this idea. In 1911, he developed a theory and technique that would profoundly affect the fledgling field of genetics in his day, and the medical genetics of today. Sturtevant proposed that the farther apart two genes are on a chromosome, the more likely a crossover is to occur between them—simply because there is more space between the two genes.

Linkage Maps

Geneticists use the correlation between crossover frequency and the distance between genes to construct **linkage maps,** diagrams of gene order and spac-

ing on chromosomes. By determining the proportion of recombinant offspring, investigators can infer the frequency with which two genes cross over. By determining this frequency, they can then infer how far apart the genes are on the chromosome. Genes that occupy opposite ends of the same chromosome would cross over often, generating a large recombinant class. But genes lying very close on the chromosome would rarely be separated by a crossover (fig. 14.3).

In 1913, Sturtevant published the first genetic linkage map, depicting the order of five genes on the X chromosome of the fruit fly. As the twentieth century progressed, genetic researchers rapidly mapped genes on all four fruit fly chromosomes, and along the human X chromosome. Mapping genes on the X chromosome was easier than on the autosomes, because human males have only a single X chromosome. This means that X chromosome alleles in the human male are never masked.

By 1950, genetic researchers began to contemplate the daunting task of mapping genes on the 22 human autosomes. The gene mapper needs a clue to begin matching a certain gene to its chromosome. Often, this clue is an individual who has a particular trait and a particular chromosome abnormality. Matching phenotypes to chromosomal variants is a field called cytogenetics.

Finding a chromosomal abnormality and using it to detect linkage to another gene is a valuable but rare clue.

More often, researchers have to rely on the sorts of experiments Sturtevant conducted on his flies—calculating the percentage of recombination (crossovers) between two genes. Because individual humans do not produce hundreds of offspring, as fruit flies do, cytogeneticists must observe traits in many families and pool the data to obtain enough information to establish gene linkages.

Today, the journal *Science* publishes the latest human chromosome maps every October. Many of these maps originated as classical linkage maps. Researchers are supplementing linkage maps with molecular approaches, discussed in chapter 16, to chart the human genome. A **genome** is the collection of all of the genes of an organism.

Sex Linkage and Sex Chromosomes

Constructing a linkage map is tedious because mappers must set up crosses, or find individuals whose allele configurations can be deciphered. The mappers must also observe large numbers of offspring to map accurately. The first linkage maps, in flies and later in humans, were maps of the X chromosome because it is easier to follow the inheritance of sex-linked traits, or traits on the X chromosome. Before we examine how sex-linked traits are inherited, it helps to understand the role that genes and chromosomes play in determining sex.

Sex Determination in Humans

In mammals, the sexes have equal numbers of autosomes, but males have one X and one Y chromosome, and females have two X chromosomes. The sex with two different sex chromosomes is the **heterogametic** sex, and the other, with two of the same sex chromosomes, is the **homogametic** sex. In all mammals, the male is the heterogametic sex. In birds, moths, and butterflies, the female is the heterogametic sex. Biology in Action 14.1 explores sex-determination mechanisms in different species.

The human X chromosome carries more than 1,000 genes. The Y chromosome is much smaller than the X and carries far fewer genes. A quarter of the genes on the Y chromosome correspond to genes on the X chromosome. As researchers fill in specific genes on the

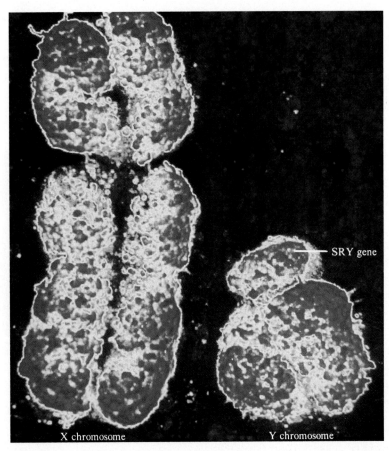

FIGURE 14.4

The human X and Y chromosomes. The SRY gene, at the top of the short arm of the Y chromosome, sets into motion the cascade of gene activity that directs male development.

Y chromosome, we will learn about traits that are distinctly male, and about how sex evolved.

Researchers identified the part of the Y chromosome that determines maleness by studying some interesting people—men who have two X chromosomes, and women who have one X and one Y chromosome, the reverse of normal. A close look at these individuals' sex chromosomes revealed that each of the XX males actually had a small piece of a Y chromosome, and each of the XY females lacked a small part of the Y chromosome. The part of the Y chromosome present in the XX males was the same part that was missing in the XY females. It was only a tiny part of the Y chromosome, about half a percent of its total structure. Somewhere within this stretch of 300,000 DNA base pairs, investigators in England finally found the male sex-determining gene. They named it *SRY*, for sex-determining region of the Y.

The SRY gene produces a protein that switches on other genes that direct the embryo to develop male structures. Recall from chapter 11 that before the sixth week of prenatal development, the human embryo has organs that could develop into either male or female reproductive systems. When the SRY gene is activated at about week 7, the rudimentary testes begin to secrete testosterone, the sex hormone that stimulates male structures to grow. Cascades of other gene activity follow, directing development towards maleness (fig. 14.4). Cords of breast tissue diminish, swellings develop into male sex organs, and, at about 16 weeks, immature sperm begin to form.

For a male to develop, female structures must be suppressed. The SRY protein activates a gene on chromosome 19 that encodes a protein called Mullerian-inhibiting substance. This protein destroys the rudimentary female structures. "Femaleness" is not a default

Sex Determination—Male or Female?

People thought about the origins of gender long before they knew about genes and chromosomes. Some ancient cultures thought the phase of the moon, the direction of the wind, or what the couple wore, ate, or said just prior to conception determined a child's sex. The ancient Greeks thought sex was determined by which testicle the sperm came from. Some European royal families perpetuated this idea—some of the men tied off or removed their left testicles to guarantee conception of a son! Mendel was actually the first to suggest that sex is inherited.

Sex determination mechanisms are diverse among different types of organisms. Grasshoppers, for example, have only one type of sex chromosome, designated X. If two sex chromosomes are present (XX), the insect is a female; if only one is present (XO), it is a male.

In the fruit fly *Drosophila melanogaster*, the X to autosome ratio is important. Individuals with a pair of X chromosomes and a pair of each autosome are normal females; those with one X chromosome and a pair of each autosome are normal males. Although fertile male fruit flies are XY, it is not the Y chromosome that determines maleness but the ratio of the one X chromosome to the paired autosomes. An XO fruit fly is a sterile male (in humans, an XO is female), and an XXY fruit fly is female (in humans, an XXY is male). X:autosome sex determination is also characteristic of flowering plants.

Sex determination is more complex in some other species. Certain beetles have 12 X chromosomes, 6 Y chromosomes, and 18 autosomes. In other species, the number of total sets of chromosomes determines whether an individual develops as a male or a female. In bees, a fertilized egg becomes a diploid female; an unfertilized egg develops into a haploid male. The queen bee actually determines sex: the eggs that she fertilizes become females, and those she doesn't become males.

FIGURE 1

Whiptail lizards are parthenogenetic, but they engage in matinglike behavior.

Some whiptail lizards (fig. 1) dispense with sex altogether. All individuals are females who lay eggs that hatch without male input. The lizards, however, engage in a behavior that resembles mating, perhaps a vestige from a sexual ancestor. Cichlid fish, in contrast, seem to have three sexes (fig. 2)—a few bright-colored, sexually aggressive males, females, and inhibited males who do not behave sexually unless an aggressive male dies.

Autosomal genes can influence sex determination. In fruit flies, a "transformer" mutation gives an XX individual a male phenotype. Different alleles can also determine sex. In the bread mold *Neurospora crassa*, the two sexes are called mating types, and each is specified by a different allele of a particular gene. Two individuals mate only if they have different alleles for this gene. In one species of wasp, a gene with nine alleles controls sex determination. All heterozygotes are female, all homozygotes male.

In some species, the environment determines sex. For some turtles, the temperature of the land the eggs are laid on determines sex; eggs laid in the sun hatch females, and those laid in the shade pro-

FIGURE 2

Cichlids have two types of males and one type of female.

duce males (fig. 3). In the sea-dwelling worm *Bonellia viridis*, larvae mature into females if no adult females are present. If an adult female is present, she secretes hormones that cause the larvae to develop as males. Some marine bivalves and fish can even change sex in adulthood (fig. 4). Compared to some of these species, human sex determination may prove to be a simple, straightforward matter!

option if the SRY gene is not present, but one of two possible gene-encoded choices in development.

Sex-linked Recessive Inheritance

Any gene on the X chromosome of a male mammal is expressed in his phenotype, because he lacks a second allele for that gene which would mask its expression. An allele on an X chromosome in a female may or may not be expressed, depending upon whether it is dominant or recessive and on the nature of the allele on the second X chromosome. The male is **hemizygous** for sex-linked traits because he has half the number of genes the female has.

A male always inherits his Y chromosome from his father and his X chromosome from his mother (fig. 14.5). A female inherits one X chromosome from each parent. If a mother is heterozygous for a particular sex-linked gene, her son has a 50% chance of inheriting either allele from her. Sex-linked genes are therefore passed from mother to son. Because a male does not receive an X chromosome from his father (he inherits the Y chromosome from his father), a father does not pass a sex-linked trait to his son. Punnett squares are used to depict transmission of sex-linked traits, as shown in the inset in figure 14.6.

Hemophilia A is a sex-linked recessive disorder in which absence or deficiency of a protein clotting factor greatly slows blood clotting. A cut may take a long time to stop bleeding. A mild bump can lead to a large bruise, because broken capillaries leak a great deal of blood before they heal. Internal bleeding may occur without the person's awareness. Receiving the missing clotting factor can control the illness. Before 1985, when hospitals first began screening the blood supply for the AIDS virus, many people with hemophilia contracted AIDS because the clotting factor they received was extracted from blood dona-

FIGURE 3

For some turtles, the environment determines sex.

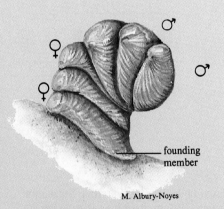

FIGURE 4

Sex reversal in the slipper limpet. To the slipper limpet *Crepidula fornicata*, position is literally everything in life. These organisms aggregate in groups anchored to rocks and shells on an otherwise muddy sea bottom. The founding member of a group is a small male. As he grows, he becomes a she, and a new, small male comes in at the top. When this second member grows large enough to become female, a third recruit enters. Thus, one's size and position in the pile determines sex. The slipper limpet shown here is in the intermediate stage.

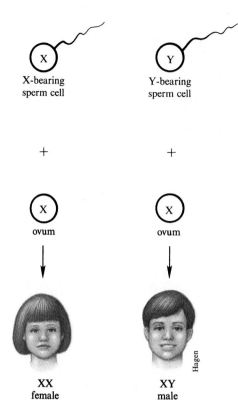

FIGURE 14.5

Sex determination in humans. An egg cell typically contains a single X chromosome. A sperm cell contains either an X chromosome or a Y chromosome. If a Y-bearing sperm cell with an SRY gene fertilizes an egg cell, the zygote is a male (XY). If an X-bearing sperm cell fertilizes an egg cell, then the zygote is a female (XX).

tions. Today, this is unlikely, both because the blood supply is more carefully scrutinized, and because the clotting factor is obtained in pure form by genetic engineering.

The risk that a carrier mother will pass hemophilia A to her son is 50%, because he can inherit either her normal allele or the mutant one. The risk that a daughter will inherit the hemophilia allele and be a carrier like her mother is also 50%.

A daughter can inherit a sex-linked recessive disorder or trait if her father is affected and her mother is a carrier. She inherits one affected X chromosome from each parent. Unless she undergoes a biochemical test, a woman usually doesn't know she is a carrier of a sex-linked recessive trait unless she has an affected son. A genetic counselor can, however,

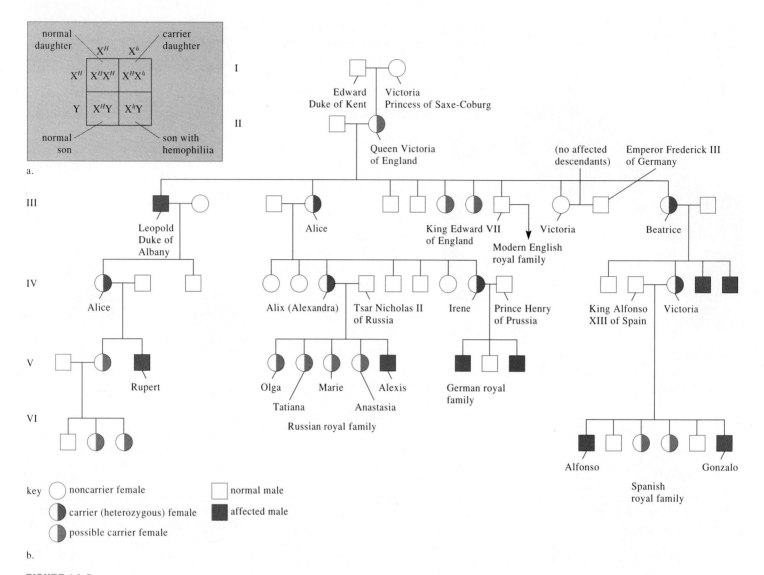

a. Sex-linked hemophilia A is usually transmitted from a heterozygous woman (designated $X^H X^h$, where X^h is the hemophilia-causing allele) to heterozygous daughters or hemizygous sons. b. The disorder has appeared in the royal families of England, Germany, Spain, and the former Soviet Union. The mutant allele apparently arose in Queen Victoria, who was either a carrier or produced an oocyte that mutated to carry the allele. She passed the allele to daughters Alice and Beatrice, who were carriers, and to Leopold, who had a case mild enough that he fathered children. In the fourth generation, Alexandra was a carrier and married Nicholas II, Tsar of Russia, passing the allele to that family. Irene married Prince Henry of Prussia, passing the allele to the German royal family, and Beatrice's descendants passed it to the Spanish royal family. This figure depicts only part of the enormous pedigree. Hemophilia is not present or carried in the modern royal family in England.

FIGURE 14.6

estimate the risk that a woman is a carrier by examining the family history and using probabilities derived from Mendel's laws.

Consider a woman whose brother has hemophilia A but whose parents are both healthy. The woman's chance of being a carrier is ½ (or 50%), based on the chance that she inherited the hemophilia allele on the X chromosome she received from her mother. (We know her mother is a carrier because the woman's

brother is affected.) The chance that the woman will conceive a son with hemophilia is ½ multiplied by ½. This is because the chance that she is a carrier is ½, and if she is a carrier, the chance that she will transmit the X chromosome bearing the hemophilia allele to a son is also ½. According to the product rule, the risk that her son will inherit hemophilia is ¼, or 25%.

Females usually don't exhibit sex-linked traits because they inherit two X chromosomes, so the wild type allele

masks the mutant. A female has the trait only if she inherits a recessive mutant allele from each parent. A sex-linked recessive trait appears more often in females if it isn't serious enough to prevent a man from surviving to father children. Color-blindness, for example, affects 8% of Caucasian men and 0.5% of Caucasian women, which is much more prevalent than other sex-linked disorders. Table 14.2 lists some sex-linked disorders, and the questions at the end of the chapter discuss others.

Table 14.2 Some Disease-related Genes on the Human X Chromosome*

Condition	Description
Eye	
Green color blindness	Abnormal green cone pigments in retina
Megalocornea	Enlarged cornea
Norrie disease	Abnormal growth of retina, eye degeneration
Ocular albinism	Eye lacks pigment
Red color blindness	Abnormal red cone pigments in retina
Retinitis pigmentosa	Constriction of visual field, night blindness, clumps of pigment in eye
Retinoschisis	Retina degenerates and splits
Inborn Errors of Metabolism	
Agammaglobulinemia	Lack of certain antibodies
Chronic granulomatous disease	Skin and lung infections, enlarged liver and spleen
Diabetes insipidus	Copious urination
Fabry disease	Abdominal pain, skin lesions, kidney failure
Gout	Inflamed, painful joints
G6PD deficiency and favism	Hemolytic anemia after eating fava beans
Hemophilia A	Absent clotting factor IX
Hemophilia B	Absent clotting factor VIII
Hypophosphatemia	Vitamin D-resistant rickets
Hunter syndrome	Deformed face, dwarfism, deafness, mental retardation, heart defects, enlarged liver and spleen
Ornithine transcarbamylase deficiency	Mental deterioration, ammonia accumulation in blood
Primary adrenal hypoplasia	Great disorganization of adrenal glands and resulting hormone deficiencies
Severe combined immune deficiency	Lack of immune system cells
Wiskott-Aldrich syndrome	Bloody diarrhea, infections, rash, too few platelets
Nerves and Muscles	
Charcot-Marie-Tooth disease	Loss of feeling in ends of arms and legs
Fragile X syndrome	X chromosome with extra constrictions; mental retardation, characteristic face, large testicles
Hydrocephalus	Excess fluid in brain
Lesch-Nyhan syndrome	Mental retardation, self-mutilation, urinary stones, spastic cerebral palsy
Menkes disease	Kinky hair, abnormal copper transport, brain atrophy
Muscular dystrophy, Becker and Duchenne forms	Progressive muscle weakness
Spinal and bulbar muscular atrophy	Muscle weakness and wasting
Other	
Amelogenesis imperfecta	Abnormal tooth enamel
Alport syndrome	Deafness, inflamed kidney tubules
Cleft palate	Opening in roof of mouth
Hypohidrotic ectodermal dysplasia	Absence of teeth, hair, and sweat glands
Ichthyosis	Rough, scaly skin on scalp, ears, neck, abdomen, and legs
Incontinentia pigmenti	Swirls of skin color, hair loss, seizures, abnormal teeth
Kallmann syndrome	Inability to smell, underdeveloped testes
Testicular feminization	Male embryo does not respond to male hormones, appears female

*Some of these conditions can also be transmitted through genes on the autosomes.

Sex-linked Dominant Inheritance

How is a dominant allele on the X chromosome expressed in each sex? A female who inherits a dominant allele has the associated trait or illness, but a male who inherits the allele is usually more severely affected. An example of a sex-linked dominant condition is incontinentia pigmenti. The name comes from the major symptom in females who have the disorder—swirls of pigment in the skin that resemble swirls of paint or marble cake. Males with the condition are so severely affected that they die in the uterus. Women with the disorder thus have a high rate of miscarriage, about 25%.

Y Linkage

In Y-linked inheritance, a trait is passed from father to son on the Y chromosome. This mode of inheritance is rare because the Y chromosome carries so few genes. Apparently, a "hairy ear" trait seen in some men in Sri Lanka, India, Israel, and aboriginal Australia is passed along the Y chromosome because it never appears in females. The trait's appearance in fathers and sons suggested Y linkage. If the trait was passed on the X chromosome, a more common pattern of inheritance would be for the trait to appear in maternal grandfathers and grandsons.

KEY CONCEPTS

The human female is the homogametic sex, with two X chromosomes; males, who are heterogametic, have one X and one Y chromosome. The SRY gene on the Y chromosome determines maleness in humans. It triggers a cascade of gene action that stimulates development of rudimentary male structures while suppressing development of female structures. Because a male is hemizygous, he expresses the genes on his X chromosome, whereas a female only expresses recessive alleles on the X chromosome if she is homozygous. Sex-linked recessive traits pass from carrier mothers to sons at a 50% rate. Sex-linked dominant conditions are expressed in females, but are more severe in males. Most sex-linked traits are inherited on the X chromosome. The Y chromosome contains very few genes. A few Y-linked traits are passed from father to son.

X Inactivation— Equalizing the Sexes

Female mammals have two alleles for every gene on the X chromosome, while males have only one. A mechanism called **X inactivation** helps balance this inequality in the number of sex-linked genes. Early in female development, one X chromosome in each cell is inactivated. This occurs at about the third week of development in humans. In most mammals, which X chromosome is turned off—the one inherited from the mother or the one from the father—is a matter of chance. As a result, a female mammal expresses the paternal X chromosome genes in some cells and the maternal genes in others. The exception is marsupials, the pouched mammals that include kangaroos. For unknown reasons, these species always shut off the X chromosome inherited from the father.

By studying rare human females who have only part of one X chromosome, researchers have identified a specific part of the X chromosome, the **X-inactivation center,** that shuts off the chromosome. X inactivation is believed to occur one gene at a time. A few genes, however, escape inactivation and remain active. The inactivation process seems to be under the control of a gene called XIST, which has so far been identified in humans, cats, dogs, mice, cows, and rabbits.

We do not know precisely how X inactivation occurs, but we do know some of its implications. X inactivation means that a female is a genetic mosaic of any heterozygous genes on the X chromosome because some cells express one allele, and other cells express the other allele. Heterozygosity can still offer a female protection from sex-linked disorders. If the female inherits one allele that specifies a vital enzyme, and another allele that specifies an inactive version, she will probably still be healthy, because some of her cells will manufacture the enzyme. A male who has the defective allele would not survive.

X inactivation may be obvious. The brown swirls of skin color in incontinentia pigmenti patients reflect cells in the deeper skin layers where the wild type allele is shut off.

Rarely, a female who is heterozygous for a sex-linked gene expresses the associated condition. This can happen in a female carrier of hemophilia. If the X chromosome carrying the normal allele for the clotting factor is, by chance, turned off in many of her immature blood platelet cells, her blood will take longer than normal to clot—a mild symptom of hemophilia. When a female carrier of a sex-linked trait expresses the phenotype, she is called a **manifesting heterozygote.**

Once an X chromosome is inactivated in one cell, all the cells that form when that cell divides have the same inactivated X chromosome. Because the inactivation occurs early in development, the adult female has patches of tissue that are phenotypically different in their expression of sex-linked genes. Now that each cell in her body has only one active X chromosome, she is numerically equivalent to the male in genetic makeup.

Biologists can easily observe X inactivation in female cells. The turned-off X chromosome absorbs a stain much more readily than the active X chromosome. In the nucleus of a female cell in interphase, the dark-staining, inactivated X chromosome is called a **Barr body.** Murray Barr, a Canadian researcher, first noticed the bodies in 1949 in the nerve cells of female cats. A normal male cell has no Barr bodies because the one X chromosome remains active (fig. 14.7).

In 1961, English geneticist Mary Lyon proposed that the Barr body is the inactivated X chromosome, and that X inactivation occurs early in development and is irreversible. She reasoned that for homozygous sex-linked genes, X inactivation would have no functional effect. No matter which X chromosome was turned off, the same allele would be expressed. But for heterozygous genes, X inactivation causes one allele or the other to be expressed in each cell.

For most sex-linked genes, X inactivation is not obvious because the cells of the body work together, and the overall phenotype reflects the mixture of cell types. (An exception is the manifesting

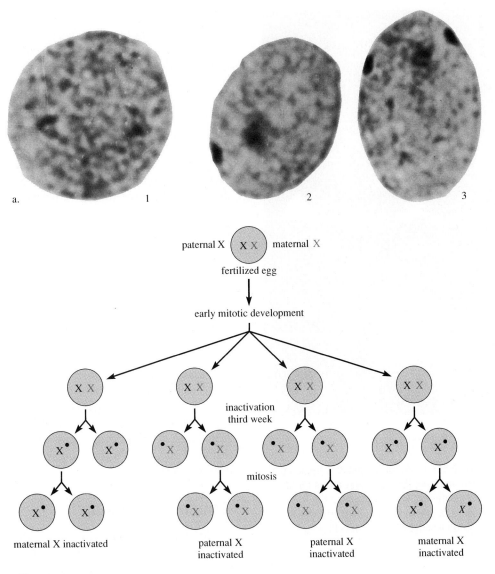

a.

paternal X (X X) maternal X

fertilized egg

early mitotic development

inactivation third week

mitosis

maternal X inactivated   paternal X inactivated   paternal X inactivated   maternal X inactivated

Key
X = paternally derived X chromosome
X = maternally derived X chromosome
• = inactivated X chromosome

b.

FIGURE 14.7

A Barr body marks an inactivated X chromosome. One X chromosome is inactivated in the cells of a female mammal. The turned-off X chromosome absorbs a stain called Giemsa faster than the active X chromosome, forming a dark spot called the Barr body, which is visible at the edge of the nucleus. *a.* A normal male cell has no Barr body (*1*), and a normal female cell has one Barr body (*2*). Individual (*3*) has two Barr bodies and three X chromosomes. She is, in appearance, behavior, and intellect, normal but has a lower IQ than her siblings, a slight deficit possibly due to her extra X chromosome. How many X chromosomes does she have? *b.* At about the third week of embryonic development in the human female, one X chromosome in each diploid cell is inactivated, and all daughter cells of these cells have the same X chromosome turned off. The inactivated X may come from the mother or father, resulting in a female who is mosaic at the cellular level for X chromosome gene expression.

heterozygote.) But one strikingly obvious example of X inactivation appears in the coat color pattern of the calico cat. Two sex-linked alleles, one specifying orange and one black, control the shapes and positions of the cat's characteristic orange and black patches. Cells containing inactivated orange alleles develop into black patches, and cells containing inactivated black alleles develop into orange patches. The earlier X inactivation occurs, the larger the patches are (fig. 14.8). The only way in which a male calico cat can arise is through a rare meiotic mishap that results in an animal with an XXY sex chromosome constitution. Only one in 10,000 calico cats is a male.

X inactivation has a valuable medical application in detecting carriers of some sex-linked disorders.

Geneticists can identify carriers of Duchenne muscular dystrophy (DMD) by demonstrating inactivation. A woman who has a 50% chance of being a carrier—for example, if her sister has had an affected son—has a muscle biopsy. The tissue is stained for the presence of dystrophin, the muscle protein missing in people with the condition. If she is a carrier, some muscle cells (cells with the mutant allele inactivated) manufacture dystrophin, while others (cells with the wild type allele inactivated) do not.

A carrier for DMD has truly hybrid muscle structure. She may have mildly weak muscles and produce enzymes that indicate that some of her muscle tissue is breaking down.

KEY CONCEPTS

In mammals, X inactivation evens the differences in the number of genes males and females carry on the X chromosome. Early in a female's development, one X chromosome in each cell is turned off. X inactivation can cause noticeable effects when heterozygous alleles are each expressed in different tissues.

a.

FIGURE 14.8

A striking exhibition of X inactivation occurs in the calico cat. Each orange patch is made up of cells descended from a cell in which the X chromosome carrying the coat color allele for black was inactivated; each black patch is made of cells descended from a cell in which the X chromosome carrying the orange allele was turned off. *a.* X inactivation happened early in development, resulting in large patches. *b.* X inactivation occurred later in development, producing smaller patches.

Gender Effects on Phenotype

A sex-linked trait generally is more prevalent in males than females because males carry only one X chromosome. Other situations can also cause different gene expressions in the sexes.

Sex-limited Traits

A **sex-limited trait** affects a body structure or function present in only one sex. The gene that controls such a trait may be sex-linked or autosomal, and it may be difficult to distinguish between the two.

Sex-limited inheritance is important in animal breeding. In cattle, for example, traits such as milk yield and horn development affect only one sex, but either parent may transmit the genes that control these traits. In humans, beard growth and breast size are sex-limited traits. A woman cannot grow a beard because she does not manufacture the hormones needed to grow facial hair. She can, however, pass the genes specifying heavy beard growth to her sons.

Sex-limited inheritance is important to consider in diagnosing breast cancer. One young woman had a cousin and aunt with breast cancer, but did not consider herself at high risk because no one in her immediate family was affected. Because the cousin's and aunt's cancer was found to be caused by an autosomal dominant gene, the young woman had a mammogram. She indeed had a very small cancerous tumor. The causative gene had been passed by her father, who did not express the trait.

A sex-limited trait affecting only males may present the same family pattern as a Y-linked trait. In this case, only identifying and mapping the gene can distinguish between these alternatives.

Sex-influenced Traits

In a **sex-influenced trait,** an allele is dominant in one sex but recessive in the other. Hormonal differences can cause this difference in expression. For example, a gene for hair growth pattern has two alleles, one that produces hair all over the head and another that causes pattern baldness (fig. 14.9). The baldness allele is dominant in males but recessive in females, which is why more men than women are bald. A heterozygous male is bald, but a heterozygous female is not. What is the genotype of a bald woman?

Genomic Imprinting

In **genomic imprinting,** the expression of a disorder differs depending upon which parent transmits the disease-causing gene or chromosome. The phenotype may differ in degree of severity, age of onset, or even in the nature of the symptoms.

X inactivation in mammals is a broad example of genomic imprinting—many genes on the X chromosome are inactivated in a female, but not in a male. Genomic imprinting also appears in the Angelman and Prader-Willi syndromes, discussed in the previous chapter. In these two disorders affecting the same region of chromosome 15, different sets of symptoms arise depending upon the sex of the parent transmitting the gene. Genomic imprinting also plays a role in juvenile diabetes, some forms of asthma and hay fever, Huntington disease, and certain childhood cancers.

a.

b.

c.

d.

FIGURE 14.9

Pattern baldness is a sex-influenced trait and was a genetic trademark of the illustrious Adams family. *a.* John Adams (1735–1826) was the second president of the United States. He was the father of John Quincy Adams (1767–1848) (*b*), the sixth president. John Quincy was the father of Charles Francis Adams (1807–86) (*c*), a diplomat and the father of Henry Adams (1838–1918) (*d*), who was a historian.

Genetic Heterogeneity

The four modes of Mendelian inheritance reflect the location of genes on chromosomes. They are:

   autosomal recessive

   autosomal dominant

   sex-linked recessive

   sex-linked dominant

The same phenotype may result from genes inherited in different ways. A trait or disorder exhibits **genetic heterogeneity** if a different single gene causes the trait in different individuals. For example, we know of nearly 200 forms of hereditary deafness. Of these, 132 forms are transmitted as autosomal recessive traits, 64 as autosomal dominant, and the remaining four as sex-linked recessive traits. Many of the sex-linked conditions listed in table 14.2 are also inherited as autosomal disorders, including cleft palate, albinism, and diabetes insipidus.

   Genetic heterogeneity can occur when genes' products act at different points in the same biochemical pathway, ultimately affecting the same biological function. For example, 11 different genes, each specifying a different clotting factor may cause hemophilia when absent or abnormal. Similarly, an abnormal flower color may result from mutations in any of several genes whose products participate in the biosynthetic pathway for the wild type pigment. Mutant petunia plants with whitish leaves may have a mutation in any of 16 genes controlling chlorophyll synthesis.

KEY CONCEPTS

The same genes may be expressed differently in each sex, and, conversely, different genes may cause the same trait or disorder in different individuals. A sex-limited trait affects only one gender. A sex-influenced allele is dominant in one sex but recessive in the other. In genomic imprinting, the phenotype differs depending on whether a gene is inherited from the mother or the father. Because of genetic heterogeneity, different mutant genes can result in the same phenotype.

   A full explanation of Mendel's experiments awaited the discovery of chromosomes. Today, chromosomes provide valuable information to such diverse fields as human health care, agriculture, environmental pollution, and evolution. The remainder of the chapter considers cytogenetics, the area of genetics that links chromosomal aberrations to phenotypes.

Awaiting the Amnio Result

Impending parenthood makes many people acutely aware of chromosomes, because chromosomes are windows into the health of a fetus. In fact, some people do not form an emotional attachment to the unborn child until they pass a medical milestone: ascertaining that the fetus has the normal 46 chromosomes. Once they learn that the chromosome count is normal, many parents-to-be focus on one particular chromosome—the Y, which carries the sex-determining SRY gene. Fetal chromosome tests are performed for medical reasons, such as when a pregnant women is over age 35, or when a parent is otherwise at risk for transmitting an abnormal chromosome.

   A fetal chromosome check doesn't always bring prospective parents good news. Genetic health is a matter of balance—too much or too little genetic material, particularly of the autosomes, usually impairs health. The majority of chromosomal abnormalities are so harmful that prenatal development ceases after only a few weeks. Some do not kill the fetus, but do cause specific syndromes that affect the child for life. At the other end of the clinical spectrum, some chromosomal variants rearrange genetic material without causing specific symptoms.

Chromosome Structure

A chromosome is more than a very long molecule of DNA. Several biochemicals emanate from the DNA, including RNA and various proteins that serve as DNA scaffolds, help replicate DNA, and help transcribe RNA. Chromosomes also include proteins specific to certain cell types.

The DNA in chromosomes is highly coiled (fig. 14.10). During metaphase, when DNA coils tightly, key physical landmarks distinguish the chromosome pairs. Chromosomes are categorized by size, although this originally resulted in generalized groupings because many chromosomes are of similar size. Chromosomes are numbered from largest to smallest.

Each chromosome has a characteristically located constriction, the centromere. A chromosome is **metacentric** if the centromere divides it into two arms of approximately equal length, **acrocentric** if the centromere establishes one long arm and one short arm, and **telocentric** if it pinches off only a small amount of material (fig. 14.11). The centromere consists of DNA, and may help orient chromosomes during cell division. The long arm of a chromosome is designated "q," and the short arm "p." Some chromosomes are distinguished further by bloblike ends called **satellites** that extend from a stalklike bridge from the rest of the chromosome.

Researchers use stains to distinguish chromosomes. Dark-staining genetic material, called **heterochromatin,** is more highly coiled than the lighter-staining **euchromatin** (fig. 14.12). Heterochromatin's tighter conformation may entrap dye particles so that it stains more intensely. Stained chromosomes appear striped, with long stretches of heterochromatin near the centromere

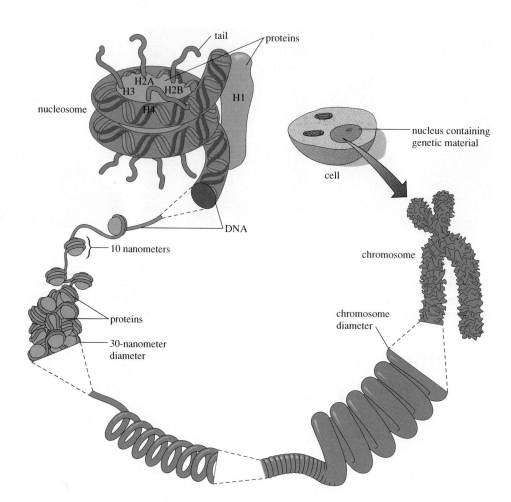

FIGURE 14.10

A chromosome consists of a very tightly wound molecule of DNA, plus associated proteins.

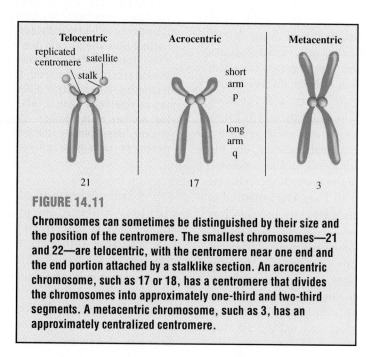

FIGURE 14.11

Chromosomes can sometimes be distinguished by their size and the position of the centromere. The smallest chromosomes—21 and 22—are telocentric, with the centromere near one end and the end portion attached by a stalklike section. An acrocentric chromosome, such as 17 or 18, has a centromere that divides the chromosomes into approximately one-third and two-third segments. A metacentric chromosome, such as 3, has an approximately centralized centromere.

and at the **telomeres,** or tips. Different chromosomes stain in different patterns.

Interestingly, DNA sequences are identical in the telomeres of many species of fish, amphibians, reptiles, birds, and mammals, indicating that these structures are very important. Heterochromatin, in the telomeres and elsewhere, may be crucial for maintaining the structural integrity of chromosomes. Euchromatin, on the other hand, encodes many proteins.

KEY CONCEPTS

Chromosomes consist of DNA, RNA, and proteins. Chromosomes are distinguishable by size, centromere location, the presence of satellites, and heterochromatin and euchromatin staining. Heterochromatin maintains structural integrity, and euchromatin mostly encodes proteins.

Abnormal Chromosome Numbers

Each species has a characteristic number of chromosomes (table 14.3). An altered chromosome number drastically affects the phenotype because a chromosome consists of so many genes. Chromosomes can be abnormal in a variety of ways. Charts called **karyotypes** are used to display chromosomes (fig. 14.13)

Extra Chromosome Sets—Polyploidy

Sometimes an error in meiosis produces a sperm or egg cell that has an extra complete set of chromosomes, a condition called **polyploidy** (many sets). For example, if a sperm with the normal one copy of each chromosome fertilizes an egg with two copies of each chromosome, the resulting zygote will have three copies of each chromosome, a type of polyploidy called triploidy (fig. 14.14). Most human polyploids die as embryos or fetuses, but occasionally one survives

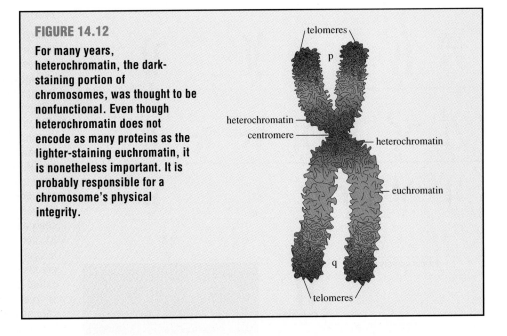

FIGURE 14.12

For many years, heterochromatin, the dark-staining portion of chromosomes, was thought to be nonfunctional. Even though heterochromatin does not encode as many proteins as the lighter-staining euchromatin, it is nonetheless important. It is probably responsible for a chromosome's physical integrity.

Table 14.3	Number of Chromosomes in Various Species
Species	**Diploid Number of Chromosomes**
Human	46
Dog	78
Horse	64
Cat	38
Mouse	40
Hamster	44
Rat	42
Guinea pig	64
Alligator	32
Housefly	12
Grasshopper	24
Mosquito	6
Barley	14
Rice	24
Onion	16
Tobacco	48
Corn	20
Potato	48
Kidney bean	22
Pine	24

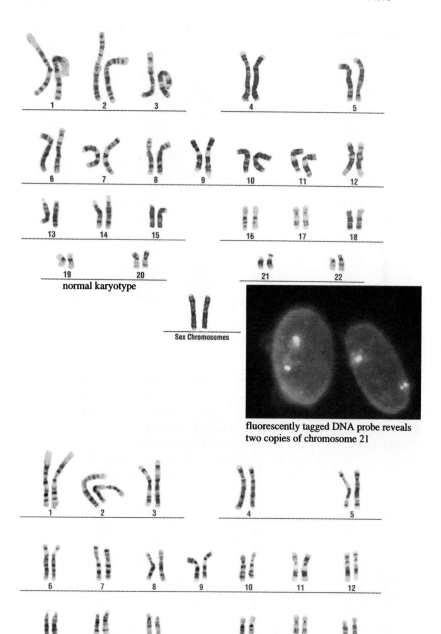

fluorescently tagged DNA probe reveals
two copies of chromosome 21

abnormal karyotype
revealing trisomy 21

fluorescently tagged DNA probe reveals
three copies of chromosome 21

a.

FIGURE 14.13

A karyotype (*a*) displays chromosomes arranged in pairs in descending size order. Computers scan cells on microscope slides, focusing on one cell in which the chromosomes do not overlap, and manipulating the images to construct the karyotype. Until recently, cytogeneticists constructed karyotypes by taking photographs of cells and cutting and pasting the chromosome images in size-order pairs! Karyotypes are refined by locating DNA sequences known to be on specific chromosomes, tagging them with fluorescent chemicals, and using them as probes. A probe placed on a chromosome preparation binds to its normal site on the chromosome, emitting a fluorescent signal. The insets show a normal karyotype with two probes to chromosome 21, and a karyotype from a fetus who has an extra flash—indicating Down syndrome. *b*. Genes are added, as they are discovered or localized, to depictions of chromosomes called ideograms. On this ideogram, *p* designates the short arm, and *q* the long arm of this metacentric chromosome. The numbers indicate subdivisions of the chromosome, and the brackets indicate localized genes causing the conditions listed on the right. It is fascinating to compare ideograms published in *Science* magazine each October, as they grow more and more crowded with new discoveries!

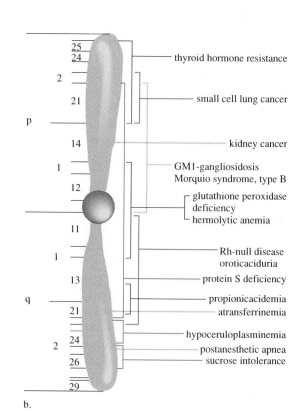

b.

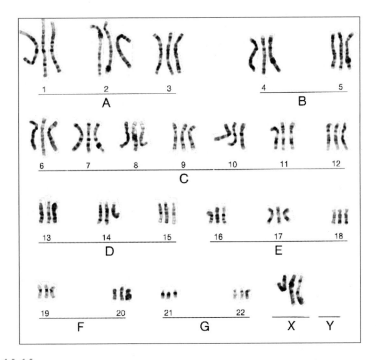

FIGURE 14.14

Individuals with three copies of each chromosome (triploids) account for 17% of all spontaneous abortions and 3% of stillbirths and newborn deaths.

for a few days after birth, with defects in nearly all organs. However, about 30% of flowering plant species tolerate polyploidy well. Many agriculturally important plant variants are derived from polyploids. For example, the type of wheat used to make pasta is a tetraploid—it has four sets of seven chromosomes. The wheat species used to make bread is a hexaploid—it has six sets of seven chromosomes.

Extra and Missing Chromosomes—Aneuploidy

A **euploid** ("true set") cell has a normal chromosome number. An **aneuploid** ("not true set"), by contrast, has a missing or extra chromosome. Aneuploidy symptoms depend upon which chromosome is affected. Autosomal aneuploidy often results in mental retardation, because so many genes contribute to brain function. Sex chromosome aneuploidy is less severe.

Extra genetic material is apparently less dangerous than missing material. This is why most children born with the wrong number of chromosomes

have an extra one (called a **trisomy**) rather than a missing one (a **monosomy**). Trisomies and monosomies are named according to the affected chromosome, and the syndromes they cause were originally named for the investigator who first described them. Down syndrome, for example, is also known as trisomy 21.

Aneuploidy occurs due to a meiotic mishap called **nondisjunction** (fig. 14.15). In normal meiosis, pairs of homologous chromosomes separate so that each of the resulting sperm or egg cells contains only one member of each pair. In nondisjunction, a chromosome pair fails to separate, either at the first or second meiotic division. The result is a sperm or egg with either two copies of a particular chromosome or none at all, rather than the normal one copy. In humans, when such a gamete fuses with its opposite at fertilization, the resulting zygote has either 45 or 47 chromosomes instead of 46, the normal number for a human. Because there are 23 types of human chromosomes, there are 49 ways in which a chromosome can be aneuploid for one chromosome—any of

the 22 pairs of autosomes may have an extra or missing chromosome, plus the sex chromosomes may occur in five abnormal, aneuploid combinations (YO, XO, XXX, XXY, or XYY). However, only 9 of these 49 possibilities appear in newborns; the others halt development before birth. In fact, extra or missing chromosomes cause about 50% of all spontaneous abortions. Similarly, aneuploidy in plants may produce a seed (an embryo and its food supply) that does not germinate.

Aneuploidy and polyploidy can also arise during mitosis, producing groups of somatic cells with a chromosomal aberration. If only a few cells have an abnormal number of chromosomes, health may not be affected. For example, the human liver often contains patches of polyploid cells, but they have no obvious effect on the organ's activity. However, if such a mitotic abnormality occurs in a very early embryo, many cells descend from the original defective one, causing more serious problems. Some people who have mild Down syndrome, for example, are really chromosomal mosaics—that is, some of their cells carry the extra chromosome 21, and some are normal. Let's take a closer look at Down syndrome and some of the other more common human aneuploids.

Down Syndrome—A Closer Look

The most common autosomal aneuploid is **trisomy 21,** or Down syndrome. Sir John Langdon Haydon Down first described the syndrome in the 1880s. As the medical superintendent of a facility for the profoundly mentally retarded, Down noted that about 10% of his patients had characteristically slanted eyes and flat faces, superficially resembling people of the Mongolian race. This prompted him to coin the inaccurate term "mongolism" to describe the disorder. Characteristic facial features are associated with many inherited disorders. In actuality, males and females of all races can have Down syndrome.

Along with the distinctive facial features Down noted, a person with Down syndrome is generally short and has straight, sparse hair, a protruding

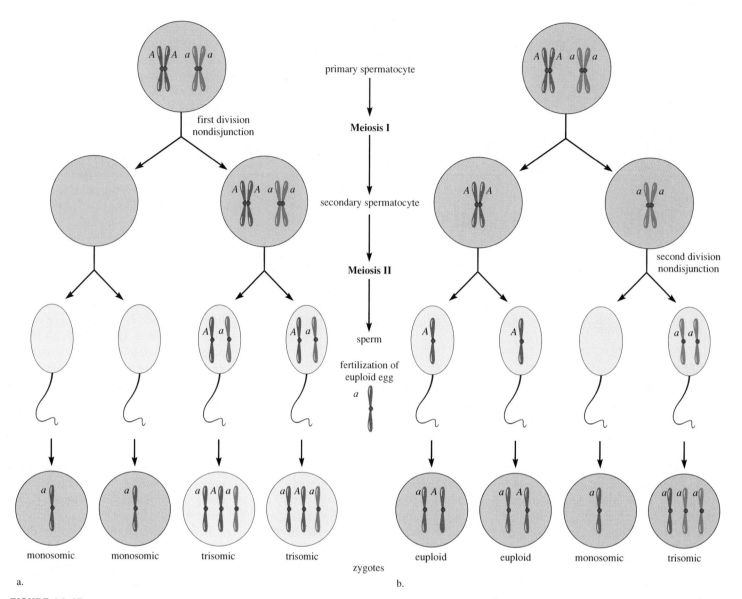

primary spermatocyte

Meiosis I

first division
nondisjunction

secondary spermatocyte

Meiosis II

second division
nondisjunction

sperm

fertilization of
euploid egg

monosomic   monosomic   trisomic   trisomic

zygotes

euploid   euploid   monosomic   trisomic

a.   b.

FIGURE 14.15

Extra and missing chromosomes—aneuploidy. Unequal division of chromosome pairs into sperm and egg cells can occur at either the first or the second meiotic division. *a.* A single pair of chromosomes unevenly partitioned into the two cells arising from the first division of meiosis in a male. The result: two sperm cells that have two copies of the chromosome, and two sperm cells that have no copies of that chromosome. When a sperm cell with two copies of the chromosome fertilizes a normal egg cell, the zygote is trisomic for that chromosome; when a sperm cell lacking the chromosome fertilizes a normal egg cell, the zygote is monosomic for that chromosome. Symptoms depend upon which chromosome is involved. *b.* This nondisjunction occurs at the second meiotic division. Because the two products of the first division are unaffected, two of the mature sperm are normal and two aneuploid. Egg cells can undergo nondisjunction as well, leading to zygotes with extra or missing chromosomes when they are fertilized by normal sperm cells.

tongue, and thick lips. The individual has an abnormal pattern of hand creases, loose joints, and poor reflex and muscle tone, creating a "floppy" appearance. Children with Down syndrome reach developmental milestones (such as sitting, standing, and walking) slowly, and toilet training may take several years. Their intelligence varies greatly; some have profound mental impairment, while others can follow simple directions, read, and use a computer.

Although tests can determine that a fetus has trisomy 21, they cannot predict the severity of the syndrome before the child is born (fig. 14.16). However, nearly 50% of people with Down syndrome die before their first birthdays, often of heart or kidney defects, or a suppressed immune system that can make a bout of influenza deadly. Digestive system blockages are common and must be corrected surgically shortly after birth. A child with Down syndrome is 15 times

FIGURE 14.16

Medical genetics can present difficult choices. Down syndrome is usually caused by an extra chromosome 21. Unfortunately, the test that diagnoses Down syndrome in a fetus cannot detect how severely affected the child will be. The decision of whether to terminate such a pregnancy or raise a child with Down syndrome is very difficult.

more likely to develop leukemia (a white-blood-cell cancer) than a healthy child. Individuals with Down syndrome who live past age 40 develop Alzheimer disease, which greatly impairs memory and reasoning.

The likelihood of giving birth to a child with Down syndrome increases dramatically as a woman ages. For women under 30, the chances of conceiving a child with the syndrome are 1 in 3,000. For a woman of 48, the probability jumps to 1 in 9. This increase may be related to the fact that an egg cell completes meiosis after fertilization. The older a woman is, the longer her oocytes have been arrested on the brink of completing meiosis. During this time, the oocytes may have been exposed to chromosome-damaging chemicals or radiation. Other trisomies are more likely to occur among the offspring of older women too. As table 14.4 indicates, after Down syndrome, the two most frequently seen autosomal aneuploids are trisomy 13 (Patau syndrome) and trisomy 18 (Edward syndrome).

Sex Chromosome Aneuploids

People with the same chromosomal abnormality often share strikingly similar physical characteristics, as in Down syndrome. At a 1938 medical conference, endocrinologist Henry Turner reported on seven young women, aged 15 to 23, who were short and sexually undeveloped, had folds of skin on the backs of their necks, and had malformed elbows. Other physicians soon began spotting patients with this disorder, Turner syndrome, in their practices. Everyone assumed the odd collection of traits reflected a hormonal insufficiency. They were right, but there was more to the story—a chromosomal imbalance lay behind the hormone deficit.

In 1954, P. E. Polani, a British medical researcher, discovered that the cells of Turner syndrome patients lack Barr bodies, the dark spots that indicate a second X chromosome in each cell. Might the symptoms, particularly the failure to mature sexually, be caused by the absence of a sex chromosome? By

1959, karyotyping confirmed that the cells of Turner syndrome patients lack an X chromosome. Today geneticists hypothesize that the symptoms may result from certain genes normally exempt from X inactivation and also present on the Y. Healthy individuals would have two copies of these genes; people with Turner syndrome have only one.

Like the autosomal aneuploids, Turner syndrome occurs more frequently among spontaneously aborted fetuses than among liveborns—99% of affected fetuses die before birth. The syndrome affects about 1 in 2,500 to 10,000 female births. However, the only abnormal expression in individuals born with the condition may be short stature and a flap of skin at the back of the neck. Many Turner's patients do not know they have a chromosomal abnormality until they lag behind their age-mates in sexual development. Turner syndrome patients are usually of normal intelligence and, if treated with hormone supplements, lead fairly normal lives, but they remain infertile. Interestingly, Turner syndrome is the only aneuploid condition that is no more prevalent among offspring of older mothers.

About 1 in every 1,000 to 2,000 females is born with an extra X chromosome in each cell, a condition called **triplo-X.** The only symptoms seem to be tallness, menstrual irregularities, and a normal-range IQ that is nevertheless lower than that of other family members. However, a woman with triplo-X may produce some oocytes bearing two X chromosomes, which increases her risk of giving birth to triplo-X daughters or XXY sons. Can you see why?

Males with an extra X chromosome (XXY) have Klinefelter syndrome. Individuals with Klinefelter are sexually underdeveloped, with rudimentary testes and prostate glands and no pubic or facial hair. They also have very long arms and legs, large hands and feet, and may develop breast tissue. Individuals with Klinefelter syndrome may be slow to learn, but they are usually not mentally retarded unless they have more than two

Table 14.4 Aneuploids Surviving Birth

Chromosome Constitution	Syndrome	Incidence	Phenotype

Trisomy 21	Down	1/770	Mental retardation
			Abnormal pattern of palm creases
			Flat face
			Sparse, straight hair
			Short stature
			High risk of:
			  cardiac anomalies
			  leukemia
			  cataracts
			  digestive blockages
			  Alzheimer disease

Trisomy 13	Patau	1/15,000	Mental and physical retardation Skull and facial abnormalities Defects in all organ systems Cleft lip Large, triangular nose Extra digits
Trisomy 18	Edward	1/4000–8000	Mental and physical retardation Skull and facial abnormalities Defects in all organ systems Excess muscle tone
Trisomy 22		Very few reported	Multiple defects Large ears and nose Narrow face Small jaw Excess muscle tone
Monosomy 21		Very few reported	Mental and physical retardation Skull, jaw, and facial abnormalities
XO	Turner	1/2500–10,000 females	No sexual maturity Short stature Webbed neck Wide-spaced nipples Narrow aorta Pigmented moles Malformed elbows
XXY	Klinefelter	1/500–2000 males	No secondary sexual characteristics Breast swelling Long arms and legs Large hands and feet No sperm
XXX	Triplo-X	1/1000–2000	Tall and thin Menstrual irregularity Slightly lower IQ
XYY	Jacob	1/1000 males	Tall Acne Possible speech and reading difficulties

X chromosomes, which rarely happens. Klinefelter syndrome occurs in 1 out of every 500 to 2,000 male births.

One male in 1,000 has an extra Y chromosome. In 1965, researcher Patricia Jacobs published results of a survey among inmates at Carstairs, a high-security mental facility in Scotland. Of 197 men, 12 had chromosome abnormalities, 7 of them an extra Y! Jacobs hypothesized that the extra Y chromosome might cause these men to be violent and aggressive and perhaps accounted for their incarceration. When other facilities reported similar numbers of XYY men, *Newsweek* magazine ran a cover story on "congenital criminals." In 1968, attorneys in France and Australia defended violent clients on the basis of **Jacobs syndrome,** the inherited extra Y chromosome. Meanwhile, the National Institute of Mental Health in Bethesda, Maryland, held a conference on the condition, lending legitimacy to the hypothesis that an extra Y predisposes the affected men to violent behavior.

In the early 1970s, hospital nurseries in England, Canada, Denmark, and Boston began screening for Jacobs syndrome. Social workers and psychologists offered "anticipatory guidance" to the parents of baby boys with XYY karyotypes, instructing them in how to deal with their toddling criminals. However, a correlation between two events does not necessarily signify cause and effect. By 1974, geneticists and others halted the program, pointing out that singling out these boys on the basis of a few statistical studies could lead to a self-fulfilling prophecy.

Today, we know that 96% of XYY males are apparently normal. The only symptoms attributable to the extra chromosome may be great height, acne, and perhaps speech and reading problems. An explanation for the high prevalence of XYY among prison populations may be more psychological than biological. Teachers, employers, parents, and others may expect more of these physically large boys and men, and a small percentage of them may cope by becoming aggressive.

Medical researchers have never reported a sex chromosome constitution of one Y and no X. When a zygote lacks any X chromosome, so much genetic material is missing that it probably cannot sustain more than a few cell divisions.

KEY CONCEPTS

Abnormal numbers of chromosomes cause a variety of problems. Polyploids have whole extra sets of chromosomes, while aneuploids have a single missing (monosomic) or extra (trisomic) chromosome. Meiotic nondisjunction leads to aneuploidy. Mitotic nondisjunction produces chromosomal mosaics. Trisomics are more likely to survive than monosomics, and sex chromosome aneuploidy is less severe than autosomal aneuploidy. Down syndrome (trisomy 21) is the most common human aneuploid. Sex chromosome aneuploids in humans include Turner syndrome (XO), Klinefelter syndrome (XXY), triplo-X females, and XYY males (Jacobs syndrome).

Chromosome Rearrangements

Sometimes only part of a chromosome is missing (a **deletion**) or extra (a **duplication**). Geneticists can visualize deletions and duplications as missing or extra bands on stained chromosomes.

Translocations

In a **translocation,** different (nonhomologous) chromosomes exchange parts or combine. Certain viruses (such as the virus that causes the mumps), drugs (anticancer drugs), and radiation (medical X rays or ultraviolet radiation) can cause translocations.

In a **nonreciprocal translocation,** a piece of one chromosome breaks off and attaches to another chromosome. In 97% of patients with chronic myeloid leukemia, for example, the tip of one chromosome 22 breaks off and attaches to chromosome 9. In a **reciprocal translocation,** two different chromosomes exchange parts. If the chromosome exchange does not break any genes, a

person who has both translocated chromosomes is healthy, although he or she is a **translocation carrier.** This person has the normal amount of genetic material, but it is rearranged.

A translocation carrier produces some "unbalanced gametes"—sperm or oocytes with deletions or duplications of genes in the translocated chromosomes. This occurs when the sperm or egg receives one reciprocally translocated chromosome but not the other, resulting in a genetic imbalance. One common reciprocal translocation occurs between chromosomes 14 and 21; figure 14.17 shows the pedigree for a family with the associated chromosomal abnormalities. When a family has many specific birth defects and spontaneous abortions, a physician or genetic counselor would suspect a translocation carrier. A translocation of chromosome 21 causes about 6% of Down syndrome cases.

Inversions

In an **inversion,** part of the chromosome flips around, rearranging the gene sequence (fig. 14.18).

An adult who is heterozygous for an inversion in one chromosome can still be healthy but is likely to have reproductive problems. This is because the inverted chromosome and its non-inverted homolog twist around each other during meiosis in a way that generates chromosomes with missing or extra genes, if crossing over occurs. The missing or extra genes may cause repeated spontaneous abortions or birth defects in a family.

Other Chromosomal Abnormalities

Another meiotic error that disrupts the balance of genetic material is the formation of an **isochromosome,** a chromosome with identical arms. This occurs when the centromere splits in the wrong plane (fig. 14.19).

Ring-shaped chromosomes form in 1 out of 25,000 human conceptions and may affect any chromosome. Ring

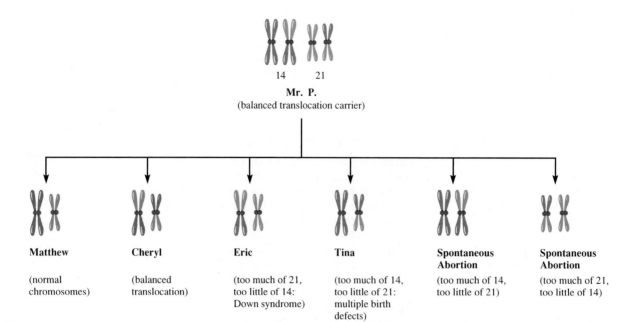

Mr. P.
(balanced translocation carrier)

Matthew	Cheryl	Eric	Tina	Spontaneous Abortion	Spontaneous Abortion
(normal chromosomes)	(balanced translocation)	(too much of 21, too little of 14: Down syndrome)	(too much of 14, too little of 21: multiple birth defects)	(too much of 14, too little of 21)	(too much of 21, too little of 14)

FIGURE 14.17

Families with many birth defects may have a translocation. After two spontaneous abortions, Mr. and Mrs. P. were overjoyed to have two healthy children, Matthew and Cheryl. Their third child, Eric, had Down syndrome. When the doctor told them it was unlikely they would have another child with Down syndrome, they decided to have a fourth child. Tina was born with multiple medical problems. Chromosome studies revealed that Mr. P. was a translocation carrier. His 14th largest and 21st largest chromosomes had exchanged parts. This had no effect on him, because his cells still had all the genes normally found in a human cell. But whenever one mixed-up chromosome was packaged into a sperm cell without the other, and that sperm fertilized a normal egg, an abnormal child was conceived. Tina inherited extra material from chromosome 14 and too little from chromosome 21. Eric inherited the opposite condition, resulting in Down syndrome. Cheryl too had abnormal chromosomes, but hers were balanced, so she was a translocation carrier like her father. Matthew had normal chromosomes. Mrs. P.'s two spontaneous abortions were also probably caused by her husband's translocation—the embryos lacked too much genetic material to develop.

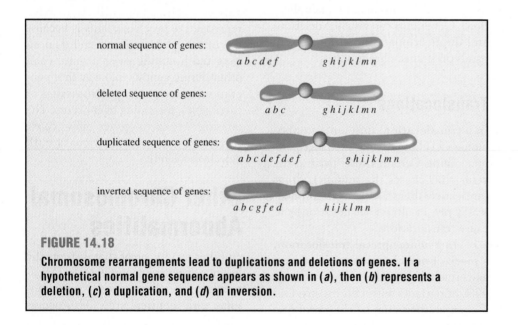

FIGURE 14.18

Chromosome rearrangements lead to duplications and deletions of genes. If a hypothetical normal gene sequence appears as shown in (*a*), then (*b*) represents a deletion, (*c*) a duplication, and (*d*) an inversion.

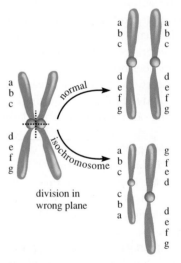

FIGURE 14.19

Isochromosomes have identical arms. They form when chromatids divide along an axis offset by 90 degrees.

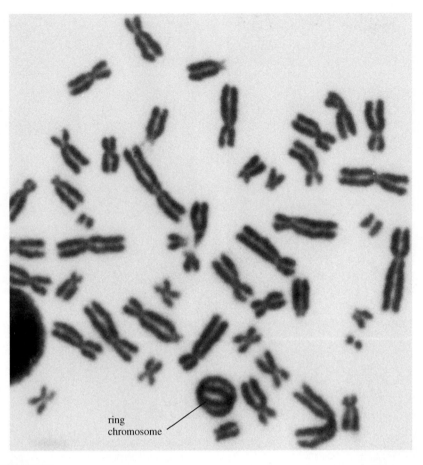

ring
chromosome

FIGURE 14.20

Ring chromosomes. Exposure to toxic chemicals and radiation can break the tips off of chromosomes. The tipless chromosomes sometimes attract one another and fuse, forming rings. The appearance of ring chromosomes or chromosome fragments in a karyotype is a sign of possible exposure to a toxin. The ring chromosome in this photograph arose from X ray exposure. This chromosome is derived from two chromosomes that fused, as evidenced by its two centromeres.

KEY CONCEPTS

Abnormal chromosome rearrangements cause gene deletions and duplications. In a nonreciprocal translocation, a piece of one chromosome breaks off and attaches to another chromosome. In a reciprocal translocation, two chromosomes exchange parts. If a translocation leads to a deletion or duplication, the affected individual will exhibit symptoms. Gene duplications and deletions also occur in isochromosomes and ring chromosomes, and when crossovers occur within chromosomal inversions. In fragile X syndrome, an expanding gene destabilizes the X chromosome, causing mental impairment and characteristic facial features.

Why Study Chromosomes?

Karyotypes are useful at several levels. For example, a chromosome chart can be used to diagnose a chromosomal "accident," such as Down syndrome, which occurs in a child of chromosomally normal parents. This helps the parents assess the risk that they will give birth to other affected offspring. If the karyotype reveals trisomy 21, the syndrome is unlikely to recur in another child. But if the Down syndrome is caused by a translocation, it may repeat. Sometimes recognizing an abnormal chromosome can save lives. In one New England family, several adults died from a very rare form of kidney cancer. Because the cancer was so unusual, medical investigators studied the family's health closely. Karyotypes showed a translocation between chromosomes 3 and 8 in every relative who had cancer. When two healthy young family members proved to have the translocation, physicians examined their kidneys. They both exhibited very early stages of the cancer! Subsequent surgeries prolonged their lives.

Karyotypes can be compared between species to clarify evolutionary relationships. The more recently two

chromosomes may arise when telomeres are lost, leaving "sticky" ends that tend to fuse to one another (fig. 14.20). Radiation exposure can cause ring chromosomes to form.

Ring chromosomes can produce symptoms when they duplicate genetic material. For example, a small ring chromosome consisting of DNA from chromosome 22 causes cat eye syndrome. Affected children are mentally retarded and have vertical pupils, heart and urinary tract anomalies, and skin-covered anuses. These individuals have 47 chromosomes—the normal two chromosome 22s, and the ring.

A chromosomal aberration in a class by itself lies behind fragile X syndrome, a form of mental retardation affecting 1 in 1,000 individuals, mostly males. When X chromosomes from affected people are cultured in media low in folic acid, an X chromosome tip dangles, making it prone to breakage (fig. 14.21a). A person with fragile X syndrome characteristically has protruding ears and a long jaw and is mentally impaired (fig. 14.21b). This syndrome is one of a handful of disorders caused by an expanding gene, discussed in chapter 16.

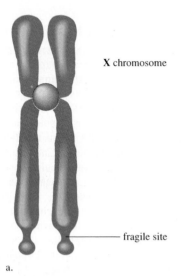

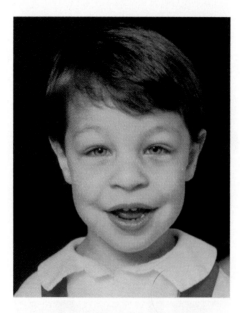

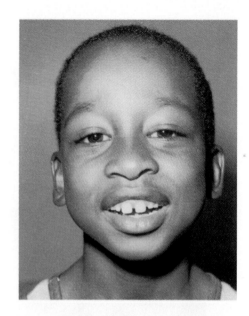

FIGURE 14.21

a. A constriction appears in the X chromosome of individuals who have symptoms of fragile X syndrome when their cells are grown in a medium that lacks thymidine and folic acid. *b.* The characteristic facial structure and features of fragile X syndrome sufferers become more pronounced with age.

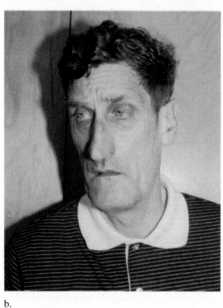

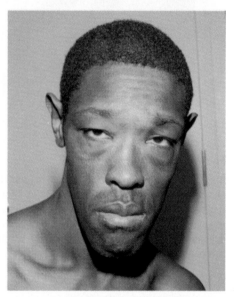

b.

species diverged from a common ancestor, the more closely related they are, and the more alike their karyotypes should be. Indeed, human chromosome bands are strikingly similar to those of our closest relative, the chimpanzee.

In the past, people usually learned their chromosomal makeup only when something was wrong—when they had a history of reproductive problems, developed cancer, had a prenatal test that revealed a chromosomal anomaly in a fetus, or had symptoms that resembled those associated with a known chromosomal disorder. Because our ability to

peer into the biochemical makeup of our chromosomes is rapidly improving, and because researchers are daily identifying the genes that cause common disorders, it is very likely that chromosome testing will soon be an integral and routine part of medical care—beginning before birth (Biology in Action 14.2).

KEY CONCEPTS

Karyotypes provide information on genetic disease, exposure to environmental toxins, and evolutionary relationships between species. Soon, chromosome testing may become an integral part of health care from the prenatal period on.

Biology in ACTION

Prenatal Peeks

Prenatal diagnosis techniques give us unprecedented glimpses of the human embryo and fetus. In many instances, doctors can now predict whether a child will be born with a genetic (inherited) or congenital (noninherited) birth defect long before the baby is born. Although prenatal diagnosis cannot reveal all potential medical problems, it dispels some of the mystery surrounding prenatal health and development.

Ultrasound

Mrs. F. is 38 years old and is 15 weeks pregnant with her first child. She mentions to her doctor that her sister just gave birth to a child with Down syndrome. Should she be concerned about her child? Because of her age and family background, the doctor suggests a few tests, beginning with an ultrasound exam. In this exam, sound waves bounced off the fetus are converted to produce an image on a monitor (fig. 1).

The next day, a nervous Mr. and Mrs. F. arrive for the ultrasound. After Mrs. F. drinks several glasses of water, a nurse spreads a cold jelly on her abdomen, then sweeps a hand-held device called a transducer across it. The procedure is painless, and the parents-to-be are so excited to see fingers and toes that they almost forget the doctor is looking for organ abnormalities that sometimes signify Down syndrome. All looks well. The beating heart is clear, and the lengths of the tiny arms and legs are about right for a 15-week fetus. When the fetus's body parts are too large or too small for its age of gestation, it may indicate a problem.

Amniocentesis

The patient's relief following the ultrasound exam is short-lived. The doctor advises her to undergo amniocentesis a week later. In this procedure, the doctor removes fetal cells and fluids from the uterus with a needle (fig. 2). The cells are cultured in the laboratory, and karyotypes are constructed

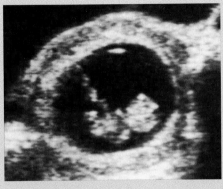 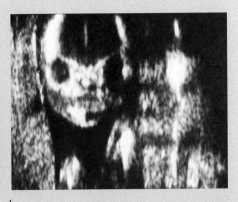

a.   b.

FIGURE 1

In an ultrasound exam, sound waves are bounced off the embryo or fetus, and the pattern of deflected sound waves is converted into an image. Not very much detail is visible in the 6-week embryo (*a*), but by 13 weeks (*b*), the face can be discerned.

from a few cells. The fluid for telltale biochemicals that could indicate a metabolic disorder. Amniocen-tesis can detect about 400 of the more than 5,000 known chromosomal and biochemical abnormalities. Ultrasound helps the physician insert the needle without harming the fetus, and the procedure only hurts the mother momentarily. Amniocentesis causes spontaneous abortion in about 1 in 200 cases.

Mrs. F.'s amnio results will come back in about 10 days. The cytogeneticist who examines their sample will look specifically for the extra or translocated chromosome 21 that identifies Down syndrome. Although she is apprehensive about the procedure, Mrs. F. is most concerned that the test results will not arrive until she is 17 weeks pregnant. She knows that a fetus can survive at the age of 24 weeks.

Chorionic Villus Sampling

Mrs. F. is not the only one concerned about the delays in performing amniocentesis. Researchers are experimenting with amniocentesis as early as 12 weeks. Had the doctor known of the family's Down syndrome when Mrs. F. was in the tenth week of pregnancy, an experimental procedure, chori-

onic villus sampling, could have been performed. In this technique, the physician snips cells from the fingerlike projections that develop into the placenta and runs chromosomal tests on the tissue (fig. 3).

In chorionic villus sampling, the doctor removes the cells through the vagina. A karyotype is prepared directly from the collected cells, rather than culturing them first, as in amniocentesis. Results are ready in days. Because chorionic villus cells descend from the fertilized egg, their chromosomes are identical to those of the embryo. Occasionally, however, chromosomal mosaicism occurs; the karyotype of a villus cell differs from that of an actual embryonic cell. The procedure is thus slightly less accurate than amniocentesis, which probes actual fetal cells. Also, the sampling procedure does not obtain any amniotic fluid, so other biochemical tests cannot be run.

Fetal Cell Sorting

In 1957, a pregnant woman died when cells from a very early embryo lodged in a major blood vessel in her lung, blocking blood flow. How did a cell from the embryo find its way into her bloodstream? To answer this question, researchers began studying

Biology in Action—*Continued*

the blood cells of pregnant women. They found that, sometimes, a few of the woman's body cells contained a Y chromosome! These women went on to deliver boys. Cells from the male embryo or fetus appeared to have found their way into the maternal bloodstream.

Today, two technologies allow researchers to extract fetal cells from a pregnant woman's blood as an alternative to more invasive prenatal tests. A device called a fluorescence-activated cell sorter separates out fetal cells by distinguishing surface characteristics that differ from the woman's cells. Then, a gene amplification technique allows the researcher to copy the fetal genetic material, creating enough to construct a karyotype and run biochemical tests to assess fetal health.

Although fetal cell sorting was not available to Mr. and Mrs. F., they received good news a week after amniocentesis—their unborn child did not have Down syndrome or any other detectable medical problem.

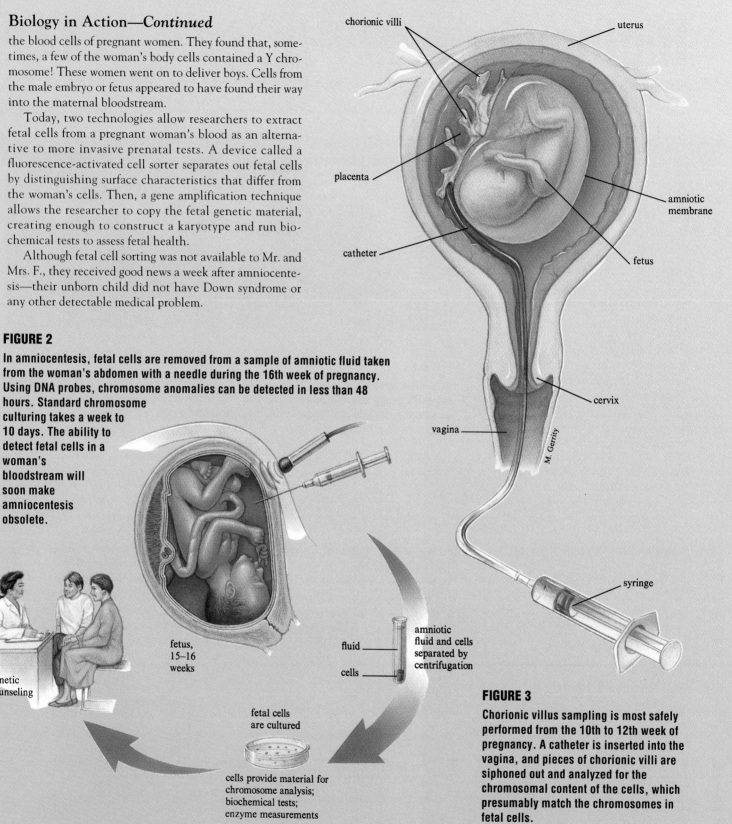

FIGURE 2

In amniocentesis, fetal cells are removed from a sample of amniotic fluid taken from the woman's abdomen with a needle during the 16th week of pregnancy. Using DNA probes, chromosome anomalies can be detected in less than 48 hours. Standard chromosome culturing takes a week to 10 days. The ability to detect fetal cells in a woman's bloodstream will soon make amniocentesis obsolete.

genetic counseling

Hagen

fetus, 15–16 weeks

fluid

cells

amniotic fluid and cells separated by centrifugation

fetal cells are cultured

cells provide material for chromosome analysis; biochemical tests; enzyme measurements

chorionic villi

uterus

placenta

amniotic membrane

catheter

fetus

cervix

vagina

M. Gerrity

syringe

FIGURE 3

Chorionic villus sampling is most safely performed from the 10th to 12th week of pregnancy. A catheter is inserted into the vagina, and pieces of chorionic villi are siphoned out and analyzed for the chromosomal content of the cells, which presumably match the chromosomes in fetal cells.

SUMMARY

Mapping localizes a gene to a specific region on a chromosome. Genes on the same chromosome are linked; rather than demonstrating independent assortment, they produce a large number of parental genotypes and a small number of recombinant genotypes. When geneticists know whether linked alleles of two genes are in coupling or repulsion, and when they can determine crossover frequencies by pooling data, they can predict the probabilities that certain genotypes will appear in the F1 generation. Crossover frequency between two linked genes is directly proportional to the distance between them.

The first linkage maps were established for genes on the X chromosome because it is easier to follow the inheritance of sex-linked traits. In humans, the male is the heterogametic and hemizygous sex, with one X and one Y chromosome. The female, with two X chromosomes, is homogametic. The SRY gene on the Y chromosome determines sex—it controls other genes that stimulate the development of male structures and suppress the development of female structures. A sex-linked trait passes from mother to son because the male inherits his X chromosome from his mother and his Y chromosome from his father. A sex-linked allele may be dominant or recessive.

X inactivation shuts off one X chromosome in the tissues of female mammals, equalizing the number of sex-linked genes each sex carries. Early in development, each female cell inactivates one X chromosome. When a female is heterozygous for genes on the X chromosome, this makes her cells mosaic. A heterozygous female who expresses the phenotype of a sex-linked gene is a manifesting heterozygote.

Sex-limited traits may be autosomal or sex-linked, but they only affect one sex because of anatomical or hormonal differences. A sex-influenced gene is dominant in one sex but recessive in the other. In genomic imprinting, the phenotype corresponding to a particular genotype differs depending on the parent who passes the gene. A genetically heterogeneic trait may be transmitted by more than one gene or mode of inheritance.

A chromosome consists of DNA, RNA, and proteins. Chromosomes are distinguishable by size, centromere position, satellites, and staining patterns. Dark-staining DNA, called heterochromatin, maintains the chromosome's structure. It surrounds the centromere, comprises the telomeres, and is interspersed elsewhere with light-staining, protein-encoding euchromatin.

Chromosomal abnormalities may affect number of chromosomes, sex chromosome constitution, and gene rearrangements. A euploid human cell has 23 pairs of chromosomes. Polyploid cells have extra chromosome sets, and aneuploids have extra or missing individual chromosomes. Trisomies (containing an extra chromosome) are less harmful than monosomies (involving a missing chromosome), and a sex chromosome aneuploidy is less severe than autosomal aneuploidy. Nondisjunction, an uneven division of chromosomes in meiosis, causes aneuploidy. Most aneuploids die before birth.

Down syndrome is the most common autosomal aneuploid. Turner (XO) and Klinefelter (XXY) syndromes are common sex chromosome aneuploids characterized by lack of sexual development.

When chromosomal rearrangements disrupt meiotic pairing, gene deletions and duplications occur. A nonreciprocal translocation occurs when part of a chromosome breaks off and attaches to another chromosome. A reciprocal translocation occurs when two chromosomes exchange parts. Translocation carriers have reproductive problems because they produce a predictable proportion of unbalanced gametes as a result of the unequal division of chromosome regions during meiosis. A heterozygote for an inversion may have reproductive problems if a crossover occurs between the inverted region and the noninverted homolog, generating deletions and duplications. Isochromosomes have two identical chromosome arms, deleting some genes and duplicating others. They form when the centromere divides in the wrong plane during meiosis. Ring chromosomes form when telomeres break off and the remnants fuse.

Chromosome studies provide useful information on genetic disease and evolutionary relationships. Soon chromosome testing may be an integral part of routine medical care—even before birth.

KEY TERMS

acrocentric   310
aneuploid   313
Barr body   306
deletion   317
duplication   317
euchromatin   310
euploid   313
genetic heterogeneity   309
genome   300
genomic imprinting   308

hemizygous   303
heterochromatin   310
heterogametic sex   301
homogametic sex   301
inversion   317
isochromosome   317
Jacobs syndrome   317
karyotype   311
linkage map   300
linked genes   298

manifesting heterozygote   306
metacentric   310
monosomy   313
nondisjunction   313
nonreciprocal translocation   317
parental   299
polyploidy   311
reciprocal translocation   317
recombinant   299
satellite   310
sex-influenced trait   308

sex-limited trait   308
telocentric   310
telomere   311
translocation   317
translocation carrier   317
triplo-X   315
trisomy   313
trisomy 21   313
X inactivation   306
X-inactivation center   306

REVIEW QUESTIONS

1. How are linked genes inherited differently than genes that are located on different chromosomes?

2. How are sex chromosome genes inherited differently in male and female humans?

3. How is sex determined in different organisms?

4. A normal-sighted woman with a normal-sighted mother and a color-blind father marries a color-blind man. What are the chances that their son will be color blind? Their daughter?

5. The tribble is a cute, furry creature ideal for embryonic experimentation. Cells from tribble embryos genetically destined to become brightly colored adults can be transplanted into tribble embryos that would otherwise develop into albino (pure white) adults. Investigators can follow the numbers and positions of cells descended from the original transplants by observing the pattern of colored patches in the adult tribble.

A gene in the tribble confers coat color. The brown allele *B* is dominant to the orange allele *b*. A single cell from a *Bb* female embryo at the 8-cell stage is transplanted into an albino embryo. The resulting adult tribble has both brown and orange patches on an albino background. But when a single cell from a *Bb* embryo at the 64-cell stage is transplanted into an albino embryo, only orange patches, or only brown patches, appear in the adult.

State the genetic principle demonstrated. What is the difference between the two experiments?

6. For an exercise in a college genetics laboratory course, a healthy student takes a drop of her blood, separates out the white cells, stains her chromosomes, and constructs her own chromosome chart. She finds only one chromosome 3 and one chromosome 21, plus two unusual chromosomes that do not seem to have matching partners.

   a. What type of chromosomal abnormality do you think she has?
   b. Why doesn't she have any symptoms?

   c. Would you expect any of her relatives to have any particular medical problems? If so, which ones?

7. What happens during meiosis to produce each of the following?

   a. an aneuploid
   b. a polyploid
   c. a translocation
   d. isochromosomes

8. Scientists once thought heterochromatin was nonfunctional because it does not encode proteins. Give two examples of roles heterochromatin fulfills.

9. Describe an individual with each of the following chromosome constitutions. Mention the person's sex and possible phenotype.

   a. 47, XXX
   b. 45, XO
   c. 47, XX, trisomy 21
   d. 47, XXY

10. Herbert is 58 years old and bald. His wife Sheri also has pattern baldness. What is the risk that their son Frank will lose his hair?

TO THINK ABOUT

1. Why doesn't the inheritance pattern of linked genes disprove Mendel's laws?

2. A fetus dies in the uterus. Several of its cells are examined for chromosomal content. Approximately 75% of the cells are diploid, and 25% are tetraploid (containing four copies of each chromosome). What has happened, and when in development did it probably occur?

3. Which chromosomal anomaly might you expect to find more frequently among the members of the National Basketball Association than in the general population? Cite a reason for your answer.

4. Two babies are born with identical blockages in the esophagus, the tube leading from the mouth to the stomach. One baby has no other medical problems; the other has Down syndrome and an associated heart defect, which will later require surgery. Do you think that each child has the same right to receive the esophageal surgery? Cite a reason for your answer. (A landmark legal case concerned parents of a baby with Down syndrome who refused such surgery on the basis of the child's Down syndrome).

5. A fetus is found to have an inverted chromosome. What information might reveal whether the expected child will have health problems stemming from the inversion?

6. Patricia Jacobs concluded in her 1965 *Nature* article on XYY syndrome, "It is not yet clear whether the increased frequency of XYY males found in this institution is related to aggressive behavior or to their mental deficiency or to a combination of these factors." Why is this wording unscientific? What harm could it (or did it) do?

7. In the fruit fly *Drosophila melanogaster*, the contribution from the sex chromosomes in the two sexes is equalled by overexpression of X-linked genes in the male. How is this equalization accomplished in mammals?

8. Stella Walsh won a gold medal in the 100 meter dash in the 1932 Olympics. In 1980, she was killed in a robbery. When a physician examining the body noticed that she had ambiguous genitalia, she ordered a chromosome test. Stella had an XY karyotype. The same was true for another athlete, Ewa Klobukowska, who set a 1965 world record for the 100 meter dash—she, too, turned out to be XY. What must have been unusual about the Y chromosomes of Stella Walsh and Ewa Klobukowska?

SUGGESTED READINGS

Aasen, Eric, and Juan Fernando Medrano. December 1990. Amplification of the Zfy and Zfx genes for sex identification in humans, cattle, sheep, and goats. *Bio/Technology*, vol. 8. The sex chromosomes are very similar in all mammals.

Bird, Thomas D. July 1993. Are linkage studies boring? *Nature Genetics*, vol. 4. Linkage studies were the backbone of the fledgling field of genetics early in the century. Today, they are key to gene discoveries. But they can be tedious to carry out.

Foote, S., D. Vollrath, A. Hilton, and D. C. Page. October 2, 1992. The human Y chromosome: Overlapping DNA clones spanning the euchromatic region. *Science*, vol. 258. Using the genetic material from an unusual man with four Y chromosomes, these researchers developed the first map for the gene-encoding portion of the Y chromosome.

Gebhart, Fred. February 1991. Lawrence Livermore laboratory transfers whole chromosome paints technology. *Genetic Engineering News*. In situ hybridization distinguishes between chromosomes, tracking abnormalities, cancer, and toxic exposure.

Gillis, Anna Maria. March 1994. Turning off the X chromosome. *BioScience*, vol. 44. Researchers are working out the molecular details of X inactivation.

Gorman, Christine. January 20, 1993. Sizing up the sexes. *Time*. Whether one is male or female depends upon the species and its sex determination mechanism.

Lewis, Ricki. January 1991. Genetic imprecision. *BioScience*. Chromosome checks can reveal abnormalities—but they are not necessarily linked to symptoms.

Lyon, Mary F. 1993. Controlling the X chromosome. *Current Biology*, vol. 3, no. 4. The discoverer of X inactivation updates her hypothesis.

Moyzis, Robert K. August 1991. The human telomere. *Scientific American*. The dark-staining tips of chromosomes are vital to chromosomal integrity.

Rothman, Barbara Katz. 1986. *The tentative pregnancy*. New York: Penguin books. Many expectant couples do not accept their impending parenthood until the amnio results are in.

Science, vol. 255. February 28, 1992. Experts slam Olympic gene test. Sex chromosome tests are no longer adequate proof of sex.

Suarez, Brian K., and Carol L. Hampe. March 1994. Linkage and association. *The American Journal of Human Genetics*, vol. 54. A short history of linkage analysis.

 **Interactive Software**

Heredity in Families

This interactive exercise allows students to explore how the character and location of a gene influences how it is inherited within families. The exercise presents a pedigree of hemophilia, a classic genetic disorder affecting the royal families of Europe. Students can investigate how altering the dominance/recessiveness of the inherited allele creates different patterns of affected individuals as the allele is inherited from one generation to the next. Students can then move the gene from an X chromosome (its true location) to an autosome and learn how this influences the pedigree. These alterations of pedigrees are often all a human geneticist has to work with in attempting to assess the dominance and chromosomal location of a trait.

1. In analyzing family pedigrees, why are sex-linked alleles expressed so much more often in male offspring than in females?

2. Why do you imagine some genetic disorders like sickle cell disease and hemophilia are common, while others are rare?

3. In a pedigree, what effect does sex linkage have on your ability to determine whether a trait is being influenced by a single gene?

4. What is the minimal evidence that you would accept that a trait appearing in a family is indeed hereditary (that is, caused by an allele), rather than environmentally induced (that is, not due to an allele)?

CHAPTER 15

•••

DNA Structure and Replication

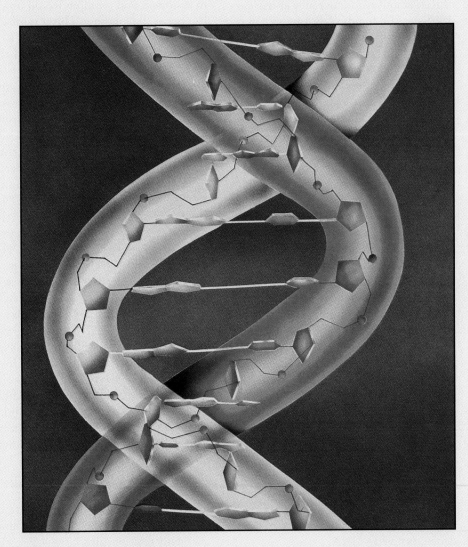

DNA is the hereditary material.

LEARNING OBJECTIVES

By the chapter's end, you should be able to answer these questions:

1. What kinds of experiments revealed the structure of the genetic material?

2. How did investigators demonstrate DNA replication?

3. Why is it necessary to replicate genetic material?

4. How does DNA structure provide a replication mechanism?

DNA Makes History

One night in July 1918, Tsar Nicholas II of Russia and his family met gruesome deaths at the hands of Bolsheviks in a Ural mountain town called Ekaterinburg. Captors led the tsar, tsarina, their four daughters and one son, the family physician, and three servants to a cellar and shot them, bayoneting those who did not immediately die. The executioners then stripped the bodies and loaded them onto a truck, planning to hurl them down a mine shaft. But the truck broke down, and the killers instead placed the bodies in a shallow grave, then damaged them with sulfuric acid so that they could not be identified.

In another July—many years later, in 1991—two Russian amateur historians found the grave. Because they were aware that the royal family had spent its last night in Ekaterinburg, they alerted the government that they might have unearthed the long-sought bodies of the Romanov family. An official forensic examination soon determined that the skeletons represented nine individuals. The sizes of the skeletons indicated that three were children, and the porcelain, platinum, and gold in some of the skeletons' teeth suggested that they were royalty. Unfortunately, the acid had so destroyed the facial bones that some conventional forensic tests were not feasible. But one very valuable type of evidence survived—DNA.

British researchers eagerly went to work examining the molecules of heredity from cells in the skeletal remains. By finding DNA sequences specific to the Y chromosome, the investigators were able to distinguish males from females. Then the DNA detectives delved into the genetic material of mitochondria. Because these organelles pass from mother to offspring, finding an identical mitochondrial DNA pattern in a woman and children would establish her as their mother. This was indeed the case for one of the women (who had impressive dental work) and the three children.

But a mother, her children, and some companions does not make a royal family. The researchers had to find a connection between the skeletons and the royal family—and to do so, they again turned to DNA. Genetic material from one of the male skeletons shared certain rare DNA sequences with DNA from living descendants of the Romanovs. This skeleton also had aristocratic dental work and shared DNA sequences with the children! The mystery of the fate of the Romanovs was apparently solved—thanks to the help of a most fascinating molecule, deoxyribonucleic acid.

The complex but intriguing DNA molecule is affecting many peoples' lives in many ways. DNA gives doctors the means to run sophisticated new medical tests, and DNA's control over cellular activities has enabled pharmaceutical researchers to create and mass-produce many safer and more effective drugs. The ability to trace unique DNA sequences allows forensic investigators to unravel historical mysteries such as the fate of the Romanovs, solve crimes, and untangle family relationships.

It took the entire twentieth century to learn the secrets of the DNA molecule and to apply that knowledge (fig. 15.1). This chapter describes how scientists discovered the chemical basis of heredity, and how the structure of DNA enables it to make copies of itself—a quality that not only has many practical uses but is essential for the perpetuation of life. Recognition of DNA's vital role in life was a long time coming.

Identifying and Describing the Genetic Material

Swiss physician and biochemist Friedrich Miescher was the first investigator to chemically analyze the contents of a cell's nucleus. In 1869, he isolated the nuclei of white blood cells obtained from pus in soiled bandages. In the nuclei, he discovered an unusual acidic substance containing nitrogen and phosphorus. Miescher and others went on to find it in cells from a variety of sources. Because the material resided in cell nuclei, Miescher called it nuclein in his 1871 paper; subsequently it was called a nucleic acid.

Miescher's discovery, like those of his contemporary Gregor Mendel, was not appreciated for some years. Although investigators were researching inheritance, their work focused for several decades on the association between hereditary diseases and proteins.

In 1909, English physician Archibald Garrod was the first to link inheritance and protein. Garrod noted that certain inherited "inborn errors of metabolism" correlated with missing enzymes. Other researchers added supporting evidence: they linked abnormal or missing enzymes to unusual eye color in fruit flies and nutritional deficiencies in bread mold variants. Although evidence was mounting that proteins are keys to trait expression, new questions arose. Why did enzyme deficiencies occur in some cells? What controlled protein synthesis? Eventually, investigators found that the path led back to Miescher's discovery of nucleic acids.

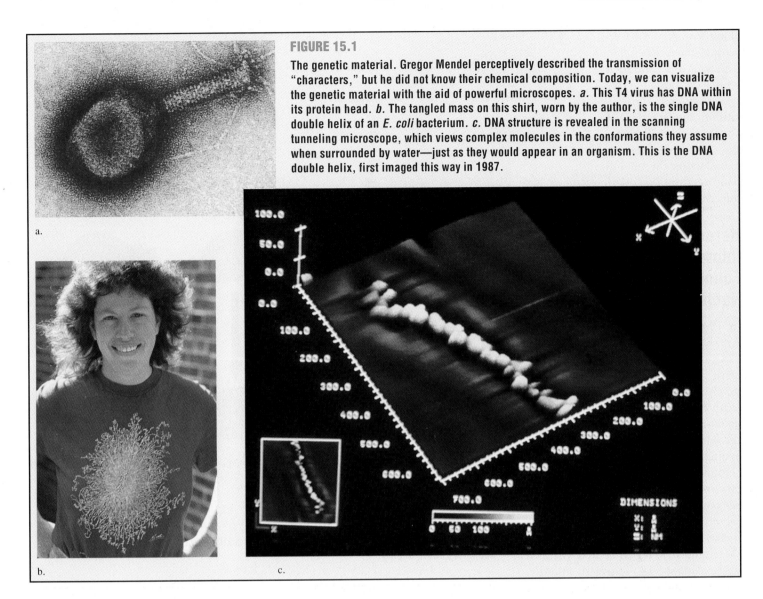

FIGURE 15.1

The genetic material. Gregor Mendel perceptively described the transmission of "characters," but he did not know their chemical composition. Today, we can visualize the genetic material with the aid of powerful microscopes. *a.* This T4 virus has DNA within its protein head. *b.* The tangled mass on this shirt, worn by the author, is the single DNA double helix of an *E. coli* bacterium. *c.* DNA structure is revealed in the scanning tunneling microscope, which views complex molecules in the conformations they assume when surrounded by water—just as they would appear in an organism. This is the DNA double helix, first imaged this way in 1987.

a.

b.

c.

DNA—The Hereditary Molecule

By 1929, scientists had discovered deoxyribose and differentiated RNA from DNA. At about the same time, in 1928, English microbiologist Frederick Griffith made an observation that would later prove to be the first step in identifying DNA as the genetic material. Griffith noticed that animals with a certain type of bacterial pneumonia harbored one of two types of *Diplococcus pneumoniae* bacteria. Type R bacteria are rough in texture. Type S bacteria are smooth, because they are surrounded by a polysaccharide capsule. Griffith saw

that mice injected with type R bacteria did not develop pneumonia, but mice injected with type S did. The polysaccharide coat seemed to be necessary for infection.

When Griffith heated type S bacteria, they no longer caused pneumonia in mice. However, when he injected mice with a mixture of type R bacteria plus heat-killed type S bacteria—neither capable of causing pneumonia on its own—the mice died of pneumonia (fig. 15.2). Their bodies contained live type S bacteria encased in polysaccharide. What was happening?

In the 1930s, Rockefeller University physicians Oswald Avery, Colin

MacLeod, and Maclyn McCarty offered an explanation. They hypothesized that something in the heat-killed type S bacteria transformed the normally harmless type R strain into a killer. Experiments showed that injecting the mice with a protein-dismantling enzyme (a protease) failed to inhibit the type R strain from transforming into a killing strain. However, injecting them with a DNA-dismantling enzyme disrupted the transformation. Could DNA control the transformation?

Avery, MacLeod, and McCarty confirmed that DNA transformed the bacteria by isolating DNA from heat-killed type S bacteria and injecting it along with type R bacteria into mice (fig. 15.3). The mice died, and their bodies contained active type S bacteria. The conclusion: the type S DNA altered the type R bacteria, enabling them to manufacture the smooth coat necessary to cause infection.

Hereditary Information— Found in the Nucleus

In 1943, a year before Avery, MacLeod, and McCarty published their work, Danish biologist Joachim Hammerling conducted a clever experiment that demonstrated the role the nucleus plays in directing development. Hammerling removed different parts of *Acetabularia*, a large, single-cell green alga, to investigate whether the nucleus controls inheritance. He experimented with two *Acetabularia* species—one with a flowery cap (*A. crenulata*) and one with a disk-like cap (*A. mediterranea*). Hammerling cut individuals of the two species into three pieces—cap, stalk, and base. He found that when he grafted a stalk from one species to a base from the other, the regenerated cap matched the base species (fig. 15.4). Therefore, something in the base—the nucleus, perhaps?—directed the cap's development. In the nucleus were the nucleic acids.

Despite the accumulating evidence, scientists still hesitated to accept DNA as the chemical of heredity. Proteins, they knew, consisted of 20 different building blocks and performed many important functions.

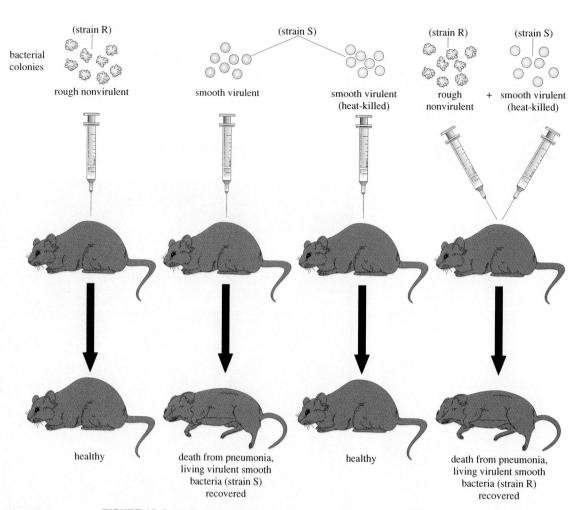

FIGURE 15.2

Griffith's experiments showed that a biochemical in a killer strain of bacteria can make a nonkilling strain deadly.

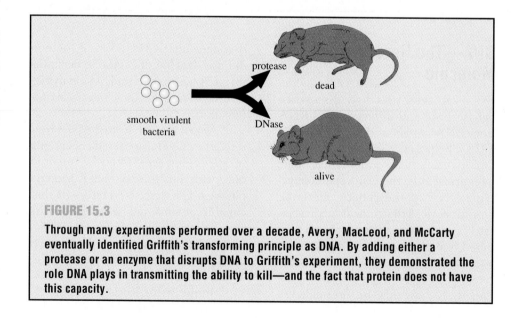

FIGURE 15.3

Through many experiments performed over a decade, Avery, MacLeod, and McCarty eventually identified Griffith's transforming principle as DNA. By adding either a protease or an enzyme that disrupts DNA to Griffith's experiment, they demonstrated the role DNA plays in transmitting the ability to kill—and the fact that protein does not have this capacity.

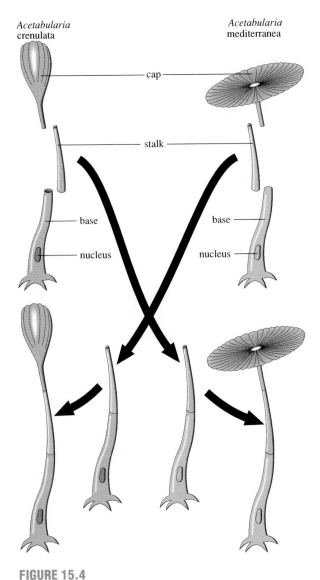

Acetabularia crenulata

Acetabularia mediterranea

cap

stalk

base

nucleus

base

nucleus

FIGURE 15.4

Hammerling's experiments swapping body parts in *Acetabularia* showed that the nucleus contains the genetic material.

Wasn't protein more complex—and therefore capable of carrying more information—than the simpler, less known nucleic acids? Some researchers even suggested that protein had contaminated the Rockefeller group's DNA and controlled the bacterial transformation. However, before long, another set of researchers would definitively answer the question: was DNA or protein responsible for inheritance?

The Hereditary Material—DNA, Not Protein

In 1950, American microbiologists Alfred Hershey and Martha Chase confirmed that DNA—not protein—is the genetic material. To do so, they infected *E. coli* bacteria with a virus. They knew that some part of the virus would enter the bacteria and commandeer them, directing them to produce more virus. By carefully designing their experiment, Hershey and Chase could determine which part of the simple virus controlled replication—the nucleic acid, or the protein coat that surrounded it.

Hershey and Chase already knew that virus grown with radioactive sulfur became radioactive, and the radioactivity was emitted from the protein coats. When Hershey and Chase repeated the experiment with radioactive phosphorus, the viral nucleic acid, DNA, emitted the radioactivity instead. This showed that sulfur exists in protein but not in nucleic acid, and that phosphorus exists in nucleic acid but not in protein. (Recall that Miescher had identified phosphorus in "nuclein" nearly a century earlier.)

Next, the researchers "labeled" two batches of virus, one with radioactive sulfur (which marked protein), and the other with radioactive phosphorus (which marked DNA). They used each type of labeled virus to infect a separate batch of bacteria, allowing several minutes for the virus particles to bind to the bacteria and inject their genetic material into them. Then they agitated each mixture in a blender,

poured them into test tubes, and centrifuged them (spun them at high speed). This settled the heavier infected bacteria at the bottom of each test tube.

Hershey and Chase examined the contents of the bacteria that had settled to the bottom of each tube. In the test tube containing sulfur-labeled virus, the virus-infected bacteria were not radioactive. But in the other tube, where the virus contained radioactive phosphorus, the infected bacteria were radioactive. This meant that the part of the virus that could enter the bacteria and direct them to mass-produce more virus was the part with the phosphorus label—the DNA. The genetic material, therefore, was DNA, and not protein (fig. 15.5).

Deciphering the Structure of DNA

In the early decades of the twentieth century, Russian-American biochemist Phoebus Levene, also at Rockefeller University, picked up where Miescher had left off by continuing to chemically analyze nucleic acids. In 1909, Levene identified the 5-carbon sugar **ribose** in some nucleic acids; in 1929, he discovered a new, similar sugar called **deoxyribose.** Levene's discovery revealed a major chemical distinction between the two types of nucleic acid, RNA (**ribonucleic acid**) and DNA (**deoxyribonucleic acid**).

Levene then determined that the three parts of a nucleic acid—carbohydrates, nitrogen-containing groups, and phosphorus-containing components—occur in equal proportions. He deduced that the building blocks that make up nucleic acids—called **nucleotides**—must each include one of each component. Levene also found that while nucleotides always contain the same carbohydrate and phosphate portions, they may contain any one of four different nitrogen-containing bases. For several years thereafter, scientists erroneously thought that the nitrogen-containing bases occur

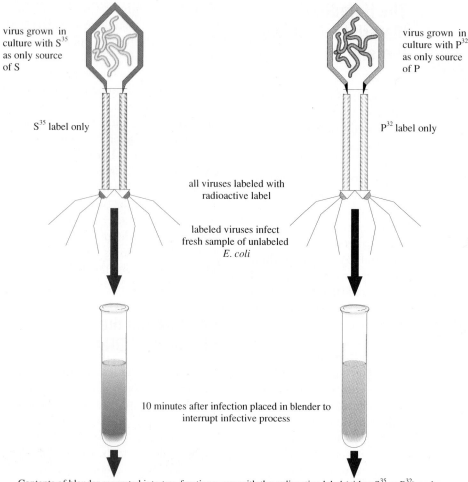

virus grown in
culture with S³⁵
as only source
of S

virus grown in
culture with P³²
as only source
of P

S³⁵ label only

P³² label only

all viruses labeled with
radioactive label

labeled viruses infect
fresh sample of unlabeled
E. coli

10 minutes after infection placed in blender to
interrupt infective process

Contents of blender separated into two fractions; one with the radioactive label (either S³⁵ or P³²) and one
fraction with no radioactivity. Each fraction put in petri dish with fresh media and scored for growth of *E. coli*
and liberation of new virus.

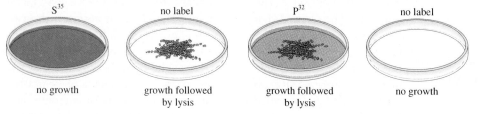

S³⁵   no label   P³²   no label

no growth   growth followed   growth followed   no growth
                   by lysis   by lysis

FIGURE 15.5

**By labeling either sulfur (found in protein but not in DNA) or phosphorus (found in DNA but not
in protein) in viruses used to infect *E. coli* bacteria, Hershey and Chase proved that DNA is the
hereditary material—and protein is not.**

in equal amounts. If this were the case, DNA bases could not occur in as many combinations and therefore could not encode as much information.

In the 1930s and 1940s, Scottish chemist Alexander Todd confirmed and extended Levene's work by synthesizing DNA nucleotides. This set the stage for others to decipher the DNA molecule's complex three-dimensional conformation and so to explain its unique role in perpetuating life.

In the early 1950s, two lines of experimental evidence finally converged to elucidate DNA's structure. Austrian-American biochemist Erwin Chargaff showed that DNA in several species contains equal amounts of the bases **adenine** and **thymine** and equal amounts of the bases **guanine** and **cytosine.** For example, if a species's DNA contains 20% adenine, it would also contain 20% thymine, meaning that the remaining 60% would be evenly split as

well—30% each of guanine and cytosine. Next, two English researchers, physicist Maurice Wilkins and chemist Rosalind Franklin, bombarded DNA with X rays using a technique called X-ray diffraction. The resulting X-ray deflection pattern revealed a regularly repeating structure of nucleotides.

In 1953, American biochemist James Watson and English physicist Francis Crick drew these two clues together to build a replica of the DNA molecule using ball-and-stick models (fig. 15.6). Their model included equal amounts of guanine and cytosine and equal amounts of adenine and thymine, and it demonstrated the symmetry the X-ray diffraction pattern had revealed. Watson and Crick's model, which was based upon so many others' experimental evidence, was the now famous and familiar double helix.

KEY CONCEPTS

DNA, the genetic material, contains the information the cell requires to synthesize protein and to replicate itself. Many scientists helped unravel DNA's secrets in nearly a century of research. Miescher first isolated DNA in 1869, naming it nuclein. Hammerling confirmed that the nucleus contains the hereditary material and Garrod first linked heredity to enzymes. Griffith identified a substance capable of transmitting a trait, which Avery, MacLeod, and McCarty later showed was DNA. In 1950, Hershey and Chase confirmed that DNA—not protein—is the genetic material. By combining Chargaff's discovery that DNA contains equal numbers of paired bases with Franklin's finding that DNA has a symmetrical structure, Watson and Crick finally deciphered the double-helix conformation of DNA in 1953, coming full circle to Miescher's original discovery.

Gene and Protein— An Important Partnership

Genes and proteins encode information because they consist of sequences—genes consist of four types of DNA bases in different sequences, and proteins of 20

FIGURE 15.6

James Watson (left) and Francis Crick. Watson was the first director of the Human Genome Project, an ongoing worldwide effort to systematically determine the sequences of all human genes.

amino acids arranged in chains. Proteins provide the connection between a gene and the trait it produces. Pigment proteins provide pea color; protein hormones control plant height. In the human body, the amino acids that assemble proteins ultimately come from the diet. Enzymes catalyze the chemical reactions of metabolism. Proteins such as collagen and elastin provide structural support in connective tissues. Tubulin builds the cytoskeleton, myoglobin and hemoglobin transport oxygen, and antibodies protect against infection. Malfunctioning or inactive proteins, which reflect genetic defects, can be devastating to health.

The genetic material—a molecule of DNA—consists of two long strands entwined to form a double helix, which resembles a twisted ladder. The rungs of the ladder are pairs of nitrogenous (nitrogen-containing) bases held together by hydrogen bonds. The ladder's rails, which make up the DNA's sugar-phosphate "backbone," consist of alternating units of deoxyribose and phosphate (PO_4) held together by covalent bonds. As Levene deduced, a single building block of DNA—a nucleotide—consists of one deoxyribose, one phosphate, and one nitrogenous base. Figure 15.7 shows depictions of the DNA molecule.

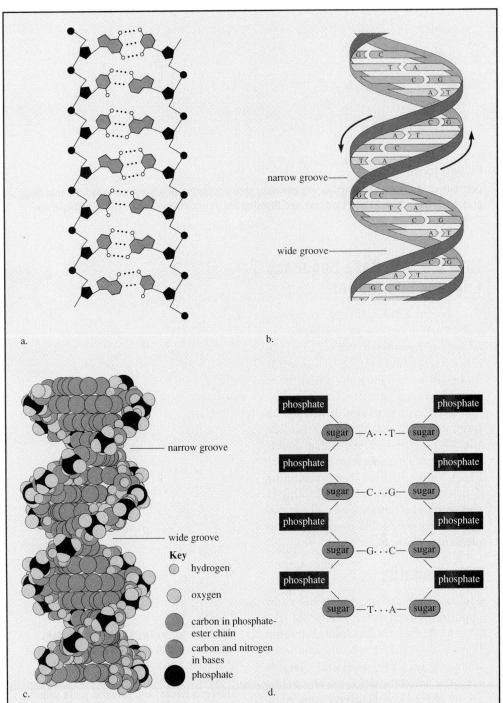

FIGURE 15.7

Different ways to represent the DNA double helix. *a.* The helix is unwound to show the base pairs in color and the sugar-phosphate backbone in black. *b.* The sugar-phosphate backbone is emphasized, along with the fact that the two strands run in opposite directions. *c.* This representation shows the relationships of all of the atoms. *d.* A schematic representation of an unwound section of the double helix outlines the relationship of the sugar-phosphate rails to the base pairs. The informational content of DNA lies in the sequence of bases. The sugar-phosphate rails are identical in all DNA molecules.

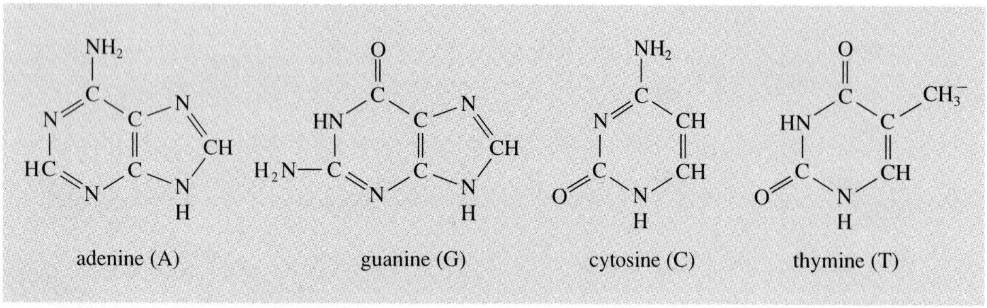

FIGURE 15.8

DNA bases. Adenine and guanine are purines, each composed of a six-membered organic ring plus a five-membered ring. Cytosine and thymine are pyrimidines, each built of a single six-membered ring.

How the DNA Base Sequence Encodes Information

The key to DNA's ability to convey information lies in the sequence of adenines, thymines, cytosines, and guanines (A,T,C, and G for short). Adenine and guanine are **purines,** which have a double organic ring structure. Cytosine and thymine are **pyrimidines,** which have a single organic ring structure (fig. 15.8). The sleek, symmetric double helix forms when nucleotides containing A bases pair with those containing T bases, and nucleotides containing G pair with those containing C. These specific purine-pyrimidine couples are **complementary base pairs** (fig. 15.9).

Directionality

The two chains of the double helix run opposite to each other, somewhat like artist M. C. Escher's depiction of drawing hands (fig. 15.10a). This head-to-tail arrangement, called **antiparallelism,** is apparent when the deoxyribose carbons are numbered consecutively from right to left, according to chemical convention (fig. 15.10b). Where one chain ends in the 3′ carbon, the opposite chain ends in the 5′ carbon (fig. 15.10c). (The 3′ and 5′ designate the positions of the carbon atoms in the DNA molecule.)

The Highly Coiled Structure of DNA

DNA molecules are incredibly long. A single stretched-out molecule, repre-

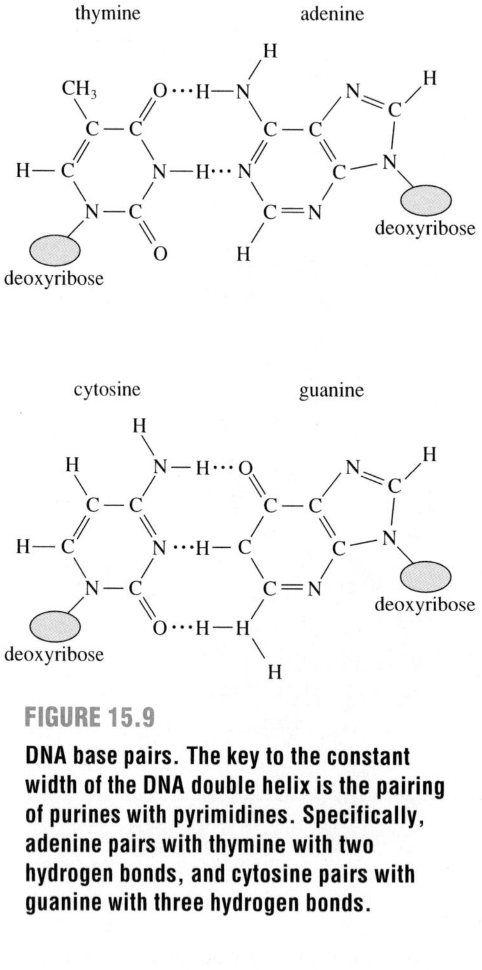

FIGURE 15.9

DNA base pairs. The key to the constant width of the DNA double helix is the pairing of purines with pyrimidines. Specifically, adenine pairs with thymine with two hydrogen bonds, and cytosine pairs with guanine with three hydrogen bonds.

senting a typical human chromosome, would be about an inch long. If the DNA bases of all 46 human chromosomes were typed as A,C,T, and G, the 3 billion letters would fill 4,000 books of 500 pages each! How can a cell only one millionth of an inch across contain so much material?

The explanation is that DNA is wrapped around proteins, much as a very long length of thread is wound around a wooden spool (see fig. 14.10). A length of DNA 146 nucleotides long wraps twice around a structure of eight histone proteins. DNA entwined around such a histone octet is called a **nucleosome;** a nucleosome is 10 nanometers (nm, billionths of a meter) in diameter. Nucleosomes are connected by a continuous thread of DNA like beads on a string, and are in turn folded to form a structure 30 nm in diameter. Just like thread, the DNA must unwind for it to function.

As sections of DNA unwind, the DNA in the more widely separated nucleosomes may be expressed. Whether the DNA rolls off the histone to be transcribed into RNA or remains tightly rolled, it maintains its structural integrity and sequence. Together, the DNA and histones constitute **chromatin,** the material that makes up chromosomes.

For many years, biologists thought histone proteins were mere structural backdrops for DNA. However, the five types of histones are remarkably consistent in their amino acid sequences, even in species as different as people and peas. If a protein is very similar in unrelated modern species, we can deduce that it has not changed much through evolutionary time. This means its function must be very important—certainly more than just serving as a spool to hold DNA, which any other protein of the same general shape could do. Histones, it turns out, also actively participate in controlling gene expression—turning on some genes and repressing others. Gene expression is the subject of the next chapter.

KEY CONCEPTS

Gene structure determines protein production, and proteins account for the expression of traits. A gene is a long section of DNA. The DNA double helix is a backbone of alternating deoxyribose and phosphate groups, with rungs formed by complementary A-T and G-C base pairs. A and G are purines; T and C are pyrimidines. The DNA double helix is directional and antiparallel. Though a single DNA molecule is incredibly long, enabling it to encode abundant genetic information, it is tightly wound to fit into living cells.

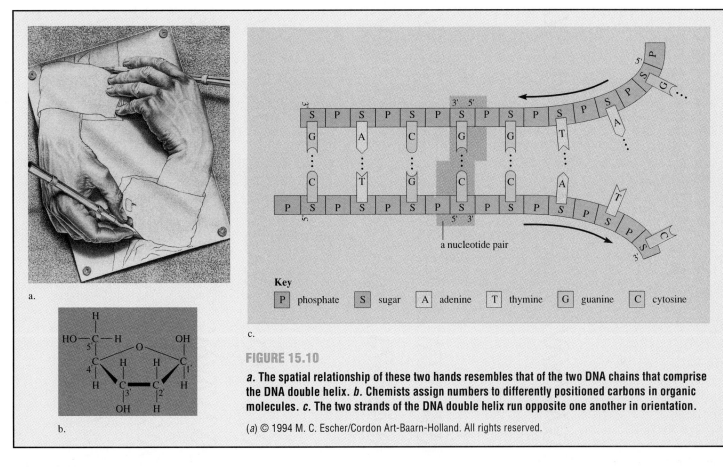

Key

P	phosphate	S	sugar	A	adenine	T	thymine	G	guanine	C	cytosine

FIGURE 15.10

a. The spatial relationship of these two hands resembles that of the two DNA chains that comprise the DNA double helix. *b.* Chemists assign numbers to differently positioned carbons in organic molecules. *c.* The two strands of the DNA double helix run opposite one another in orientation.

DNA Replication— Maintaining Genetic Information

Once Watson and Crick deciphered DNA's structure, its replication mechanism became obvious. In their classic 1953 paper describing the genetic material, the two scientists wrote that "the specific pairing we have postulated immediately suggests a possible copying mechanism for the genetic material." But experimentation had to provide the evidence that DNA actually replicates as Watson and Crick predicted.

Semiconservative Replication

Watson and Crick envisioned a double helix disentangling, with each half serving as a template, or mold, for the assembly of a new half. Every cell contains unattached nucleotides supplied by diet and metabolism. The bases in these nucleotides, Watson and Crick proposed, dock opposite complementary bases on the single parental DNA strand, forming hydrogen bonds. Covalent bonds then cement the sugar-phosphate backbone in the new, or daughter, DNA strand. This mechanism demonstrates **semiconservative replication** because half ("semi") of each double helix was part of a previous double helix.

However logical this hypothesized mechanism for DNA replication, scientists needed experimental evidence to prove the semiconservative mode of DNA replication and to rule out other possible mechanisms. What other ways might DNA replicate?

One hypothesis, **conservative replication,** proposed that the double helix did not permanently separate but instead somehow directed the construction of an entirely new double helix. A third suggestion, the **dispersive replication** strategy, postulated that the double helix broke into pieces, joining with unattached nucleotides to build two new double helices from one (fig. 15.11).

In 1957, Matthew Meselson and Franklin Stahl, at the California Institute of Technology, devised an experiment to demonstrate the mode of DNA replication. Meselson and Stahl added nitrogen that contained an extra neutron in its nucleus to *E. coli* growing on media in the laboratory. Newly synthesized bacterial DNA incorporating this "heavy" nitrogen, or ^{15}N, would be distinguishable from the previous genration's "light" nitrogen or ^{14}N-containing DNA.

Meselson and Stahl tracked the results using a technique that entailed spinning DNA in a centrifuge to separate it by density. Very dense DNA would indicate that the DNA had incorporated ^{15}N in both halves of the double helix; less dense DNA would only contain ^{14}N. An intermediate form would contain ^{15}N in one strand of the double helix and ^{14}N in the other.

First, the researchers grew *E. coli* on medium containing heavy nitrogen, or ^{15}N. After several generations, the bacteria had completely incorporated ^{15}N into both DNA strands (fig 15.12). Meselson and Stahl then shifted the bacteria to medium containing ^{14}N, allowing enough time for the bacteria to divide only once (about 20 minutes).

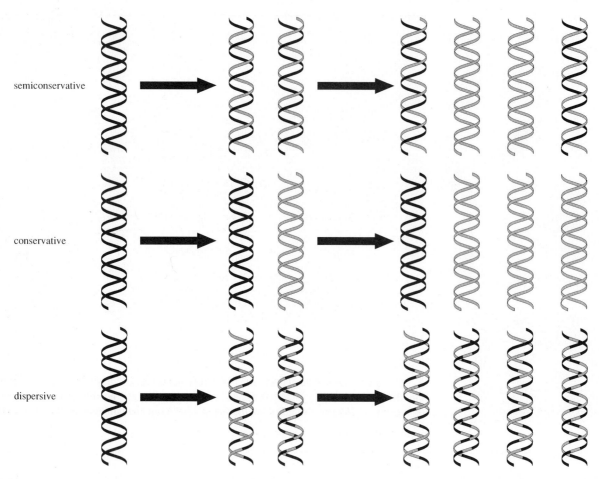

semiconservative

conservative

dispersive

FIGURE 15.11

Three routes to DNA replication—which one is correct?

When the researchers collected the DNA and centrifuged it, the double helices were all of intermediate density, indicating that they contained half ^{14}N and half ^{15}N. This pattern was consistent with semiconservative DNA replication—but it could also be consistent with a dispersive mechanism. The result did rule out conservative replication; it would have produced two density bands, one of DNA labeled only with ^{15}N and one containing only "light" ^{14}N.

By extending the experiment one more generation, Meselson and Stahl were able to disprove dispersion as well. They shifted the hybrid DNA—half ^{14}N and half ^{15}N—to another medium containing ^{14}N. If semiconservative replication occurred, the hybrids would part and assemble new halves from the bases labeled with ^{14}N. This would produce two hybrid double helices with one ^{15}N (heavy) and one ^{14}N (light) chain, plus two completely ^{14}N double helices.

The experiment did produce two density gradient bands—one heavy-light and one light-light band. This supported the semiconservative hypothesis. If a dispersive form of replication had occurred, a single large band of relatively light DNA would have appeared, since the double helices would have randomly reassembled, incorporating the new ^{14}N.

Other experimenters confirmed the semiconservative mode of DNA replication in other species. These experiments extended Meselson and Stahl's results by demonstrating semiconservative replication in the cells of more complex organisms and at the whole-chromosome level.

KEY CONCEPTS

DNA replication could theoretically be conservative, semiconservative, or dispersive. Watson and Crick's double helix model suggested a semiconservative mechanism. Meselson and Stahl, along with their successors, grew cells with DNA labels to test replication. Their experiments showed that DNA replication is semiconservative.

Steps and Participants in DNA Replication

Once researchers proved the semiconservative nature of DNA replication, the next challenge was to decipher the steps of the process.

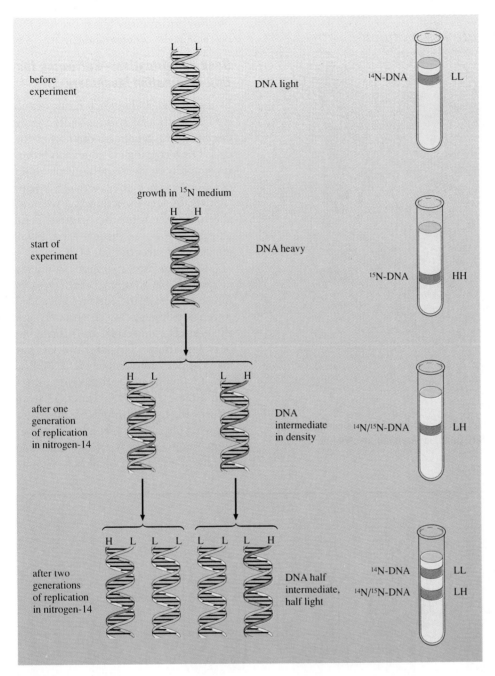

FIGURE 15.12

Meselson and Stahl found that newly synthesized DNA incorporated a label in a pattern consistent with the semiconservative mechanism suggested by the three-dimensional conformation of the molecule.

Unraveling and Separating the Strands

Even before they elucidated DNA's structure, biologists had pondered how the long molecule of heredity might replicate. To put the problem of DNA replication into perspective, imagine sewing a hem on a pair of jeans using a doubled thread the length of a football field! The doubled thread would rapidly become

hopelessly tangled. To maintain the length of the thread, you would have to continuously untwist the tangles or periodically cut the thread, untwist it locally, and reattach the cut ends.

Conceptually speaking, this is what happens when a long length of DNA replicates. A contingent of enzymes carries out the process, as figure 15.13 shows in rather whimsical form. Enzymes called

topoisomerases break apart, untwist, and mend DNA. While the DNA strands are apart, other enzymes guide the assembly of a new DNA strand.

Replication Proceeds at Several Points Simultaneously

A human chromosome replicates at hundreds of points along its length, and then the individual sections rejoin—a little like the fence Tom Sawyer whitewashed by convincing several children to each paint a small part. DNA in the process of replicating resembles a fork in a road—in fact, the open portion of the double helix is called a **replication fork.** A replicating chromosome has many replication forks.

DNA replication begins when an unwinding protein breaks the hydrogen bonds that connect a base pair. The location where this occurs is called an **initiation site.** An enzyme called RNA primase then attracts complementary RNA nucleotides to build a short piece of RNA, called an **RNA primer,** at the start of each DNA segment to be replicated. Next, the RNA primer attracts **DNA polymerase,** which draws in DNA nucleotides to complement the exposed bases on the template strand. The new DNA strand grows as hydrogen bonds form between the complementary bases.

DNA polymerase "proofreads" as it goes, excising mismatched bases and inserting the correct ones. At the same time, another enzyme removes the RNA primer and replaces it with the correct DNA bases. (RNA nucleotides differ in that they contain ribose rather than deoxyribose and **uracil** rather than thymine.) Enzymes called **ligases** (ligase means "to tie") catalyze the formation of the covalent bonds that hold together the sugar-phosphate backbone. Figure 15.14 depicts the replication process.

Replication is Discontinuous

DNA polymerase works directionally, adding new nucleotides to the exposed 3′ end of the sugar in the growing strand. Replication proceeds in a 5′ to 3′ direction. In order to follow this 5′ to 3′ "rule," DNA replication proceeds continuously (in one piece) on one strand,

Molecular Biology made simpler

QUADRANT

FIGURE 15.13

DNA replication is carried out by an army of enzymes. This advertisement for a company that sells DNA-cutting enzymes depicts the number of participants (enzymes) involved in replicating and repairing DNA.

but discontinuously (in short 5´ to 3´ pieces) on the other strand (fig. 15.15).

Gene Amplification—Borrowing the DNA Replication Machinery

Every time a cell divides, it replicates all its DNA. In a technique called **gene amplification,** geneticists can borrow the cell's DNA copying machinery and focus it to rapidly replicate millions of copies of DNA sequences they particularly want to study. Biology in Action 15.1 recounts the history behind the original gene amplification technique—the polymerase chain reaction, or PCR.

PCR requires two short, lab-made pieces of DNA that bracket the gene of interest. Replicating enzymes are added to repeatedly replicate the primers and the genetic material of interest in between them. The technology has many applications (table 15.1). PCR is used to amplify genes that cause inherited disease and DNA sequences from viruses that cause infectious diseases, allowing physicians to run highly sensitive diagnostic tests. PCR can also amplify the DNA in small bits of tissue

FIGURE 15.14

DNA replication takes several steps. Parental strands are depicted as being thicker, to distinguish them from newly replicated DNA. Replication begins as the helix unwinds locally at many sites of origin, and RNA primers are synthesized. DNA polymerase extends the primers, DNA replaces the RNA primer, and the small, replicated portions of the chromosome join as ligase cements the sugar-phosphate backbone.

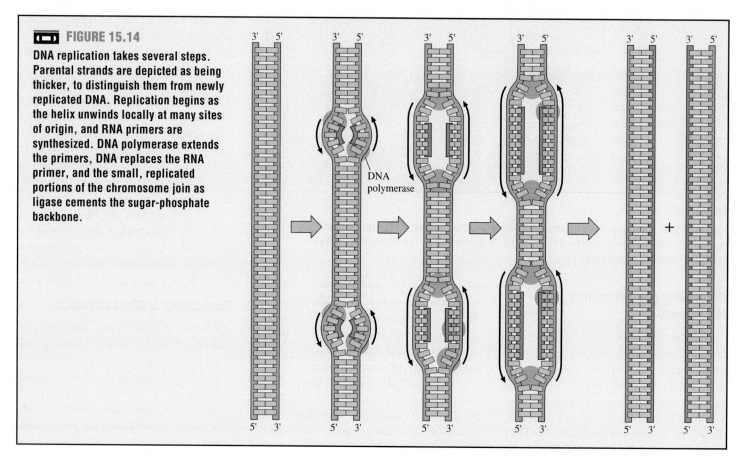

Table 15.1　PCR Applications

PCR Has Been Used to Amplify:

HIV genetic material in a human blood sample when infection is so recent that antibodies are not yet detectable.

A bit of DNA in a preserved quagga (a relative of the zebra) and a marsupial wolf, that recently became extinct.

DNA in sperm cells obtained from the body of a rape victim. Specific sequences were compared to DNA sequences from a suspect in the crime.

Genes from microorganisms that cannot be grown or maintained in culture for study.

Mitochondrial DNA from various modern human populations. Comparisons of mitochondrial DNA sequences indicate that *Homo sapiens* originated in Africa, supporting fossil evidence.

Genes from several organisms that are very similar in sequence. Comparing the extent of similarity reveals evolutionary relationships between species.

DNA from the brain of a 7,000-year-old human mummy. The DNA indicated that native Americans were not the only people to dwell in North America long ago.

Genetic material from saliva, hair, skin, and excrement of organisms that we cannot catch to study. The prevalence of a rare DNA sequence among all of the bird droppings from a certain species in an area can be extrapolated to estimate the population size.

DNA in the digestive tracts of carnivores. This can reveal food web interactions.

DNA in deteriorated road kills and carcasses washed ashore. This helps identify locally threatened species.

DNA in products illegally made from endangered species, such as powdered rhinoceros horn, which is sold as an aphrodisiac.

DNA sequences that are unique to the bacteria that cause Lyme disease. When these sequences are found in animals, they provide clues to how the disease is transmitted.

DNA from genetically altered microbes that are released in field tests, to trace their dispersion.

DNA from a cell of an 8-celled human preembryo, to diagnose cystic fibrosis.

Y-chromosome-specific DNA from a human egg fertilized in the laboratory, to determine the fertilized ovum's sex.

A papilloma virus DNA sequence present in, and possibly causing, an eye cancer.

An unusual form of HIV that an infected person does not manufacture antibodies against.

so that enough material is available from a crime scene to either establish or rule out the identity of a suspect. Other gene amplification methods use ligases and other types of replication enzymes.

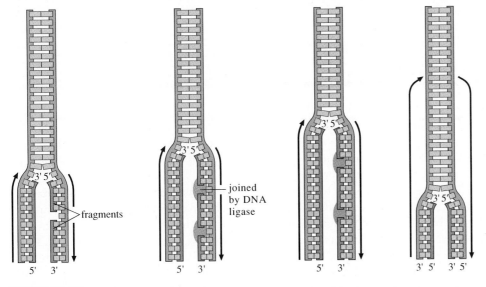

FIGURE 15.15

To maintain the 5′ to 3′ directionality of DNA replication, at least one strand must be replicated in short pieces, from the fork outwards. The pieces are joined by ligase.

KEY CONCEPTS

DNA replication occurs in several steps. Topoisomerases locally unwind the double helix. In humans, this occurs simultaneously from several points on each chromosome. At each initiation site, RNA primase directs the synthesis of a short RNA primer; DNA eventually replaces the RNA primer. DNA polymerase extends complementary DNA bases to the RNA primer, building a new strand against each template. Then ligase joins the sugar-phosphate backbone. DNA is synthesized in a 5′ to 3′ direction, discontinuously on one strand. The polymerase chain reaction can replicate large amounts of selected genes.

Using PCR to Find DNA Needles in a Haystack

Inspiration on a Starry Night

The polymerase chain reaction (PCR) formed in Kary Mullis's mind on a moonlit night in northern California in 1983, when knowledge of DNA structure was 30 years old. As he drove up and down the hills, Mullis, a modern-day Thomas Edison, was thinking about the incredible precision and power of DNA replication. Suddenly, a way to tap into that power popped into his mind. He excitedly explained his idea to his girlfriend and then went home to think it through further. "It was difficult for me to sleep with deoxyribonuclear bombs exploding in my brain," he wrote much later, after PCR went on to revolutionize the life sciences.

The idea behind PCR was so stunningly straightforward that Mullis had trouble convincing his superiors at Cetus Corporation that he was really onto something. He spent the next year using the technique to amplify a well-studied gene so he could prove that his brainstorm was not just a flight of fancy. One by one, other researchers glimpsed Mullis's vision, born that starry night. After he convinced his colleagues at Cetus, Mullis published his landmark 1985 paper and filed patent applications, launching the era of gene amplification. The technology is a direct application of the DNA replication mechanism.

Surprisingly Simple

PCR rapidly replicates a selected sequence of DNA in a test tube. The requirements include:

A known target DNA sequence to be amplified.

Two types of primers, both lab-made, single-stranded, short pieces of DNA. Their sequences must be complementary to opposite ends of the target sequence. A hefty supply of the four types of DNA nucleotide building blocks. Taq1—a DNA polymerase produced by *Thermus aquaticus,* a microbe that inhabits hot springs. This enzyme is adapted to its host's hot surroundings and does not fall apart when DNA is heated. (Other heat-tolerant polymerases can be used, too.)

In the first step of PCR, heat is applied to separate the two strands of the target DNA. Next, the temperature is lowered and the two short DNA primers are added. The primers bind by complementary base pairing to the separated target strands. In the third step, Taq1 DNA polymerase and bases are added. The enzyme adds bases to the primers and builds a sequence complementary to the target sequence (fig. 1). The newly synthesized strands then act as templates in the next round of replication, which is immediately initiated by raising the temperature. All of this happens in an automated device called a thermal cycler that controls the key temperature changes.

The pieces of DNA accumulate geometrically. The number of amplified pieces of DNA equals 2^n, where n equals the number of temperature cycles. After just 20 cycles, one million copies of the original sequence float in the test tube. Table 15.1 lists some of the diverse applications of PCR.

PCR's greatest strength is that it works on crude samples of rare and minute DNA sequences, such as a bit of brain tissue on the bumper of a car, which led to identification of a missing person in one criminal case. PCR's greatest weakness, ironically, is its exquisite sensitivity. If a blood sample submitted for diagnosis of an infection is

DNA Repair

Any manufacturing facility tests a product in several ways to see whether it has been put together correctly. Production mistakes are rectified before the item goes on the market—most of the time. The same is true of a cell's DNA production.

DNA replication is incredibly accurate—only about one in 100,000 bases is incorporated incorrectly. Just as genes control replication by encoding enzymes such as polymerases and ligases, they oversee the fidelity of replication by specifying repair enzymes.

The discovery of DNA repair systems began with observations made in the late 1940s. Investigators noted that when fungi were exposed to ultraviolet radiation, the cultures nearest a window grew best. Exposure to light remedied the DNA-damaging effects of ultraviolet radiation in a variety of organisms.

Ultraviolet radiation damages DNA by causing an extra covalent bond to form between adjacent pyrimidines, particularly thymines. The linked thymines are called thymine dimers. This extra attachment forms a kink in the otherwise sleek double helix. It can disrupt replication and lead to the insertion of a noncomplementary base. However, **photoreactivation enzymes** absorb energy from visible light and use it to detect and bind to pyrimidine dimers, breaking the extra bond (fig. 15.16). This type of repair mechanism enables ultraviolet-damaged fungi to recover when placed in sunlight.

In the early 1960s, scientists discovered another type of DNA self-

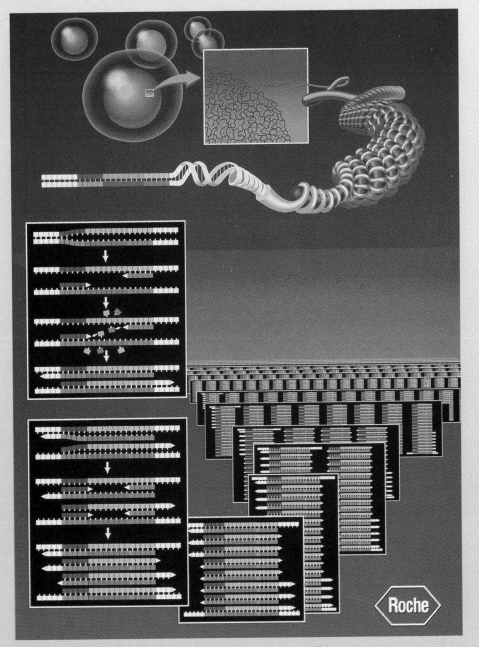

contaminated by leftover DNA from a previous run, or if a stray eyelash drops from the person running the reaction, it can yield a false positive result. The technique is also limited because a user must know the sequence to be amplified.

As for Kary Mullis, he prefers the moonlight to the limelight. As PCR mania surged, he left the world of corporate biotechnology to become an independent consultant. Some bitterness surrounds the way he was treated. Cetus paid him $10,000 for what may be the greatest invention of the century—then, a few years later, sold the technology to another company for $300 million without further compensating Mullis. However, he won both the Japan prize and the Nobel Prize in 1993. We probably haven't heard the last from Kary Mullis. He continues to invent, and says he has projects on the drawing board that will dwarf even the incredible impact of PCR.

mending, **excision repair,** in mutant *E. coli* bacteria that were extra-sensitive to ultraviolet radiation. The enzymes that carry out excision repair cut the bond between the DNA sugar and base and snip out the pyrimidine dimer and surrounding bases. Then, a DNA polymerase fills in the correct nucleotides, using the exposed template as a guide. Another collection of enzymes "proofreads" the newly replicated DNA, spotting and replacing mismatched bases in a process called **postreplication repair.**

Dozens of different genes specify the enzymes for DNA repair. When even one such enzyme is impaired, the health consequences can be devastating. DNA repair disorders are usually characterized by chromosome breaks and a high susceptibility to cancer following exposure to ionizing radiation or chemicals that affect cell division. Biology in Action 15.2 discusses some of these disorders, and figure 15.17 shows a child with one of them—a condition called xeroderma pigmentosum. DNA repair systems may also be damaged by excessive sun exposure, leading to skin cancer.

KEY CONCEPTS

DNA replication is incredibly accurate. Many genes encode enzymes that search through replicating DNA for errors and correct them. Ultraviolet radiation can cause the formation of thymine dimers and can lead to noncomplementary base insertion. Photoreactivation enzymes, excision repair, or postreplication repair mechanisms can replace mismatched bases. Abnormal repair genes cause disorders characterized by chromosome breaks and predisposition to cancer.

Biology in *ACTION*

15.2

DNA Replication and Repair Disorders

Life is far from normal for a child with *xeroderma pigmentosum* (XP). While other youngsters burst out of their houses on a sunny day, a child who has XP must wear long sleeves and pants even in midsummer and must apply sunscreen to every bit of exposed skin. Moderate sun exposure leads easily to skin sores or cancer and, even with precautions, the child's skin is a sea of freckles.

XP was first described in Vienna in 1874, and in the 1960s was linked to defects in DNA repair. XP appears in only one out of 250,000 births. Because it—like the other repair disorders—is inherited as an autosomal recessive trait, parents who have one affected child face a 25% risk of producing a child with the problem each pregnancy. The impaired ability to repair DNA predisposes the person with XP to a variety of cancers.

Ataxia telangiectasia (AT) appears first as a lack of muscle control (ataxia) in infancy, which worsens as the child begins to walk.

Table 1	**DNA Replication and Repair Disorders**	
Disorder	**Frequency**	**Defect**
Ataxia telangiectasia	1/40,000	Deficiency in topoisomerases (cannot break DNA strands)
Bloom syndrome (two types)	100 cases since 1950	DNA ligase is inactive or heat sensitive, slowing replication
Fanconi anemia (several types)	As high as 1/22,000 in some populations	Deficient excision repair
Xeroderma pigmentosum (nine types)	1/250,000	Deficient excision repair

An uncontrollable gait soon leads to the need for a wheelchair. The child's eyes dart uncontrollably. AT affects a part of the brain's cerebellum that monitors balance, coordination, and posture. Red marks (telangiectasia) appear on the face and neck, caused by dilations of the smallest blood vessels just under the skin. The child is prone to frequent lung infections.

The most serious symptom of AT is impaired ability to repair chromosome breaks caused by extreme sensitivity to ion-

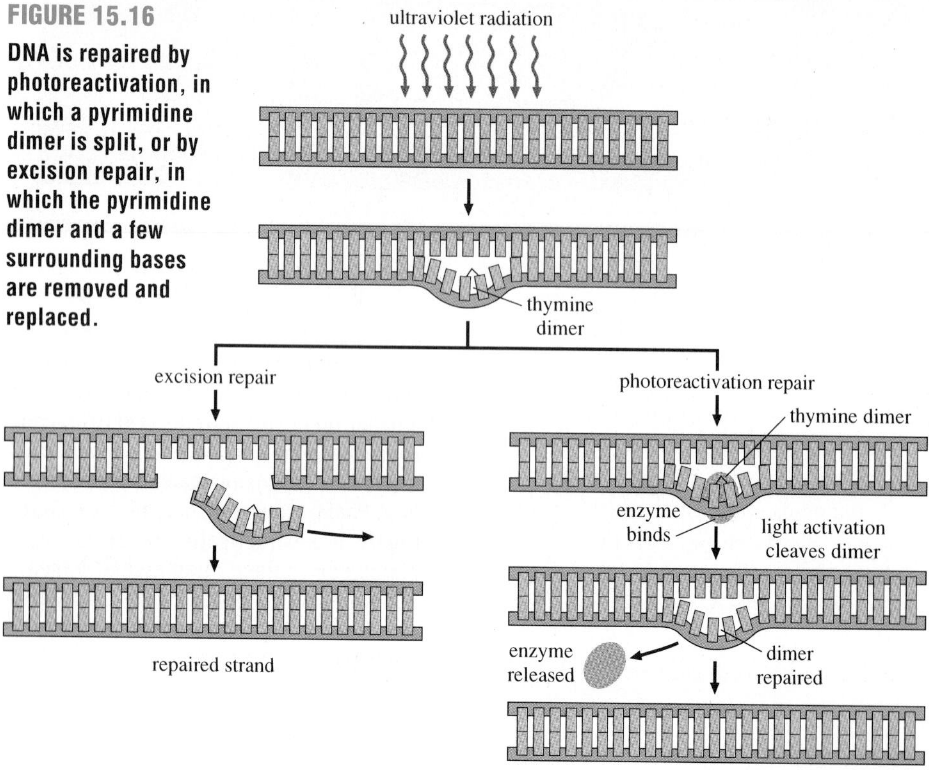

FIGURE 15.16

DNA is repaired by photoreactivation, in which a pyrimidine dimer is split, or by excision repair, in which the pyrimidine dimer and a few surrounding bases are removed and replaced.

ultraviolet radiation

thymine dimer

excision repair

photoreactivation repair

thymine dimer

enzyme binds

light activation cleaves dimer

repaired strand

enzyme released

dimer repaired

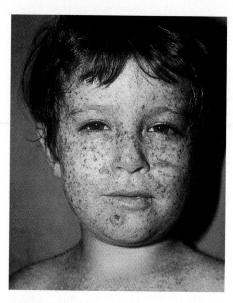

FIGURE 15.17

The marks on this child's face are a result of sun exposure. The child is highly sensitive to sunlight because he has inherited xeroderma pigmentosum. The large lesion on his chin is a skin cancer.

342

izing radiation such as X rays. Studies on cells from AT patients support this idea. In normal cells, X-ray exposure delays DNA replication. But in X-rayed cells from AT patients, DNA replication continues, apparently without pausing to repair the damage.

The clinical consequence of impaired DNA repair is a cancer risk 61 to 184 times higher than that of the general population. About 15% of AT patients develop cancer, usually leukemia or lymphoma. Ironically, a common treatment for many cancers—ionizing radiation—severely damages tissue in AT patients.

The 1% of the United States white population who are AT carriers may have mild symptoms. They face a 2- to 6-fold greater risk than the general population for many cancers. The risk is even greater—a 7-fold increase—for breast cancer, and 9 to 18% of all breast cancer patients are carriers of AT. Like the person whose body cannot normally repair damage caused by sun exposure, both the AT homozygote and heterozygote may be unable to recover from low doses of X rays that pose no threat to others.

A person with *Bloom syndrome* confronts some of the problems faced by XP and AT patients. A rash of telangiectases appears on the face as early as infancy, in response to extreme sensitivity to the sun. The rash, which worsens in the summer, affects males to a greater extent than females. The child with Bloom syndrome is characteristically small and has a narrow face, large nose, and small chin. Respiratory and digestive infections are common. As in the other repair disorders, in Bloom syndrome, chromosome breaks go unrepaired, and the result is a predisposition to cancer, particularly leukemia.

The culprit behind Bloom syndrome is ligase, the enzyme that glues the sugar-phosphate backbone of the DNA double helix. Abnormal ligase causes replication forks to move too slowly.

Symptoms of *Fanconi anemia* result from impaired ability to excise ultraviolet radiation-induced thymine dimers (fig. 1). The anemia affects all blood cell types. It can be corrected only by a bone marrow transplant. A child who has Fanconi anemia is short and has a triangular face, dark skin,

and stunted or absent thumbs. Learning disabilities or mental retardation are often part of the syndrome. Like people with other repair disorders, the child with Fanconi anemia cannot heal chromosome breaks and faces a high risk of developing leukemia.

FIGURE 1

The chromosomes of a patient with a DNA repair disorder appear characteristically fragmented upon treatment with certain chemicals. The patient who donated these cells has Fanconi anemia.

SUMMARY

The genetic material—DNA—must have the capacity to encode the information necessary for a cell's survival and specialization, and to replicate. An impressive set of experimenters discovered and described DNA. Miescher identified DNA in white blood cell nuclei, and later Hammerling localized it to the nucleus in *Acetabularia*. Garrod conceptually connected heredity to symptoms caused by enzyme abnormalities. Griffith identified a substance that transmits infectiousness in pneumonia-causing bacteria; Avery, MacLeod, and McCarty discovered that the transforming principle is DNA; Hershey and Chase confirmed that the genetic material is DNA and not protein. Levene described the three components of a nucleotide and found that

they appear in DNA in equal amounts. Chargaff discovered that the amount of adenine equals the amount of thymine, and the amount of guanine equals that of cytosine. Watson and Crick put all these clues together to propose the double helix conformation of DNA.

The rungs of the DNA double helix consist of hydrogen-bonded complementary base pairs (A with T, and C with G). The rails are chains of alternating sugars and phosphates, which run antiparallel to each other. DNA is highly coiled.

Meselson and Stahl proved the semiconservative nature of DNA replication with density shift experiments. During replication, the DNA unwinds locally at several initiation points.

Replication forks form as the hydrogen bonds break between an initial base pair. RNA polymerase builds a short RNA primer, which is eventually replaced with DNA. Next, DNA polymerase fills in DNA bases, and ligase seals the sugar-phosphate backbone. Replication proceeds in a 5′ to 3′ direction, necessitating that the process be discontinuous in short stretches on at least one strand.

Gene amplification techniques, such as PCR, utilize the power and specificity of DNA replication enzymes to selectively amplify certain sequences. DNA can repair itself in a variety of ways, relying on DNA polymerase proofreading, excision repair, photoreactivation, and postreplication repair.

KEY TERMS

adenine 332
antiparallelism 334
chromatin 334
complementary base pairs 334
conservative replication 335
cytosine 332
deoxyribonucleic acid 331

deoxyribose 331
dispersive replication 335
DNA polymerase 337
excision repair 341
gene amplification 338
guanine 332
initiation site 337

ligase 337
nucleosome 334
nucleotide 331
photoreactivation enzyme 340
postreplication repair 341
purine 334
pyrimidine 334

replication fork 337
ribonucleic acid 331
ribose 331
RNA primer 337
semiconservative replication 335
thymine 332
uracil 337

REVIEW QUESTIONS

1. DNA specifies and regulates the cell's synthesis of protein. If a cell contains all the genetic material it must have to carry out protein synthesis, why must the DNA also be replicated?

2. How are the designs of the experiments conducted by Avery, MacCleod, and McCarty; Hershey and Chase; and Meselson and Stahl similar?

3. What part of the DNA molecule encodes information?

4. Place the following proteins in the order they begin to function in DNA replication:
ligase
DNA polymerase
RNA polymerase
topoisomerases

5. Write the complementary sequence of a daughter strand of DNA replicated from each of the following parental base sequences:
  a. T C G A G A A T C T C G A T T
  b. C C G T A T A G C C G G T A C
  c. A T C G G A T C G C T A C T G

6. A person with deficient or abnormal ligase or excision repair may have an increased cancer risk and chromosomes that cannot heal breaks. The person is, nevertheless, alive. How long would an individual lacking DNA polymerase be likely to survive?

7. Why don't we see drastic mutations in the genes encoding histone proteins?

8. Choose an experiment mentioned in the chapter and analyze how it follows the scientific method.

TO THINK ABOUT

1. To diagnose a rare form of encephalitis (brain inflammation), a researcher needs a million copies of a viral gene. She decides to use the polymerase chain reaction on a sample of cerebrospinal fluid, the fluid that bathes the person's infected brain. If one cycle of PCR takes two minutes, how long will it take the researcher to obtain her million-fold amplification?

2. Give an example from the chapter when different types of experiments were used to address the same hypothesis. Why might this be necessary?

3. The experiments that revealed DNA structure and function utilized a variety of organisms—from peas, barley, fruit flies, and *E. coli* to yeast, molds, mice, and humans. How can such diverse organisms demonstrate the same genetic principles?

4. Avery, MacLeod, and McCarty had a difficult time proving that DNA is the genetic material because scientists were so convinced that protein served this function. Today, a few scientists, PCR inventor Kary Mullis among them, insist that HIV is not the sole cause of AIDS, and they face some of the same obstacles

Avery, MacLeod, and McCarty encountered. How should the scientific community treat scientists who espouse very unpopular ideas? What criteria should a hypothesis have to meet to be considered sufficiently proven? Should the public be made aware of such controversies? Should researchers whose ideas are outside the mainstream of current scientific thought receive funding?

SUGGESTED READINGS

Bootsma, D., and J. H. J. Hoeigmakers. May 13, 1993. DNA repair: Engagement with transcription. *Nature*, vol. 363. Although geneticists discovered DNA repair 30 years ago, they are still working out the details.

Cook, Peter R. August 23, 1991. The nucleoskeleton and the topology of replication. *Cell*. Preparing for replication is an awesome physical challenge.

Grunstein, Michael. October 1992. Histones as regulators of genes. *Scientific American*. Once thought to be mere bystanders to gene action, histones are now known to initiate and repress transcription and to serve as scaffolds for the DNA to wind around.

Jaroff, Leon. March 15, 1993. Happy birthday, double helix. *Time*. Watson and Crick shared a rare visit on the fortieth anniversary of their monumental deciphering of DNA structure.

Kornberg, Arthur. 1992. *DNA replication*. 2d ed. New York: W. H. Freeman. Kornberg made many of the original discoveries concerning DNA replication. This volume is an update of a classic.

Lewis, Ricki. January 21, 1991. Innovative alternatives to PCR technology are proliferating. *The Scientist*. Replication enzymes are used to selectively amplify genes.

Lewis, Ricki. July 26, 1993. PCR innovations. *The Scientist*. PCR continues to fit an incredible diversity of applications.

Mullis, Kary B. April 1990. The unusual origin of the polymerase chain reaction. *Scientific American*. How the technique of gene amplification arose from a brainstorm.

Richards, Robert I., and Grant R. Sutherland. February 1994. Simple repeat DNA is not replicated simply. *Nature Genetics*, vol. 6. When a gene contains extensive regions of short repeats, replication can be a problem.

Strauss, Bernard S. November 25, 1993. DNA repair: The nick in time. *Nature*, vol. 366. Researchers continue to unravel the steps of DNA repair in yeast.

Watson, James D. 1968. *The double helix*. New York: New American Library. An exciting, personal account of the discovery of DNA structure.

Watson, James D., and F. H. C. Crick. April 25, 1953. Molecular structure of nucleic acids: A structure for deoxyribose nucleic acid. *Nature*, vol. 171, no. 4356. The original paper describing the structure of DNA.

CHAPTER 16

Gene Function

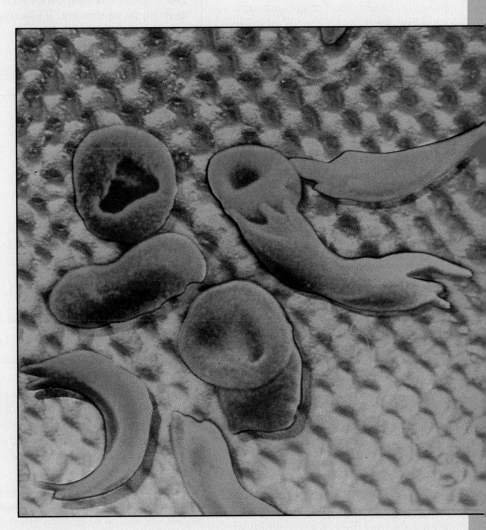

A mutation in a single DNA base causes these red blood cells to form a sickle shape.

LEARNING OBJECTIVES

By the chapter's end, you should be able to answer these questions:

1. Why is DNA transcribed into RNA?

2. How does RNA differ chemically from DNA?

3. What are RNA's functions?

4. How is transcription selectively activated in certain cells?

5. What are the steps of transcription?

6. What types of experiments eventually deciphered the genetic code?

7. What are the steps in building an amino acid chain from information transcribed into RNA?

8. How can a gene mutate in different ways at different sites?

9. What causes a mutation?

10. Why don't all mutations alter the phenotype?

Anatomy of an Illness

The sisters at first puzzled their doctors. At seven and three years of age, the girls clearly suffered from the same disorder. Their too-soft leg bones bowed outwards, bearing so much strain that their knees protruded like knobs. The girls' rib cages caved in. They had a history of other bone abnormalities since infancy, and both had lost most of their hair. They had the unmistakable symptoms of rickets, a bone-weakening disease that usually results from vitamin D deficiency—but these children followed a healthy diet.

The doctors had one intriguing hint to the nature of the girls' medical problem: their parents were second cousins. The parents could have both inherited the same autosomal recessive disease-causing gene from a common ancestor. Each girl would have then inherited two copies of the disease-causing allele—one from each carrier parent.

A second definitive clue appeared when physicians analyzed the relative amounts of intermediate biochemicals the girls' cells produced as they used vitamin D. The diagnosis: a very rare, inherited form of vitamin D refractory rickets. Even though the girls ate foods rich in vitamin D, took vitamin supplements, and got plenty of sunshine (needed to activate vitamin D precursors in the skin), they were still sick because their bodies could not use the abundantly supplied vitamin. But at what point did the cell's utilization of vitamin D go wrong?

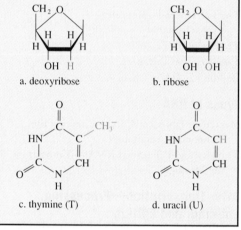

FIGURE 16.1

DNA and RNA differ structurally from each other in three ways. First, DNA nucleotides contain the sugar deoxyribose (*a*), whereas RNA contains ribose (*b*). Second, DNA nucleotides include the pyrimidine thymine (*c*), whereas RNA has uracil (*d*). And third, DNA is double-stranded, while RNA is generally single-stranded. The two types of nucleic acids also have different functions.

a. deoxyribose

b. ribose

c. thymine (T)

d. uracil (U)

The small intestine may have failed to absorb vitamin D, so that it never entered the bloodstream. The girls might have lacked an enzyme that modifies the vitamin into an accessible form. Or perhaps the problem was in the receptor protein that normally binds the vitamin D derivative in the cell and ferries it to the DNA, where it turns on genes that encode bone-building proteins. The receptor might not have been properly binding to the vitamin derivative, or to the DNA.

The answer finally came from laboratory studies of the girls' connective tissue cells (fibroblasts). The doctors isolated their vitamin D receptors and mixed them with activated vitamin D. Like a ferry able to pick up passengers but unable to dock at the destination, the receptors picked up the activated vitamin and transported it within the cell, but could not bind to the DNA. The activated vitamin D, in plentiful supply because of the girls' healthy lifestyle, never interacted with their DNA.

Transcription—The Genetic Crossroads

The sisters with hereditary rickets illustrate the importance of the middle stage of gene expression—the **transcription** of an RNA intermediate from DNA. RNA differs from DNA in three ways: it contains ribose instead of deoxyribose, it contains the nitrogenous base uracil in place of thymine, and it is single- rather than double-stranded (fig. 16.1).

DNA replication preserves genetic information. **Translation** literally translates the information in a DNA sequence into a protein's amino acid sequence. Between replication and translation lies transcription, a genetic crossroads that determines which genes are expressed. Vitamin D refractory rickets is one of many clinical consequences of transcription gone awry.

Transcription is far more complex than biologists originally realized. The central dogma of molecular biology, a

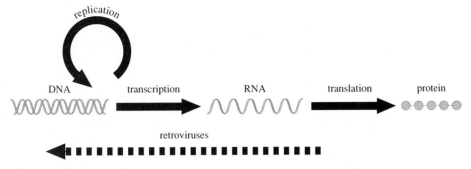

FIGURE 16.2

The central dogma of molecular biology, circa 1965, envisioned DNA being replicated and transcribed into RNA, some of which was then translated into protein. All organisms and most viruses exhibit this directional flow of genetic information. A class of viruses called retroviruses, which includes the human immunodeficiency virus thought to cause AIDS, uses RNA as its genetic material. Once in a host cell, these viruses use an enzyme, reverse transcriptase, to manufacture DNA from RNA, confounding the early theories. The original central dogma also oversimplified the role of RNA. This versatile nucleic acid assists in DNA replication, and we are still discovering how RNA helps control gene expression.

paradigm envisioned shortly after Watson and Crick elucidated DNA structure, portrayed RNA as simply a stepping-stone between a gene's nucleotide base language and a protein's amino acid language (fig. 16.2). Although essentially correct, the central dogma only explains the end result of transcription—how a gene's information enables a cell to manufacture protein. It does not explain how a cell "knows" which genes to express and when to express them. What, for example, directs a bone cell to transcribe the genes controlling collagen synthesis rather than muscle-specific proteins? The central dogma also does not explain the "backward" flow of genetic information in the retroviruses, such as the human immunodeficiency virus (HIV). These viruses inject RNA into a cell, then reverse transcribe it into DNA that lodges in the host's genome. RNA is far more than a go-between that connects gene to protein—it participates in all aspects of gene expression.

The Lac Operon—Revealing Transcriptional Complexity

Scientists gained an inkling of the complexity of gene expression back in 1961. French biologists François Jacob and Jacques Monod described the remarkable ability of *E. coli* bacteria to produce enzymes needed to metabolize the sugar lactose only when lactose was present in the cell's surroundings. What was "telling" a simple bacterial cell to transcribe the genes whose products metabolize lactose at precisely the right time?

The lactose itself proved to be the trigger. It binds to a protein that normally sits atop a certain DNA sequence. This sequence contains the signal that tells the cell to begin transcribing the three enzymes needed to break down the sugar. The lactose removes the protein, and the cell synthesizes the enzymes. Lactose, in a sense, causes its own dismantling.

Jacob and Monod named the genes (and their controls) that produce the enzymes needed for lactose metabolism an **operon.** Soon, geneticists discovered operons for the metabolism of other nutrients, and in other bacteria. Some, like the lac operon, negatively control transcription by removing a block. Others act positively, producing factors that turn on transcription.

Transcription Factors

Bacterial operons function like switches, turning gene transcription on or off. In multicellular eukaryotes, genetic control is more complex because different cell types express different subsets of genes. To manage such complexity, groups of proteins called **transcription factors** come together, bind DNA, and initiate transcription. The transcription factors, activated by extracellular signals, begin transcription by forming a pocket for **RNA polymerase (RNAP)**—the enzyme that actually builds an RNA chain.

Four or more transcription factors are required to transcribe a eukaryotic gene. Genes for transcription factors may be located near the genes they control, or as far away as 40,000 bases. The DNA may be able to form loops so that the transcription factor genes encoding proteins that act together come near each other. **Nuclear matrix proteins** organize the cell's nucleus so that genes actively being transcribed, and transcription factors in use, are partitioned off together.

Scientists have identified hundreds of transcription factors. Many of them have common localized regions called **motifs** that fold into similar three-dimensional shapes or conformations. The abnormality in the vitamin D receptor protein that caused the rickets described earlier is located in a motif called a **zinc finger** (fig. 16.3). The zinc fingers are the sections of the receptors that bind DNA by extending outward, resembling fingers. The affected vitamin D receptors can bind the activated vitamin D derivative, but they have "broken" zinc fingers. Biology in Action 16.1 describes yet another type of transcription factor and its effects.

Transcription Initiation

A section of DNA may contain thousands of genes. How do transcription factors and RNA polymerase "know" where to bind to DNA to initiate transcription for a specific gene? Transcription factors and RNA polymerase are attracted to certain control sequences located near the start of the gene, which collectively form a region called the **promoter.** The first transcription factor to bind is attracted to a sequence called a TATA box, which consists of the base sequence TATA (thymine, adenine, thymine, adenine). Once the first transcription factor is bound, it attracts others, and finally RNA polymerase joins the complex

FIGURE 16.3

Vitamin D refractory rickets can be caused by a mutation in a gene encoding a zinc finger region of the steroid receptor protein, which prevents the protein from binding to DNA. The fingers form as the result of four strategically located cysteines. These amino acids attract each other because they contain sulfur. The attraction creates a pocket to entrap zinc, which in turn stabilizes the finger formation.

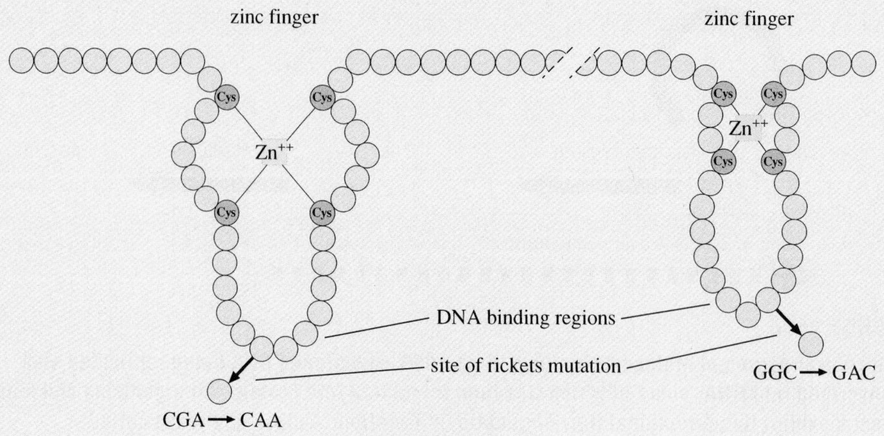

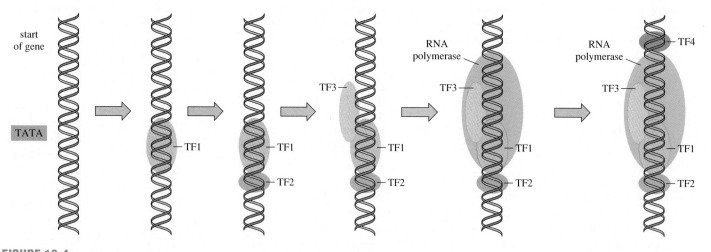

FIGURE 16.4

Transcription factors come together to start transcription of a gene. To begin transcription, four or more protein transcription factors bind near the start of the gene, guided by a "TATA" box. The bound transcription factors then enable an RNA polymerase to bind. The entire region where transcription factors and RNA polymerase bind is the promoter.

(fig. 16.4). Other sequences in or near the promoter regulate the frequency of transcription. The transcription factor complex guides RNA polymerase so that it binds just before the gene sequence starts.

Building an RNA Chain

Transcription, like DNA replication, is guided by complementary base pairing. First, enzymes unwind the DNA double helix. RNA nucleotides bond with the exposed bases on one strand (fig. 16.5). RNA polymerase knits together the RNA nucleotides in the sequence specified by the DNA, moving along the DNA strand in a 3′ to 5′ direction. For example, the DNA sequence GCG-

TATG is transcribed into RNA with the sequence CGCAUAC. (Remember, RNA contains uracil in place of thymine.) A terminator sequence in the DNA indicates the end of the gene's RNA-encoding region.

RNA is transcribed from only one strand of the DNA double helix for a particular gene. The transcribed DNA strand is called the **sense strand,** and the DNA strand that isn't transcribed is called the **antisense strand.** In antisense technology, researchers introduce synthetic DNA strands that are complementary to the sense strand for a particular gene, which "silences" that gene's expression. For example, silencing the gene that controls a ripening enzyme in tomatoes extends shelf life by delaying ripening.

KEY CONCEPTS

Transcription is the intermediate step between gene replication and gene expression (protein synthesis). A gene sequence is transcribed into RNA, a single-strand nucleic acid with ribose instead of deoxyribose and uracil instead of thymine. Transcription is highly controlled; in prokaryotes, genes are expressed as needed, and in multicellular organisms, specialized cell types express the genes required to perform specific functions. Transcription factors recognize sequences near a gene and bind sequentially, creating a binding site for RNA polymerase. RNA polymerase then begins transcription. Transcription proceeds as RNAP inserts complementary RNA bases opposite the sense strand of the DNA double helix.

Biology in ACTION

Mutant Mice Reveal the Role of a Muscle-Specific Transcription Factor

Muscle development in mice begins when early embryo structures called somites begin to specialize into myoblasts, or immature muscle cells. Individual myoblasts migrate and divide, cease dividing, then fuse to form the multinucleate fibers of skeletal (voluntary) muscle.

Researchers have identified four types of muscle-specific transcription factors. Each prompts a variety of cell types growing in culture to develop the long, spindly contractile fibers characteristic of muscle cells. But there is a big difference between cells nurtured in laboratory glassware and cells in a living organism.

Two groups of investigators, in Texas and Japan, decided to study muscle formation in mice. They bred mice with inactivated forms of both alleles encoding a muscle transcription factor called myogenin. The newborn homozygous recessive mice (–/–) were strikingly different from their heterozygous (+/–) and homozygous wild type (+/+) siblings—but the researchers didn't realize how different until they dissected the mutant mice. The mutant newborn mice could not move (although their hearts beat because heart muscle is not skeletal muscle). They could not take a breath; their breathing muscles were practically nonexistent. This

explained their bluish color. Skeletal muscle was greatly diminished throughout their oddly shaped bodies.

Figure 1 shows the abnormalities that afflict mice lacking this transcription factor. Mice that do not produce myogenin have curved spines, deformed ribcages, thickened tissue on the backs of their necks, and weak skeletons because of the lack of muscular support. Curiously, the muscle-depleted regions of their bodies have normal numbers of nuclei. This indicates that while cells were in place, they never specialized into the characteristic multinucleate cells of skeletal muscle.

FIGURE 1

Absence of a transcription factor leads to death at birth. The newborn mice in the top panel of (*a*) never took their first breaths, because their breathing muscles are nearly absent. Note also their abnormal shapes. The mouse on the bottom to the left is a heterozygote, and to the right, a homozygous wild type. In (*b*), red and blue stains highlight skeletal abnormalities in two of the mice. These defects result from the failure of immature muscle cells to specialize, which in turn is due to absence of a transcription factor.

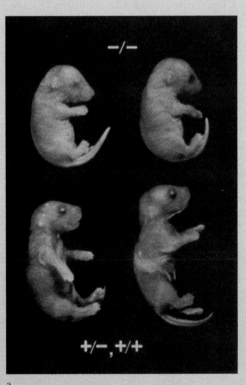

a.

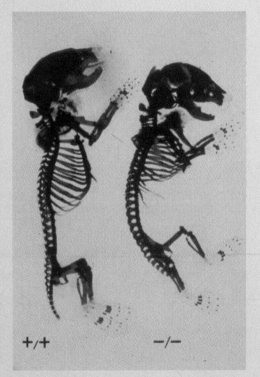

b.

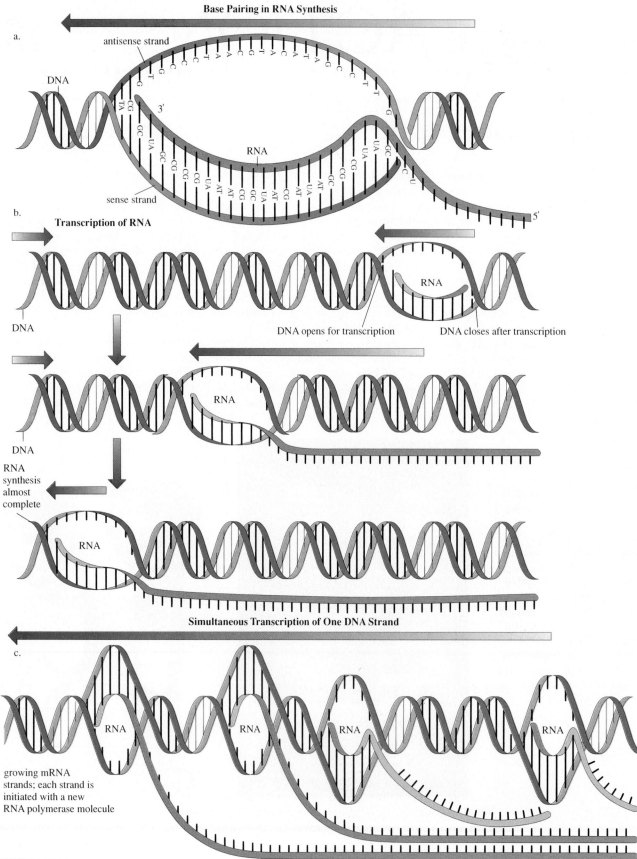

Base Pairing in RNA Synthesis

a.

antisense strand

DNA

sense strand

RNA

b.

Transcription of RNA

DNA

DNA opens for transcription   DNA closes after transcription

DNA

RNA

synthesis

almost

complete

RNA

Simultaneous Transcription of One DNA Strand

c.

growing mRNA
strands; each strand is
initiated with a new
RNA polymerase molecule

FIGURE 16.5

Transcription of RNA from DNA. (*a*) The base sequence of one side of
the DNA double helix dictates the sequence of the RNA. (*b*) The DNA
double helix opens and closes, allowing RNA polymerase access
to the DNA sequence RNA is transcribed from. (*c*) The same DNA
sequence can be transcribed several times simultaneously.

Types of RNA

As RNA is synthesized along its DNA template, it curls into conformations determined by complementary base pairing within the same RNA molecule. These three-dimensional shapes determine how RNA functions, as we shall soon see. Several types of RNA exist (table 16.1).

Messenger RNA (mRNA)—Carrying the Genetic Information

Messenger RNA (mRNA) carries the information specifying a particular protein product. Each three mRNA bases in a row forms a genetic code word, or **codon,** that specifies a particular amino acid. Because genes vary in length, mRNA molecules do also. Most mRNAs are 500 to 1,000 bases long.

Transfer and Ribosomal RNAs—Translating the Message

Two other types of RNA help utilize the information carried in an mRNA.

Ribosomal RNA (rRNA) is 100 to nearly 3,000 nucleotides long. It associates with certain proteins to form a **ribosome,** a structural support for protein synthesis (fig. 16.6). A ribosome has two subunits that are separate in the cytoplasm, but which join at the site of protein synthesis.

Transfer RNA (tRNA) molecules are "connectors" that bind to an mRNA codon at one end and to a specific amino acid at the other. A tRNA molecule is small, only 75 to 80 nucleotides long. Some of its bases form weak (hydrogen) bonds with each other, folding the tRNA into a characteristic cloverleaf shape (fig. 16.7). One loop of the tRNA has three bases in a row that form the **anticodon,** which is complementary to an mRNA codon. The end of the tRNA opposite the anticodon forms a strong (covalent) bond to a specific amino acid. A tRNA with a particular anticodon always carries the

Table 16.1	**Major Types of RNA**	
Molecule	**Size (number of nucleotides)**	**Function**
mRNA	500–1,000	Codons encode amino acid sequence
tRNA	75–80	Forms cloverleaf shape that binds mRNA codon on one end, amino acid on the other, physically connecting a gene's message to the amino acid sequence it encodes
rRNA	100–3,000	Associates with proteins to form ribosomes, which provide structural support for protein synthesis

FIGURE 16.6

A ribosome consists of about half protein and half ribosomal RNA. In *E. coli,* the larger subunit is built of two rRNA molecules and 35 proteins, and the smaller subunit is one rRNA molecule and 21 proteins. A ribosome from a eukaryote also has two subunits, but they are each slightly larger than the corresponding prokaryotic subunits and contain 80 proteins and four rRNA molecules altogether.

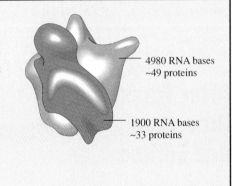

4980 RNA bases ~49 proteins

1900 RNA bases ~33 proteins

FIGURE 16.7

Transfer RNA. Certain nucleotide bases within a tRNA molecule hydrogen bond with each other to give the molecule a conformation that can be represented in two dimensions. The dotted lines indicate hydrogen bonds, and the filled-in bases at the bottom form the anticodon. Each tRNA terminates with the sequence CCA, and a particular amino acid covalently bonds with the RNA at this end.

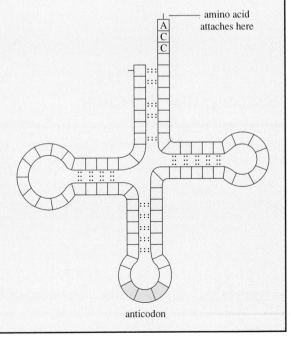

amino acid attaches here

anticodon

same amino acid. For example, a tRNA with the anticodon sequence AAG always picks up the amino acid phenylalanine. Special enzymes attach amino acids to the rRNAs that bear the appropriate anticodons.

After Transcription— Processing, Splicing, and Editing

In prokaryotes, RNA is translated into protein at the same time it is transcribed from DNA. In eukaryotic cells, mRNA must first exit the nucleus to enter the cytoplasm, where protein synthesis occurs. RNA undergoes great changes before it "matures" enough to participate in protein synthesis in these more complex cells.

Messenger RNA Processing

As mRNA is transcribed, a short **cap,** consisting of modified nucleotides, joins the 5′ end. A **poly A tail** of 100 to 200 adenines binds to the 3′ end as the mRNA moves away from the DNA, but is still in the nucleus. The cap and tail assist mRNA in its journey from the nucleus to the cytoplasm.

Splicing—Genes in Pieces

When nucleic acid sequencing technology made it possible to compare DNA and RNA base by base, researchers were in for a surprise. They expected to find a direct correspondence between the numbers of bases in a gene and in its RNA. Instead, the RNA was sometimes altered. RNA could either contain bases the gene did not specify or lack stretches of bases the gene did contain. A gene might be 10,000 bases long while its corresponding mRNA was only 6,000 bases long. "Extra" noncoding regions of a gene—such as the 4,000 bases in our example—are called **introns;** sequences of DNA bases in the gene that do appear in the "mature" RNA are called **exons** (fig. 16.8).

Investigators discovered introns as they studied mRNA. The introns are transcribed, but they are later cut out, and the cut ends of the remaining molecule are spliced together before the mRNA is translated. The mRNA is called **pre-mRNA** prior to intron removal. Introns are excised by small RNA molecules called **ribozymes** that function as enzymes—that is, they are not consumed in the splicing reaction. The ribozymes associate with proteins to form **small nuclear ribonucleoproteins (snRNPs),** or "snurps," as they are sometimes called. Several snurps work together to form a complex called a **spliceosome,** which actually cuts the introns out and knits the exons together to form the mature mRNA that exits the nucleus.

Introns range from 65 to 100,000 bases. While the average exon is 100 to 300 bases long, the average intron is about 1,000 bases long. Many genes are riddled with introns—the human collagen gene, for example, contains 50 of them. The number, size, and arrangement of introns varies from gene to gene. Biology in Action 16.2 discusses scientists' ideas on the functions of introns.

Translation— Expressing Genetic Information

To translate information from gene to RNA to protein, particular mRNA codons specify particular amino acids (fig. 16.9). This correspondence between the chemical languages of mRNA and protein is called the **genetic code.** In the 1960s, many researchers helped decipher which mRNA codons correspond to which amino acids using a combination of logic and experimentation. Certain questions had to be answered to understand the genetic code.

The Genetic Code— From Genetic Message to Protein Product

Question 1—How many RNA bases specify one amino acid?

Because the number of different protein building blocks (20) exceeds the number of different mRNA building blocks (four), each codon must contain more than one mRNA base. If a codon consisted of only one mRNA base, then only four different amino acids could be specified, one corresponding to each of the four bases: A, C, G, and U. If a codon consisted of two bases, then 16 different amino acids could be specified, one

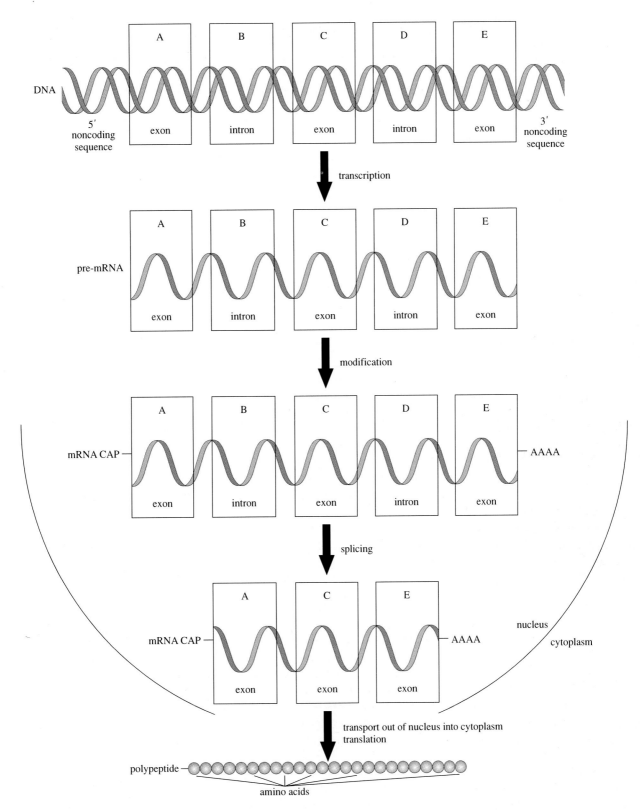

FIGURE 16.8

Messenger RNA processing—the maturing of the message. Several steps carve the mature mRNA. First, noncoding surrounding bases in the DNA are not transcribed. A cap head and poly A tail are added, and introns are spliced out.

Biology in ACTION

Introns and Ribozymes

Watson and Crick's 1953 portrait of the DNA double helix provided an elegant basis for the molecular picture of the gene. The gene, geneticists believed, was a continuous informational polymer that specified another informational polymer. By 1969, molecular biologists thought they had thoroughly described the major players in gene expression.

This straightforward view of the gene upended in 1977, when the ability to sequence DNA revealed genes far too big to account for their respective proteins. The 600-base rabbit beta globin gene, for example, encodes a chain of 146 amino acids. Since three DNA bases code for one amino acid, this protein should have been specified by a gene 438 bases long. Why did the gene contain 162 extra DNA bases?

In the 1970s, we learned that the extra bases are transcribed into RNA, but are removed before the protein is built. Soon, extra DNA was found in the genes of several eukaryotes (yeast, fruit flies, mice, chickens, frogs, humans, and others) and in their viruses. Much later, extra DNA was found in some prokaryotes (bacteria and cyanobacteria).

In February 1978, Harvard University's Walter Gilbert named the extra DNA sequences introns, for intragenic regions (also called intervening sequences). The expressed genes are termed exons. Gilbert and geneticists everywhere then struggled to shoehorn the unexpected complication of "genes in pieces" into the central dogma of molecular biology—the idea that genes are transcribed into RNA, which is translated directly into protein.

Introns explained some genetic mysteries. They accounted for the observation that messenger RNAs have longer precursor forms. Introns also explained excess DNA as compared to the number of known proteins, tRNAs, and rRNAs in well-studied bacteria. More theoretically, introns

provided a mechanism for rapid evolution of a new function. The intron could audition a new activity while the old sequence was retained, thus ensuring survival.

Some scientists advanced a very popular hypothesis—that introns act as mediators or "exon shufflers," bringing together exon sequences that code for different protein domains. This would enable a few exons to encode a great variety of proteins, much as a few items of clothing can be mixed and matched to create a varied wardrobe.

Francis Crick was among the biologists who, for a time, called introns "parasitic" or "junk" DNA, suggesting that they were molecular stowaways from past viral infections. Today our view of the role of introns—though still a matter of intense debate—has risen to a comparatively lofty height. Many believe that RNA transcribed from what was once called "junk" DNA may be descended from the primordial molecule of life. This change in thinking came when experimenters demonstrated that RNA is far more than a go-between that links DNA to protein.

Enter Ribozymes

In the early 1980s, Sidney Altman at Yale University was looking at enzymes that modify RNA. RNase P is an enzyme in *E. coli* that cleaves tRNA precursors to form the shorter, mature tRNA. Altman had discovered RNase P in the early 1970s while working in Francis Crick's lab. It was a peculiar enzyme in that it always had about 377 RNA bases tagging along. Was the persistent presence of the RNA the result of sloppy lab technique, or might it indicate something about the enzyme's function? The idea that a nucleic acid might assist an enzyme was radical, because enzymatic catalysis was thought to be restricted to proteins.

Altman showed that the RNA in RNase P was more than a stowaway. He started

by adding a nuclease to his preparations. This destroyed any lingering RNA—and it also destroyed the enzyme's function. Altman and Norman Pace of Indiana University then showed that the RNA portion of RNase P had catalytic activity.

Two years earlier, Thomas Cech at the University of Colorado in Boulder was trying to demonstrate how introns are spliced out of mRNA molecules. He studied ribosomal RNA (rRNA) in the single-celled eukaryote *Tetrahymena thermophila*. When he added an extract from the nucleus to the rRNA, splicing occurred. However, in the control experiment, he observed the activity of the RNA without the nuclear extract. Much to his surprise, the intron was still cut out and the exons still linked to form the mature mRNA that would be translated into protein. Cech had found that RNA alone could cut and paste introns and exons.

"It was quite a surprise," Cech recalled. "We had no reason to believe that RNA could catalyze a splicing reaction. We were looking for a protein enzyme, or even multiple enzymes." Cech and Altman shared the 1989 Nobel prize in chemistry after showing that these RNA enzymes, called ribozymes, are not consumed by their activity. Several companies worldwide are currently developing ribozymes as a new type of antiviral treatment.

In 1990, RNA was found to be both a catalyst and a template. Researchers at Massachusetts General Hospital built a ribozyme from part of an intron in a well-studied gene of a virus that infects *E. coli*. The ribozyme could replicate new RNA strands—using itself, unassembled and unwound, as a template!

RNA therefore carries information and has the ability to replicate itself. Such a multifunctional molecule fills the requirements for a biochemical first step to life. RNA may be the long-sought bridge between an inanimate chemical and life.

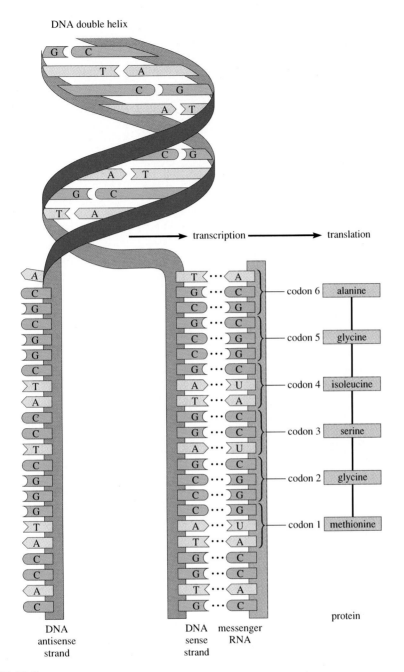

DNA double helix

transcription → translation

codon 6 — alanine
codon 5 — glycine
codon 4 — isoleucine
codon 3 — serine
codon 2 — glycine
codon 1 — methionine

protein

DNA antisense strand

DNA sense strand

messenger RNA

FIGURE 16.9

From DNA to RNA to protein. Messenger RNA is transcribed from a locally unwound portion of DNA. In translation, transfer RNA matches up mRNA codons with amino acids.

corresponding to each of the 16 possible combinations of two RNA bases (AA, CC, GG, UU, AC, CA, AU, UA, CG, GC, GU, UG, GA, AG, UC, and CU). If a codon consisted of three bases, then as many as 64 different amino acids could be specified. Because 20 different amino acids must be indicated by at least 20 different codons, the minimum number of bases in a codon is three.

Francis Crick and his coworkers conducted experiments that confirmed the triplet nature of the genetic code. They added one, two, or three bases within a gene with a known sequence and protein product. Altering the sequence by one or two bases greatly disrupted the coded order of amino acids, known as the **reading frame.** However, adding or deleting three contiguous bases caused only a localized addition or deletion of a single amino acid in the protein product.

Question 2—Does the genetic code overlap?

Consider the hypothetical mRNA sequence AUCAGUCUA. If the genetic code is triplet and does not overlap (that is, if each base is part of only one codon), then this sequence of nine bases contains only three codons: AUC, AGU, and CUA. If the code overlaps, the sequence contains seven codons: AUC, UCA, CAG, AGU, GUC, UCU, and CUA.

An overlapping code would pack maximal information into a limited number of bases. However, it would also constrain protein structure because certain amino acids would always follow certain others. For example, the amino acid specified by the first codon listed, AUC, would always be followed by an amino acid whose codon begins with UC. Experiments that sequence proteins show that no specific amino acid is always followed by another. Any amino acid can follow any other amino acid in a protein's sequence—the code does not overlap.

Question 3—Can mRNA codons signal anything other than amino acids?

Chemical analysis eventually showed that the genetic code contains directions for starting and stopping protein translation. The codon AUG signals "start," and the codons UGA, UAA, and UAG each signify "stop." A short sequence of bases at the start of each mRNA, called the **leader sequence,** also enables the mRNA to form hydrogen bonds with part of the rRNA in a ribosome. Thus, the codons do carry other signals.

Question 4—Do all species use the same genetic code?

The fact that all species use the same mRNA codons to specify the same amino acids is one piece of the abundant evidence that all life on earth evolved from a common ancestor. The genetic code is universal, with the exception of a few genes in the mitochondria of certain

Table 16.2 The Genetic Code

First Letter	Second Letter				Third Letter
	U	**C**	**A**	**G**	
U	UUU UUC } phenylalanine (phe) UUA UUG } leucine (leu)	UCU UCC UCA UCG } serine (ser)	UAU UAC } tyrosine (tyr) UAA "stop" UAG "stop"	UGU UGC } cysteine (cys) UGA "stop" UGG tryptophan (try)	U C A G
C	CUU CUC CUA CUG } leucine (leu)	CCU CCC CCA CCG } proline (pro)	CAU CAC } histidine (his) CAA CAG } glutamine (giln)	CGU CGC CGA CGG } arginine (arg)	U C A G
A	AUU AUC AUA } isoleucine (ilu) AUG methionine (met) and "start"	ACU ACC ACA ACG } threonine (thr)	AAU AAC } asparagine (asn) AAA AAG } lysine (lys)	AGU AGC } serine (ser) AGA AGG } arginine (arg)	U C A G
G	GUU GUC GUA GUG } valine (val)	GCU GCC GCA GCG } alanine (ala)	GAU GAC } aspartic acid (asp) GAA GAG } glutamic acid (glu)	GGU GGC GGA GGG } glycine (gly)	U C A G

single-celled organisms. The ability of a cell from one species to translate mRNA from another species makes recombinant DNA technology, discussed in the next chapter, possible.

Question 5—Which codons specify which amino acids?

Biochemist Marshall Nirenberg and his coworkers at the National Institutes of Health began deciphering the codons that specify particular amino acids in 1961. First they synthesized mRNA molecules in the laboratory. Then they added them to test tubes containing all the chemicals and structures needed for translation, which they had extracted from *E. coli* cells.

The first synthetic mRNA tested had the sequence UUUUUU. In the test tube, this translated into a polypeptide consisting entirely of one amino acid: phenylalanine. The first entry in the genetic code dictionary thus noted that

the codon UUU specifies the amino acid phenylalanine. The number of phenylalanines always equaled one-third of the number of mRNA bases, confirming that the genetic code is triplet and does not overlap. In the next three experiments, Nirenberg discovered that AAA codes for the amino acid lysine, GGG for glycine, and CCC for proline.

Next, the researchers synthesized chains of alternating bases. When they tested mRNA of the sequence AUAUAU, it introduced codons AUA and UAU. When translated, the mRNA yielded an amino acid sequence of alternating isoleucines and tyrosines. But was AUA isoleucine and UAU tyrosine, or vice versa?

An mRNA of sequence UUU-AUAUUUAUA encoded alternating phenylalanine and isoleucine. Because the first experiment showed that UUU codes for phenylalanine, the experimenters knew that AUA must code for

isoleucine. If AUA codes for isoleucine, they reasoned, looking back at the previous experiment, then UAU must code for tyrosine.

By the end of the 1960s, the researchers had deciphered the entire genetic code (table 16.2). Many research groups contributed to this monumental task. Crick organized an "amino acid club," inducting a new member whenever someone added a new piece to the puzzle of the genetic code.

Sixty-one of the possible 64 codons specify particular amino acids, while the others indicate "stop" or "start." (AUG specifies "start" as well as the amino acid methionine.) Because there are only 20 amino acids in biological proteins, some amino acids are specified by more than one codon. For example, UUU and UUC both encode phenylalanine. Different codons that specify the same amino acid are called **degenerate**. They often differ only in the third base. The degeneracy of the genetic code provides protection against mutation, because changes in the DNA that cause one degenerate codon to replace another would not affect the protein's amino acid sequence. We shall return to this point later in the chapter.

KEY CONCEPTS

To translate information from gene to protein, mRNA codons specify amino acids according to the genetic code. The genetic code is triplet, nonoverlapping, continuous, universal, and degenerate.

Building a Protein

Protein synthesis requires mRNA, tRNAs carrying amino acids, ribosomes, energy-storing molecules such as adenosine triphosphate (ATP), and various protein factors. These pieces come together at the beginning of translation in a stage called **initiation.** First, the mRNA leader sequence hydrogen bonds with a short sequence of rRNA in a small ribosomal subunit. The first mRNA codon to specify an amino acid is always AUG (the "start" signal). This codon attracts an initiator tRNA that carries a special form of the amino acid methionine, called formylated methionine (abbreviated fmet) (fig. 16.10). Fmet signifies the start of a polypeptide, or chain of amino acids. The small ribosomal subunit, the mRNA bonded to it, and the initiator tRNA with its attached fmet form the **initiation complex.**

In the next stage of translation, **elongation,** a large ribosomal subunit attaches to the initiation complex. The codon adjacent to the initiating codon (AUG) then bonds to its complementary anticodon. In figure 16.10, the codon following the initiation codon is GGA. The anticodon (CCU) is on one end of a free-floating tRNA that carries the amino acid glycine on the other end of its cloverleaf. The two amino acids (fmet and glycine in the example) align. A peptide bond forms between them, and the first tRNA is released. It will pick up another amino acid and be used again. The ribosome and its attached

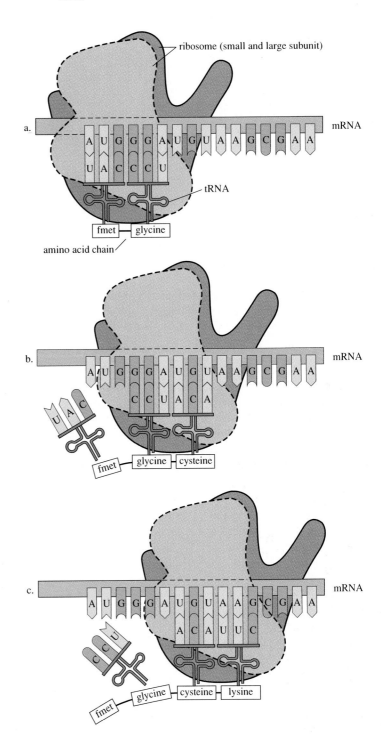

FIGURE 16.10

Translating a polypeptide. Translation begins when an mRNA molecule binds to a segment of rRNA that is part of a small ribosomal subunit. The anticodon of a tRNA bearing formylated methionine (fmet) hydrogen bonds to the initiation codon (AUG) on the mRNA. These bound structures form the initiation complex. Next, a large ribosomal subunit binds to the complex, and a tRNA bearing a second amino acid (glycine, in this example) forms hydrogen bonds between its anticodon and the second mRNA's codon (*a*). The fmet brought in by the first tRNA forms a dipeptide bond with the amino acid brought in by the second tRNA, and the first tRNA detaches and floats away. The ribosome moves down the mRNA by one codon, and a third tRNA arrives—in this example, carrying the amino acid cysteine (*b*). A fourth amino acid is linked to the growing polypeptide chain (*c*), and the process continues until a termination codon is reached.

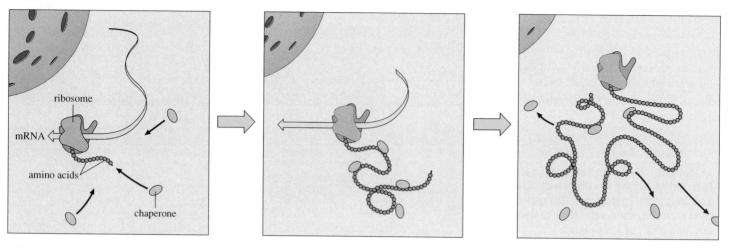

FIGURE 16.11

Protein folding. As this amino acid chain extends from the ribosome it is synthesized on, chaperone proteins bind the portions that will reside on the molecule's interior, leaving the other regions free to interact. When the final form is assumed, the chaperone proteins leave.

mRNA are now bound to a single tRNA with two amino acids extending from it. This forms the start of a polypeptide.

Next, the ribosome moves down the mRNA by one codon. A third tRNA molecule enters, carrying its amino acid (cysteine in fig. 16.10). This third amino acid aligns with the other two and forms a peptide bond to the second amino acid in the growing chain. The tRNA attached to the second amino acid is released and recycled. The polypeptide continues to build one amino acid at a time as new tRNAs attract. The tRNAs carry anticodons that correspond to the mRNA codons.

Elongation halts at an mRNA "stop" codon (UGA, UAG, or UAA) because there are no tRNA molecules that correspond to these codons. The last tRNA is released from the ribosome, the ribosomal subunits separate from each other and are recycled, and the new polypeptide floats away.

Protein Folding

For many years, biochemists thought that protein folding was straightforward; the amino acid sequence dictated specific attractions and repulsions between parts of a protein, tangling it into its final form as it emerged from the ribosome complex. But the attractions and repulsions of the amino acid sequence are not sufficient to ensure that the polypeptide assumes the highly specific form essential to its function. A protein apparently needs help to fold correctly.

An amino acid chain may start to fold as it emerges from the ribosome (fig. 16.11). Localized pockets form, and possibly break apart and form again, as translation proceeds. Experiments that isolate proteins as they are synthesized show that other proteins oversee the process. These accessory proteins include enzymes that foster chemical bonds and **chaperone proteins,** which stabilize partially folded regions important to the molecule's final form. Just as repair enzymes check newly replicated DNA for errors, accessory proteins scrutinize folding protein, detecting and dismantling incorrectly folded regions.

Certain proteins must also undergo structural modifications before they become functional. Sometimes enzymes must shorten a polypeptide chain for it to become active. Insulin, which is 51 amino acids long, for example, is initially translated as the polypeptide proinsulin, which is 80 amino acids long. Some polypeptides must join others to form larger protein molecules. The blood protein hemoglobin, for example, consists of four polypeptide chains.

Protein synthesis is economical. A cell can produce large amounts of a particular protein from just one or two copies of a gene. A plasma cell in the human immune system, for example, can produce 2,000 identical antibody molecules per second. To mass-produce on this scale, RNA, ribosomes, enzymes, and other proteins must be continually recycled. Many mRNAs can be transcribed from a single gene. Several ribosomes can simultaneously translate an mRNA, each at a different point along the message, producing polypeptides of different lengths that branch from them (fig. 16.12).

KEY CONCEPTS

As translation begins, mRNA, tRNA carrying bound amino acids, ribosomes, energy molecules, and protein factors assemble. The mRNA leader sequence binds to rRNA in the small subunit of a ribosome, and the first codon attracts a tRNA that bears fmet. Next, as elongation begins, the large ribosomal subunit attaches to the mRNA and successive mRNA codons bind to appropriate anticodons on tRNAs. Each tRNA carries an amino acid, and as the tRNAs align along the mRNA, the amino acids bind to build a polypeptide chain. Protein synthesis ceases at a stop codon. Protein folding begins during translation, as enzymes and chaperone proteins assist the amino acid chain in assuming its final functional form. Translation is efficient—many components recycle, and different steps in the process occur simultaneously.

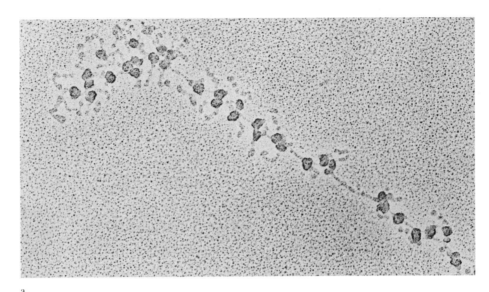

ribosome

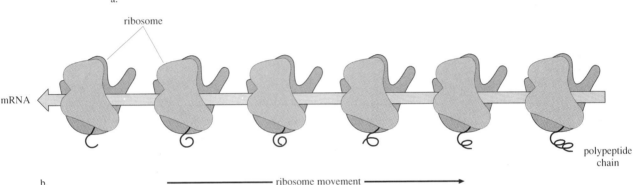

mRNA

polypeptide chain

b.

ribosome movement

FIGURE 16.12

Many polypeptides can be manufactured simultaneously from an mRNA molecule. A single mRNA molecule can be translated by several ribosomes at one time. The ribosomes have different-sized polypeptides dangling from them—the closer a ribosome is to the end of a gene, the longer its polypeptide chain. *a.* The line is an mRNA molecule, the dark round structures are ribosomes, and the short chains extending from the ribosomes are polypeptides. *b.* A schematic illustration of polyribosomes.

Genetic Misinformation—Mutation

Inherited traits and illnesses vary greatly. One person with cystic fibrosis, for example, may have severe breathing and digestive problems, while another has only asthma or male infertility. Great variation in gene expression is partly due to the fact that a gene can be changed, or mutate, in many ways. This is hardly surprising, given DNA's role as an informational polymer. Think of how many ways the words on this page can be altered!

Of Mutants and Mutations

A **mutation** is a physical change in the genetic material. The change can be a single DNA base substitution; the addition or deletion of one or more DNA bases to or from a gene; even a transfer of a stretch of bases on a chromosome. A mutation can occur in the part of a gene that encodes a protein, in a sequence that controls transcription, or at a site critical to intron removal and the joining together of exons. A mutation may or may not cause a mutant (different) phenotype. This depends upon how the alteration affects the gene's product or activity.

KEY CONCEPTS

A mutation is any change in a gene. Genes can mutate by substituting a single base or by adding, deleting, or transferring longer base sequences. A mutation may or may not affect the phenotype.

Understanding Sickle Cell Disease

The first genetic illness researchers were able to trace to a distinct mutation was sickle cell disease. In the 1940s, scientists

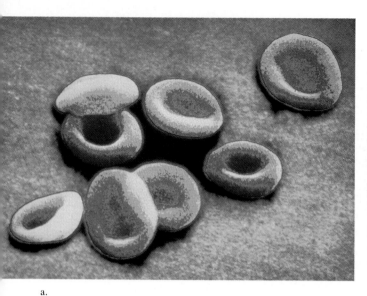

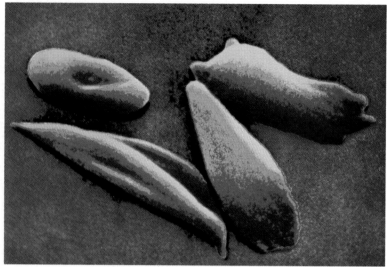

a.

b.

FIGURE 16.13

a. A normal red blood cell is a concave disc containing about 200 million molecules of the protein hemoglobin. *b.* A mutation in the beta globin gene results in abnormal hemoglobin that crystallizes when oxygen tension is low and bends the red blood cells into sickle shapes.

These abnormally shaped cells obstruct circulation, causing pain and loss of function in various organs. A single DNA nucleotide substitution on both homologs causes sickle cell disease.

showed that sickle-shaped red blood cells (fig. 16.13) accompanied this inherited form of anemia (weakness and fatigue caused by a deficient number of red blood cells). In 1949, a team led by Linus Pauling discovered that hemoglobin (the oxygen-carrying molecule in red blood cells) functioned differently depending on whether it came from healthy people or people with sickle cell disease. When Pauling placed the hemoglobin in a solution in an electrically charged field (a technique called electrophoresis), each type of hemoglobin moved to a different position. Hemoglobin from sickle cell carriers showed both movement patterns.

The researchers suspected that a physical difference accounted for this difference. But how could they identify the protein portion of hemoglobin that sickle cell disease alters? Hemoglobin is a very large molecule. It consists of four globular-shaped polypeptide chains, each surrounding an iron atom (fig. 16.14).

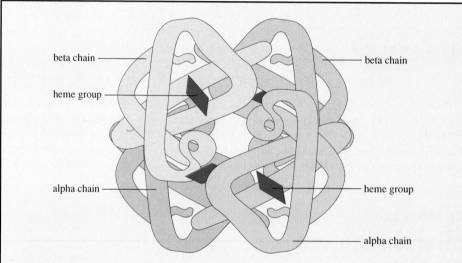

FIGURE 16.14

Defects in the beta globin genes. Adult hemoglobin is built of two beta chains and two alpha chains, each of which binds an iron atom. More than 300 different mutations of the human beta globin gene are known. Mutations can disrupt the binding of the globin chains to each other or to the iron groups; change a stop codon to one specifying an amino acid, resulting in an elongated beta globin chain; or change an amino-acid-specifying codon into a stop codon, producing a stunted beta chain. Many beta chain mutations are silent because they change a codon into one specifying the same amino acid or a similar one or they occur in a part of the chain that is not essential to function.

Protein chemist V. M. Ingram tackled the problem by cutting normal and sickle hemoglobin with a protein-digesting enzyme, then separating the pieces, staining them, and displaying them on paper. The patterns of fragments—known as peptide fingerprints—differed. This meant, Ingram deduced, that the two molecules must differ in amino acid sequence. One piece of the molecule occupied a different position in the two types of hemoglobin. Ingram concentrated on this peptide (just eight amino acids long, rather than the 146 of a full beta globin sequence) to find the site of the mutation. He found it in the sixth amino acid in the gene sequence. The amino acid valine appeared in the sickle cell hemoglobin, where glutamic acid appeared in the normal chain. At the DNA level, a single base caused the mutation—a CAC codon replaced the normal CTC. This tiny change causes devastating effects in the human body.

What Causes a Mutation?

A mutation can form spontaneously or be induced by chemical or radiation exposure. An agent that causes mutation is called a **mutagen.**

Spontaneous Mutation

Two healthy people of normal height give birth to a child with short stature. The parents are surprised to learn that an autosomal dominant mutation caused their son's achondroplasia (dwarfism). This means that each of their son's children will face a 50% chance of inheriting the condition. How could this happen when there are no other people with the condition in the family?

The boy's achondroplasia arose from a new, or *de novo,* mutation in either his mother's or father's gamete. Such a **spontaneous mutation** usually originates as a DNA replication error.

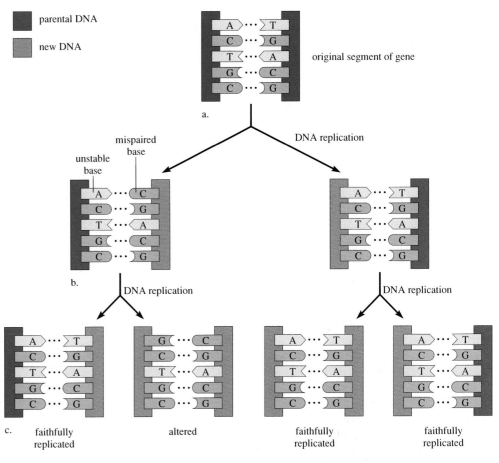

a.

b.

c.   faithfully   altered   faithfully   faithfully
       replicated   replicated   replicated

FIGURE 16.15

Spontaneous mutation. *a.* In DNA, A nearly always pairs with T, and C with G. However, DNA bases are very slightly chemically unstable, and for brief moments, they exist in altered forms. *b.* If a replication fork encounters a base in its unstable form, a mismatched base pair can result. *c.* After another round of replication, one of the daughter cells has a different base pair than the one in the corresponding position in the original DNA segment. Such a substituted base pair can alter the structure of a gene. If the gene's function is affected, the individual's phenotype may change.

Sometimes spontaneous mutation stems from chemical changes in DNA's nitrogenous bases. Free nitrogenous bases tend to exist in two slightly different forms called **tautomers.** For a short time, each base takes on an unstable tautomeric form. If such an unstable base is inserted by chance into newly-forming DNA, an error will be perpetuated (fig. 16.15).

In the 1960s, Watson and Crick suggested that tautomerism underlies many mutations. More recent chemical experiments suggest that although tau-tomerism may account for most **transition mutations,** which replace a purine (A or G) with a purine or a pyrimidine (C or T) with a pyrimidine, other mechanisms cause the more unusual **transversion mutations,** which replace a purine with a pyrimidine (or vice versa).

The spontaneous mutation rate varies for different genes and in different organisms. The dominant gene that causes neurofibromatosis type I, for example, mutates at one of the highest rates known, arising in 40 to 100 of every million germ cells. The large size of this gene may contribute to its mutability—its long sequence can change in more ways. Each human gene has about a one in 100,000 chance of mutating. Each of us probably carries at least a few spontaneously mutated genes, but most are recessive, so they do not affect the phenotype.

Human spontaneous mutation rates are difficult to assess because of the length of a single generation—usually about 30 years. In bacteria, a new generation arises every half hour or so, and genetic change is therefore much more rapid. When disease-causing bacteria mutate, they may become resistant to the antibiotic drugs we use to destroy them. For example, the antibiotic quinoline disables *E. coli* by binding to gyrase, an enzyme that unwinds replicating DNA. *E. coli* that are sensitive to the antibiotic cannot replicate, so they die. However, a few bacteria may spontaneously mutate in a way that alters the conformation of their gyrase so that quinoline cannot bind to it. A person infected with the resistant bacteria would gain no relief from the antibiotic. Genes conferring such resistance can also jump from one bacterial strain to another, creating strains that resist a variety of drugs.

The genetic material of viruses also rapidly undergoes spontaneous mutation. An influenza vaccine manufactured to fight one strain may be ineffective when a new, resistant strain appears the next flu season. Genetic changes can so alter the virus's surface that the vaccine no longer recognizes it. The virus that causes AIDS mutates at a very rapid rate, which has stymied efforts to create a vaccine.

A gene's sequence also influences mutation rate. Genes with repeated base sequences, such as GCGCGC . . . , are more likely to mutate because bases from a single strand may pair as the double helix unwinds to replicate. This within-one-strand base pairing blocks replication enzymes and causes sequence errors.

The **alpha globin genes** that comprise part of the hemoglobin molecule are also prone to mutate. Two alpha globin genes are repeated in their entirety next to each other on each human chro-

2 copies of alpha globin gene

chromosome 16

chromosome 16

α α

α α

a. normal chromosome arrangement of alpha globin genes

α α

α α

crossover

b. misalignment of alpha globin genes during meiosis I

α α α

c. chromosome resulting from crossover with three alpha globin genes

α

d. chromosome resulting from crossover with one alpha globin gene

FIGURE 16.16

The repeated nature of the alpha globin genes makes them prone to mutation by mispairing during meiosis.

mosome 6. If mispairing occurs in meiosis, chromosomes may have more or fewer than the normal two copies of the gene (fig. 16.16).

KEY CONCEPTS

Spontaneous mutations result when an error occurs in DNA replication. Unstable base tautomers may insert into replicating DNA, disrupting the normal DNA sequence. This is more likely to happen when the gene is large or when the DNA sequence is repetitive. Spontaneous mutations occur more frequently in microorganisms and viruses because they reproduce often. Different genes have different mutation rates.

Induced Mutations

Geneticists often use the classical tool of comparing mutant to wild type phenotypes. Gregor Mendel used tall and short pea plants with wrinkled or round, green or yellow seeds to derive the laws of inheritance. Odd eye and body colors in fruit flies helped scientists decipher the chromosomal basis of inheritance.

Researchers can often infer a gene's normal function by observing what happens when the gene is altered. But the spontaneous mutation rate is far too low to provide enough genetic variants for experiments.

To obtain mutants for study, geneticists often use mutagens. Alkylating agents, for example, cause a DNA base to drop away; since any one of the four bases may replace it, and three of them will be a mismatch, an alkylating agent is very likely to produce a mutation. Researchers often use these compounds in fruit fly experiments—they can induce up to five mutants per 100 exposed flies. Other chemical mutagens act in different ways. X rays and other forms of radiation break DNA, causing both large- and small-scale mutations in DNA.

Some commonly encountered substances may cause mutations in cells growing in culture, such as sodium nitrite in smoked meats. Nuclear accidents, bomb tests, and mustard gas have caused mutations. Natural mutagens include radiation from rocks, cosmic rays, and the ultraviolet light in sunlight.

Natural Protection against Mutation

The natural repair systems discussed in chapter 15 represent only some of the built-in protections against mutation.

The Genetic Code

When scientists first deciphered the genetic code, it seemed to have too much information—61 codons to specify only 20 amino acids. However, the redundancy of the genetic code helps protect against mutation. Redundant codons ensure that many DNA alterations involving the third codon position are "silent." For example, both CAA and CAG specify glutamine. A change from one to the other does not alter the designated amino acid, so it would not alter a protein containing that amino acid.

Mutations in the second codon position often cause one amino acid to replace another with a similar conformation. This does not always disrupt the protein's form too drastically. For example, if a GCC codon mutates to GGC, glycine replaces alanine; both are very small amino acids.

Position in the Protein

A mutation's effect on the phenotype depends upon where the change affects the gene and how it alters the conformation of the gene product. A mutation may replace an amino acid with a very dissimilar substitute, but this may not affect the phenotype if the change occurs in a part of the protein that is not critical to its function. Certain mutations in the beta globin gene, for example, do not cause anemia or otherwise affect how a person feels, but may slightly alter how the protein migrates in an electric field.

Some proteins are more vulnerable to disruption by mutation than others. Collagen, a major constituent of connective tissue, is unusually symmetrical. The slightest change in its DNA sequence can greatly disrupt its overall structure, leading to a disorder of connective tissue.

Types of Mutations

Mutations can be considered according to whether they affect somatic cells or sex cells, and by the extent of the alteration.

Somatic versus Germinal Mutations

A **germinal mutation** occurs during the DNA replication period preceding meiosis. The change appears in the resulting gamete and all the cells that descend from it following fertilization. A **somatic mutation** can occur during the DNA replication period just before mitosis. In this case, the genetic change is perpetuated in the daughter cells of the original mutated cell. Because a somatic mutation occurs in somatic rather than sex cells, it affects a subset of a multicellular organism's cells.

Point Mutations

A **point mutation** is a change in a single DNA base. There are two types of point mutations, distinguished by their consequences—missense and nonsense mutations.

A **missense mutation** changes a codon that normally specifies a partic-ular amino acid into one that codes for a different amino acid. If the substituted amino acid alters the protein's conformation sufficiently or at a site critical to its function, the phenotype changes. A missense mutation can profoundly affect a gene's product if it alters a site normally recognized by the ribozymes that remove introns. The resulting gene product could contain stretches of additional nucleotides or lack blocks of nucleotides.

A **nonsense mutation** is a single base change that changes a codon specifying an amino acid into a "stop" codon. This shortens the protein product, which can profoundly influence the phenotype. Can you predict what happens when a normal stop codon mutates into a codon that specifies an amino acid?

Alterations in the Number of DNA Bases

In genes, the number three is very important, because triplets of DNA bases specify amino acids. Adding or deleting bases by any number other than a multiple of three devastates a gene's function. It disrupts the reading frame, and therefore, also disrupts the sequence of amino acids. Such a change is called a **frameshift mutation.** Sometimes, even adding or deleting multiples of three can alter a phenotype.

Missing Genetic Material

Missing genetic material—sometimes even a single base—is a mutation that can greatly alter gene function. A DNA deletion can be so large that detectable sections of chromosomes are missing, or it can be so small that only a few genes or parts of a gene are lacking. The mutation that causes severe cystic fibrosis deletes only a single codon, and the resulting protein lacks just one amino acid.

Expanding Genes

Just as deleting a DNA base can upset a gene's reading frame, adding a base can, too. A recently discovered type of mutation is an expanding gene, in which the number of repeats of a short nucleotide sequence increases over several generations. Consider myotonic dystrophy, which until 1992 was a mystery. This autosomal dominant form of muscular dystrophy seemed to worsen from one generation to the next. A grandfather might experience only mild weakness in his forearms. His daughter might have moderate arm and leg weakness. Her children, if they inherited the genes, might have severe muscle impairments.

For many years, doctors attributed the worsening symptoms to "anticipation," hypothesizing that people worried themselves into reporting increasingly severe symptoms. However, once researchers could sequence genes, they found startling evidence that the disease was indeed worsening with each generation because the gene was expanding! The gene, on chromosome 19, has an area rich in the trinucleotide CTG. This area repeats over and over, almost like a molecular stutter. A person who does not have myotonic dystrophy has from two to 28 repeats, whereas a person with the disorder has from 28 to 50 copies of the sequence.

Researchers have discovered that expanding genes underlie several well-known inherited disorders, including Huntington disease and fragile X syndrome. They all have phenotypes involving brain function, but the precise connection between genotype and phenotype is not yet well understood.

FIGURE 16.17

Corn kernels are splotched because their cells contain jumping genes. Normally, a dominant gene inhibits the production of pigment in the outer layer of the kernel. But when a jumping gene inserts itself into the dominant pigment-inhibiting gene, normally turned-off pigment genes turn on. That cell and all of its descendant cells then produce a pigment until the jumping gene jumps out. The result is a speckled kernel. The earlier in development the jumping gene turns pigment production on, the larger the colored section resulting from the event.

Jumping Genes

Many types of organisms have some genes that move from one chromosomal location to another, a phenomenon discovered in corn by Barbara McClintock in the 1940s. These "jumping genes," or **transposable elements,** can block gene expression by inserting into a gene and preventing its transcription (fig. 16.17).

Table 16.3 summarizes the different types of mutations using an English sentence analogy.

KEY CONCEPTS

Altering the number of bases in a gene is one form of mutation. Inserting or deleting bases upsets the DNA reading frame, causing a frameshift mutation. Several disorders are associated with expanded triplet repeats. Transposable elements can move and insert into genes, disrupting their expression.

The Expanding Genetic Universe

Nearly a half century after Watson and Crick elucidated DNA structure, the genetic material still holds many surprises. At first we thought that every three nucleotides in a gene encodes an amino acid; later, we found that introns interrupt the informational parts of some genes. At first we thought that most genes encode protein; now we know that many do not—and we do not yet know exactly what these non-protein-encoding sequences do. We thought we understood all the ways mutations disrupt gene function—until we discovered expanding genes in the early 1990s. And we have barely scratched the surface of understanding how genes interact with each other, and with the environment.

It seems that the more we learn about genetics, the more we learn that there is to know.

Table 16.3 Types of Mutations

A sentence of three-letter words can serve as an analogy to demonstrate the effects of mutations on gene sequence:

Wild type	THE ONE BIG FLY HAD ONE RED EYE
Missense	THQ ONE BIG FLY HAD ONE RED EYE
Nonsense	THE ONE BIG
Frameshift	THE ONE QBI GFL YHA DON ERE DEY
Deletion	THE ONE BIG HAD ONE RED EYE
Duplication	THE ONE BIG FLY FLY HAD ONE RED EYE
Insertion	THE ONE BIG WET FLY HAD ONE RED EYE
Expanding mutation	**P1** THE ONE BIG FLY HAD ONE RED EYE
	F1 THE ONE BIG FLY FLY FLY HAD ONE RED EYE
	F2 THE ONE BIG FLY FLY FLY FLY FLY FLY HAD ONE RED EYE

SUMMARY

For a gene to function, the information it contains must first be transcribed into ribonucleic acid (RNA). RNA is a single-stranded nucleic acid similar to DNA, but containing uracil and ribose rather than thymine and deoxyribose.

Transcription factors regulate which genes are transcribed in a particular cell type. These factors have certain common regions called motifs. Transcription begins when transcription factors help RNA polymerase (RNAP) to bind to a gene's starting region, called the promoter. RNAP then adds RNA nucleotides to a growing chain, in a sequence complementary to the DNA sense strand.

Several types of RNA participate in protein synthesis (translation). Messenger RNA (mRNA) carries a protein-encoding gene's information. Ribosomal RNA (rRNA) associates with certain proteins to form ribosomes, which provide physical support for protein synthesis. Transfer RNA (tRNA) is cloverleaf-shaped, with a base sequence complementary to mRNA on one end. It bonds to a particular amino acid at the other end.

Each three consecutive mRNA bases forms a codon that specifies a particular amino acid. The correspondence between each codon and the amino acid it specifies constitutes the genetic code. Of the 64 different possible codons, 61 specify amino acids and three signal the end of polypeptide synthesis. Because 61 codons can specify the 20 amino acids, more than one type of codon may encode a single amino acid type. The genetic code is nonoverlapping, triplet, and identical in nearly all genes of all species. Scientists deciphered the code by constructing synthetic RNA molecules, exposing them to the contents of *E. coli* cells, and noting which amino acids were strung into peptide chains. Other experiments confirmed the triplet nature of the code.

Translation requires tRNA, ribosomes, energy storage molecules, enzymes, and protein factors. First, an initiation complex forms when mRNA, a small ribosomal subunit, and a tRNA carrying formylated methionine (fmet) join. Then amino acid chain elongation begins when a large ribosomal subunit joins the small one already in place. Next, a second tRNA binds by its anti-codon to the next mRNA codon, and its amino acid forms a dipeptide bond with the fmet the first tRNA brought in. t RNAs continue to add amino acids, and a polypeptide forms as each of them bonds to the existing chain. The ribosome moves down the mRNA as the chain grows. When the ribosome reaches a "stop" codon, it falls apart into its two subunits and is released, and the new polypeptide breaks free. After translation, some polypeptides are cleaved and some aggregate to form larger proteins. The cell either uses or secretes the protein. Various proteins, including chaperones, help fold a new polypeptide.

A mutation is a change in DNA (genotype) and a mutant is the corresponding phenotype. A gene can mutate spontaneously when its DNA adds, deletes, substitutes, or transfers nucleotides. Large or repetitive DNA sequences are more likely to mutate. Mutagens are chemicals or radiation that increase the mutation rate. These agents may be used to intentionally induce mutation in genetic studies.

Some mutations are silent. A mutation in the third position of a degenerate codon can substitute the same amino acid. A mutation in the second codon position can replace an amino acid with a similarly shaped one. A mutation in a nonvital part of a protein may not affect function.

Researchers classify mutations by the type of tissue they affect, and by their specific nature. A germinal mutation originates in meiosis and affects all cells of an individual. A somatic mutation originates in mitosis and affects only a subset of cells. A point mutation alters a single DNA base. This mutation may be a transition (purine replacing purine or pyrimidine replacing pyrimidine) or a transversion (purine replacing pyrimidine, or vice versa). It may also be missense (substituting one amino acid for another) or nonsense (substituting a "stop" codon for an amino acid-coding codon). Altering the number of bases in a gene is another type of mutation. This type of mutation may entail the deletion or addition of genetic material, which may disrupt the reading frame. Some inherited illnesses are caused by expanding genes. Mobile DNA, such as a transposable element, is also a mutation because it can disrupt the function of a gene that it enters.

KEY TERMS

alpha globin gene 364
anticodon 353
antisense strand 350
cap 354
chaperone protein 360
codon 353
degenerate codon 358
elongation 359
exon 354
frameshift mutation 366
genetic code 354
germinal mutation 365

initiation 359
initiation complex 359
intron 354
leader sequence 357
messenger RNA (mRNA) 353
missense mutation 365
motif 349
mutagen 363
mutation 361
nonsense mutation 365
nuclear matrix protein 349
operon 349

point mutation 365
poly A tail 354
pre-mRNA 354
promoter 349
reading frame 357
ribosomal RNA (rRNA) 353
ribosome 353
ribozyme 354
RNA polymerase (RNAP) 349
sense strand 350
small nuclear ribonucleoproteins (snRNPs, or snurps) 354

somatic mutation 365
spliceosome 354
spontaneous mutation 363
tautomer 363
transcription 348
transcription factor 349
transfer RNA (tRNA) 353
transition mutation 363
translation 348
transposable elements 366
transversion mutation 363
zinc finger 349

REVIEW QUESTIONS

1. Refer to the figure at the right to answer the following questions.

  a. Label the mRNA and the tRNA molecules, and draw in the ribosomes.
  b. What are the next four amino acids to be added to peptide (*b*)?
  c. Fill in the correct codons in the mRNA opposite the sense strand (*c*).
  d. What is the sequence of the DNA antisense strand (as much as can be determined from the figure)?
  e. Is the end of the peptide encoded by this gene indicated in the figure? How can you tell?
  f. What might happen to peptide (*b*) after it is terminated and released from the ribosome?

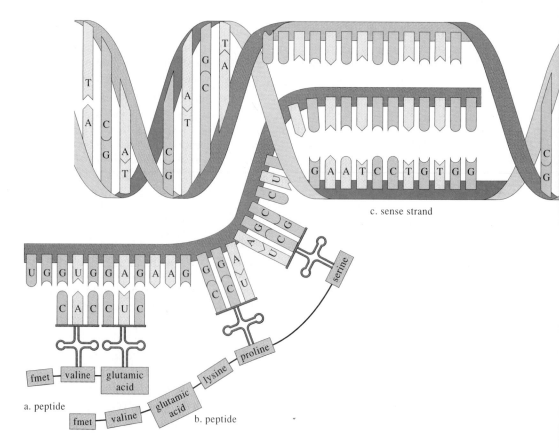

c. sense strand

a. peptide

b. peptide

2. To answer the following questions, refer to this DNA sequence that is transcribed:

GCAAAACCGCGATTATCATGCTTC

a. What is the sequence of the corresponding DNA strand that was not transcribed?

b. What is the mRNA sequence?

c. What amino acid sequence is specified by the transcribed DNA sequence?

d. What would the amino acid sequence be if the genetic code were completely overlapping (that is, if a new codon began with each single consecutive base)?

e. If the fifteenth DNA base mutates to thymine, what happens to the amino acid chain?

3. How is complementary base pairing responsible for

a. the structure of the DNA double helix?

b. DNA replication?

c. transcription of RNA from DNA?

d. mRNA's attachment to a ribosome?

e. codon/anticodon pairing?

f. tRNA conformation?

4. Cite two ways in which the central dogma is an oversimplification.

5. Many antibiotic drugs work by interfering with bacterial protein synthesis. Explain how each of the following antibiotic mechanisms disrupts genetic function in bacteria.

a. tRNAs misread mRNA codons, binding with the incorrect codon and bringing in the wrong amino acid

b. fmet is released from the initiation complex before translation can begin

c. tRNA cannot bind to the ribosome

d. ribosomes cannot move

e. a tRNA picks up the wrong amino acid

6. Cite two ways that proteins assist in their own synthesis. How does RNA assist in its own synthesis?

7. A protein-encoding region of a gene has the following DNA sequence:

GTAGCGTCACAAACA
AATCAGCTC

Determine how each of the following mutations alters the amino acid sequence.

a. substitution of a T for the C in the tenth DNA base

b. substitution of a G for the C in the nineteenth DNA base

c. insertion of a T between the fourth and fifth DNA bases

d. insertion of a GTA between the twelfth and thirteenth DNA bases

e. deletion of the first DNA base

TO THINK ABOUT

1. Many articles on genetics in the popular press state, "We are now beginning to crack the genetic code," or "Everyone has his or her own unique genetic code." What is inaccurate about these statements?

2. How can a mutation alter the sequence of DNA bases in a gene but not produce a noticeable change in the gene's polypeptide product?

3. Why do different cell types have different rates of transcription and translation?

4. Why would two-nucleotide codons not be able to encode the number of amino acids in biological proteins?

5. How could a mutation involving one DNA base be more devastating than a mutation involving three contiguous DNA bases?

6. Outline the steps of the scientific method used to demonstrate the role of the transcription factor myogenin (Biology in Action 16.1).

SUGGESTED READINGS

Akhtar, Saghir, and Adrian J. Ivinson. July 1993. Therapies that make sense. *Nature Genetics*, vol. 4. Biotechnologists are using the concept of sense and antisense nucleic acids to stifle expression of particular genes.

Ashizawa, T., et al. March 1994. Characteristics of intergenerational contractions of the CTG repeat in myotonic dystrophy. *The American Journal of Human Genetics*, vol. 54. This mild form of muscular dystrophy worsens with each generation because of an expanding gene.

Cooper, D. N., and M. Krawczak. 1993. *Human genetic mutation*. New York: BIOS Scientific Publisher. A look at how a variety of spontaneous mutations affect human health.

Gething, Mary-Jane, and Joseph Sambrook. January 2, 1992. Protein folding in the cell. *Nature*, vol. 355. Proteins help other proteins to fold.

Gilbert, Walter. February 9, 1978. Why genes in pieces? *Nature*, vol. 271. A classic and insightful look at the enigma of introns.

Gusella, J. F. August 19, 1993. Elastic DNA elements—boon or blight? *The New England Journal of Medicine*. Expanding DNA repeats are useful tools for identifying individuals—but they are also implicated in disease.

Hasty, Paul, et al. August 5, 1993. Muscle deficiency and neonatal death in mice with a targeted mutation in the myogenin gene. *Nature*, vol. 364. The vital role of one transcription factor is revealed in mice that lack it.

The Huntington disease collaborative research group. March 26, 1993. A novel gene containing a trinucleotide repeat that is expanded and unstable on Huntington's disease chromosomes. *Cell*, vol. 72. This long-awaited paper details the hunt for a very elusive gene.

Ingram, V. M. 1957. Gene mutations in human hemoglobin: The chemical difference between normal and sickle cell hemoglobin. *Nature*, vol. 180, p. 1326. The classic paper explaining the molecular basis for sickle cell disease.

Kunkel, Thomas A. September 16, 1993. Nucleotide repeats: Slippery DNA and diseases. *Nature*, vol. 365. Trinucleotide repeats lie behind Huntington disease, fragile X syndrome, myotonic dystrophy, and many other disorders of the human brain.

Lewis, Ricki. April 1992. Transcription factors—the genetic crossroads. *Biology Digest*. The coordinated functioning of protein transcription factors controls gene expression.

Morell, Virginia. June 4, 1993. The puzzle of the triple repeats. *Science*, vol. 260. Expanding DNA triplets cause disease—but how?

Rosenthal, Nadia. July 7, 1994. DNA and the genetic code. *The New England Journal of Medicine*, vol. 331. A very concise overview of gene function.

Steitz, Joan Argetsinger. June 1988. Snurps. *Scientific American*. How the nonsense is deleted from genetic messages.

 # EXPLORATIONS   **Interactive Software**

Cystic Fibrosis

This interactive exercise allows students to explore the way in which transport proteins influence the passage of water in and out of cells by examining the effects of a mutation that disables a particular transport protein, the one responsible for chloride ion transport. Students can explore the consequences of changing extracellular chloride ion concentrations on transport of water into lung cells, learning that high extracellular concentrations lead to inhibition of water transport into cells because of osmosis. Students can then invoke the cf mutation, which prevents the chloride channel from opening and thus leads to a buildup of chloride ion outside lung cells. The resulting failure of water transport into lung cells is the direct cause of cystic fibrosis symptoms.

1. What is the effect of increased extracellular chloride ion concentration on water movement into the cell?

2. What is the effect of disabling the chloride channel upon water movement?

3. Can artificially decreasing the chloride ion concentration in extracellular fluids counteract cystic fibrosis?

CHAPTER 17

Genetic Technology

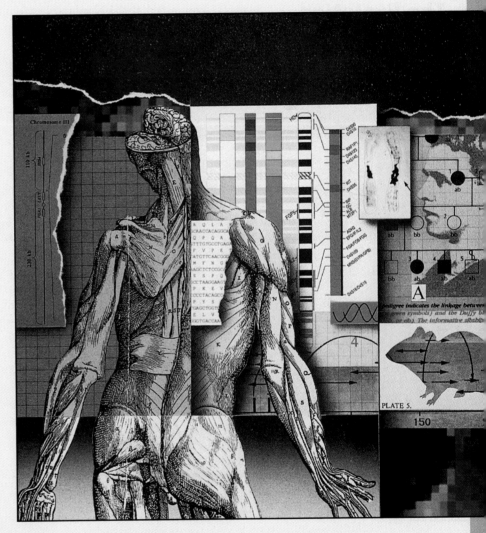

Each October, *Science* magazine publishes an updated human gene map—it actually grows more crowded every week as researchers discover new genes.

LEARNING OBJECTIVES

By the chapter's end, you should be able to answer these questions:

1. How are genetic technologies based on the structure and function of the genetic material?

2. What issues concerned researchers as they considered the implications of recombinant DNA technology in 1975?

3. What tools and steps are needed to construct recombinant DNA molecules?

4. What are some applications of recombinant DNA technology?

5. How are transgenic organisms used?

6. Why is gene targeting more precise than transgenic technology?

7. How are DNA probes used to diagnose disease?

8. How do geneticists detect differences in individuals' DNA sequences?

9. What gene therapies are being tested in humans?

The Evolution of a Cure

Crystal Emery and Leonard Gobea had already lost a five-month-old baby to the ravages of severe combined immune deficiency (SCID), a disorder caused by a missing enzyme. When Crystal became pregnant again late in 1992, amniocentesis showed that this child, too, had inherited SCID. The anxious parents knew this meant the baby would be unable to fight off life-threatening infections. However, doctors could now partially restore an affected child's immunity by replacing the missing enzyme, adenosine deaminase (ADA). In fact, physicians told the couple, new technologies could allow their son to be the first recipient of a more lasting treatment—replacing the errant gene in blood cells from his umbilical cord, and then sending these bolstered cells into his bloodstream before he was born.

The parents agreed, and in May 1993, newborn Andrew Gobea made medical history (fig. 1.1b). His story reflects the culmination of several other pioneering treatments for SCID.

Laura Cay Boren spent her first four years, from 1982 to 1986, in and out of hospitals, battling frequent infections. Like Andrew, her immune system lacked both T and B cells due to the missing enzyme (fig. 17.1). Laura Cay was clinging to life when she became the first recipient of a long-acting form of ADA in 1986.

SCID treatment reached another milestone on September 14, 1990, at 12:52 P.M. A four-year-old Ohio girl sat up in bed at the National Institutes of Health in Bethesda, Maryland, intravenously receiving her own white blood cells, which doctors had earlier removed and patched with normal ADA genes. Even as the girl recovered her immunity over the ensuing months, researchers were onto the next step. Wouldn't the treatment last longer, they reasoned, if they could treat immature blood cells? These stem cells would remain in the circulatory system longer, providing a longer-lasting effect. The researchers knew that stem cells reside in the bone marrow, but there, they number literally one in a million; however, they also knew that stem cells are abundant in umbilical cord blood. Eventually, their reasoning led to Andrew Gobea's experimental stem cell gene therapy for SCID. Medical researchers are currently exploring ways to store umbilical cord blood and use it to treat a variety of disorders later in life.

Gene therapies are making headlines now as an increasing number are tested in humans. But more mature genetic technologies have been used for years—for example, **recombinant DNA technology** has provided abundant and pure supplies of several important drugs since 1978. Recombinant DNA technology combines genetic material from different species. This is possible because all life uses the same genetic code (fig. 17.2). When researchers place a human gene in bacteria, for example, the bacteria divide and yield many copies, or clones, of the "foreign" DNA. They also produce many copies of the protein that DNA specifies.

Single-cell recombinant DNA technology is used to produce valuable proteins. In the 1980s, scientists began genetically engineering multicellular organisms by altering DNA at the one-cell stage of development (in a gamete or fertilized ovum). The resulting individuals, who carry the genetic alteration in every cell, are **transgenic.** Presently, human gene therapy is performed on somatic cells to correct problems in specific tissues, but is not performed on all cells, as it is in transgenic organisms.

Gene therapy, recombinant DNA technology, and transgenic technologies are revolutionizing such diverse fields as human health care, veterinary medicine, agriculture, food processing, anthropology, paleontology, and forensics. This chapter considers some of the fascinating ways scientists apply their rapidly increasing knowledge of the chemistry of the gene.

Recombinant DNA Technology

In February 1975, 140 molecular biologists convened at Asilomar, a seaside conference center on California's Monterey Peninsula, to discuss the safety and implications of a new type of experiment. Investigators had found a simple way to combine the genes of two species, and they were initially concerned about experiments requiring the use of a cancer-causing virus. At the conference, the biologists expressed general concerns about the promising new field of recombinant DNA technology.

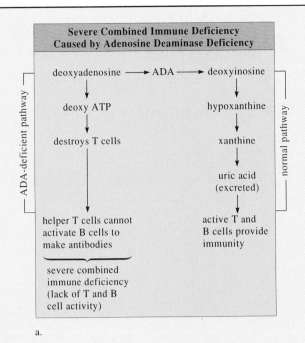

**Severe Combined Immune Deficiency
Caused by Adenosine Deaminase Deficiency**

deoxyadenosine → ADA → deoxyinosine

deoxy ATP → hypoxanthine

destroys T cells → xanthine

→ uric acid (excreted)

helper T cells cannot activate B cells to make antibodies → active T and B cells provide immunity

severe combined immune deficiency (lack of T and B cell activity)

ADA-deficient pathway / normal pathway

a.

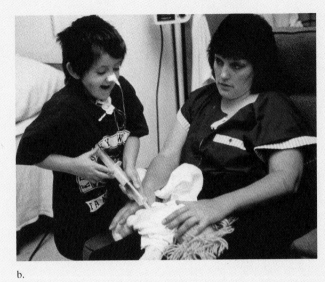

b.

FIGURE 17.1

Neither Andrew Gobea nor Laura Cay Boren can manufacture an enzyme, ADA; as a result, both suffer from profound immunity breakdown. *a.* Lack of ADA blocks a biochemical pathway (right) that converts biochemical intermediates to uric acid, which is then excreted. As toxic deoxy ATP builds up (left), T cells die. They therefore cannot secrete their protective biochemicals nor activate antibody-producing B cells. Andrew was treated with gene therapy on umbilical cells (see fig. 1.1*b*); Laura (*b*) was helped by an earlier and more temporary treatment, a chemically altered form of ADA. Until she was treated, Laura spent much of her life in hospitals. Here, she pretends to inject her doll as her mother looks on.

a.

b.

FIGURE 17.2

The genetic code is universal. Recombinant DNA technology is built on the fact that all organisms utilize the same DNA codons to specify the same amino acids. A striking illustration of the universality of the code is shown in this tobacco plant (*a*), which has been genetically engineered to contain genes from the firefly that specify the "glow" enzyme luciferase. The plant is bathed in a chemical that allows the enzyme to be expressed, causing the plant to glow (*b*).

The scientists discussed restrictions they could place on the sorts of organisms used in recombinant DNA research and what they could do to prevent escape of a recombinant organism from the laboratory. The Asilomar guidelines outlined physical containment measures, such as using specialized hoods and airflow systems to keep the recombinant organisms inside the laboratory. They also discussed biological containment measures such as using organisms so weakened that they could not survive outside of the laboratory.

A decade after the Asilomar meeting, many members of the original group reconvened at the meeting site to assess progress in the field. Nearly all agreed on two things: recombinant DNA technology had proven safer than most had predicted, and the technology had spread from the research laboratory to industry far faster and in more diverse ways than anyone had imagined (Biology in Action 17.1).

Do We Need Recombinant DNA Drugs? It Depends on the Drug

EPO

The invention of kidney hemodialysis in 1961 to simulate normal kidney function was a milestone in modern medicine. The technique began saving 120,000 lives per year in the United States alone. However, a serious side effect, anemia, was usually so severe that dialysis patients required frequent transfusions. This was because dialysis, in addition to cleansing the blood, also depleted **erythropoietin (EPO),** a small protein hormone manufactured in the kidneys.

The kidneys sense the oxygen level in the blood. If it is too low, they secrete EPO, which travels to the bone marrow and signals it to produce more oxygen-carrying red blood cells. Boosting patients' EPO levels can counteract dialysis-induced anemia. In 1970, the federal government initiated a program to find ways to produce large, pure amounts of the substance.

EPO levels in human plasma are too low to make pooling from donors feasible. A more likely potential source was people suffering from aplastic anemia or anemia caused by hookworm infection, disorders that cause them to secrete large amounts of EPO. The National Institutes of Health ran a program in which they extracted EPO from the urine of South American farmers who had hookworm infections. The material was transported on government planes in diplomatic pouches! In 1976, the National Heart, Lung and Blood Institute began a grant program in search of ways to purify EPO. In 1977, 2,550 liters of urine from Japanese aplastic anemia patients supplied the EPO for the research.

Problems loomed. Was it ethical to obtain a scarce substance from the urine of sick, usually poor, people from one country, to treat comparatively wealthy people from the United States? Then AIDS arose. Extracting any biochemical from human body fluids was simply no longer safe.

Recombinant DNA technology solved the EPO problem. Two companies engineered bacteria to produce the human protein; clinical trials were successful, and despite some quibbling between the manufacturers over patent rights, dialysis patients can now obtain pure, human EPO—though at a cost of $10,000 per year (fig. 1).

Scientists continue to learn about EPO. A large Scandinavian family has an inherited condition, benign erythrocytosis, that causes them to overproduce EPO. The symptom? Great athletic skill. An Olympic skier in the family may owe his gold medal to his genes.

tPA

Using recombinant DNA technology to produce EPO was a breakthrough, because there seemed to be no other way to obtain this one-of-a-kind biomolecule. In the mid 1980s, another human biochemical, tissue plasminogen activator (tPA), was also widely heralded as a wonder drug available courtesy of genetic engineering. tPA is a "clot buster." If injected within four hours after a heart attack, it dramatically limits damage to the heart muscle by restoring blood flow.

As was the case for EPO, several companies genetically engineered tPA, with the usual corporate infighting. Also like EPO, tPA is expensive—$2,200 per life-saving shot. The two drugs differ, however, in the crucial area of need. Unlike EPO, tPA is not unique—it is not the only clot-dissolving drug. Others are nearly as effective and far cheaper. One of them, streptokinase, is a bacterial protein used as a standard of cardiac care for many years. Streptokinase costs $300 per injection.

At first, researchers predicted that tPA would prove to be a better clot buster because of its human origin. Many extensive clinical trials compared tPA and streptokinase. Surprisingly, tPA and streptokinase have proven about equal in efficacy. Some physicians who were quick to deliver tPA are now returning to the old standby, streptokinase, reserving the more expensive tPA for patients who have already received the other drug and could therefore have an allergic reaction to it.

The lesson learned from EPO and tPA is that the value of genetic engineering cannot be judged in a scientific vacuum—economics, marketing, and plain common sense also enter the picture.

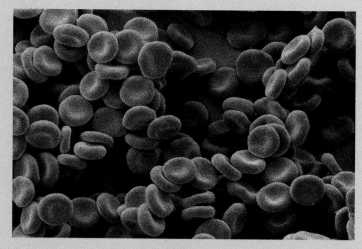

FIGURE 1

These immature red blood cells are mass-produced in the bone marrow of a patient being treated with erythropoietin.

Constructing Recombinant DNA Molecules

To manufacture recombinant DNA molecules, scissorlike biochemicals called **restriction enzymes** cut a gene from its normal location. Other biochemicals insert the gene into a circular piece of DNA, which is transferred into cells of another species. Each of the hundreds of types of restriction enzymes cuts DNA at a specific base sequence. Cutting double-stranded DNA can generate single-stranded ends that "stick" to each other due to complementary base pairing (fig. 17.3).

The natural function of restriction enzymes is to protect bacteria by cutting and thereby inactivating the DNA of infecting viruses. The bacterium's own DNA contains protective methyl (CH_3) chemical groups that shield it from its restriction enzymes. Geneticists use restriction enzymes from bacteria to cut DNA at specific base sequences. A restriction enzyme that recognizes a sequence of seven bases, for example, cuts DNA into larger fragments than a restriction enzyme that recognizes a sequence of four bases. It also cuts it into fewer pieces, since each fragment is larger.

Another natural "tool" geneticists use in recombinant DNA technology is a piece of DNA called a **vector**. DNA from one type of organism can attach to a vector and then transfer into another organism's cell. One commonly used type of vector is a **plasmid,** a small circle of double-stranded DNA found in some bacteria, yeasts, plant cells, and other types of organisms (fig. 17.4).

Viruses that infect bacteria (bacteriophages) are another type of vector. Scientists can manipulate them so that they transport genetic material but do not cause disease. Disabled retroviruses (that use RNA as their genetic material) can act as vectors, too. Later in the chapter, we'll discuss how geneticists introduce vectors into cells.

To create a recombinant DNA molecule, scientists isolate DNA from a donor cell and cut it with a restriction enzyme (fig. 17.5). They choose an

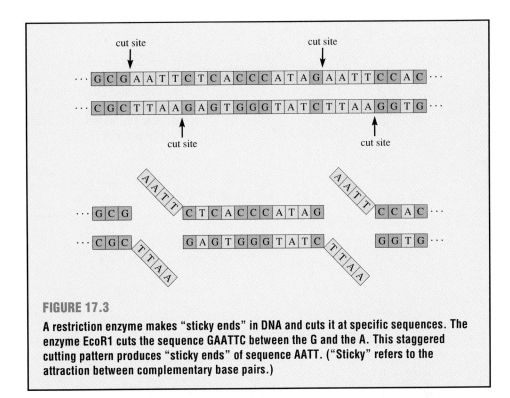

FIGURE 17.3

A restriction enzyme makes "sticky ends" in DNA and cuts it at specific sequences. The enzyme EcoR1 cuts the sequence GAATTC between the G and the A. This staggered cutting pattern produces "sticky ends" of sequence AATT. ("Sticky" refers to the attraction between complementary base pairs.)

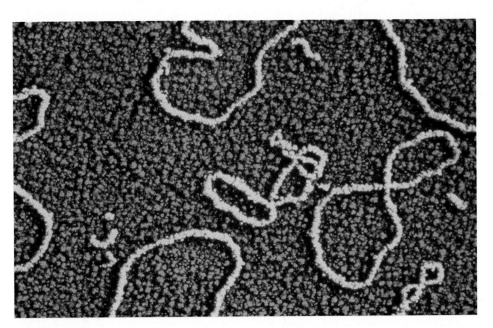

FIGURE 17.4

Plasmids are small circles of DNA found naturally in the cells of some organisms. A plasmid can replicate itself, as well as any other DNA inserted into it. For this reason, plasmids make excellent "cloning vectors"—structures that carry DNA from one species into the cells of another.

enzyme that cuts at sequences known to bracket the gene they want to study. The enzyme leaves single-stranded ends dangling from the cut DNA, each bearing a characteristic base sequence. Next, the researchers isolate a plasmid and use the same restriction enzyme they used to cut the donor DNA to cut it. Because both the donor DNA and the plasmid DNA are cut by the same restriction enzyme, the same single-stranded base sequences extend from the cut ends of each. When

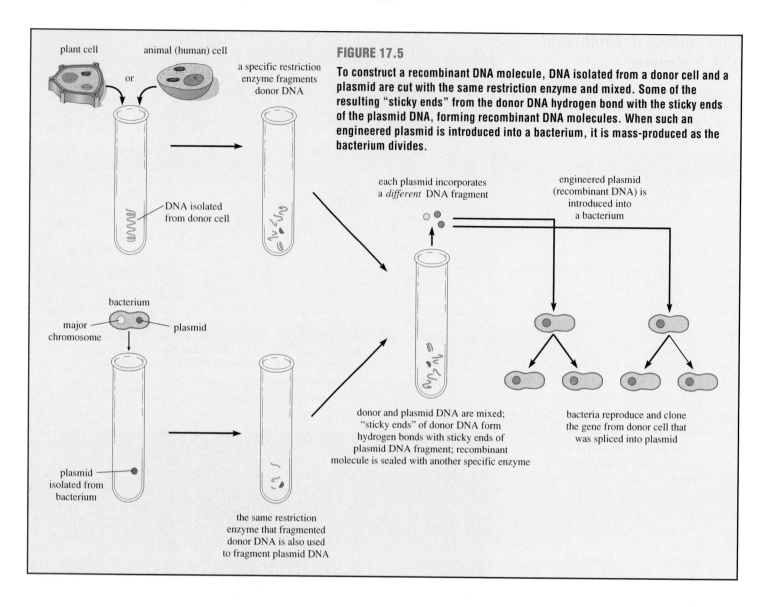

plant cell   animal (human) cell

a specific restriction
enzyme fragments
donor DNA

FIGURE 17.5

To construct a recombinant DNA molecule, DNA isolated from a donor cell and a plasmid are cut with the same restriction enzyme and mixed. Some of the resulting "sticky ends" from the donor DNA hydrogen bond with the sticky ends of the plasmid DNA, forming recombinant DNA molecules. When such an engineered plasmid is introduced into a bacterium, it is mass-produced as the bacterium divides.

DNA isolated
from donor cell

each plasmid incorporates
a *different* DNA fragment

engineered plasmid
(recombinant DNA) is
introduced into
a bacterium

bacterium

major
chromosome   plasmid

donor and plasmid DNA are mixed;
"sticky ends" of donor DNA form
hydrogen bonds with sticky ends of
plasmid DNA fragment; recombinant
molecule is sealed with another specific enzyme

bacteria reproduce and clone
the gene from donor cell that
was spliced into plasmid

plasmid
isolated from
bacterium

the same restriction
enzyme that fragmented
donor DNA is also used
to fragment plasmid DNA

the plasmid and the donor DNA are mixed together, the single-stranded "sticky ends" of some plasmids match base pairs with the sticky ends of the donor DNA. The result is a recombinant DNA molecule, such as a plasmid carrying the human insulin gene. Researchers can now transfer the plasmid with its stowaway human gene to the cell of an individual of another species.

First, however, the recombinant molecules must be separated from molecules containing just donor DNA or just plasmid DNA. To separate them, the researchers use a plasmid containing two genes that make it resistant to two particular antibiotics, so that a cell

containing the plasmid is able to grow in the presence of those particular antibiotic drugs. When the piece of DNA the scientists want to clone (mass-produce) inserts into the plasmid, it inactivates one of the antibiotic genes. A researcher can tell which cells have taken up a plasmid containing the foreign DNA by exposing cells to each antibiotic. Cells that grow in the presence of both antibiotics contain one of the original plasmids—one that does not carry the inserted foreign DNA, and thus is still antibiotic resistant. Cells that die from exposure to either antibiotic contain only donor DNA—since they lack the plasmid, they are not resistant to

antibiotics. Finally, cells that die in the presence of one antibiotic but grow in the presence of the other harbor the plasmid containing the recombinant foreign DNA.

When bacteria containing the recombinant plasmid divide, so does the plasmid. Within hours, the original bacterium gives rise to a culture of cells containing the recombinant plasmid. Because the enzymes, ribosomes, energy molecules, and factors necessary for protein synthesis are also present in the bacterial cells, the cells transcribe and translate the plasmid DNA and its stowaway human gene. The bacterial culture thus produces a human protein.

Applications of Recombinant DNA Technology

Recombinant DNA technology allows geneticists to isolate individual genes from complex organisms and observe their functions on the molecular level. It also has practical applications. A protein with medical uses can be mass-produced through recombinant DNA technology. Often a genetically engineered drug is safer than drugs extracted directly from organisms. This is true of human insulin, which was the first recombinant drug.

Two million people in the United States have diabetes. Before 1982, the insulin they injected daily came from the pancreases of cattle in slaughterhouses. Cattle insulin is so similar to the human protein, differing in only two of 51 amino acids, that most people with diabetes can use it. However, about one in 20 patients is allergic to cow insulin because of the slight chemical difference. Before recombinant DNA technology was feasible, allergic patients had to use expensive combinations of insulin from a variety of other animals or from human cadavers.

Normally, humans carry a gene that instructs beta cells in the pancreas to produce insulin. Medical researchers cut this gene from its chromosome and inserted it into the genetic material of an *E. coli* bacterium. When the bacterium reproduced, the human insulin gene was duplicated along with the bacterium's own genetic material. When the bacterial genes directed the bacteria to synthesize *E. coli* proteins, the transplanted human gene directed the cells to manufacture its protein product—human insulin. Today, a major pharmaceutical company grows huge vats of *E. coli* bacteria engineered to manufacture human insulin. A person with diabetes can now purchase genetically engineered human insulin with a prescription at local drugstores.

Human growth hormone is another drug produced with recombinant DNA technology. Doctors use it to treat pituitary dwarfism in children, and some-

Table 17.1   Drugs Produced Using Recombinant DNA Technology

Drug	Use
Atrial natriuretic factor	Dilates blood vessels, promotes urination
Epidermal growth factor	Accelerates healing of wounds and burns; treats gastric ulcers
Erythropoietin	Stimulates production of red blood cells to alleviate anemia
Factor VIII	Promotes blood clotting in patients with hemophilia
Fertility hormones (follicle stimulating hormone, luteinizing hormone, human chorionic gonadotropin)	Promotes fertility
Human growth hormone	Promotes muscle and bone growth in pituitary dwarfs
Insulin	Allows cells to take up glucose in treatment of diabetes
Interferon	Destroys some cancer cells and some viruses
Lung surfactant protein	Helps lung alveoli to inflate in infants with respiratory distress syndrome
Renin inhibitor	Lowers blood pressure
Somatostatin	Slows muscle and bone growth in pituitary giants
Superoxide dismutase	Prevents further damage to heart muscle after heart attack
Tissue plasminogen activator	Dissolves blood clots related to heart attacks and arterial blockages

times to make normal short children grow taller, though this use is controversial. The hormone used to come from pituitary glands of human cadavers. This source is costly because each gland contains very little of the substance, and it is dangerous because it can transmit serious viral infections of the central nervous system. In contrast, the hormone that genetically engineered *E. coli* produces is pure and plentiful. Table 17.1 lists this and other drugs produced using recombinant DNA technology.

Products of recombinant DNA technology are also used in the food industry. The enzyme rennin, normally produced in calves' stomachs, is used to

make cheese. Scientists can insert the gene coding for the enzyme into plasmids and transfer the plasmids to bacteria, which are then mass-cultured to produce large quantities of pure rennin.

Recombinant DNA technology may help clean the environment in a process called **bioremediation.** This process endows harmless microbes with genes that enable them to detoxify pollutants. For example, a gene from the bacterium *Pseudomonas mendocina* specifies toluene monooxygenase, an enzyme that degrades trichloroethylene (TCE), an industrial degreaser and solvent found in many polluted areas. Toluene monooxygenase lowers TCE contamination levels from 20 parts per million to two parts per billion. However, *Pseudomonas mendocina,* the natural host, requires a constant supply of toluene (another solvent) to activate the enzyme that digests TCE.

Recombinant DNA technology overcomes this problem. *E. coli* engineered to harbor the *Pseudomonas* toluene monooxygenase gene effectively degrades TCE without toluene, leaving behind harmless chemicals and cellular debris.

KEY CONCEPTS

Recombinant DNA technology combines genetic material from different species. When a cell is given a vector that contains foreign DNA, the cell transcribes and translates the foreign gene, producing pure, abundant supplies of particular proteins. Genetic engineering technologies are possible because of the universality of the genetic code and the ability of restriction enzymes to cut DNA at specific sequences, creating sticky ends that bind by complementary base pairing. Products of recombinant DNA technology are used in such varied fields as health care, food technology, agriculture, forensics, and bioremediation.

Transgenic Organisms

A genetically altered gamete or fertilized ovum develops into a transgenic organism—that is, an individual who displays the change in each cell. Transgenic plants can derive from somatic tissue as well as from gametes or fertilized ova, but animals usually cannot. Scientists use different vectors and gene transfer techniques in plants because their cell walls (which are not present in animals) are difficult to penetrate.

Genetic engineers have several ways to introduce DNA into cells (table 17.2). Chemicals such as polyethylene glycol and calcium phosphate are used to open transient holes in cell membranes. **Liposomes,** fatty bubbles that

Table 17.2 Gene Transfer Techniques

Approach	How It Works
Virus	A human gene is inserted into a herpes virus and transferred by infection into a human cell, where it is expressed.
Retrovirus	An RNA virus carrying an RNA version of a human gene infects a somatic cell. The gene is reverse transcribed to DNA and inserts into a human chromosome. Here it may produce a missing or abnormal protein.
Liposome transfer	A fatty bubble called a liposome, which carries a gene, is engulfed by a somatic cell. The delivered gene may then replace an abnormal one.
Chemical	Calcium phosphate or dextran sulfate opens transient holes in a cell membrane, admitting replacement DNA.
Electroporation	Electrical current opens transient holes in a cell's membrane, admitting replacement DNA.
Microinjection	A tiny needle injects DNA into a cell that lacks that DNA sequence.
Particle bombardment	Metal pellets coated with DNA are shot with explosive force or air pressure into recipient cells.

can carry genetic cargo, are enveloped by cell membranes and drawn into cells. In **electroporation,** a brief jolt of electricity opens transient holes in cell membranes so that foreign DNA may enter. Microscopic needles can inject DNA into cells (fig. 17.6).

Another way to introduce DNA into cells is **particle bombardment,** nicknamed "biolistics" because it uses a "gene gun." A gunlike device shoots tiny metal particles, usually gold or tungsten, coated with foreign DNA. Some of the projectiles enter target cells. Gene guns were pioneered on plant cells to blast through their tough cell walls, but they also effectively send DNA—at speeds up to 4,500 feet per second—into mitochondria, chloroplasts, bacteria, animal somatic cells, embryos, and eggs. In a variation on the gene gun, genetic engineers prepare a mist of two types of microscopic droplets, one consisting of DNA in solution, and the other of microprojectiles.

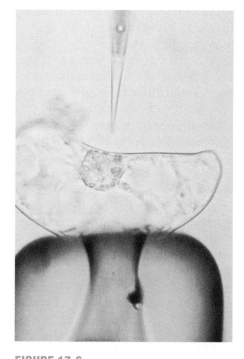

FIGURE 17.6

One way to move foreign DNA into a plant cell nucleus is by direct injection with a microscopic glass needle.

When the projectiles blast through a cell membrane, they open up a pathway for the DNA.

Once foreign DNA moves into a target cell, it must enter the nucleus, replicate along with the cell's own DNA, and be transmitted when the cell divides in order to produce a transgenic individual. Finally, a mature individual must develop from the engineered cell and express the desired trait in the appropriate tissues at the right time in development. Then the organism must pass the characteristic on to the next generation. Figure 17.7 indicates the complexity of engineering a transgenic organism.

Transgenic organisms are used in two basic ways—to secrete valuable medical substances (table 17.3), and as research tools to study the functions of individual genes. Transgenic mice are very useful research tools, because they are, as one geneticist puts it, "a remarkable experimental surrogate for us." Such mice often provide valuable information that geneticists use to design the clinical trials for gene therapies. For example, researchers studying Duchenne muscular dystrophy knew the gene and its product well and were close to developing ways to deliver normal dystrophin protein into boys with the affliction. But what would happen if the dystrophin entered tissues other than muscle, or if too much built up in the boys' bodies? When scientists engineered transgenic mice to contain more than 50 times the amount of dystrophin in human muscle, and to produce it in other tissues as well, the dystrophin caused no ill effects. The researchers then could be reasonably confident that human trials would be safe.

Transgenic mice have been vital in studying single-gene disorders, many human cancers, and even multifactorial disorders. For example, a transgenic mouse used in obesity research develops fat-encrusted arteries, even on a low-fat diet, because it carries a mutant human gene for a cholesterol-carrying lipoprotein in each of its cells.

Gene Targeting

Transgenic technology is not very precise; since the introduced DNA is not directed to a particular chromosomal locus, the transferred gene can disrupt another gene's function or come under another gene's control sequence. Even if a transgene does insert into a chromosome and is expressed, its effect may be overshadowed by the host's version of the same gene.

A more precise method of genetic engineering is **gene targeting,** in which the introduced gene exchanges places with its counterpart on a host cell's chromosome. Gene targeting uses a natural process called **homologous recombination,** in which a DNA sequence locates and displaces a very similar or identical sequence in a chromosome. By swapping an inactivated gene for an active one in mice, researchers are able to "knock out" a gene and observe the effects of its absence, even in embryos and fetuses. The technique is helping us understand immune system protein function, how oncogenes interact to produce cancer, and how genetic disorders develop.

Gene targeting entails genetic engineering plus complex developmental manipulations (fig. 17.8). It does not work on mammals' fertilized eggs, so the intervention must occur later in development. Using electroporation or microinjection, researchers can send a gene into an embryonic cell before the embryo implants in the uterus. This cell is called an **embryonic stem (ES) cell.** ES cells are not specialized, which means that many of their genes have not yet been expressed.

In the procedure depicted in figure 17.8, an engineered ES cell with inactivated pigment genes is injected into a blastocyst (a ball of cells that does not yet show the layered structure of an embryo) from a colored mouse. The blastocyst is implanted into a surrogate mother mouse, where it continues to develop into an embryo that has some cells bearing the targeted gene. The eventual newborn mouse is a chimera (mosaic), with patches of tissue whose cells bear the introduced gene and patches with cells the gene did not enter. The mosaic mouse is then bred to another mosaic. If all goes well, some of the offspring will be homozygous for the targeted gene, demonstrating successful gene targeting.

Gene targeting is useful in developing animal models of genetic diseases. First, researchers identify the animal's version of a human disease-causing allele. Then they transfer the corresponding human mutant allele into ES cells and eventually breed an animal homozygous for the inactivated gene.

Animals with "knocked out" genes are also useful in studying complex disorders that involve more than one gene. For example, researchers are studying atherosclerosis by inactivating combinations of genes whose products oversee lipid metabolism. Similarly, researchers can study multiple genetic changes responsible for some cancers, such as colon cancer, by targeting the genes in various combinations.

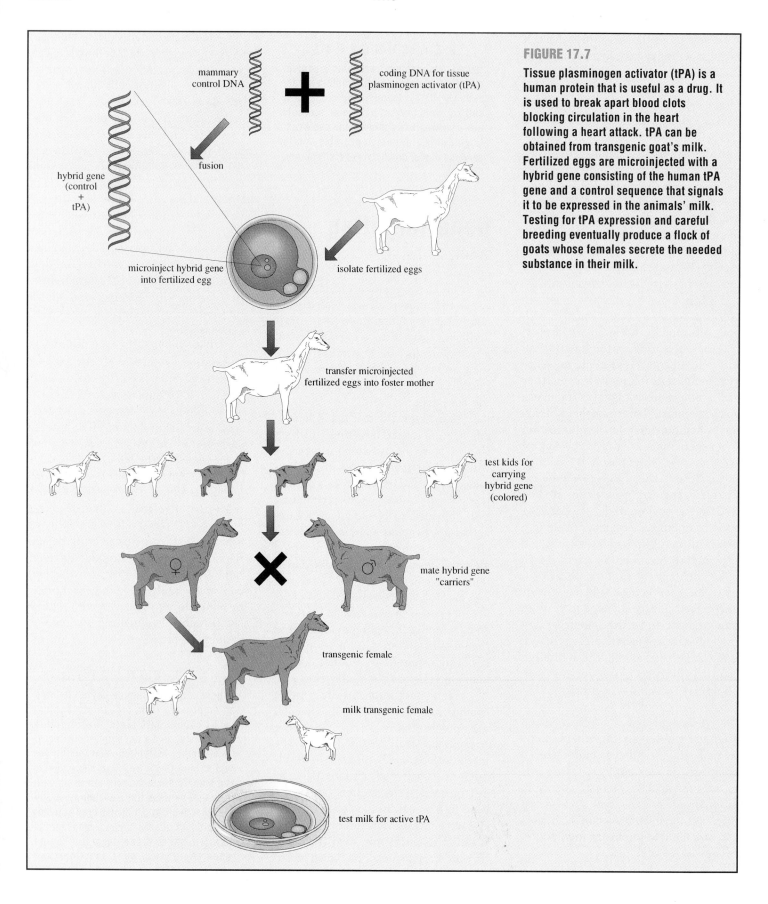

FIGURE 17.7

Tissue plasminogen activator (tPA) is a human protein that is useful as a drug. It is used to break apart blood clots blocking circulation in the heart following a heart attack. tPA can be obtained from transgenic goat's milk. Fertilized eggs are microinjected with a hybrid gene consisting of the human tPA gene and a control sequence that signals it to be expressed in the animals' milk. Testing for tPA expression and careful breeding eventually produce a flock of goats whose females secrete the needed substance in their milk.

Table 17.3   Transgenic Pharming

Host	Product	Potential Use
Cows	Lactoferrin	Added to infant formula to bind iron and prevent bacterial infection
Goats	tPA	Breaks up blood clots
Mice	tPA	
	CFTR	Treats cystic fibrosis
Pigs	Hemoglobin	Blood substitute
Rabbit	Erythropoietin	Treats anemia from dialysis
Rat	Human growth hormone	Treats pituitary dwarfism
Sheep	Alpha-1-antitrypsin	Treats hereditary emphysema

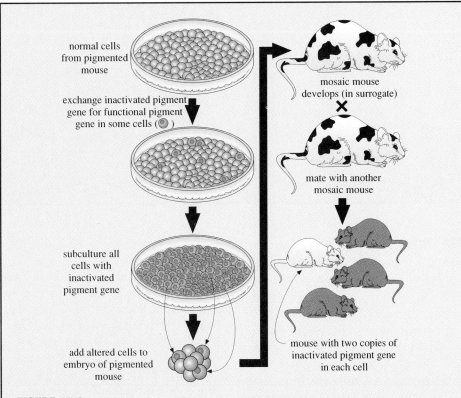

FIGURE 17.8

Gene targeting. Inactivated pigment-encoding genes are inserted into mouse embryonic stem cells, where they trade places with functional pigment-encoding alleles. The engineered ES cells are cultured and injected into early mouse embryos. Mosaic mice develop, with some cells heterozygous for the inactivated allele. These mice are bred and, if all goes well, they will yield some offspring homozygous for the knocked-out allele. It's easy to tell which mouse this is in the pigment gene example, but gene targeting is particularly valuable for revealing unknown gene functions by inactivating targeted alleles.

Genetic Technology in Human Health Care

Genetic researchers in many nations are cooperating to discover the sequences and functions of all human genes in an effort called the **human genome project.** Scientists are analyzing genomes from several organisms that have been important in genetic research and in agriculture (table 17.4). For example, researchers completed a genome map for the cow in 1994, to help breeders understand how certain complex traits are inherited. Table 17.5 lists goals for the human genome project.

Italian-American virologist Renato Dulbecco came up with the idea for the human genome project in 1986. Dulbecco shared in the 1975 Nobel Prize for physiology or medicine because he developed a way to study the genes of cancer-causing viruses. Later, he proposed that systematically dissecting the human genome would reveal how cancer arises. During the summer of 1986, geneticists convened at the Cold Spring Harbor Laboratory on Long Island to debate the idea. Enthusiasm ran so high that just two years later, a worldwide effort was underway.

Since then, participants have announced new gene discoveries on a nearly weekly basis. With each revelation that a specific gene is responsible for a set of symptoms, researchers learn how to more precisely attack and treat health problems. For example, in early 1993, researchers discovered that the gene behind amyotrophic lateral sclerosis (ALS, or Lou Gehrig disease), a degenerative neuromuscular disease that affects adults, encodes a well-known enzyme—superoxide dismutase. This enzyme detoxifies free radicals, highly reactive byproducts of oxygen metabolism that can harm tissues. Researchers now know to study free radical damage in their search for treatments for ALS.

Table 17.4 Sequencing Genomes of Model Organisms

Organism	Genome Size (millions of base pairs)
Bacterium (*Escherichia coli*)	4.8
Yeast (*Saccharomyces cerevisiae*)	14.4
Roundworm (*Caenorhabditis elegans*)	100.0
Mustard weed (*Arabidopsis thaliana*)	100.0
Fruit fly (*Drosophila melanogaster*)	165.0
Mouse (*Mus musculus*)	3,000.0
Human (*Homo sapiens*)	3,000.0

Table 17.5 Goals of the Human Genome Project

1. To map and sequence the genomes of model organisms, such as bacteria, fruit flies, mice, algae, and mustard weed, in order to:

   estimate how long it will take to sequence the human genome,

   establish evolutionary relationships between species, and

   identify highly conserved genes by sequence homologies.

2. To discover the relative chromosomal locations of genes, markers, and other DNA sequences to better understand gene structure, function, and interaction.

3. To develop new diagnostic tests and, eventually, treatments for Mendelian and multifactorial disorders.

4. To develop new computer and database capabilities to handle large amounts of dense information, and to detect meaning in nucleic acid and amino acid sequences.

5. To address and set policy on legal, medical, social, and ethical problems and concerns that could result from others' knowledge of individual genome information.

From a clinical viewpoint, disease gene discoveries yield three tiers of advances:

1. New diagnostic techniques
2. New drug treatments targeted to the exact biochemical anomaly that underlies the illness
3. More lasting treatment in the form of gene therapy

Diagnostics

To diagnose a hereditary or infectious illness at the DNA level, researchers detect mutant DNA sequences or genetic material of infectious organisms or viruses. Traditional medicine usually detects only the phenotypic changes or symptoms of infection.

DNA Probes

The chemical attraction between complementary strands of DNA is the basis of DNA probe technology. A **DNA probe** is a piece of DNA, usually 1,000 to 6,000 bases long, whose base sequence is complementary to the sequence of the DNA segment under study, such as a gene that causes an inherited disorder or a gene from an infecting virus or bacterium. A probe applied to cells from a preembryo, embryo, or fetus can provide a prenatal diagnosis (fig. 17.9).

DNA probe technology requires the geneticist to know the sequence of the gene of interest. To locate a gene, probes containing radioactive or fluorescent atoms are applied to pieces of DNA from a patient's sample.

(Restriction enzymes have cut the DNA at specific points in the sequence.) The probe binds wherever it locates a complementary sequence, and the researcher can detect the probe by detecting radioactivity or fluorescence, by exposing photographic film laid over the preparation. The DNA segment the probe binds to is identified by its size using a technique called **Southern blotting** (fig. 17.10). (Southern is the last name of the inventor of the technique.)

DNA probes are useful in a number of ways. They can reveal molecular mechanisms behind biological functions. For example, geneticists used DNA probes to identify the three human genes that provide color vision. DNA probes are also used to diagnose infections by detecting DNA sequences unique to the

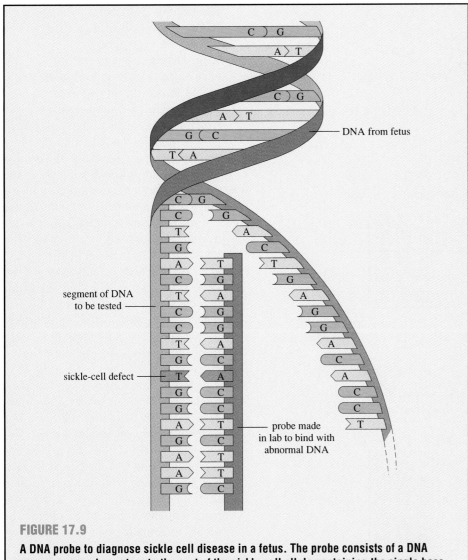

DNA from fetus

segment of DNA
to be tested

sickle-cell defect

probe made
in lab to bind with
abnormal DNA

FIGURE 17.9

A DNA probe to diagnose sickle cell disease in a fetus. The probe consists of a DNA sequence complementary to the part of the sickle cell allele containing the single base difference that causes the disorder. Fetal cells are collected from a sample of amniotic fluid. The DNA is separated into single strands and unwound, and the radioactively labeled probe is applied. DNA from a fetus who has inherited two copies of the disease-causing allele (and will therefore have sickle cell disease) shows twice as much fluorescence as DNA from a fetus who has only one disease-causing allele (a carrier).

with a paper point, then sends it to a lab where DNA probes detect the three microorganisms that cause periodontal disease. DNA probes detect fungal infections in turf grasses on golf courses. A very valuable DNA probe tests for *Salmonella* bacteria in food samples. The test takes two days, compared to a week to detect the bacteria by growing them the traditional way.

KEY CONCEPTS

Scientists are attempting to sequence entire genomes for a number of species that have played important roles in the study of genetics. Sequencing the human genome will provide information about many diseases and improve diagnosis and treatment. DNA probes and polymorphisms help diagnose genetic disease. DNA probes are labeled pieces of DNA that bind to DNA sequences of interest.

DNA Polymorphisms

It is unusual to know enough about a gene's product or sequence to manufacture an exact probe. More often, probes attract large fragments of DNA that contain the complementary sequence.

The same restriction enzyme applied to the same chromosomal region in two individuals sometimes yields different-sized DNA pieces. This means that their DNA sequences vary. Varying DNA sequences at corresponding chromosomal sites is termed **polymorphism** ("many forms"). A polymorphism may be a single base difference, or a different number of repeated short sequences.

Polymorphisms can occur in DNA that encodes protein or in DNA with a regulatory or unknown function. However, polymorphisms do not always affect the phenotype. Can you see why?

The following hypothetical DNA sequence illustrates how a DNA polymorphism can alter the size of a restriction enzyme-generated fragment:

Sequence 1 CGT TAA GCT <u>A</u>AT
CGC CTA

Sequence 2 CGT TAA GCT <u>G</u>AT
CGC CTA

infecting organism or virus. Consider the viral liver infection hepatitis B. A single strand of DNA complementary to one strand of the viral DNA is attached to a molecule of fluorescent dye and added to a blood sample from a person with hepatitis B symptoms. Then the sample is heated, which separates the DNA into single strands. If the hepatitis B virus is present, the DNA probe hydro-gen bonds to it, reconstituting a short double helix. Next, the geneticist separates these double strands from single strands. The double-strand DNA fluoresces (under lighting that detects the dye) only if the hepatitis B virus is present.

DNA probes are also used in dental offices, on golf courses, and in the food industry. A dentist swabs a patient's gums

FIGURE 17.10

Southern blotting displays DNA fragments. In this procedure, DNA pieces that have been separated by size are blotted onto a nitrocellulose filter (which looks like a square of paper) and exposed to radioactively tagged pieces of DNA, or probes. When the probed filter is placed next to X-ray film, the sites at which the probes bind to the DNA fragments expose the film. The completed Southern blot shows the probed DNA fragments in size order. (An actual Southern blot looks more like a line of smears than the neat lines shown in the illustration.)

Source: National Institute of General Medical Sciences, "The New Human Genetics," in *NIH Publication No. 84-662,* September 1984, page 29.

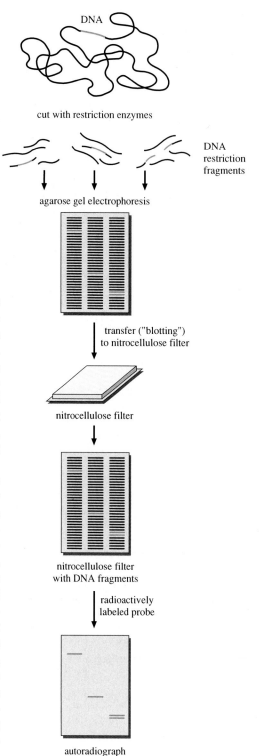

If a restriction enzyme cuts at the sequence AAT, it cuts at a site in sequence 1 that does not exist in sequence 2 because of a single base difference. These variations in cutting sites are called **restriction fragment length polymorphisms,** or **RFLPs** (pronounced "riflips"). Geneticists use RFLPs to detect genes of interest, directly or indirectly, by describing the sequences surrounding those genes.

In direct genetic disease diagnosis using RFLPs, a disease-causing allele may be disrupted in a way that alters the recognition site for a particular restriction enzyme. If the allele contains a cutting site not present in the wild type, that restriction enzyme can be used to distinguish between the defective and normal alleles. In other words, if the normal allele yields different DNA fragment sizes than the mutant allele when each is cut with the same restriction enzyme, physicians can detect the defective allele and diagnose the disease. Geneticists detect sickle cell disease by using restriction enzymes, because normal cutting sites are altered in the mutant beta globin gene (fig. 17.11).

Differences in the sizes of restriction fragments also occur when homologs have different numbers of repeats of a short sequence, or a **variable number of tandem repeats (VNTRs)** (fig. 17.12). VNTRs are often used in DNA fingerprinting, a technique that distinguishes between individuals (Biology in Action 17.2).

When the gene that causes a disorder has not been identified, restriction enzymes can be used for diagnosis if a piece of DNA located near the disease-causing gene has an extra or missing restriction enzyme cutting site. This indirect approach uses a **genetic marker,** a sequence of DNA bases, constituting a RFLP, that is physically close to the disease-causing gene on the chromosome. Detecting a marker infers the presence of the disease-causing allele.

Unlike direct DNA probes or RFLPs within a disease-causing gene, which can be used with any individual, genetic markers must be traced in several family members. This is because a DNA polymorphism that travels on the same chromosome as a disease-causing allele in one family may accompany the normal allele in another. The marker-disease gene association, not just the presence of a RFLP, must be detected.

A gene and its marker are on the same homolog. A RFLP on the other homolog—the one that does not carry the abnormal allele—cannot be a marker in that family, because it would separate from the disease-causing allele during meiosis (fig. 17.13).

To be useful as a genetic marker, the marked DNA sequence must always be present in ill family members but never in healthy adult relatives. For families with inherited disease, a marker test can predict who will develop the illness. For researchers, a marker is literally a toehold on a chromosome, a starting point that allows them to use other techniques to actually pinpoint the disease-causing gene.

KEY CONCEPTS

Many genetic diagnostic techniques work by detecting polymorphisms, or variations in DNA fragment size. RFLPs are differences in restriction enzyme cutting sites. If RFLPs occur in certain gene sequences, scientists can directly detect those genes. They can also indirectly detect them if RFLPs serve as closely linked genetic markers. Polymorphisms are variations in sequence or in the number of repeats of a short sequence. VNTR sequences alter the sizes of restriction fragments.

Gene Therapy

Once researchers developed ways to transfer genes to specific cells, where the genes were transcribed and translated, they turned their attention to deciding

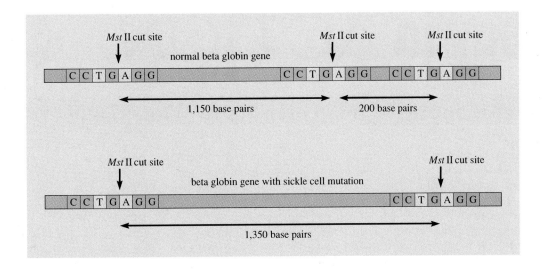

FIGURE 17.11

Restriction enzymes can be used to diagnose sickle cell disease. The normal beta globin gene contains a cutting site for the restriction enzyme *Mst* II that is lacking in the allele that causes sickle cell disease. Therefore, cells of a fetus with sickle cell disease have two unusually long segments of DNA that are 1,350 bases long. A fetus who has inherited two normal copies of the gene has pieces for that gene of 1,150 bases long and 200 bases long. What size pieces would you expect for a carrier of sickle cell disease?

Source: Data from Bob Conrad, *Newsweek,* March 5, 1984.

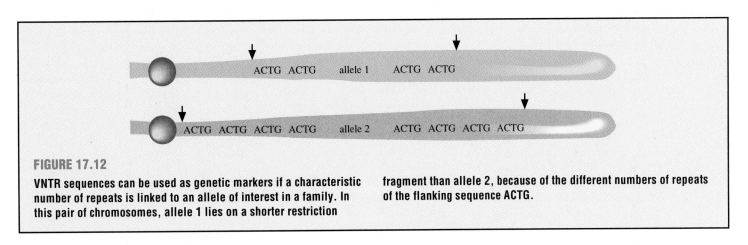

FIGURE 17.12

VNTR sequences can be used as genetic markers if a characteristic number of repeats is linked to an allele of interest in a family. In this pair of chromosomes, allele 1 lies on a shorter restriction fragment than allele 2, because of the different numbers of repeats of the flanking sequence ACTG.

which genetic disorders should be treated first. Should they begin with those that affect the greatest number of people, or those that would be easiest to treat? Many felt that gene therapy should be perfected on the disorders that we know the most about—those for which we know the affected gene, protein, and tissue. As a result, some researchers focused on gene therapy for ADA deficiency, described at the beginning of the chapter, even though only a few dozen youngsters are born with the disorder each year. More recent gene therapy projects address conditions that affect many people, such as hypertension, very high blood cholesterol levels, diabetes, and

DNA sequence A is a marker for a particular family when it is located on the same homolog as the gene of interest:

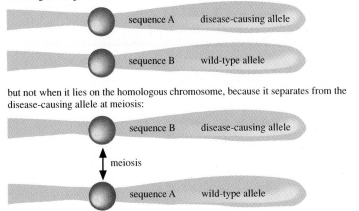

but not when it lies on the homologous chromosome, because it separates from the disease-causing allele at meiosis:

FIGURE 17.13

Markers are based on meiosis.

385

DNA Fingerprinting—From Forensics to Wine Making

DNA fingerprinting distinguishes between individuals by comparing restriction fragment length polymorphisms (RFLPs). RFLPs reflect differences in DNA sequence or in the number of repeats of short DNA sequences. The technology first aroused public interest in 1988, when it entered courtrooms as a new way to match a suspect to DNA recovered from a crime scene, or to vindicate a falsely accused person. Today, DNA fingerprinting continues to be a valuable forensic tool, but it has become an even more important research tool. The technique gives evolutionary biologists powerful views into the taxonomic relationships between species. It also has commercial applications. Let's look at two arenas DNA fingerprinting has entered.

In the Courtroom

Tommie Lee Andrews was a very meticulous rapist. He picked out his victims months before he attacked, watching them so that he knew exactly when they would be home alone. On a balmy Sunday night in May 1986, Andrews lay in wait for Nancy Hodge, a young computer operator at nearby Disney World in Orlando, Florida. The burly man surprised her when she was in the bathroom removing her contact lenses, then repeatedly raped and brutalized her.

Andrews was very careful not to leave fingerprints, threads, hairs, or any other indication that he had been in Hodge's home. But he had not counted on the new technology of DNA fingerprinting. This technique allows forensic specialists to establish (or disprove) identity by identifying differences in individual DNA sequences. If the DNA fingerprint of a forensic sample such as sperm, blood, or hair matches the DNA fingerprint of a suspect's cells, the person's presence at the crime scene is established.

Thanks to Nancy Hodge's quick arrival at the hospital following the attack, a medical team was able to obtain a vaginal secretion sample containing the rapist's sperm cells. Two district attorneys who had read about DNA fingerprinting then sent some of the sperm to Lifecodes, a Westchester, New York biotechnology company. There, scientists extracted DNA from the sperm cells and cut it with restriction enzymes. The DNA pieces were then mixed with laboratory-synthesized DNA probes labeled with radioactive isotopes. The probes bound to the DNA segments containing the complementary base sequence.

After Tommie Lee Andrews was apprehended as a suspect in Nancy Hodge's assault, the same procedure of extracting, cutting, and probing DNA was performed on white blood cells taken from both Hodge and Andrews. When the radioactive DNA pieces from each sample were separated out and displayed according to size, the resulting pattern of bands—the DNA fingerprint—for the sperm sample and Andrews's blood matched exactly. There was no doubt that he was the attacker.

The sizes of the DNA pieces in a DNA fingerprint vary from person to person because of differences in DNA sequence in the regions surrounding the probed genes, or in the number of repeats of short sequences. The power of the technology stems from the fact that there are far more ways for the 3 billion bases of the human genome to vary than there are people on earth. Figure 1 shows the steps of DNA fingerprinting used in forensics. However, the technique can be used on any species.

In the Garden

Vitis vinifera—better known as the grape—was domesticated more than 6,000 years ago. Traditionally, grape tasters distinguish cultivated strains, or cultivars, based on phenotypic traits, including growth characteristics; vine branching pattern; leaf, fruit, and flower shape; and pollen grain diameter. The number of cultivars identified by these traits ranges from 14,000 to 24,000, but geneticists estimate the number of genetically distinct types at 5,000 to 8,000. Researchers are currently cataloging grape cultivars using RFLPs derived from differences in repeated DNA sequences. The information will clarify the evolutionary relationships between different types of grapes, as well as provide a more objective form of identification in patent disputes and in developing grapes with new combinations of characteristics.

Beets have also been the focus of DNA fingerprinting. RFLPs showing differences in the numbers of repeated DNA sequences reveal the close evolutionary relationship of sugar beets (*Beta vulgaris*) to three cultivars (fodder beets, leaf beets, and garden beets) and also to a wild subspecies (*B. vulgaris maritima*). Not surprisingly, DNA fingerprinting shows that sugar beets are all very similar genetically, whereas the wild strain is much more variable. Researchers are using this information to identify valuable traits in the wild strain, such as cold tolerance and disease resistance, that could be bred (or genetically engineered) into sugar beets.

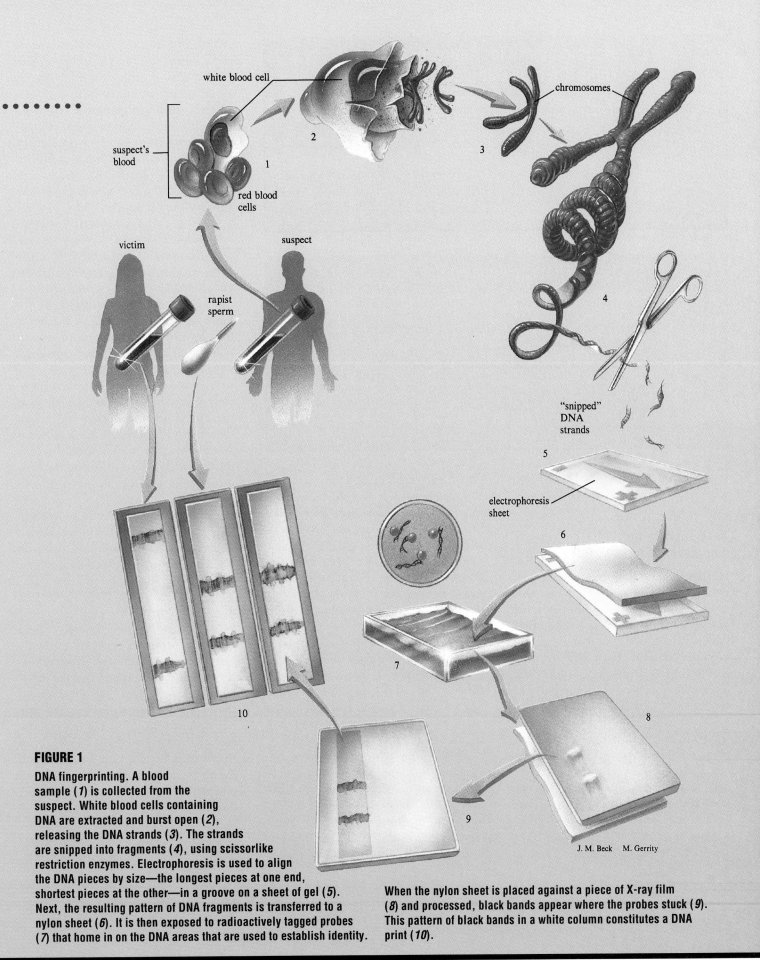

white blood cell

chromosomes

suspect's blood

red blood cells

1

2

3

4

victim

rapist sperm

suspect

"snipped" DNA strands

5

electrophoresis sheet

6

7

8

9

10

J. M. Beck   M. Gerrity

FIGURE 1

DNA fingerprinting. A blood sample (*1*) is collected from the suspect. White blood cells containing DNA are extracted and burst open (*2*), releasing the DNA strands (*3*). The strands are snipped into fragments (*4*), using scissorlike restriction enzymes. Electrophoresis is used to align the DNA pieces by size—the longest pieces at one end, shortest pieces at the other—in a groove on a sheet of gel (*5*). Next, the resulting pattern of DNA fragments is transferred to a nylon sheet (*6*). It is then exposed to radioactively tagged probes (*7*) that home in on the DNA areas that are used to establish identity.

When the nylon sheet is placed against a piece of X-ray film (*8*) and processed, black bands appear where the probes stuck (*9*). This pattern of black bands in a white column constitutes a DNA print (*10*).

Table 17.6 Requirements for Approval of Gene Therapy Clinical Trials

1. Knowledge of defect and how it causes symptoms
2. An animal model
3. Success in human cells growing in vitro
4. Either no alternate therapies, or a group of patients for whom existing therapies are not possible or have not worked
5. Safety

Table 17.7 Gene Therapy Concerns

1. Which cells should be treated?
2. What proportion of the targeted cell population must be corrected to alleviate or halt progression of symptoms?
3. Is overexpression of the therapeutic gene dangerous?
4. If the engineered gene "escapes" and infiltrates other tissues, is there danger?
5. How long will the affected cells function?
6. Will the immune system attack the newly introduced cells?

several types of cancer. Tables 17.6 and 17.7 list requirements and concerns for gene therapy.

Gene Therapy for Mendelian Disorders

Heritable gene therapy is transgenic; it alters all cells of an organism. It is not a goal for humans. **Nonheritable (somatic) gene therapy,** on the other hand, corrects only the somatic cells a genetic condition affects. A bone marrow transplant from a donor's healthy marrow or from genetically altered cells reinfused into the recipient, or clearing cystic fibrosis-congested lungs with a spray containing the normal gene are nonheritable gene therapies because they affect only certain tissues. Nonheritable gene therapy may help individuals, but it doesn't benefit future generations because the defective gene is unchanged in gametes.

Unfortunately, we do not yet know enough about most genetic disorders to attempt nonheritable gene therapy. In Tay-Sachs disease, for example, we know which enzyme is missing and the consequence of the deficit, but we do not know which nerve cells become buried in fat. Even when we know which tissues are affected, delivering a healthy gene is often difficult.

Despite these limitations, genetic researchers are making great progress in treating a variety of disorders with gene therapy (fig. 17.14). Geneticists know how to remove several cell types, give them genes that produce needed biochemicals, and then reimplant them.

Researchers take inventive approaches to target gene-carrying vectors to specific tissues. A vector called adenovirus-associated virus (AAV), for example, naturally infects only those human bone marrow cells that develop into red blood cells, which lack nuclei

in their mature form in the bloodstream. AAV tagged to a functional beta globin gene might treat sickle cell disease by carrying the functional gene in to correct the defect in immature red blood cells, filling them with normal hemoglobin before they enter the circulation. Similarly, a disabled common cold virus can deliver a functional cystic fibrosis gene to cells lining the lungs of a person who has the illness.

If a cell type seems resistant to genetic manipulation, genetic researchers can alter nearby cells in the hope that they will have a "bystander effect" and alter the abnormal cells.

Consider a gene treatment for mice injured in the part of the brain that degenerates in Alzheimer disease in humans. Scientists can genetically engineer fibroblasts to contain a gene for nerve growth factor. When the genetically altered fibroblasts are implanted at the wound site, they secrete the growth factor, which neurons need in order to enlarge. The traumatized neurons then grow and secrete their neurotransmitters, stimulated by the nerve growth factor bath. This type of treatment may one day boost neurotransmitter levels in Huntington disease, Parkinson disease, and clinical depression, all of which involve neurotransmitter deficits.

Gene Therapy for Cancer

Researchers are enlisting genes to fight cancer. They can remove cells from a tumor and transfer genes that encode cancer-fighting immune system biochemicals, such as interferons and interleukins, into them. Once reimplanted, the engineered tumor cells secrete the biochemicals, killing themselves and the surrounding tumor cells. Figure 17.15 shows such an approach. In the figure, the physician is injecting an immune system gene, encased in a liposome, into a skin cancer.

A different genetic approach is being tried to fight glioma, a rapidly fatal cancer in which certain cells called glia,

FIGURE 17.14

Sites of gene therapy experiments.

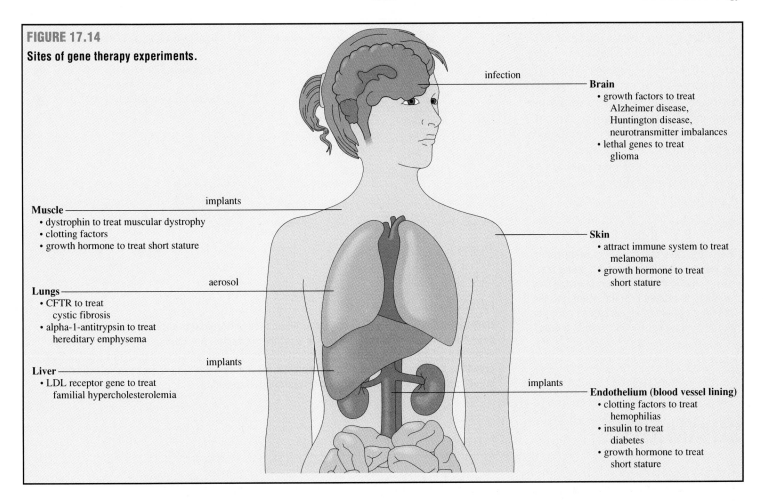

infection → **Brain**
- growth factors to treat Alzheimer disease, Huntington disease, neurotransmitter imbalances
- lethal genes to treat glioma

Muscle — implants
- dystrophin to treat muscular dystrophy
- clotting factors
- growth hormone to treat short stature

Skin
- attract immune system to treat melanoma
- growth hormone to treat short stature

Lungs — aerosol
- CFTR to treat cystic fibrosis
- alpha-1-antitrypsin to treat hereditary emphysema

Liver — implants
- LDL receptor gene to treat familial hypercholesterolemia

implants → **Endothelium (blood vessel lining)**
- clotting factors to treat hemophilias
- insulin to treat diabetes
- growth hormone to treat short stature

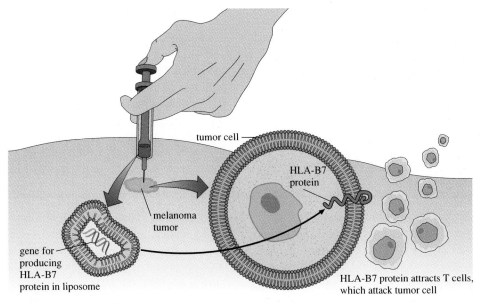

tumor cell

HLA-B7 protein

melanoma tumor

gene for producing HLA-B7 protein in liposome

HLA-B7 protein attracts T cells, which attack tumor cell

FIGURE 17.15

A gene encoding a cell surface protein that attracts the immune system's tumor-killing T cells is injected directly into a melanoma tumor.

which surround nerve cells, divide very rapidly (fig. 17.16). First, the scientists use a retrovirus vector to place a herpes virus gene that encodes an enzyme called thymidine kinase (TK) into mouse connective tissue cells. Thymidine kinase renders the cells susceptible to a drug called ganciclovir.

The retrovirus with its herpes gene stopped glioma from growing in culture and slowed tumor progression in mice and rats. In ongoing human trials, physicians inject the engineered mouse fibroblasts into the brain tumors of patients who have a short time to live. There, the implanted cells produce the viruses, which infect neighboring tumor cells. The tumor cells then produce thymidine kinase. When the patient takes ganciclovir, the tumor cells die. The technique may also treat ovarian cancer.

Not long ago, the idea of gene therapy was just that—an idea. Today, gene therapy is rapidly becoming an important part of medical practice. As researchers discover new gene functions and put genes to use to fight not only rare inborn errors of metabolism, but also multifactorial disorders and cancer, it becomes ever more likely that some form of gene therapy will be part of your future.

KEY CONCEPTS

Heritable gene therapy alters every cell in an individual. In nonheritable gene therapy, somatic cells incorporate genes that produce a needed protein, stimulate the immune system to attack cancer cells, or render cancer cells vulnerable to a drug.

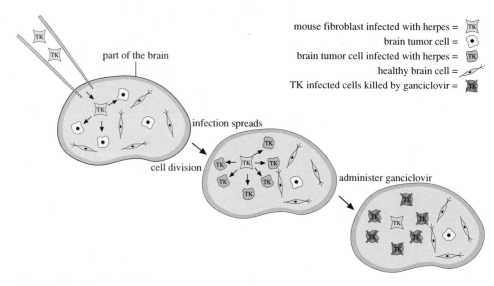

FIGURE 17.16

Herpes hitches a ride into cancer cells. When mouse fibroblasts harboring a thymidine kinase gene from the herpes simplex virus, in a retrovirus vector, are implanted near the site of a brain tumor, the engineered viruses infect the rapidly dividing tumor cells. When the patient is given the antiviral drug ganciclovir, the cells producing thymidine kinase are selectively killed, providing a gene-based cancer treatment from within.

SUMMARY

Genetic technologies allow geneticists to alter cells to cure disease or provide a useful product. Recombinant DNA technology combines genetic material from different species. This is possible because of the universality of the genetic code. Recombinant DNA is used to mass-produce proteins in bacteria or other single cells, and to design transgenic multicellular organisms by altering them at the single-cell stage. Recombinant DNA technology, begun hesitantly in 1975, has matured into a valuable method to mass-produce drugs and other useful proteins.

To construct a recombinant DNA molecule, geneticists use restriction enzymes to cut the gene of interest as well as a vector. Because the foreign DNA and the vector are cut at the same sequence, their ends complementary base pair, enabling some vectors to incorporate foreign DNA. Recombinant plasmids are then introduced into host cells, where foreign gene is propagated and expressed.

Recombinant DNA on a multicellular level produces a transgenic organism. A single cell—a gamete in an animal or plant, or a somatic cell in a plant—is genetically altered. As it divides, it includes the change in each new cell. There are many ways to introduce DNA into cells.

Gene targeting uses the natural attraction of a DNA sequence for its complementary sequence to swap one gene for another. This approach is more precise than transgenic technology, which inserts a foreign gene but does not direct it to a specific chromosomal site. Gene targeting requires complex breeding schemes to achieve homozygous expression of the targeted gene. It is used to inactivate genes and then observe the consequence.

DNA probes and RFLPs (differences in sequence, or polymorphisms, which alter restriction enzyme cutting sites) are used to directly detect genes. RFLPs and variable numbers of tandem repeats (VNTRs) are used to detect marker sequences closely linked to genes of interest.

Gene therapies obtain tissue, deliver new genes, implant the tissue, and encourage production of a needed substance. Heritable gene therapy, which produces transgenic organisms, transmits the genetic change to future generations. This type of therapy is not done in humans. Researchers perform nonheritable gene therapy on a variety of cell types to supply proteins from within or to fight cancer. A herpes virus gene renders cancer cells susceptible to an antiviral drug. Tumor cells incorporate genes with the ability to synthesize immune system biochemicals, and tumor cells also receive genes that create cell surface markers that attract the immune system. As geneticists learn more about genetic function and structure, we will undoubtedly see many more applications of genetic technology.

KEY TERMS

bioremediation   377
DNA probe   382
electroporation   378
embryonic stem (ES) cell   379
erythropoietin (EPO)   374
gene targeting   379
genetic marker   384

heritable gene therapy   388
homologous recombination   379
human genome project   381
liposome   378
nonheritable gene therapy   388
particle bombardment   378

plasmid   375
polymorphism   383
recombinant DNA
   technology   372
restriction enzyme   375
restriction fragment length
   polymorphism (RFLP)   384

Southern blotting   382
transgenic   372
variable number of tandem
   repeats (VNTRs)   384
vector   375

REVIEW QUESTIONS

1. List the components of an experiment to produce recombinant human insulin in *E. coli* cells.

2. Would recombinant DNA technology be impossible if the genetic code were not universal? Why or why not?

3. Why must manipulations designed to create a transgenic organism take place at the single-cell stage? Why couldn't you engineer an adult human to become transgenic?

4. What roles do restriction enzymes play in genetic engineering?

5. Mouse models for cystic fibrosis have been developed by inserting a human transgene, and by gene targeting to inactivate the mouse counterparts of the normal alleles. How do these methods differ? Which method do you think produces a more accurate model of human cystic fibrosis, and why?

6. Describe three ways to insert foreign DNA into cells.

7. Collagen is a large protein often used in cosmetics and skin care products. Would collagen be purer if extracted from human cadavers or manufactured in recombinant bacteria? Cite a reason for your answer.

8. How would each of the following phenomena make it difficult to find a RFLP marker for a disease-causing gene?

a. incomplete penetrance
b. genetic heterogeneity
c. a phenocopy

9. Figures 17.9 and 17.11 depict ways to prenatally diagnose sickle cell disease. How do these techniques differ?

10. A student reads this chapter and decides to rid herself of her curly red hair, which she inherited from her mother, by undergoing heritable gene therapy. Why is her plan doomed to failure? Suggest a better therapy.

TO THINK ABOUT

1. Catfish are extensively farmed in the southern United States, where they are a popular food. Rex Dunham, a biologist at Auburn University in Alabama, has approached the federal government's Agricultural Biotechnology Research Advisory Committee with a proposal to transfer the growth hormone gene from rainbow trout into catfish in an attempt to engineer faster-growing, larger catfish. What factors do you think the committee should consider as they decide whether Dunham's research should proceed?

2. In recent years, people have objected to giving dairy cows genetically engineered growth hormone to increase milk output. This treatment supplies cows with extra amounts of a hormone they

already manufacture. People have also objected to a tomato that stays ripe longer because it lacks a gene encoding a ripening enzyme. Do you think public fear of genetic engineering is justified? What can be done to address the public's concerns?

3. In 1980, Mario Capecchi, now of the Howard Hughes Medical Institute at the University of Utah in Salt Lake City, submitted a grant proposal to the National Institutes of Health. In it, Capecchi suggested inactivating genes in cells by sending in an inactivated allele that would switch places with the functional allele. The NIH turned the proposal down because the scientists who reviewed it felt that the technique was too difficult, although the reasoning was

sound. Capecchi found funds elsewhere, perfected the technique, and has pioneered the exciting technology of gene targeting. Given the current climate of financial restrictions, yet considering Capecchi's story, what factors do you think advisory committees should consider when they decide which genetic engineering proposals should receive funding?

4. Gene therapies to treat glioma and melanoma are being tested on people who have unsuccessfully tried existing treatments and are very close to losing their battles. Do you think this is the best sort of patient for these pioneering treatments? What might we learn by giving them to healthier patients?

5. Rats have helped scientists develop a nasal spray gene therapy for cystic fibrosis. Mice manufacture many human proteins in their skin and blood. Dogs were important in developing treatments for hemophilia; without earlier research performed on them, bone marrow transplants would not be possible. Rabbits are models of human familial hypercholesterolemia. Yet, many people object to using mammals in experiments.

Do you feel that animal models are necessary to fight human genetic disease? Which treatments described in this chapter do you think are safe enough to try on humans without running animal tests first?

6. Cancer of the pancreas is especially difficult to detect and treat because the gland is located in a very inaccessible place. Suggest a way to use gene therapy to treat pancreatic cancer.

7. Cirrhosis of the liver, emphysema, and heart disease are all conditions that can be caused by a faulty gene or by a dangerous lifestyle habit (drinking alcohol, smoking, following a poor diet). When gene therapies become available for these conditions, should people with gene-caused disease be given treatment priority? If not, what other criteria should be used to decide who should receive a limited medical resource?

SUGGESTED READINGS

Barendse, W., et al. March 1994. A genetic linkage map of the bovine genome. *Nature Genetics,* vol. 6. A gene map is the latest chapter in the centuries-old study of the domestic cow.

Carmen, Ira H. February 1992. Debates, divisions, and decisions: Recombinant DNA advisory committee (RAC) authorization of the first human gene transfer experiments. *American Journal of Human Genetics,* vol. 50. Before geneticists attempted gene therapy, they had to perfect protocols for gene delivery.

Culliton, Barbara J. August 26, 1993. A home for the mighty mouse. *Nature,* vol. 364. Transgenic and "knockout" mice are now affordable, thanks to several disease-related research organizations.

DiTullio, Paul, Seng H. Cheng, John Marshall, Richard J. Gregory, Karl M. Ebert, Harry M. Meade, and Alan E. Smith. January 1992. Production of cystic fibrosis transmembrane conductance regulator in the milk of transgenic mice. *Bio/Technology.* Transgenic technology has enabled us to mass-produce complex human gene products.

Garver, Kenneth L. January 1994. New genetic technologies: Our added responsibilities. *The American Journal of Human Genetics,* vol. 54. How will we implement new genetic technology wisely?

Hayes, Catherine V., M. A. November 12, 1992. Genetic testing for Huntington's disease—A family issue. *The New England Journal of Medicine,* vol. 327, no. 20. A personal view of the feelings associated with presymptomatic diagnosis of an adult-onset, untreatable genetic disorder.

Hoffman, Eric P. January 1994. The evolving genome project: Current and future impact. *The American Journal of Human Genetics,* vol. 54. Physicians must learn a new field—genetics.

Jaroff, Leon. May 31, 1993. Brave new babies. *Time.* Stem cell therapy can treat immune deficiency, and the first recipients are newborns.

Leahy, J. L., and A. E. Boyd, III. January 7, 1993. Diabetes genes in non-insulin-dependent diabetes mellitus. *The New England Journal of Medicine,* vol. 328. A marker for this common form of diabetes may make earlier treatment possible.

Lewis, Ricki. February 1987. History of recombinant DNA technology. *Biology Digest.* From ancient Mexican Indians to today's scientific researchers, we have learned to manipulate the genetic material.

Lewis, Ricki. Fall 1987. Genetic marker testing. *Issues in Science and Technology.* Presymptomatic diagnosis of untreatable inherited illness presents many thorny issues. This article brings up challenges that are even more compelling today than when it was written.

Lewis, Ricki. June 28, 1993. Gene discovery: The giant first step toward therapy. *The Scientist.* In early 1993, researchers discovered several disease-causing genes.

Longmore, Gregory D. June 1993. Erythropoietin receptor mutations and Olympic glory. *Nature Genetics,* vol. 4. Inheriting a tendency to overproduce EPO has athletic benefits.

Maddox, John. July 8, 1993. New genetics means no new ethics. *Nature,* vol. 364. A review of the history of recent genetic technology and its implications for the future.

Mulligan R. C. May 14, 1993. The basic science of gene therapy. *Science,* vol. 260. A review of the essentials of this new type of therapy.

Microscopy

As the study of life has progressed steadily from observing organisms to probing the molecules of life, the technology scientists use to view living things has grown accordingly. Today's biologist has a range of microscope types to choose from, and the instrument used depends upon the nature of the biological material being observed (table 1). An *ultraviolet microscope* might be used to highlight stained chromosomes; a *polarizing microscope* to focus in on protein arrays of a cytoskeleton; a *phase contrast microscope* to view cells while they are still alive. A *scanning electron microscope* reveals the topography of cell and organelle surfaces. A *confocal microscope* presents startlingly clear peeks at biological structures in action, and a *scanning probe microscope* reveals surfaces of individual atoms.

All microscopes provide two types of power—*magnification* and *resolution* (also called resolving power). A microscope produces an enlarged, or magnified, image of an object. Magnification is defined as the ratio between image size and object size. Resolution refers to the smallest degree of separation at which two objects appear distinct, rather than as a blurry, single image. Resolution is important in distinguishing structures from one another. A *compound microscope* commonly used in college biology teaching laboratories can resolve objects that are 0.1 to 0.2 micrometers (4 to 8 millionths of an inch) apart. The resolving power of an electron microscope is 10,000 times greater.

Table 1	Compound Microscopes		
Method	**Basis**	**Advantages**	**Disadvantages**
Phase contrast microscopy	Converts differences in the velocity of light through different parts of specimen into observable contrasts	Can be used on live cells	Not all subcellular structures are visible; halos seen around structures
Interference microscopy	Two beams of light hit specimen and join in image plane	No halos on structures; fine detail	Cumbersome to use; expensive
Differential-interference (Nomarski-optics) microscopy	Detects localized differences in velocities at which light passes through specimen	Fine transparent detail visible	
Polarizing microscopy	Ray of plane-polarized (unidirectional) light hits specimen, splits into two directions at two different velocities, creating image of ordered molecular detail	Highlights detail at molecular level	Works best on highly oriented, crystalline or fibrous structures
Fluorescence microscopy	Light of one wavelength is selectively absorbed by certain molecules that reemit light of a longer wavelength		Only creates images of structures that absorb the wavelength of light used
Ultraviolet (uv) microscopy	Ultraviolet light used with quartz lens	High resolving power; excellent for viewing proteins and nucleic acids	

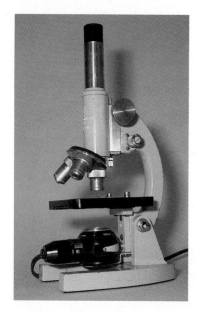

Figure 1
Light microscope

The Light Microscope

The compound light microscope focuses visible light through a specimen. Different regions of the object scatter the light differently, producing an image. In modern microscopes, three sets of lenses contribute to the generation of an image (fig. 1). The *condenser lens* focuses light through the specimen. The *objective lens* receives light that has passed through the specimen, generating an enlarged image. The *ocular lens,* or eyepiece, magnifies the image further. Total magnification is calculated by multiplying the magnification of the objective lens by that of the ocular lens. The user can manipulate the coarse and fine adjustment knobs to bring the magnified image into sharp focus. The mirror directs the light into the condenser lens, and the *diaphragm* controls the amount of light the specimen is exposed to.

A limitation of a light microscope is that the user can view only one two-dimensional plane of the specimen at a time. Thus, when a light microscope focuses on the top of a specimen, it reveals different structures than when it focuses at a deeper level. It can be difficult to envision the three-dimensional nature of the specimen from the two-dimensional views the light microscope affords. The problem is similar to focusing on particular parts of a scene with a camera. If the photographer focuses on his children in the foreground of a shot, he may entirely miss the antics of a cat and dog several feet behind them. Similarly, light microscope views at different cell depths reveal different structures.

Electron Microscopes

Electron microscopes provide greater magnification, better resolution, and a better sense of depth than light microscopes. Instead of focusing visible light, the *transmission electron microscope* (TEM) sends a beam of electrons through the specimen, using a magnetic field rather than a glass lens to focus the beam (fig. 2). Different parts of the specimen absorb different numbers of electrons. These contrasts are rendered visible to the human eye by a fluorescent screen coated with a chemical that gives off visible light rays when electrons from the specimen excite it.

Although the TEM has provided some spectacular glimpses into the microscopic structures of life, it does have limitations. The specimen must be killed, treated with chemicals, cut into very thin sections, and placed in a vacuum. This treatment can distort natural structures. A close cousin of the TEM eliminates these drawbacks. The *scanning electron microscope* (SEM) bounces electrons off of a three-dimensional specimen, generating a three-dimensional image on a device similar to a television screen. The resulting depth of field highlights crevices and textures. Not all specimens to be examined with the SEM need to be subjected to harsh treatment. Although many SEM specimens are coated with a heavy metal to highlight their surfaces, some living specimens (such as fruit flies) can undergo SEM examination with no apparent harm.

A variation of the electron microscope is the *photoelectron microscope* (PEM), originally used to probe metal surfaces but now used to examine cells

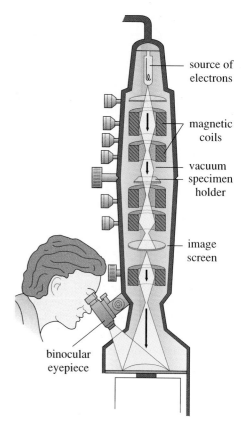

Figure 2
Electron microscope

as well. The PEM bombards a specimen with ultraviolet light, ejecting the valence shell electrons of molecules on a cell or organelle surface. These electrons are accelerated and focused by an electron lens system. The excited electrons are quite sensitive to the surface detail of the specimen, and their deflection pattern provides a high-resolution view of minute surface details. The PEM is especially useful for zeroing in on specific molecules that have been labeled with fluorescent antibodies. PEM is an electron-based version of fluorescence microscopy.

While the SEM highlights large surface features, the PEM provides a closer look. It is like comparing a topographic map of a mountain (SEM) to a picture of a bump in the terrain of the mountain (PEM). A light microscope and all three electron microscopes can be used in conjunction to paint a detailed portrait of biological structures, which can clarify functions at the organelle or even the molecular level.

 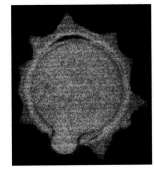 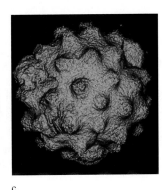

a.   b.   c.

Figure 3

Three views of pollen, as seen through *(a) a* conventional flourscence microscope, *(b) a* confocal laser scanning microscope, and *(c)* a tandem-scanning microscope.

The Confocal Microscope

A limitation of light microscopy is that light reflected from regions of the sample near the object of interest can interfere with the image, making it blurry or hazy. A *confocal microscope* avoids this interference and greatly enhances resolution by passing white or laser light through a pinhole and a lens to the object (fig. 3). The light is then reflected through a beam splitter and through another pinhole, a detector, and, finally, a photomultiplier. The result is a scan of highly focused light on one tiny part of the specimen at a time, usually an area 0.25 µm in diameter and 0.5 µm deep. The microscope is called "confocal" because the objective and the condenser lenses are both focused on the same small area.

Marvin Minsky patented the idea of a confocal microscope in 1961, but it was not developed until the mid 1980s, when computers became able to integrate many scans of different sites and at different depths and to translate them into a dynamic image. By using fluorescent dyes that label specific cell parts and that are activated by the incoming light, confocal microscopes can distinguish different structures. The microscope can capture the first division of a fertilized sea urchin egg; the spindle fibers appear green, and the chromosomes pulled in opposite directions appear a vibrant blue. Confocal microscopy has also revealed changing concentrations of calcium ions in a neuron receiving a biochemical message; the cytoskeleton in action; platelets aggregating at the scene of an injury to form a clot; sperm fertilizing an ovum; and nerve cells infiltrating the developing brain of an embryo.

Scanning Probe Microscopes

The world of microscopy was again revolutionized in 1981 with the invention of the *scanning tunneling microscope* (STM) by Gerd K. Binnig and Heinrich Rohrer. This device reveals detail at the atomic level. A very sharp metal needle, its tip as small as an atom, is scanned over a molecule's surface. Electrons "tunnel" across the space between the sample and the needle, thereby creating an electrical current. The closer the needle, the greater the current. An image is generated as the scanner continually adjusts the space between needle and sample, keeping the current constant over the topography of the molecular surface. The needle's movements over the micro-scopic hills and valleys are expressed as contour lines, which a computer converts and enhances to produce a colored image of the surface.

Electrons do not pass readily from many biological samples, which limits the use of STM. However, the same principle of adjusting a probe over a changing surface is used in *scanning ion-conductance microscopy* (SICM), developed by Paul K. Hansma and Calvin Quate. An SICM uses ions instead of electrons—useful in the many biological situations where ions travel between cells. The probe is made of hollow glass filled with a conductive salt solution, which is also applied to the sample. When voltage is passed through the sample and the probe, ions flow to the probe. The rate of ion flow is kept constant, and a portrait is painted by the compensatory movements of the probe. SICM is useful for studying cell membrane surfaces and muscle and nerve function.

Another type of scanning probe microscope, the *atomic force microscope* (AFM), was developed in 1986 by the inventors of the SICM. It uses a diamond-tipped probe that resembles the stylus on a phonograph but that presses a molecule's surface with a force millions of times gentler. As the force is kept constant, the probe moves, generating an image. AFM is especially useful for recording molecular movements, such as those involved in blood clotting and cell division.

New and improved microscopes do not always replace existing models but do complement the information other models provide. Many researchers today create their own versions of microscopes to suit their particular experiments. All modern microscopes though, some of them quite technologically sophisticated, support the cell theory the early microscopists advanced. It is hard to believe they had only very crude light microscopes to work with.

Units of Measurement Metric/English Conversions

Length

1 meter = 39.4 inches = 3.28 feet = 1.09 yards

1 foot = 0.305 meters = 12 inches = 0.33 yard

1 inch = 2.54 centimeters

1 centimeter = 10 millimeter = 0.394 inch

1 millimeter = 0.001 meter = 0.01 centimeter = 0.039 inch

1 fathom = 6 feet = 1.83 meters

1 rod = 16.5 feet = 5 meters

1 chain = 4 rods = 66 feet = 20 meters

1 furlong = 10 chains = 40 rods = 660 feet = 200 meters

1 kilometer = 1,000 meters = 0.621 miles = 0.54 nautical miles

1 mile = 5,280 feet = 8 furlongs = 1.61 kilometers

1 nautical mile = 1.15 miles

Area

1 square centimeter = 0.155 square inch

1 square foot = 144 square inches = 929 square centimeters

1 square yard = 9 square feet = 0.836 square meters

1 square meter = 10.76 square feet = 1.196 square yards = 1 million square millimeters

1 hectare = 10,000 square meters = 0.01 square kilometers = 2.47 acres

1 acre = 43,560 square feet = 0.405 hectares

1 square kilometer = 100 hectares = 1 million square meters = 0.386 square miles = 247 acres

1 square mile = 640 acres = 2.59 square kilometers

Volume

1 cubic centimeter = 1 milliliter = 0.001 liter

1 cubic meter = 1 million cubic centimeters = 1,000 liters

1 cubic meter = 35.3 cubic feet = 1.307 cubic yards = 264 U.S. gallons

1 cubic yard = 27 cubic feet = 0.765 cubic meters = 202 U.S. gallons

1 cubic kilometer = 1 million cubic meters = 0.24 cubic mile = 264 billion gallons

1 cubic mile = 4.166 cubic kilometers

1 liter = 1,000 milliliters = 1.06 quarts = 0.265 U.S. gallons = 0.035 cubic feet

1 U.S. gallon = 4 quarts = 3.79 liters = 231 cubic inches = 0.83 imperial (British) gallons

1 quart = 2 pints = 4 cups = 0.94 liters

1 acre foot = 325,851 U.S. gallons = 1,234,975 liters = 1,234 cubic meters

1 barrel (of oil) = 42 U.S. gallons = 159 liters

Mass

1 microgram = 0.001 milligram = 0.000001 gram

1 gram = 1,000 milligrams = 0.035 ounce

1 kilogram = 1,000 grams = 2.205 pounds

1 pound = 16 ounces = 454 grams

1 short ton = 2,000 pounds = 909 kilograms

1 metric ton = 1,000 kilograms = 2,200 pounds

Temperature

Celsius to Fahrenheit
$°F = (°C \times 1.8) + 32$

Fahrenheit to Celsius
$°C = (°F - 32) \div 1.8$

Energy and Power

1 erg = 1 dyne per square centimeter

1 joule = 10 million ergs

1 calorie = 4.184 joules

1 kilojoule = 1,000 joules = 0.949 British Thermal Units (BTU)

1 kilocalorie = 1,000 calories = 3.97 BTU = 0.00116 kilowatt-hour

1 BTU = 0.293 watt-hour

1 kilowatt-hour = 1,000 watt-hour = 860 kilocalories = 3,400 BTU

1 horsepower = 640 kilocalories

1 quad = 1 quadrillion kilojoules = 2.93 trillion kilowatt-hours

A P P E N D I X   C

Metric Conversion

	Metric Quantities	Metric to English Conversion	English to Metric Conversion

Length	1 kilometer (km) = 1,000 (10^3) meters
	1 meter (m) = 100 centimeters
	1 centimeter (cm) = 0.01 (10^{-2}) meter
	1 millimeter (mm) = 0.001 (10^{-3}) meter
	1 micrometer* (μm) = 0.000001
	  (10^{-6}) meter
	1 nanometer (nm) = 0.000000001
	  (10^{-9}) meter
	*formerly called micron
		1 km = 0.62 mile
		1 m = 1.09 yards = 39.37 inches
		1 cm = 0.394 inch
		1 mm = 0.039 inch
			1 mile = 1.609 km
			1 yard = 0.914 m
			1 foot = 0.305 m
			     = 30.5 cm
			1 inch = 2.54 cm

Area	1 square kilometer (km^2) = 100 hectares
	1 hectare (ha) = 10,000 square meters
	1 square meter (m^2) = 10,000 square centimeters
	
	1 square centimeter (cm^2) = 100 square
	  millimeters
		1 km^2 = 0.3861 square mile
		1 ha = 2.471 acres
		1 m^2 = 1.1960 square yards
		    = 10.764 square feet
		1 cm^2 = 0.155 square inch
			1 square mile = 2.590 km^2
			1 acre = 0.4047 ha
			1 square yard = 0.8361 m^2
			1 square foot = 0.0929 m^2
			1 square inch = 6.4516 cm^2

Mass	1 metric ton (t) = 1,000 kilograms 1 metric ton (t) = 1,000,000 grams 1 kilogram (kg) = 1,000 grams 1 gram (g) = 1,000 milligrams 1 milligram (mg) = 0.001 gram 1 microgram (μg) = 0.000001 gram	1 t = 1.1025 ton (U.S.) 1 kg = 2.205 pounds 1 g = 0.0353 ounce	1 ton (U.S.) = 0.907 t 1 pound = 0.4536 kg 1 ounce = 28.35 g

Volume (Solids)	1 cubic meter (m^3) = 1,000,000
	  cubic centimeters
	1 cubic centimeter (cm^3) = 1,000
	  cubic millimeters
		1 m^3 = 1.3080 cubic yards
		    = 35.315 cubic feet
		1 cm^3 = 0.0610 cubic inch
			1 cubic yard = 0.7646 m^3
			1 cubic foot = 0.0283 m^3
			1 cubic inch = 16.387 cm^3

Volume (Liquids)	1 liter (l) = 1,000 milliliters 1 milliliter (ml) = 0.001 liter 1 microliter (μl) = 0.000001 liter	1 l = 1.06 quarts (U.S.) 1 ml = 0.034 fluid ounce	1 quart (U.S.) = 0.94 l 1 pint (U.S.) = 0.47 l 1 fluid ounce = 29.57 ml
Time	1 second (sec) = 1,000 milliseconds 1 millisecond (msec) = 0.001 second 1 microsecond (μsec) = 0.000001 second		

TAXONOMY

We can group the millions of living and extinct species that have dwelled on earth according to many schemes. Taxonomists group organisms to reflect both anatomical similarities and descent from a common ancestor. Two, three, four, five, and most recently, six kingdom classifications have been proposed, as well as a system of three "domains" that supercede kingdoms. The five-kingdom scheme is outlined here with all phyla briefly described. Short statements explaining the rationale behind phyla groupings are given wherever possible, and indented subheadings reflect these groupings. Scientific names are followed by more familiar names of organisms. Figures from the text accompany the listing to help you visualize the wide diversity of life. A separate supplement on biodiversity visits the kingdoms in detail.

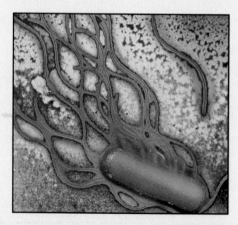

Kingdom Monera The monerans are unicellular prokaryotes that obtain nutrients by direct absorption, by photosynthesis, or by chemosynthesis. Most monerans reproduce asexually, but some can exchange genetic material in a primitive form of sexual reproduction. (The six-kingdom classification system divides the monerans into two kingdoms, recognizing bacteria and other single-celled organisms called archaebacteria.) They differ in certain key molecular structures and environmental requirements, though neither has nuclei or organelles.

Phylum Bacteria The bacteria and cyanobacteria.

Phylum Archaea Unicellular organisms lacking nuclei and organelles that share biochemical characteristics with some eukaryotes.

Kingdom Protista Protists are the structurally simplest eukaryotes, and they can be unicellular or multicellular. They can absorb or ingest nutrients or photosynthesize. Reproduction is asexual or sexual. Some forms move by ciliary or flagellar motion, and others are nonmotile. The protists' early development differs from that of the fungi, plants, and animals. The kingdom includes the protozoans, algae, and the water molds and slime molds.

Protozoans Unicellular, nonphotosynthetic, lack cell walls.

Phylum Sarcomastigophora Locomote by flagella and/or pseudopoda and include the familiar *Amoeba proteus*.

Phylum Labyrinthomorpha Aquatic, live on algae.

Phylum Apicomplexa Parasitic, with characteristic twisted structure on anterior end at some point in life cycle.

Phylum Myxozoa Parasitic on fish and invertebrates.

Phylum Microspora Parasitic on invertebrates and primitive vertebrates.

Phylum Ciliophora The ciliates have cilia at some stage of the life cycle.

Algae Unicellular or multicellular, photosynthetic, some have cell walls. Distinguished by pigments.

Phylum Euglenophyta Unicellular and photosynthetic, with a single flagellum and contractile vacuole.

Phylum Chrysophyta Diatoms, golden-brown algae, and yellow-green algae. Unicellular and photosynthetic.

Phylum Pyrrophyta Dinoflagellates. Unicellular and photosynthetic.

Phylum Chlorophyta Green algae. Unicellular or multicellular, photosynthetic.

Phylum Phaeophyta Brown algae (kelps). Multicellular and photosynthetic.

Phylum Rhodophyta Red algae. Multicellular and photosynthetic.

Water and Slime Molds

Phylum Oomycota   The water molds. Unicellular or multinucleate, with cellulose cell walls. Live in freshwater.

Phylum Chytridiomycota   The chytrids. Multicellular, with chitinous cell walls. Aquatic.

Phylum Myxomycota   Multinucleated, "acellular" slime molds.

Phylum Acrasiomycota   Multicellular "cellular" slime molds.

Kingdom Fungi

With the exception of the yeasts, fungi are multicellular eukaryotes that decompose organisms to obtain nourishment. Fungal cells have chitinous cell walls. Phyla are distinguished by their mode of reproduction.

Phylum Zygomycota   Reproduce with sexual resting spores.

Phylum Ascomycota   Yeasts, morels, truffles, molds, lichens. Reproduce with sexual spores carried in asci. Some ascomycetes cause food spoilage and some plant diseases; others are used in the production of certain foods, beverages, and antibiotic drugs.

Phylum Basidiomycota   Mushrooms, toadstools, puffballs, stinkhorns, shelf fungi, rusts, and smuts. Reproduce by spore-containing basidia.

Kingdom Plantae

Plants are multicellular, land dwelling, photosynthetic, and reproduce both asexually and sexually in an alternation of generations. Their cells have cellulose cell walls. Plants have specialized tissues and organs but lack nervous and muscular systems.

Nonvascular Plants (Bryophytes)   Lack specialized conducting tissues and true roots, stems, and leaves. The gamete-producing reproductive phase predominates.

Division Bryophyta   Liverworts, hornworts, mosses.

Vascular Plants (Tracheophytes)   Xylem and phloem transport water and nutrients, respectively, throughout the plant body of roots, stems and leaves. The spore-producing reproductive phase predominates.

Primitive Plants   Sperm cells travel in water to meet egg cells.

Division Pterophyta   Ferns.

Division Psilophyta   Whisk ferns.

Division Lycophyta   Club mosses and others.

Division Spenophyta   Horsetails.

Seed Plants   Sperm cells and egg cells enclosed in protective structures.

Gymnosperms   (naked seed plants)   Male and female cones produce pollen grains and ovules.

Division Coniferophyta   Conifers.

Division Cycadophyta   Cycads.

Division Ginkgophyta   Ginkgos.

Division Gnetophyta   Gnetophytes.

Angiosperms (seeds in a vessel)   The flowering plants.

Division Anthophyta   Flowering plants.

Kingdom Animalia

The animals are multicellular with specialized tissues and organs, including nervous and locomotive systems. Cells lack cell walls. Animals obtain nutrients from food. Phyla are distinguished largely on the basis of body form and symmetry, characteristics generally established in the early embryo.

Mesozoa Simplest animals.

Phylum Mesozoa   Very simple, wormlike parasites of marine invertebrates. Consist of only 20 to 30 cells.

Parazoa   A separate branch from the evolution of protozoa to metazoa.

Phylum Placozoa   A single species, *Trichoplax adhaerens,* characterized by two cell layers with fluid in between.

Phylum Porifera   The sponges. Specialized cell types organized into canal system to transport nutrients in and wastes out.

Eumetazoa   Animal phyla descended from protozoa.

Radiata   Radially symmetric body plan. Sedentary, saclike bodies with two or three cell layers and a diffuse nerve net.

Phylum Cnidaria   Hydroids, sea anemones, jellyfish, horny corals, hard corals.

Phylum Ctenophora   Sea walnuts, comb jellies.

Bilateria Bilaterally symmetric body plan.

Protostomia (first mouth)
Embryonic characteristics:
1. Mouth forms close to area of initial inward folding in very early embryo.
2. Spiral cleavage: At third cell division, second group of four cells sits atop first group of four cells, rotated 45°.
3. Determinate cleavage: Cell fates determined very early in development. If a cell from a four-celled embryo is isolated, it will divide and differentiate to form only one-quarter of an embryo.
4. The protostomes are further grouped by the way in which the body cavity (coelom) forms. A true coelom is a body cavity that develops within mesoderm, the middle layer of the embryo.

Acoelomates No coelom.
Phylum Platyhelminthes Flatworms.
Phylum Nemertina Ribbonworms.
Phylum Gnathostomulida Jawworms.

Pseudocoelomates Body cavity derived from a space in the embryo between the mesoderm and endoderm. The body cavity is called "pseudo" because it does not form within mesoderm. In the adult, the pseudocoelom is a cavity but it is not lined with mesoderm-derived peritoneum (seen in more advanced forms).
Phylum Rotifera The rotifers. Small (40 μm–3 mm), intricately shaped organisms that have a structure on their anterior ends that resembles rotating wheels. The rotifers occupy a variety of habitats.
Phylum Gastrotricha Aquatic, microscopic, flattened organisms with a scaly outer covering.
Phylum Kinorhyncha Marine worms less than 1 mm long.
Phylum Nematoda Roundworms. Found everywhere, many parasitic.

Phylum Nematomorpha Horsehair worms. Juveniles are parasitic in arthropods; adults are free-living.
Phylum Acanthocephala Spiny-headed worms. Spiny projection from anterior end used to attach to intestine of host vertebrate. Range in size from 2 mm to more than a meter.
Phylum Entoprocta Nonmotile, sessile, mostly marine animals that look like stalks anchored to rocks, shells, algae, or vegetation on one end, with tufted growth on the other.

Eucoelomates Coelom forms in a schizocoelous fashion, in which the body cavity forms when mesodermal cells invade the space between ectoderm and endoderm, and then proliferate so that a cavity forms within the mesoderm.
Major Eucoelomate Protostomes
Three phyla, with many species.
Phylum Mollusca Snails, clams, oysters, squids, octopuses.
Phylum Annelida Segmented worms.
Phylum Arthropoda Spiders, scorpions, ticks, mites, crustaceans, millipedes, centipedes, insects.
The Lesser Protostomes Seven phyla, including many extinct species. Little-understood offshoots of annelid-arthropod line.
Phylum Pripulida Bottom-dwelling marine worms.
Phylum Echiurida Marine worms.
Phylum Sipunculida Bottom-dwelling marine worms.
Phylum Tardigrada "Water bears." Less than 1 mm, live in water film on mosses and lichens.
Phylum Pentastomida Tongue worms. Parasitic on respiratory system of vertebrates, mostly reptiles.
Phylum Onychophora Velvet worms. Live in tropical rain forest and resemble caterpillars, with 14 to 43 pairs of unjointed legs and a velvety skin.
Phylum Pogonophora Beard worms. Live in mud on ocean bottom.

Lophophorates Three phyla distinguished by a ciliary feeding structure called a lophophore.
Phylum Phoronida Small, wormlike bottom-dwellers of shallow, coastal temperate seas. Live in a tube that they secrete.
Phylum Ectoprocta Bryozoa, or "moss animals." Aquatic, less than 1/2 mm long, live in colonies but each individual lives within a chamber secreted by the epidermis. Bryozoa look like crust on rocks, shells, and seaweeds.
Phylum Brachiopoda Lampshells. Attached, bottom-dwelling marine animals that have two shells and resemble mollusks, about 5 to 8 mm long.

Deuterostomia (second mouth)
Embryonic characteristics:
1. Mouth forms far from area of initial inward folding in very early embryo.
2. Radial cleavage: At third cell division, second group of four cells sits directly atop first group.
3. Indeterminate cleavage: Cell fates of very early embryo not determined. If a cell from a four-cell embryo is isolated, it will develop into a complete embryo.
4. Coelom formation is enterocoelous. Body cavity forms from outpouchings of endoderm that become lined with mesoderm.
Phylum Echinodermata Sea stars, brittle stars, sea urchins, sea cucumbers, sea lilies. Radial symmetry in adult but larvae are bilaterally symmetric. Complex organ systems, but no distinct head region.
Phylum Chaetognatha Arrow worms. Marine-dwelling with bristles surrounding mouth.
Phylum Hemichordata Acorn worms and others. Aquatic, bottom-dwelling, nonmotile, wormlike animals.
Phylum Chordata Tunicates, lancelets, hagfishes, lampreys, sharks, bony fishes, amphibians, reptiles, birds, mammals. Chordates have a notochord, dorsal nerve cord, gill slits, and a tail. Some of these characteristics may only be present in embryos.

GLOSSARY

●●

A

abortion *a-BOR-shun* Termination of a pregnancy. 260

accessory pigments *ax-ES-or-ee PIG-mentz* Plant pigments other than chlorophyll that extend the range of light wavelengths useful in photosynthesis. 149

acetyl CoA formation *AS-eh-til FOR-MAY-shun* The first step in aerobic respiration, occurring in the mitochondrion. Pyruvic acid loses a carbon dioxide and bonds to coenzyme A to form acetyl CoA. 130

acid *AS-id* A molecule that releases hydrogen ions into water. 43

acidophile *a-SID-o-file* Member of an acid-loving species. 45

acidosis *a-sid-O-sis* Abnormal human blood pH of 7.0 to 7.3, causing disorientation, fatigue, and labored breathing. 43

acquired immune deficiency syndrome (AIDS) *ak-KWY-erd im-MUNE dah-FISH-en-see SIN-drome* Infection caused by the human immunodeficiency virus (HIV), which kills a certain class of helper T cells. (This causes profound immune suppression, which results in susceptibility to opportunistic infections and cancer. 261

acrocentric *ak-ro-SEN-trik* A chromosome whose centromere divides the chromosome into a long arm and a short arm. 310

acrosome *AK-ro-som* A protrusion on the anterior end of a sperm cell containing digestive enzymes that enable the sperm to penetrate the protective layers around the oocyte. 203

actin *AK-tin* A type of protein in the thin myofilaments of skeletal muscle cells.

action potential *AK-shun po-TEN-shel* The measurement of an electrochemical change caused by ion movement across the cell membrane of a neuron. The message formed by this change is the nerve impulse.

active site *AK-tiv SITE* The portion of an enzyme's conformation that directly participates in catalysis. 50

active transport *AK-tiv TRANZ-port* Movement of a molecule through a membrane against its concentration gradient, using a carrier protein and energy from ATP. 98

adaptation *AD-ap-TAY-shun* An inherited trait that enables an organism to survive a particular environmental challenge. 7

adaptive radiation *ah-DAP-tiv RAID-ee-AY-shun* The divergence of several new types of organisms from a single ancestral type.

adenine *AD-eh-neen* One of two purine nitrogenous bases in DNA and RNA. 52, 332

adenosine triphosphate (ATP) *ah-DEN-o-seen tri-FOS-fate* A molecule whose three high-energy phosphate bonds power many biological processes. 98, 118

adhesion *ad-HE-jhun* The tendency of water to hydrogen bond to other compounds. 43

adipose tissue *AD-eh-pose TISH-ew* Tissue consisting of cells laden with lipid. 48

adrenal glands *ad-REE-nal GLANZ* Paired two-part glands that sit atop the kidneys and produce catecholamines, mineralocorticoids, glucocorticoids, and sex hormones.

aerobic respiration *air-O-bik res-per-A-shun* Cellular respiration that requires oxygen as a final electron acceptor and that generates ATP. 126

alcoholic fermentation *AL-ko-HALL-ik FER-men-TAY-shun* An anaerobic step that yeast utilize after glycolysis. Pyruvic acid is converted to ethanol and carbon dioxide. 138

alkalosis *al-ka-LO-sis* Abnormal human blood pH of 7.5 to 7.8, producing agitation and dizziness. 43

allele *ah-LEEL* An alternate form of a gene. 200, 272

allosteric inhibition *al-o-STARE-ic in-hi-BI-shun* Suppression of an enzyme's function caused when a substance binds at a site other than its active site. 121

amino acid *ah-MEEN-o-AS-id* An organic molecule consisting of a central carbon atom bonded to a hydrogen atom, an amino group, a carboxylic acid, and an R group. A polymer of amino acids is a peptide. 48

amino group *a-MEEN-o GROOP* A nitrogen atom single-bonded to two hydrogen atoms. 49

anabolism *eh-NAB-o-liz-um* Synthetic metabolic reactions that use energy. 116

anaerobes *AN-air-robes* Organisms that can live in an environment lacking oxygen. 138

anaerobic respiration *an-air-RO-bic res-per-A-shun* Cellular respiration in the absence of oxygen.

anaphase *AN-ah-faze* The stage of mitosis when centromeres split and the two sets of chromosomes move to opposite ends of the cell. 171

anaphase I *AN-ah-faze I* Anaphase of meiosis I; homologs separate. 200

anaphase II *AN-ah-faze II* Anaphase of meiosis II, when centromeres separate and chromatids move to opposite poles. 200

aneuploid *AN-you-ploid* A cell with one or more extra or missing chromosomes.

angiogenesis *AN-gee-o-GEN-e-sis* The formation of new blood vessels. 176

anion *AN-i-on* A negatively charged ion. 40

anticodon *AN-ti-ko-don* A three-base sequence on one loop of a transfer RNA molecule that is complementary to an mRNA codon and therefore brings together the appropriate amino acid and its mRNA instructions. 353

antiparallelism *ANT-i PAR-a-lel-izm* The head-to-tail relationship of the two rails of the DNA double helix. 334

antisense strand *AN-ti-SENSE STRAND* The side of the DNA double helix not transcribed into mRNA. 350

apoptosis *a-pop-TOE-sis* Programmed cell death. 233

aqueous solution *AWK-kwee-us so-LEW-shun* A solution in which water is the solvent. 43, 92

artificial insemination *AR-teh-FISH-el in-SEM-eh-NAY-shun* Placing donated sperm in a woman's reproductive tract to start a pregnancy. 245

asexual reproduction *A-sex-yu-al re-pro-DUK-shun* Reproduction in which a cell doubles its contents and then splits in two to yield two identical cells. 8, 192

aster *AS-ter* A starlike outgrowth of the mitotic spindle that determines the site of the cleavage furrow. 171

atom *AT-um* A chemical unit, composed of protons, neutrons, and electrons, that cannot be further broken down by chemical means. 36

atomic number *a-TOM-ic NUM-ber* The number in the lower left hand corner of an element's symbol in the periodic table, indicating the number of protons or electrons in one atom of the element. 36

ATP synthase *ATP SIN-thaze* An enzyme that allows protons to move through the mitochondrial membrane and trigger phosphorylation of ADP to ATP. 134

autosomal dominant *AW-toe-soe-mal DOM-i-nent* Controlled by an allele that masks the expression of another allele. 280

autosomal recessive *AW-toe-soe-mal re-SESS-ive* Controlled by an allele whose expression is masked by another allele. 280

autosome *AW-toe-soam* A nonsex chromosome.

autotroph(ic) *AW-toe-trof* An organism that manufactures nutrient molecules using energy harnessed from the environment. 113, 144

B

Barr body *BAR BOD-ee* The dark-staining body in the nucleus of a female mammal cell, corresponding to the inactivated X chromosome. 306

base *BASE* A molecule that releases hydroxide ions into water. 43

benign tumor *bee-NINE TOO mer* A noncancerous tumor. 179

binary fission *BI-nair-ee FISH-en* A type of asexual reproduction in which a cell divides into two identical cells. 192

biochemical reactions *bi-o-KEM-i-kal re-AK-shunz* Photosynthetic reactions that do not require light, instead using products of photochemical reactions to synthesize organic molecules. 152

biodiversity *bi-o-di-VER-city* The spectrum of different life forms. 8

bioenergetics *bi-o-n-er-JET-ix* The study of energy in living things. 112

biological evolution *bi-o-LOJ-ik-kal ev-o-LEW-shun* The process by which the genetic structure of populations change over time.

bioluminescence *bi-o-loom-in-ES-ents* A biochemical phenomenon that causes an organism to glow. 119

bioremediation *bi-o-ree-meed-e-AY-shun* Using organisms that metabolize toxins to clean the environment. 377

blastocyst *BLAS-toe-syst* The preembryonic stage of human development when the organism is a hollow, fluid-filled ball of cells. 219

blastomere *BLAS-toe-mere* A cell in a preembryonic organism resulting from cleavage divisions. 219

Bohr model *BOR MAH-del* A depiction of atomic structure. 37

bond energy *BOND EN-er-gee* The energy required to form a particular chemical bond. 116

bone *BONE* A connective tissue consisting of bone-building osteoblasts, stationary osteocytes, and bone-destroying osteoclasts, embedded in a mineralized matrix infused with spaces and canals (lacunae, canaliculi, and Haversian canals).

buffer system *BUFF-er SIS-tem* Pairs of weak acids and bases that maintain body fluid pH. 44

bulbourethral glands *BUL-bo-u-REE-thral GLANZ* Small glands near the male urethra that secrete mucus. 197

bulk element *BULK EL-e-ment* An element humans need in large amounts. 36

bundle sheath cells *BUN-dull SHEETH SELLS* Thick-walled plant cells surrounding veins that function in C₄ photosynthesis. 158

C

Calvin cycle *CAL-vin SI-kel* A biochemical reaction of photosynthesis that fixes carbon from carbon dioxide into small organic molecules.

capacitation *cah-PASS-eh-TAY-shun* Activation of sperm cells in the human female reproductive tract. 218

carbohydrate *CAR-bo-HI-drate* Compounds containing carbon, hydrogen, and oxygen, with twice as many hydrogens as oxygens. Carbohydrates include the sugars and starches. 46

carboxyl group *kar-BOX-ill GROOP* A carbon atom double-bonded to an oxygen and single-bonded to a hydroxyl group (OH). 48

cardiac muscle *CAR-dee-ak MUS-sel* Striated, involuntary, single-nucleated contractile cells found in the mammalian heart.

cartilage *CAR-teh-lij* A supportive connective tissue consisting of chondrocytes embedded in collagen and proteoglycans.

catabolism *cah-TAB-o-liz-um* Metabolic degradation reactions, which release energy. 116

catalysis *kat-AL-i-sis* Speeding of a chemical reaction. 50

cation *KAT-i-on* A positively charged ion. 40

cdc 2 kinase *see-dee-see-TOO ki-NASE* An enzyme that controls the cell division cycle. 177

cell *SEL* The structural and functional unit of life. 6, 64

cell culture *SEL KUL-chur* Scientific technique of growing cells in glass dishes. 181

cell cycle *SEL CY-kel* The life of a cell, in terms of whether it is dividing or in interphase. 168

cell membrane (plasmalemma) *SEL MEM-brane* An oily structure consisting of proteins embedded in a lipid bilayer, which forms the boundary of cells. 68

cell population *SEL pop-u-LAY-shun* A group of cells with characteristic proportions in particular stages of the cell cycle. 178

cell theory *SEL THER-ee* The ideas that all living matter consists of cells, cells are the structural and functional units of life, and all cells come from preexisting cells. 66

cellular respiration *SEL-u-ler res-per-A-shun* Biochemical reactions involved in energy extraction in the mitochondrion. 115

cell wall *SEL WALL* A rigid boundary consisting of peptidoglycans in prokaryotic cells and cellulose in plant cells. 68

central vacuole *SEN-tral VAC-u-ole* A membrane-bound storage organelle in plant cells. 75

centrioles *SEN-tre-olz* Paired, oblong structures consisting of microtubules and found in animal cells, where they organize the mitotic spindle. 78, 174

centromere *SEN-tro-mere* A characteristically located constriction in a chromosome. 170

centrosome *SEN-tro-soam* A region near the cell nucleus that contains the centrioles. 174

cervical cap *SIR-vik-al CAP* A contraceptive device that blocks a woman's cervix. 260

cervix *SER-viks* In the female human, the opening to the uterus. 198

chaperone protein *shap-er-one pro-teen* A protein that stabilizes a growing amino acid chain. 360

chemical bond *KEM-e-cal BOND* Attachments atoms form by sharing or exchanging electrons. 38

chemical equilibrium *KEM-e-cal e-kwil-IB-ree-um* When a chemical reaction proceeds in both directions at the same rate. 39, 117

chemical reaction *KEM-e-cal re-AC-shun* Interactions in which atoms exchange or share electrons, forming new chemicals. 38

chemiosmosis *KEM-ee-oss-MOE-sis* Phosphorylation of ADP to ATP occurring when protons that are following a concentration gradient contact ATP synthase. 128

chemoautotroph *KEEM-o-awt-o-trofe* An organism that metabolizes chemicals from nonliving surroundings to obtain energy. 113

chlorophyll a *KLOR-eh-fill A* A green pigment used by plants to harness the energy in sunlight. 78, 148

chloroplast *KLOR-o-plast* A plant cell organelle housing the reactions of photosynthesis. 78, 149

cholesterol *kole-ES-tir-awl* A sterol lipid found in cell membranes and used to synthesize certain hormones. 48

chorion *KORE-ee-on* A membrane that develops into the placenta. 219

chorionic villi *KOR-ee-ON-ik VIL-i* Fingerlike projections extending from the chorion to the uterine lining. 225

chromatid *CRO-mah-tid* A continuous strand of DNA comprising an unreplicated chromosome or one-half of a replicated chromosome. 170

chromatin *KRO-ma-tin* Colored material in stained cell nuclei, including the genetic material. 168, 334

chromosome *KRO-mo-soam* A dark-staining, rod-shaped structure in the nucleus of eukaryotic cell consisting of a continuous molecule of DNA wrapped in protein. 70

cilia *SIL-ee-ah* Protein projections from cells. Cilia beat in unison, moving substances.

cleavage *KLEEV-ij* A period of rapid cell division following fertilization but before embryogenesis. 219

cleavage furrow *KLEEV-ij FUR-o* The initial indentation between two daughter cells in mitosis. 171

clitoris *CLIT-or-is* A small, highly innervated bit of tissue at the juncture of the labia that is the female anatomical equivalent of the penis. 198

codominant alleles *ko-DOM-eh-nent ah-LEELZ* Alleles that are both expressed in the heterozygote. 284

codon *KO-don* A continuous triplet of mRNA that specifies a particular amino acid. 353

coefficient of relationship *co-eff-FISH-ent of re-LAY-shun-ship* A measurement of how closely related two individuals are, based upon the proportion of genes they have in common. 291

coelom *SEE-loam* A central body cavity in an animal.

coenzyme *coe-EN-zime* An organic cofactor necessary for the function of certain enzymes. 120

cofactor *COE-fac-tor* A substance necessary for an enzyme to function. 120

cohesion *co-HEE-jhun* The strong attraction of water molecules to each other. 43

competitive inhibition *cum-PET-it-iv in-hi-BI-shun* Inhibition of an enzyme's function when a molecule other than the substrate reversibly binds to the active site. 121

complementary base pairs *kom-ple-MENT-ah-ree BASE PAIRZ* The tendency of adenine to hydrogen bond to thymine and guanine to cytosine in the DNA double helix. 334

complete penetrance *kum-PLEET PEN-e-trants* A genotype that produces a phenotype every time it is inherited. 284

complex carbohydrates *KOM-plex kar-bo-HI-drates* The polysaccharides, which are chains of sugars. Polysaccharides include starch, glycogen, cellulose, and chitin. 46

compound *KOM-pound* A molecule consisting of different atoms. 38

concentration gradient *kon-sen-TRA-shun GRAY-dee-ent* The phenomenon of ions passively diffusing from an area of high concentration to an area where they are less concentrated.

concordance *kon-KOR-dance* A measure of the inherited component of a trait, consisting of the number of pairs of either monozygotic or dizygotic twins in which both members express a trait, divided by the number of pairs in which at least one twin expresses the trait. 292

condom *CON-dum* A sheath worn over the penis to prevent pregnancy and protect against sexually transmitted diseases. 260

conformation *KON-for-MAY-shun* The three-dimensional shape of a protein. 49

conjugation *con-ju-GAY-shun* A form of gene transfer in bacteria. 193

connective tissue *kon-NECK-tiv TISH-ew* Tissues consisting of cells embedded or suspended in a matrix, including loose and fibrous connective tissues, cartilage, bone, and blood.

conservative replication *con-SER-va-tive rep-li-KAY-shun* Hypothesized, incorrect mode of DNA replication in which two parental strands specify two daughter strands. 335

continuously varying *con-TIN-you-us-lee VARE-e-ing* A trait that is especially variable because it is specified by more than one gene. 289

contraception *con-tra-SEP-shun* Techniques and devices that prevent pregnancy. 257

contraceptive sponge *con-tra-SEP-tive SPUNJ* A device a woman places in her vagina to prevent conception. 260

contractile vacuole *KON-tract-till VAK-u-ol* An organelle in a paramecium that pumps water out of the cell. 94

coupled reactions *CUP-uld re-AC-shuns* Two chemical reactions that occur simultaneously and have a common intermediate. 118

covalent bond *KO-va-lent BOND* The sharing of electrons between atoms. 39

C₃ plants *see-3 PLANTZ* Plants that use the Calvin cycle only to fix carbon dioxide. 155

C₄ photosynthesis *see-4 foto-SIN-the-sis* A pathway enabling plants to avoid photorespiration. 158

C₄ plants *see-4 PLANTZ* Plants that use C₄ photosynthesis. 158

cristae *KRIS-ty* The folds of the inner mitochondrial membrane along which many of the reactions of cellular respiration occur. 75

critical period *KRIT-eh-kel PER-ee-od* The time in an animal's life when it performs a particular imprinting behavior. 253

crossing over *KROS-ing O-ver* The exchange of genetic material between homologous chromosomes during prophase of meiosis I. 199

cyclin *SI-klin* A type of protein involved in controlling the cell cycle. 177

cytochrome *SI-to-krome* An iron-containing molecule that transfers electrons in metabolic pathways. 121

cytokinesis *SI-toe-kin-E-sis* Distribution of cytoplasm, organelles, and macromolecules into two daughter cells in cell division. 171

cytoplasm *SI-toe-PLAZ-um* The jellylike fluid in which organelles are suspended in eukaryotic cells. 70

cytosine *SI-toe-seen* One of the two pyrimidine nitrogenous bases in DNA and RNA. 52, 332

cytoskeleton *SI-toe-SKEL-eh-ten* A framework consisting of arrays of protein rods and tubules found in animal cells. 72, 88

D

daughter DNA strand *DAUT-er DNA STRAND* Newly replicated DNA.

degenerate codons *de-JEN-er-at KO-donz* Different codons specifying the same amino acid. 358

dehydration synthesis *de-hi-DRA-shun SYN-theh-sis* Formation of a covalent bond between two molecules by the loss of water. 46

deletion *dee-LEE-shun* A part missing from a chromosome. 317

density-dependent inhibition *DEN-seh-tee de-PEN-dent in-hi-BI-shun* The tendency of a cell to cease dividing once it touches another cell. 176

density shift experiment *DEN-seh-tee SHIFT ex-PER-e-ment* Experiments tracing DNA replication in which bacterial cultures are shifted between media containing different nitrogen isotopes.

deoxyribonucleic acid (DNA) *de-OX-ee-RI-bo-nu-KLAY-ic AS-id* A double-stranded nucleic acid composed of nucleotides containing a phosphate group, a nitrogenous base (A, T, G, or C), and the sugar deoxyribose. 52, 331

deoxyribose *de-OX-ee-RI-bose* A five-carbon sugar that is a structural component of DNA. 331

diaphragm *DI-ah-fram* A contraceptive device worn over a woman's cervix. 260

differentiate *diff-er-N-shee-ate* Specialize. 212

diffusion *de-FUZE-jhun* Movement of a substance from a region where it is highly concentrated to an area where it is less concentrated without energy input. 94

dihybrid cross *DI-HI-brid KROS* A cross between individuals heterozygous for two particular genes. 281

diploid cell *DIP-loid SEL* A cell with two copies of each chromosome. 193

disaccharide *di-SAK-eh-ride* A sugar built of two bonded monosaccharides, including sucrose, maltose, and lactose. 46

dispersive replication *dis-PURR-sive rep-li-KAY-shun* The hypothesized incorrect mode of DNA replication, in which the double helix dissociates with each DNA generation, picking up new nucleotides at random to form daughter strands. 335

disulfide bond *di-SUL-fide BOND* Attraction between two sulfurs within a protein molecule. 49

dizygotic twins *di-zi-GOT-ik TWINZ* Fraternal twins arising from two fertilized ova. 291

DNA polymerase *DNA po-LIM-er-ase* A type of enzyme that participates in DNA replication by inserting new bases and correcting mismatched base pairs. 337

DNA probe *DNA PROBE* A DNA sequence complementary to a sequence of interest and thus used to identify it. 382

dominant *DOM-eh-nent* An allele that masks the expression of another allele. 272

double-blind *DUB-el BLIND* An experimental protocol where neither the participants nor the researchers know which subjects have received a placebo and which have received the treatment being evaluated. 20

duplication *doop-li-KAY-shun* A repeated portion of a chromosome. 317

dynein *DI-ne-in* A motor molecule that forms part of cilia. 103

E

ecosystem *E-ko-SIS-tum* A unit of interaction among organisms and their physical environments, including all living things within a defined area. 11

ectoderm *EK-TOE-derm* The outermost embryonic germ layer, whose cells become part of the nervous system, sense organs, outer skin layer, and its specializations. 222

ectopic pregnancy *ek-TOP-ik PREG-nan-see* The implantation of a zygote in the wall of a fallopian tube rather than in the uterus. 243

ectoplasm *EK-toe-PLAZ-m* In an amoeba, an outer layer of thick, gellike material.

effector *e-FEK-ter* A muscle or gland that receives input from a neuron.

ejaculation *e-JAK-u-LAY-shun* Discharge of sperm through the penis. 197

elastin *e-LAS-tin* A type of connective tissue protein.

electrolyte *e-LEK-tro-lite* Solutions containing ions whose balance is vital for health. 43

electromagnetic spectrum *e-LEK-tro-mag-NET-ik SPECK-trum* A spectrum of naturally occurring radiation. 146

electron *e-LEK-tron* A subatomic particle, carrying a negative electrical charge and a negligible mass, that orbits the atomic nucleus. 36

electronegativity *e-LEK-tro-neg-a-TIV-it-ee* The tendency of an atom to attract electrons. 40

electron orbital *e-LEK-tron OR-bit-al* The volume of space where a particular electron is found 90% of the time. 37

electron transport chain *ee-LEK-tron TRANZ-port CHANE* Linked oxidation-reduction reactions. 121, 126

electroporation *ee-LEK-tro-por-AY-shun* Applying a brief jolt of electricity to open up transient holes in cell membranes, allowing foreign DNA to be introduced. 378

element *EL-eh-ment* A pure substance consisting of atoms containing a characteristic number of protons. 35

elongation *ee-long-A-shun* The stage of translation that builds an amino acid chain. 359

embryo adoption *EM-bree-o a-DOP-shun* A procedure in which an embryo is implanted in another woman's uterus. 251

embryonic induction *EM-bree-ON-ik in-DUK-shun* The ability of a group of specialized cells to stimulate neighboring cells to specialize. 227

embryonic stage *em-bree-ON-ik STAJE* The stage of prenatal development when organs develop from a three-layered organization. 217

embryonic stem cell *em-bree-ON-ik STEM SELL* An embryonic cell that retains the ability to specialize as any cell type. 379

emergent property *e-MER-jent PROP-er-tee* A quality that appears as biological complexity increases. 6

empiric risk *em-PEER-ik RISK* Risk calculation based on population prevalence. 289

endergonic reaction *en-der-GONE-ik re-AK-shun* An energy-requiring chemical reaction. 117

endocytosis *EN-doe-si-TOE-sis* The engulfing of an extracellular substance by the cell membrane. 75, 99

endoderm *EN-doe-derm* The innermost embryonic germ layer, whose cells become the organs and linings of the digestive, respiratory, and urinary systems. 222

endometriosis *en-do-mee-tree-O-sis* Tissue buildup in and on the uterus. 243

endometrium *EN-doe-MEE-tree-um* The inner uterine lining. 198, 221

endoplasm *EN-doe-PLAZ-m* In an amoeba, an inner layer of sol-like cytoplasm.

endoplasmic reticulum *EN-doe-PLAZ-mik reh-TIK-u-lum* A maze of interconnected membranous tubules and sacs, winding from the nuclear envelope to the cell membrane, along which proteins are synthesized (in the rough ER) and lipids synthesized (in the smooth ER). 73

endosymbiont theory *EN-doe-SYM-bee-ont THER-ee* The idea that eukaryotic cells evolved from large prokaryotic cells that engulfed once free-living bacteria. 82

endothelium *end-o-THEEL-e-um* Layer of single cells that forms a capillary wall. 105

energy *EN-er-gee* The ability to do work. 6, 112

energy of activation *EN-er-gee of ac-ti-VA-shun* Energy required for a chemical reaction to begin. 121

energy shell *EN-er-gee SHEL* Levels of energy in an atom formed by electron orbitals. 37

entropy *EN-tro-pee* A state of randomness or disorder. 6, 115

enzyme *EN-zime* A protein that catalyzes a specific type of chemical reaction. 36

enzyme-substrate complex *EN-zime SUB-strate COM-plex* A transient structure that forms when a substrate binds with an enzyme's active site. 50

epidemiology *EP-eh-dee-mee-OL-o-gee* The analysis of data derived from real-life, nonexperimental situations. 23

epididymis *EP-eh-DID-eh-mis* In the human male, a tightly coiled tube leading from each testis, where sperm mature and are stored. 197

epigenesis *ep-ee-GEN-i-sis* The idea that specialized tissue arises from unspecialized tissue in a fertilized ovum. 211

epistasis *EP-eh-STAY-sis* A gene masking another gene's expression. 284

epithelial tissue *EP-eh-THEEL-e-al TISH-ew* Tissue consisting of cells that are packed close together to form linings and boundaries.

epithelium *EP-eh-THEEL-e-um* Cells that form linings and coverings.

equational division *ee-QUAY-shun-el deh-VISZ-un* The second meiotic division, when four haploid cells are generated from the two haploid cells that are the products of meiosis I by a mitosislike division. 198

euchromatin *u-KROME-a-tin* Light-staining genetic material. 310

eukaryotic cell *u-CARE-ee-OT-ik SEL* A complex cell containing organelles that carry out a variety of specific functions. 68

euploid *U-ployd* A normal chromosome number. 313

excision repair *ex-SIZ-jhun ree-PARE* Cutting pyrimidine dimers out of DNA. 341

exergonic reaction *ex-er-GONE-ik re-AK-shun* An energy-releasing chemical reaction. 116

exocytosis *EX-o-si-TOE-sis* The fusing of secretion-containing organelles that travel to the inside surface of the cell membrane, where they transport a substance out of the cell. 74, 99

exon *EX-on* The bases of a gene that code for amino acids. 354

experiment *ex-PEAR-a-ment* A test designed to prove or disprove a hypothesis. 17

experimental control *ex-PEAR-eh-MEN-tel KON-trol* An extra test that can rule out causes other than the one being investigated. 20

extraembryonic membranes *EX-tra-EM-bree-on-ik MEM-BRANZ* Structures that support and nourish the mammalian embryo and fetus, including the yolk sac, allantois, and amnion. 217

F

facilitated diffusion *fah-SIL-eh-tay-tid dif-FU-shun* Movement of a substance down its concentration gradient with the aid of a carrier protein. 96

fallopian tubes *fah-LO-pee-an TUBES* In the human female, paired tubes leading from near the ovaries to the uterus. 197

fate map *FAY-t map* Diagrams showing the correspondence between embryonic and adult structures. 222

fatty acid *FAT-ee AS-id* A hydrocarbon chain that combines with other fatty acids and glycerol to form triglyceride fats. 47

feedback loop *FEED-bak LOOP* A complex interaction between the product of a biochemical reaction and the starting material.

fermentation *fur-men-TAY-shun* An energy pathway occurring in the cytoplasm that extracts some energy from the bonds of glucose, but not as much as is possible with aerobic respiration. 138

fertilized ovum *FUR-till-ized O-vum* The first cell of a new organism that arises from sexual reproduction. 197

fetal period *FEE-tal PEER-e-od* The final stage of prenatal development, when structures grow and elaborate. 217

fetus *FEE-tus* In humans, a developing individual from three months after conception until birth. 203

fibroblast *FI-bro-blast* A connective tissue cell that secretes the proteins collagen and elastin.

fibroids *FI-broydz* Benign uterine tumors. 243

first filial generation *FURST FILL-e-al gen-er-A-shun* The second generation in a genetics problem. 274

flagella *fla-GEL-ah* Taillike appendages on prokaryotic cells.

flavin adenine dinucleotide (FAD) *FLAY-vin AD-e-neen di-NUKE-lee-o-tide* An electron carrier molecule that functions in certain metabolic pathways. 120

fluid mosaic *FLU-id mo-ZAY-ik* Description of a biological membrane, referring to the arrangement of proteins embedded in the oily lipid bilayer. 91

fluorescence *floor-ES-ents* Radiation with a wavelength longer than that of visible light, causing a glowing effect. 152

follicle cell *FOL-ik-kel SEL* Nourishing cell surrounding an oocyte. 197

frameshift mutation *FRAME-shift mew-TAY-shun* A mutation that adds or deletes one or two DNA bases, altering the reading frame. 366

free energy *FREE EN-er-gee* The usable energy in the bonds of a molecule. 116

G

gamete *GAM-eet* A sex cell. The sperm and ovum. 196

gamete intrafallopian transfer (GIFT) *GAM-eet in-tra-fall-O-pee-an TRANS-fer* A procedure in which oocytes and sperm are deposited at a site past an obstruction in the fallopian tube. 249

gametogenesis *ga-MEET-o-gen-i-sis* Meiosis and maturation; making gametes. 198

gametophyte *gah-MEE-toe-fight* The part of a plant's life cycle when sex cells are manufactured.

gap analysis *GAP an-AL-i-sis* Compiling ecological data for a geographic area to predict endangered species.

gap phase *GAP FAZE* One of the stages of interphase. 168

gas exchange *GAS ex-CHAYNJ* An animal cell's intake of oxygen and release of carbon dioxide.

gastrula *GAS-troo-la* A three-layered embryo. 222

gene *JEAN* A sequence of DNA that specifies the sequence of amino acids in a particular polypeptide. 52, 270

gene amplification *JEAN am-pli-fi-KAY-shun* Biotechnologies that mass produce copies of a gene of interest. 338

gene pool *JEAN POOL* All the genes in a population.

gene targeting *JEAN TAR-get-ing* A biotechnology that replaces one gene with another. 379

genetic code *jeh-NET-ik KODE* The correspondence between specific DNA base sequences and the amino acids that they specify. 52, 354

genetic marker *jeh-NET-ik MAR-ker* A detectable piece of DNA closely linked to a gene of interest whose precise location is not known. 384

genome *JEE-nome* All of the DNA in the cell of an organism. 300

genomic imprinting *gee-NO-mik IM-print-ing* A phenotype that varies depending upon which parent it is inherited from. 288, 308

genotype *JEAN-o-type* The genetic constitution of an individual. 272

genotypic ratio *jean-o-TIP-ik RAY-shee-o* Proportions of genotypes among offspring of a genetic cross. 275

geothermal energy *gee-o-THER-mal EN-er-gee* Heat energy within the earth. 112

germ cell *JERM SEL* Gamete or sex cell. 196

germ-line cells *JERM-line SELZ* Somatic cells that give rise to germ cells. 198

glyceraldehyde-3-phosphate *gliss-er-AL-de-hide-3-FOSS-fate* A carbohydrate product of the Calvin cycle in photosynthesis. 155

glycerol *GLI-sir-all* A 3-carbon alcohol that forms the backbone of triglyceride fats. 47

glycolysis *gli-KOL-eh-sis* A catabolic pathway occurring in the cytoplasm of all cells. One molecule of glucose is split and rearranged into two molecules of pyruvic acid. 126

glycoprotein *GLY-ko-PRO-teen* A molecule built of a protein and a sugar. 91

Golgi apparatus *GOL-gee ap-ah-RAT-tis* A system of flat, stacked, membrane-bound sacs where sugars are polymerized to starches or bonded to proteins or lipids. 74

gonads *GO-nads* Paired organs where gametes are formed. 196

G₁ phase *JEE-1 FAZE* The gap stage of interphase when proteins, lipids, and carbohydrates are synthesized. 168

G₂ phase *JEE-2 FAZE* The gap stage of interphase when membrane components are synthesized and stored. 170

granum *GRAN-um* A stack of flattened thylakoid discs comprising the inner membrane of a chloroplast. 78, 149

growth factor *GROWTH FAK-ter* Locally acting proteins that assist in wound healing. 104, 176

guanine *GWAN-een* One of the two purine nitrogenous bases in DNA and RNA. 52, 332

H

half-life *HAF-life* The time it takes for half of the isotopes in a sample of an element to decay into the second isotopic form. This measurement is used in absolute dating of fossils. 37

haploid cell *HAP-loid SEL* A cell with one copy of each chromosome. 193

heat capacity *HEET ca-PA-city* The amount of heat necessary to raise the temperature of a substance. 44

heat of vaporization *HEET of VA-por-i-ZA-shun* The amount of heat needed to convert a liquid to vapor (gas). 44

heritable gene therapy *HARE-it-a-bull JEAN THER-a-pee* A gene alteration affecting all cells of an individual. 388

heterochromatin *het-er-o-KROME-a-tin* Dark-staining genetic material. 310

heterogametic sex *HET-er-o-gah-MEE-tik SEX* The sex with two different sex chromosomes, such as the human male. 301

heterotherm *HET-er-o-therm* An animal with variable body temperature.

heterotroph(ic) *HET-er-o-TROFE* An organism that obtains nourishment from another organism. 144

heterozygous *HET-er-o-ZI-gus* Possessing two different alleles for a particular gene. 272

homeostasis *HOME-ee-o-STA-sis* The ability of an organism to maintain constant body temperature, fluid balance, and chemistry. 7

homeotherm *HOME-e-o-therm* An animal with constant body temperature.

homeotic *home-ee-OT-ik* A mutant in which body parts form in the wrong places. 215

homogametic sex *HO-mo-gah-MEE-tik SEX* The sex with two identical sex chromosomes, such as the human female. 301

homologous pairs *ho-MOL-eh-gus PAIRZ* Chromosome pairs that have the same sequence of genes. 198

homologous recombination *hom-MOL-eh-gus re-com-bin-A-shun* A DNA sequence replacing its complement in a chromosome. 379

homozygous *HO-mo-ZI-gus* Possessing two identical alleles for a particular gene. 272

hormone *HOR-moan* A biochemical manufactured in a gland and transported in the blood to a target organ, where it exerts a characteristic effect. 104, 176

human chorionic gonadotropin (HCG) *YU-man KOR-ee-on-ik go-NAD-o-TRO-pin* A hormone secreted by the preembryo and embryo that prevents menstruation. 222

human genome project *YU-man JEAN-ome PROJ-ect* A worldwide effort to sequence all human genes. 381

human immunodeficiency virus (HIV) *YU-man imm-u-no-def-FISH-en-cee VI-rus* The retrovirus believed to cause AIDS.

hydrocarbon *HI-dro-kar-bon* A molecule containing carbon and hydrogen. 40

hydrogen bond *HI-dro-gen BOND* A weak chemical bond between negatively charged portions of molecules and hydrogen ions. 42

hydrolysis *hi-DROL-eh-sis* Splitting of a molecule in two by adding water. 46

hydronium ion (H₃O⁺) *hi-DRO-ne-um I-on* A water molecule (H_2O) with an extra hydrogen (H^+). 43

hydrophilic *HI-dro-FILL-ik* Attracted to water. 42, 90

hydrophobic *HI-dro-FOBE-ik* Repulsed by water. 42, 90

hydroxide ion (OH⁻) *hi-DROX-ide I-on* A molecule consisting of one oxygen and one hydrogen. 43

hypertonic *hi-per-TON-ik* The solution on one side of a membrane where the solute concentration is greater than on the other side. 94

hypothesis *hy-POTH-eh-sis* An educated guess based on prior knowledge. 17

hypotonic *hi-po-TON-ic* The solution on one side of a membrane where the solute concentration is less than on the other side. 94

I

imbibition *IM-bah-BISH-un* The absorption of water by a seed. 43

implantation *im-plan-TAY-shun* Nestling of the blastocyst into the uterine lining. 221

incomplete dominance *IN-kum-PLETE DOM-eh-nance* A heterozygote whose phenotype is intermediate between the phenotypes of the two homozygotes. 284

incomplete penetrance *IN-kum-PLETE PEN-e-trants* The quality of a genotype that does not always produce a phenotype. 285

independent assortment *IN-deh-PEN-dent ah-SORT-ment* The random arrangement of homologs during metaphase of meiosis I. 200, 281

infertility *IN-fer-TIL-eh-tee* The inability to conceive a child after a year of trying. 241

initiation *in-ish-e-A-shun* The first stage in translating a gene's message into protein. 359

initiation complex *in-ish-e-A-shun COM-plex* A small ribosomal subunit, mRNA, and an initiator tRNA, joined. 359

initiation site *in-ish-e-A-shun SITE* The site on a chromosome where DNA replication begins. 337

intermembrane compartment *in-ter-MEM-brain cum-PART-ment* The space between a mitochondrion's two membranes. 126

interphase *IN-ter-faze* The period when the cell synthesizes proteins, lipids, carbohydrates, and nucleic acids. 168

intron *IN-tron* Bases of a gene that are transcribed but are excised from the mRNA before translation into protein. 354

inversion *in-VER-shun* A chromosome with part of its gene sequence inverted. 317

in vitro fertilization *in VEE-tro fer-ti-li-ZA-shun* Fertilization outside the woman's body as in laboratory glassware. 247

ion *I-on* An atom that has lost or gained electrons, giving it an electrical charge. 40

ionic bond *i-ON-ik bond* Attraction between oppositely charged ions. 40

ionizing radiation *I-o-nize-ing rade-e-A-shun* Radiation that causes electrons to leave atoms. 146

irritability *IR-eh-tah-BIL-eh-tee* An immediate response to a stimulus. 7

isochromosome *ice-o-KROME-o-some* A chromosome with identical arms. 317

isotonic *ice-o-TON-ik* When solute concentration is the same on both sides of a membrane. 94

isotope *I-so-tope* A differently weighted form of an element. 37

J

Jacob syndrome *JAY-kob SIN-drome* A condition in which a human male has karyotype XYY. 317

K

karyokinesis *KAR-ee-o-kah-NEE-sus* Division of the genetic material. 171

karyotype *KAR-ee-o-type* A size-order chart of chromosomes. 311

kinase *KI-nase* A type of enzyme that activates other proteins by adding a phosphate. 177

kinetic energy *kin-ET-ik EN-er-gee* The energy of motion. 113

Klinefelter syndrome *KLINE-felt-er SIN-drome* A condition in which a male human has an XXY karyotype.

Krebs cycle *KREBS SI-kle* The stage in cellular respiration that completely metabolizes the products of glycolysis. 126

L

labia majora *LAY-bee-a ma-JOR-a* Large fleshy folds of tissue protecting the vagina. 198

labia minora *LAY-bee-a mi-NOR-a* Thin inner fleshy folds of tissue protecting the vagina. 198

labor *LAY-ber* The uterine contractions leading to a baby's birth. 217

lactic acid formation *LAK-tik AS-id for-MAY-shun* The conversion of pyruvic acid from glycolysis into lactic acid, occurring in some anaerobic bacteria and tired mammalian muscle cells. 138

law of segregation *seg-re-GA-shun* The principle that alleles of a gene separate during meiosis. 271

leader sequence *LEE-der SEE-kwens* A short sequence at the start of each mRNA that enables the mRNA to form hydrogen bonds with part of the rRNA in a ribosome. 357

lens *LENZ* The structure in the eye through which light passes and is focused.

ligand *LIG-and* A messenger molecule that binds to a cell surface protein to initiate signal transduction. 100

ligase *LIG-aze* An enzyme that catalyzes the formation of covalent bonds in the DNA sugar-phosphate "backbone." 337

linkage map *LINK-ege map* Diagram of gene order on a chromosome based on crossover frequencies. 300

linked genes *LINKT JEANS* Genes on the same chromosome. 298

lipid bilayer *LIP-id BI-lay-er* A two-layered structure formed by the alignment of phospholipids, reflecting their hydrophobic and hydrophilic tendencies.

lipids *LIP-idz* Compounds containing carbon, hydrogen, and oxygen, but with less oxygen than carbohydrates. Lipids, which include fats, are insoluble in water. 47

liposome *LIP-o-some* A fatty bubble that can carry a gene or drug. 378

low-density lipoprotein *LO DEN-sit-ee LIP-o-PRO-teen* A molecule that carries cholesterol. 100

lysosome *LI-so-soam* A sac in a eukaryotic cell in which molecules and worn-out organelles are enzymatically dismantled. 75

M

macromolecule *MAK-ro-MOL-e-kuel* A very large molecule. 39

malignant tumor *mal-IG-nant TOO-mer* Literally means evil; refers to a cancerous tumor. 179

malleus *MAL-e-us* A small bone in the middle ear.

manifesting heterozygote *MAN-e-fest-ing het-er-o-ZI-goat* A female carrier of a sex-linked trait who expresses the phenotype associated with the gene. 306

mass number *MAS NUM-ber* The total number of protons and neutrons in an atom. The number in the upper left hand corner of an element's symbol in the periodic table. 36

matrix *MAY-trix* The inner compartment of a mitochondrion. 126

maturation-promoting factor (MPF) *mat-yur-A-shun pro-MOAT-ing FAC-tor* A hypothesized (and later identified) mitotic trigger molecule. 177

meiosis *mi-O-sis* Cell division that results in a halving of the genetic material. 195

Mendelian trait *men-DEEL-e-an TRAYT* Trait specified by a single gene. 276

menstrual flow *MEN-stroo-al FLO* Shedding of the endometrium when fertilization does not occur. 198

mesoderm *MEZ-o-derm* The middle embryonic germ layer, whose cells become bone, muscle, blood, dermis, and reproductive organs. 222

mesophyll cells *MEZ-o-fill SELLS* Thin-walled plant cells which take part in C_4 photosynthesis. 158

messenger RNA (mRNA) *MESS-en-ger RNA* A molecule of ribonucleic acid that is complementary in sequence to the sense strand of a gene. 72, 353

meta-analysis *ME-tah a-NAL-i-sis* A study that combines the results of several studies. 24

metabolic pathway *met-a-BOL-ik PATH-way* A series of connected, enzymatically-catalyzed reactions in a cell. 115

metabolism *meh-TAB-o-liz-um* The biochemical reactions that acquire and utilize energy. 7, 54

metacentric *met-a-SEN-trik* A chromosome whose centromere divides it into two similarly sized arms. 310

metaphase *MET-ah-faze* The second stage of cell division, when chromosomes align down the center of a cell. In mitosis, the chromosomes form a single line. 171

metaphase I *MET-ah-faze I* In meiosis I, when chromosomes line up in homologous pairs in meiosis I. 200

metaphase II *MET-a-faze II* When replicated chromosomes align down the center of the cell in meiosis II. 200

metastasis *meh-TAH-stah-sis* The spreading of cancer from its site of origin to other parts of the body. 182

microfilaments *MI-kro-FILL-ah-ments* Tiny rods composed of actin found within cells, especially contractile cells. 103

microtubules *MI-kro-TU-bules* Long, hollow tubules, built of the protein tubulin, that provide movement within cells. 78, 102

missense mutation *MISS-sentz mu-TAY-shun* A mutation changing a codon specifying a certain amino acid into a codon specifying a different amino acid. 365

mitochondria *MI-toe-KON-dree-ah* Organelles within which the reactions of cellular metabolism occur. 75

mitosis *mi-TOE-sis* A form of cell division in which two genetically identical cells are generated from one. 168

mitotic spindle *mi-TOT-ik SPIN-del* A structure built of microtubules that aligns and separates chromosomes in mitotic cell division. 170

molecular weight *mo-LEK-yu-ler WAYT* The sum of the weights of the atoms comprising a molecule. 39

molecule *MOL-eh-kuel* A structure resulting from the combination of atoms. 38

monoclonal antibodies *MON-o-KLON-al AN-tah-BOD-eez* Antibodies descended from a single B cell and therefore identical. B cells are fused with cancer cells to create hybridomas, which are artificial cells that secrete a particular antibody indefinitely. 185

monohybrid *MON-o-HI-brid* An individual heterozygous for a particular gene. 274

monomer *MON-o-mer* A single link in a multilink (polymeric) molecule. 47

monosaccharide *MON-o-SAK-eh-ride* A sugar built of one 5- or 6-carbon unit, including glucose, galactose, and fructose. 46

monosomy *MON-o-SOAM-ee* A cell missing one chromosome. 313

monounsaturated fat *MON-o-un-SAT-yer-a-tid FAT* A fatty acid with one double bond. 48

monozygotic twins *mon-o-zi-GOT-ik TWINZ* Identical twins resulting from the splitting of a fertilized ovum. 291

morphogenesis *morf-o-GEN-e-sis* The series of events that form a newborn. 222

morphogens *MORF-o-genz* Proteins that form gradients that influence development. 215

morula *MORE-u-lah* The preembryonic stage of a solid ball of cells. 219

motif *moe-teef* A common part of a transcription factor. 349

M phase *EM-faze* Mitosis. 170

multicellular *mull-tee-SEL-u-lar* Consisting of many cells. 64

multifactorial *mull-tee-fac-TORE-e-al* Traits molded by one or more genes and the environment. 289

muscular tissue *MUS-ku-lar TISH-ew* Tissue consisting of contractile cells, providing motion.

mutagen *MUTE-a-jen* An agent that causes a mutation. 363

mutant *MU-tent* A phenotype or allele that is not the most common for a certain gene in a population, or that has been altered from the "normal" condition. 274

mutation *mu-TAY-shun* A change in a gene or chromosome. 54, 274, 361

N

natural selection *NAT-rul sah-LEK-shun* The differential survival and reproduction of organisms whose genetic traits better adapt them to a particular environment. 8

negative feedback *NEG-ah-tiv FEED-bak* The turning off of an enzyme's synthesis or activity caused by accumulation of the product of the reaction that the enzyme catalyzes. 121

neonatology *NE-o-nah-TOL-eh-gee* The study of the newborn. 257

nervous tissue *NER-vis TISH-ew* A tissue whose cells (neutrons and neuroglia) form a communication network.

neural tube *NEUR-el TOOB* The embryonic precursor of the central nervous system. A neural tube defect results from failure of the neural tube to fully close. 227

neuron *NEUR-on* A nerve cell, consisting of a cell body, a long "sending" projection called an axon, and numerous "receiving" projections called dendrites.

neurotransmitter *NEUR-o-TRANZ-mit-er* A chemical passed from a nerve cell to another nerve cell, or to a muscle or gland cell, that relays an electrochemical message.

neurulation *NEUR-u-LAY-shun* Physical contact between the notochord and nearby ectoderm, triggering formation of the nervous system. 227

neutron *NEW-tron* A particle in an atom's nucleus that is electrically neutral and has one mass unit. 36

nicotinamide adenine dinucleotide (NAD⁺) *nik-o-TIN-a-mide AD-e-neen di-NEW-kle-o-tide* An electron carrier molecule important in the energy pathways. 120

nicotinamide adenine dinucleotide phosphate (NADP⁺) *nik-o-TIN-a-mide AD-e-neen di-NEW-kle-o-tide FOSS-fate* An electron carrier important in photosynthesis. 120

nitrogenous base *ni-TRODGE-eh-nus BASE* A nitrogen-containing compound that forms part of a nucleotide, giving it individuality. 52

noncompetitive inhibition *non-cum-PET-eh-tiv in-hi-BI-shun* Suppression of an enzyme's function due to a substrate binding at a site other than the active site. Also called allosteric inhibition. 121

nondisjunction *NON-dis-JUNK-shun* The unequal partition of chromosomes into gametes during meiosis. 313

nonheritable gene therapy *non-HARE-it-a-bull JEAN THER-a-pee* A gene alteration affecting somatic cells. 388

nonpolar covalent bond *non-POE-lar co-VAY-lent BOND* A covalent bond in which atoms share electrons equally. 40

nonreciprocal translocation *non-ree-SIP-prik-al TRANZ-lo-KAY-shun* When a piece of one chromosome breaks off and joins a nonreciprocal chromosome. 317

nonsense mutation *NON-sents mu-TAY-shun* A point mutation altering a codon that encodes an amino acid to one that encodes a stop codon. 365

nonshivering thermogenesis *non-SHIV-er-ing thermo-GEN-i-sis* Hormone-directed internal heat.

notochord *NO-toe-kord* A semirigid rod running down the length of an animal's body. 227

nuclear envelope *NEW-klee-ar EN-vel-ope* A two-layered structure bounding a cell's nucleus. 73

nuclear matrix *NEW-klee-ar MAY-trix* A three-dimensional network of protein fibers that forms a scaffolding in the nucleus. 73

nuclear matrix proteins *NEW-klee-ar MAY-trix PRO-teens* Proteins in the nucleus that partition genes being transcribed. 349

nuclear pore *NEW-klee-ar POOR* A hole in the nuclear envelope. 73

nuclear pore complex *NEW-klee-ar POOR KOM-plex* A group of proteins that form a channel in a cell's nuclear envelope. 73

nucleic acid *new-CLAY-ic AS-id* A biochemical that encodes a protein's mino acid sequence. 52

nucleoid *NEW-klee-oid* The part of a prokaryotic cell where the DNA is located. 68

nucleolus *new-KLEE-o-lis* A structure within the nucleus where RNA nucleotides are stored. 72

nucleosome *NEW-klee-o-some* DNA wrapped around eight histone proteins as part of chromosome structure. 334

nucleotide *NEW-klee-o-tide* The building block of a nucleic acid, consisting of a phosphate group, a nitrogenous base, and a five-carbon sugar. 52, 331

nucleus (atomic) *NEW-klee-is* The central region of an atom, consisting of protons and neutrons. 66

nucleus (cellular) *NEW-klee-is* A membrane-bound sac in a eukaryotic cell that contains the genetic material. 36

O

octet rule *OC-tet ROOL* The tendency of an atom to fill its outermost shell. 37

oncogene *ON-ko-jean* A gene that normally controls cell division but when overexpressed leads to cancer. 182

oocyte *OO-site* The female sex cell before it is fertilized. 197

oocyte banking *OO-site BANK-ing* Freezing and storing oocytes for future use.

oocyte donation *OO-site do-NA-shun* Donating healthy oocytes to an infertile woman.

oogenesis *oo-GEN-eh-sis* The differentiation of an egg cell from a diploid oogonium, to a primary oocyte, to two haploid secondary oocytes, to ootids, and finally, after fertilization, to a mature ovum. 204

oogonium *oo-GO-nee-um* The diploid cell in which oogenesis (egg formation) begins. 204

operon *OP-er-on* A series of genes with related functions and their controls. 349

organ *OR-gan* A structure consisting of two or more tissues that functions as an integrated unit.

organelles *OR-gan-NELLZ* Specialized structures in eukaryotic cells that carry out specific functions. 6, 64

organogenesis *or-GAN-o-GEN-eh-sis* The development of organs from an embryo. 227

orgasm *OR-gazz-m* A pleasurable sensation associated with sexual activity. 197

ovary *O-var-ee* One of the paired female gonads which house developing oocytes. 197

ovulation *OV-u-LAY-shun* The release of an oocyte from the largest ovarian follicle just after luteinizing hormone peaks in the blood in the middle of a woman's menstrual cycle.

ovum *O-vum* The female sex cell after fertilization. 196

oxidation *OX-e-DAY-shun* A chemical reaction in which electrons are lost. 40, 117

oxidative phosphorylation *ox-i-DAY-tiv fos-for-e-LAY-shun* Phosphorylation of ADP to ATP coupled to protons moving down their concentration gradient across the inner mitochondrial membrane. 128

P

panspermia *pan-SPUR-mee-a* The theory that life came to earth in spores from the cosmos. 35

parental generation *pa-REN-tel gen-e-RA-shun* The first generation in a genetic cross. 274

parthenogenesis *par-tho-GEN-eh-sis* Female reproduction without male fertilization. 194

particle bombardment *PAR-tik-el bom-BARD-ment* A method to shoot genes into cells. 378

parturition *par-ter-I-shun* The birth of a baby. 217

pelvic inflammatory disease *pell-vik in-FLAMM-a-to-ree diz-EAZE* Bacterial infection of the female reproductive tract. 261

penis *PEE-nis* The male sex organ. 197

pentose phosphate pathway *PEN-tose FOS-fate PATH-way* An alternative pathway to the Krebs cycle in extracting energy from glucose. 136

peptide bond *PEP-tide BOND* A chemical bond between two amino acids resulting from dehydration synthesis. 49

periodic table *peer-ee-OD-ic TA-ble* Chart that lists naturally occurring elements according to their properties. 35

peroxisome *per-OX-eh-soam* A membrane-bound sac that buds from the smooth ER and that houses enzymes important in oxygen utilization. 75

phenocopy *FEEN-o-kop-ee* An environmentally caused trait that resembles an inherited trait. 287

phenotype *FEEN-o-type* The observable expression of a genotype in a specific environment. 272

phenotypic ratio *feen-o-TIP-ik RAY-shee-o* The proportions of different phenotypes among offspring of a genetic cross. 275

phosphoglyceric acid *FOSS-foe-gli-SARE-ik AS-id* An intermediate in the Calvin cycle of photosynthesis. 155

phospholipid *FOS-fo-LIP-id* A molecule consisting of a lipid and a phosphate that is hydrophobic at one end and hydrophilic at the other end. 90

phosphorylation *foss-for-eh-LAY-shun* Adding a phosphate (PO_4) group to a molecule. 120

photochemical reactions *fo-to-KEM-i-kal re-AK-shuns* The light-requiring reactions of photosynthesis that harness photon energy and use it to phosphorylate ADP to ATP. 152

photolysis *fo-TOL-eh-sis* A photosynthetic reaction in which electrons from water replace electrons lost by chlorophyll a.

photon *FOE-ton* A packet of light energy. 113, 146

photophosphorylation *FO-toe-FOS-for-eh-LAY-shun* A photosynthetic reaction in which energy released by the electron transport chain linking the two photosystems is stored in the high-energy phosphate bonds of ATP. 154

photoreactivation enzyme *fo-to-re-ac-ti-VAY-shun en-zime* An enzyme that uses light energy to break pyrimidine dimers, thereby repairing DNA. 340

photorespiration *fo-to-res-per-A-shun* A process that counters photosynthesis. 157

photosynthate *fo-toe-SIN-thate* The carbohydrate products of photosynthesis. 157

photosynthesis *FO-toe-SIN-the-sis* The series of biochemical reactions that enable plants to harness sunlight energy to manufacture nutrient molecules. 76, 113, 144

photosystem *FO-toe-SIS-tum* A cluster of pigment molecules that enable green plants to absorb, transport, and harness solar energy. 150

pH scale *SKALE* A measurement of how acidic or basic a solution is. 43

phylum *FI-lum* A major classification of organisms. 10

placebo *pla-SEE-bo* An inert substance used as an experimental control. Its effects are compared with those of the substance under investigation. 20

placenta *pla-CEN-tah* A specialized organ that develops in certain mammals, connecting mother to unborn offspring. 203, 217

plasma *PLAZ-ma* A watery, protein-rich fluid that forms the matrix of blood.

plasmalemma *plaz-ma-LEM-a* Cell membrane.

plasmid *PLAZ-mid* A small circle of double-strand DNA, found in some bacteria in addition to their DNA, commonly used as a vector for recombinant DNA. 375

plastid *PLAS-tid* A plant organelle that encapsulates photosynthetic membranes. 149

platelet *PLATE-let* A cell fragment that is part of the blood and orchestrates clotting.

pleiotropic *PLY-o-TRO-pik* A genotype with multiple expressions. 287

pluripotency *plur-e-POE-ten-see* A cell that retains the potential to specialize in any way. 212

point mutation *POYNT mu-TAY-shun* A change in a single DNA base. 365

polar body *PO-lar BOD-ee* A small cell generated during female meiosis, enabling cytoplasm to be partitioned into just one of the four meiotic products, the ovum. 204

polar covalent bond *PO-lar co-VAY-lent BOND* A covalent bond in which electrons are attracted more toward one atom's nucleus than the other. 40

polygenic *pol-ee-JEAN-ik* A trait caused by more than one gene. 289

polymer *POL-eh-mer* A long molecule composed of similar subunits. 47

polymerase chain reaction (PCR) *pole-LIM-er-ase CHAYN re-AK-shun* A method that uses a polymerase to mass produce copies of a gene in a test tube.

polymorphism *pol-ee-MORF-iz-m* A variant of a DNA sequence. 383

polypeptide chain *pol-ee-PEP-tide CHAYN* A long polymer of amino acids. 48

polyploidy *POL-ee-PLOID-ee* A condition in which a cell has one or more extra sets of chromosomes. 311

polyunsaturated fat *POL-ee un-SAT-yer-a-tid FAT* A fatty acid with more than one double bond. 48

positive feedback loop *PAHZ-eh-tiv FEED-bak LOOP* A biochemical pathway in which the accumulation of a product stimulates its production. 122

postreplication repair *post-rep-li-KAY-shun ree-PARE* Repair of newly replicated DNA. 341

potential energy *poe-TEN-shul EN-er-gee* The energy held in the position of matter. 113

preembryonic stage *pre-em-bree-ON-ik STAYJ* The first two weeks of human prenatal development. 217

preformation *pre-for-MAY-shun* The idea that a gamete or fertilized ovum contains an entire preformed organism. 211

preimplantation genetic diagnosis *pre-im-plan-TAY-shun jeh-NET-ic di-ag-NO-sus* Testing for the presence of a disease-causing gene in one preembryo cell. 251

premature *pre-ma-CHUR* A human baby born more than four weeks early or weighing less than five pounds. 252

pre-mRNA *PRE-m-RNA* Messenger RNA before intron removal. 354

primary oocyte *PRI-mare-ee OO-site* An intermediate in ovum formation. 204

primary spermatocyte *PRI-mare-ee spur-MAT-o-site* An intermediate in sperm formation. 202

primary structure *PRI-mare-ee STRUK-sure* The amino acid sequence of a protein. 49

primitive streak *PRIM-eh-tiv STREEK* The pigmented band along the back of a three-week embryo that develops into the notochord. 227

primordial embryo *pri-MORE-dee-al EM-bree-o* The three-layered early embryo. 222

product rule *PROD-ukt ROOL* The chance of two events occurring equals the product of the chances of either event occurring. 282

progenotes *pro-JEAN-note* Collections of nucleic acid and protein that were forerunners to cells. 54

prokaryotic cell *pro-CARE-ee-OT-ik SEL* A structurally simple cell, lacking organelles. 68

promoter *pro-MOW-ter* A control sequence near the start of a gene that attracts RNA polymerase and transcription factors. 349

pronuclei *pro-NU-kle-eye* The genetic packages of gametes. 218

prophase *PRO-faze* The first stage of cell division, when chromosomes condense and become visible. 171

prophase I *PRO-faze I* Prophase of meiosis I, when synapsis and crossing over occur. 198

prophase II *PRO-faze II* Prophase of the second meiotic division. 200

protamine *PROT-a-meen* A protein found in sperm heads; DNA wraps around it. 203

protein *PRO-teen* A long molecule consisting of amino acids bonded to each other. 48

proton *PRO-ton* A particle in an atom's nucleus carrying a positive charge and having one mass unit. 36

protoplasm *PRO-tow-plaz-m* Living matter. 64

protozoans *PRO-toe-ZO-anz* Single-celled eukaryotes often classified by their mode of movement, including the familiar amoeba, euglena, and paramecium.

pseudogene *SOOD-o-jean* A nonfunctional gene similar in sequence to and located near a functional gene.

pseudopod *SOOD-o-pod* The portion of an amoeba that extends outward, enabling the organism to move.

Punnett square *PUN-et SQWARE* A diagram used to display genotypic and phenotypic ratios resulting from a genetic cross. 275

purine *PURE-een* A type of organic molecule with a double ring structure, including the nitrogenous bases adenine and guanine. 334

pyrimidine *pie-RIM-eh-deen* A type of organic molecule with a single ring structure, including the nitrogenous bases cytosine, thymine, and uracil. 334

pyruvic acid *pi-ROO-vic AS-id* One of the products of glycolysis. 130

Q

quaternary structure *QUAR-teh-nair-ee STRUK-sure* The number and arrangement of polypeptide chains of a protein. 49

R

reactant *re-AK-tant* A starting material in a chemical reaction. 38

reaction center *re-AK-shun SEN-ter* Chlorophyll and proteins clustered together that receive photon energy in photosynthesis. 150

reading frame *REED-ing FRAME* The DNA nucleotide corresponding to the first codon position in RNA. 357

recessive allele *re-SESS-ive ah-LEEL* An allele whose expression is masked by the activity of another allele. 272

reciprocal translocation *re-SIP-ro-kal tranz-lo-CAY-shun* Two nonhomologous chromosomes exchanging parts. 317

recombinant DNA technology *re-KOM-bah-nent DNA tek-NAL-eh-gee* Transferring a gene from a cell of a member of one species to the cell of a member of a different species. 299, 372

red blood cell (erythrocyte) *RED BLUD SEL* A disc-shaped cell, lacking a nucleus, that is packed with the oxygen-carrying molecule hemoglobin.

reduction *re-DUK-shun* A chemical reaction in which electrons are gained. 40, 117

reduction division *re-DUK-shun dah-VISH-un* Meiosis I, when the diploid chromosome number is halved. 198

regulatory enzyme *REG-u-la-tor-ee EN-zime* An enzyme that controls the activity of a biochemical pathway because it can be activated or deactivated by binding a compound other than its substrate under certain conditions. 121

replication fork *rep-li-KAY-shun FORK* A locally unwound portion of DNA where replication occurs. 337

respiratory chain *RES-pir-ah-TOR-ee CHANE* A series of electron-accepting enzymes embedded in the inner mitochondrial membrane. 131

respiratory distress syndrome *RES-pir-ah-tory dis-TRESS SIN-drome* A condition in which a newborn's lungs lack surfactant. 253

restriction enzyme *re-STRIK-shun EN-zime* A bacterial enzyme that cuts DNA at a specific sequence. 375

restriction fragment length polymorphism (RFLP) *re-STRIK-shun FRAG-ment LENGTH POL-e-MORF-iz-um* Differences in restriction enzyme cutting sites between individuals. 384

retrovirus *RE-tro-VI-rus* A virus that uses RNA as its genetic material.

reverse transcriptase *re-VERS tran-SCRIPT-aze* An enzyme that constructs a DNA molecule from an RNA molecule. 54

R group *R GROOP* An amino acid side chain. 49

rhythm method *RITH-im METH-ud* Contraception by controlling the timing of sexual intercourse. 260

ribonucleic acid (RNA) *RI-bo-nu-KLAY-ik AS-id* A single-strand nucleic acid consisting of nucleotides containing a phosphate, ribose, and nitrogenous bases adenine, guanine, cytosine, and uracil. 52, 331

ribose *RI-bose* The five-carbon sugar that is a structural component of RNA. 331

ribosomal RNA (rRNA) *RI-bo-SOAM-el RNA* that, along with proteins, comprises the ribosome. 72, 353

ribosome *RI-bo-soam* A structure built of RNA and protein upon which a gene's message (mRNA) anchors during protein synthesis. 68, 353

ribozyme *RI-bo-zime* Small RNAs that function as enzymes. 354

ribulose bisphosphate (RuBP) *RI-bew-lose bis-FOS-fate* A five-carbon sugar with two phosphate groups that is part of the Calvin cycle of the biochemical reactions of photosynthesis. 155

ribulose biphosphate (RuBP) carboxylase/oxygenase (rubisco) *R-U-B-P car-BOX-e-lace OX-i-gin-ays* The enzyme that catalyzes the reaction between RuBP and CO_2 in the Calvin cycle of the biochemical reactions of photosynthesis. 155

RNA polymerase *RNA poe-LIM-er-ase* An enzyme that takes part in DNA replication and RNA transcription. 349

RNA primer *RNA PRI-mer* A small piece of RNA, inserted at the start of a piece of DNA to be replicated, which attracts RNA polymerase and is later removed. 337

RU 486 *R-U-486* A drug that induces abortion early in pregnancy. 260

rubella *rew-BELL-a* A viral illness that can severely harm a human fetus. 254

S

salt *SALT* A molecule composed of cations and anions. 40

sample size *SAM-pul SIZE* The number of individuals in an experiment. 20

satellite *SAT-e-lite* Bloblike chromosome ends. 310

saturated (fat) *SAT-yur-ray-tid* A triglyceride with single bonds between the carbons of its fatty acid tails. 48

scientific method *SI-en-TIF-ik METH-id* A systematic approach to interpreting observations, involving reasoning, predicting, testing, and drawing conclusions, then putting them into perspective with existing knowledge. 16

scrotum *SKRO-tum* The sac of skin containing the male testes. 196

secondary oocyte *SEC-un-derry OO-site* A haploid cell that is an intermediate in ovum formation. 204

secondary spermatocyte *SEC-un-derry sper-MAT-o-site* A haploid cell that is an intermediate in sperm formation. 202

secondary structure *SEC-en-DAIR-ee STRUCK-sure* The shape a protein assumes when amino acids that are close together in the primary structure chemically attract one another. 49

second filial generation *SEK-und FIL-e-al jen-er-A-shun* The third generation in a genetics problem. 274

second law of thermodynamics *SEK-und LAW of THERM-o-die-NAM-ix* The tendency of matter to become random or disordered. 6

second messenger *SEK-und MESS-en-ger* A biochemical activated by an extracellular signal that transmits a message inside a cell. 104

semiconservative replication *sem-ee-con-SERV-a-tive rep-li-KAY-shun* The correct mode of DNA replication, in which each double helix is composed of one parental and one daughter strand. 335

seminal fluid *SEM-in-el FLEW-id* Secretions that carry sperm. 197

seminal vesicles *SEM-in-el VES-eh-kels.* In the human male, the paired structures that add fructose and prostaglandins to the sperm. 197

seminiferous tubule *sem-i-NIF-er-us TUBE-you-all* Tubule within the testis where sperm form and mature. 196

sense strand *SENSE STRAND* The side of the DNA double helix containing a particular gene that is transcribed. 350

sex chromosome *SEX KRO-mo-soam* A chromosome that determines sex.

sex-influenced trait *SEX IN-flu-enced TRAIT* A trait controlled by an allele that is dominant in one sex but recessive in the other. 308

sex-limited trait *SEX LIM-eh-tid TRAIT* A trait affecting a structure or function of the body that is present in only one sex. 308

sex pilus *SEX PILL-us* A bacterial cell outgrowth used to exchange genetic material. 193

sexual reproduction *SEX-u-el RE-pro-DUK-shun* The combination of genetic material from two individuals to create a third individual. 8, 193

signal transduction *SIG-null tranz-DUCK-shun* The biochemical transmission of a message from outside the cell to inside. 104

simple carbohydrates *SIM-pel KAR-bo-HI-drates* Monosaccharides and disaccharides. 46

skeletal muscle *SKEL-eh-tel MUS-sel* Voluntary, striated muscle consisting of single, multinucleated cells that are contractile due to sliding filaments of actin and myosin.

small nuclear ribonucleoproteins (SNURPs) *SMAL NEW-klee-ar RI-bo-NEW-klee-o-PRO-teens* Proteins that associate with ribozymes, forming a complex that cuts introns from mRNA. 354

smooth muscle *SMOOTH MUS-sel* Involuntary, nonstriated contractile tissue that lines the digestive tract and other organs.

sodium-potassium pump *SO-dee-um po-TAS-ee-um PUMP* A mechanism that uses energy released from splitting ATP to transport Na^+ out of cells and K^+ into cells. 98

solute *SOL-yute* A chemical that dissolves in another, forming a solution. 43, 92

solution *so-LU-shun* A homogenous mixture of a substance (the solute) dissolved in water (the solvent). 43

solvent *SOL-vent* A chemical in which others dissolve, forming a solution. 43, 92

somatic cell *so-MAT-ik SEL* A body cell; a cell other than the sperm or ovum.

somatic mutation *sew-MAT-ik mew-TAY-shun* A mutation in a somatic (body) cell. 365

Southern blotting *SOU-thern BLOT-ting* Use of DNA probes to identify specific fragments of DNA. 382

species *SPE-shez* A group of similar individuals that interbreed in nature and are reproductively isolated from all other such groups.

spectrin *SPEK-trin* Proteins that support cell membranes from beneath. 104

sperm *SPERM* The male sex cell. 196

spermatid *sper-ma-TID* An intermediate stage in sperm development. 202

spermatogenesis *sper-MAT-o-JEN-eh-sis* The differentiation of a sperm cell from a diploid spermatogonium, to primary spermatocyte, to two haploid secondary spermatocytes, to spermatids, and finally to mature spermatozoa. 202

spermatogonium *sper-mat-o-GOWN-e-um* A diploid cell that divides, yielding daughter cells that become sperm cells. 202

spermatozoa *sper-mat-o-ZO-a* Mature sperm. 203

S phase *S FAZE* The synthesis phase of interphase, when DNA is replicated and microtubules are produced from tubulin. 170

spontaneous abortion *spon-TAY-nee-us a-BOR-shun* A pregnancy that ends naturally before the embryo or fetus has fully developed. 251

spontaneous generation *spon-TAY-nee-us JEN-er-RAY-shun* The idea, proven untrue, that living things can arise from nonliving matter. 34

spontaneous mutation *spon-TAY-nee-us mew-TA-shun* A change in DNA sequence not caused by a mutagen, usually resulting from a replication error. 363

S-shaped curve *S-shapt KURV* A leveling off of a population growth curve.

stem cell *STEM SEL* An undifferentiated cell that divides to give rise to cells that specialize. 178

sterol *STAIR-ol* A lipid with four carbon rings. 48

stimulus *STIM-yew-lus* A change in an organism's external or internal environment.

stoma *STO-mah* Pore in a plant's cuticle through which water and gases are exchanged between the plant and the atmosphere.

stroma *STRO-ma* The nonmembranous inner region of the chloroplast. 149

subatomic particle *sub-a-TOM-ic PAR-ti-cle* One of the small particles that make up atoms. 36

substrate *SUB-strate* A reactant an enzyme acts upon. 50

substrate-level phosphorylation *SUB-strate LEV-ell fos-for-i-LAY-shun* ATP formation accomplished by transferring a phosphate group to ADP. 128

sulfide *SUL-fide* A sulfur-containing compound. 45

surfactant *sir-FACT-ant* A soapy substance that allows microscopic air sacs in the human lungs to inflate. 253

surrogate mother *SURR-o-get MUTH-er* A woman who carries a child in utero for another. 247

synapsis *sin-AP-sis* The gene-by-gene alignment of homologous chromosomes during prophase of meiosis I. 199

syncytium *sin-SIK-tee-um* A mass of multinucleated cells. 171

synteny *SIN-ten-ee* Comparison of gene order on chromosomes between species.

synthesis (S) phase *SIN-the-sis FAZE* The period of mitotic interphase when genetic material replicates. 168

T

tandem duplication *TAN-dem doop-li-KAY-shun* A repeat of part of a gene's sequence.

tautomer *TAWT-o-mer* An unstable, transient variant of a nitrogenous base. 363

taxonomy *tax-ON-o-mee* The branch of biology concerned with classifying organisms on the basis of evolutionary relationships. Taxonomic levels include, in order, kingdom, phylum (or division), class, order, family, genus, and species.

telocentric *tell-o-SEN-trik* A chromosome whose centromere divides the chromosome into a long arm and a very, very short arm. 310

telomere *TELL-o-meer* A chromosome tip. 311

telophase *TELL-o-faze* The final stage of cell division, when two cells form from one and the spindle is disassembled. 171

telophase I *TELL-o-faze I* When homologs arrive at opposite poles in meiosis I. 200

telophase II *TELL-o-faze II* When nuclear envelopes form around the four products of meiosis. 200

teratogen *teh-RAT-eh-jen* A chemical or other environmental agent that causes a birth defect. 254

tertiary structure *TER-she-air-ee STRUK-sure* The shape a protein assumes when amino acids far apart in the primary structure chemically attract one another. 49

testis *TES-tis* One of the paired male gonads containing the seminiferous tubules in which sperm are manufactured. 196

thalidomide *thal-ID-o-mide* A drug that causes limb birth defects. 254

thermodynamics *THERM-o-di-NAM-ix* The field of physical science concerning energy transformations in nature. 114

thermogenesis *therm-o-JEN-i-sis* Generating heat metabolically.

thylakoid *THI-lah-koid* Membranous disc comprising the inner membrane of a chloroplast. 78

thylakoid membrane *THI-lah-koid MEM-brane* The structure that encloses the thylakoid space in a chloroplast. 149

thylakoid space *THI-lah-koid SPASE* The space within thylakoid membranes. 149

thymine *THI-meen* One of the two pyrimidine bases in DNA. 52, 332

thyroid gland *THI-roid GLAND* A gland in the neck that manufactures thyroxine, a hormone that increases energy expenditure.

tissue *TISH-ew* In multicellular organisms, groups of cells with related functions. 5

trace element *TRASE EL-e-ment* An element humans require in small amounts. 36

transcription *tranz-SKRIP-shun* Manufacturing RNA from DNA. 348

transcription factor *tranz-SCRIPT-shun fac-tor* A protein that controls which genes are turned on or off in a particular cell. 72, 349

transfer RNA (tRNA) *TRANZ-fer* A small RNA molecule that binds an amino acid at one site and an mRNA codon at another site. 72, 353

transgenic organism *TRANZ-jen-ik OR-gan-niz-um* Genetic engineering of a gamete or fertilized ovum, leading to development of an individual with the alteration in every cell. 372

transition mutation *tran-ZI-shun myu-TA-shun* A mutation that alters a purine to a purine or a pyrimidine to a pyrimidine. 363

translation *tranz-LAY-shun* Assembly of an amino acid chain according to the sequence of base triplets in a molecule of mRNA. 348

translocation *TRANZ-lo-KAY-shun* Exchange of genetic material between nonhomologous chromosomes. 317

translocation carrier *TRANZ-lo-KAH-shun KARE-ee-er* A person who has two translocated chromosomes that are balanced, so that a complete genome is present. 317

transposable elements *tranz-POSE-a-bull EL-e-mentz* Jumping genes. 366

transversion mutation *tranz-VER-jhun myu-TA-shun* A mutation that alters a purine to a pyrimidine or vice versa. 363

triglyceride *tri-GLI-sir-ide* A type of fat consisting of one glycerol and three fatty acids. 47

triplo-X *TRIP-low X* A female human with an extra X chromosome. 315

trisomy *TRI-som-mee* A cell with one extra chromosome. 313

trisomy 21 *TRI-som-mee 21* An extra chromosome 21, causing Down syndrome. 313

trophoblast *TRO-fo-blast* A layer of cells in the preembryo that develops into the chorion and then the placenta. 219

tubal ligation *TOOB-al li-GA-shun* Contraception by tying off the fallopian tubes. 260

tubulin *TOOB-u-lin* Proteins that comprise microtubules. 102

tumor suppressor *TOO-mer sup-PRESS-er* A gene which, when inactivated or suppressed, causes cancer. 182

Turner syndrome *TERN-er SIN-drome* Phenotype resulting from an XO karyotype.

U

ultratrace element *UL-tra-trase EL-e-ment* An element vital to humans in very small amounts. 36

umbilical cord *um-BIL-ik-kel KORD* A ropelike structure containing one artery and two veins that connects mother to unborn child. 217

uniparental disomy *uni-pa-REN-tal DI-so-mee* Inheriting two copies of a gene from one parent. 288

unsaturated (fat) *un-SAT-yur-RAY-tid* A triglyceride with double bonds between some of its carbons. 48

uracil *YUR-eh-sil* One of the two pyrimidine bases in RNA. 52, 337

urethra *u-RETH-rah* The tube leading from the bladder through which urine exits the body. 197

uterus *U-ter-us* The muscular, saclike organ in the human female in which the embryo and fetus develop. 197

V

vagina *va-GINE-a* The birth canal in a female. 198

valence electrons *VAY-lense e-LEC-trons* The electrons in an atom's outermost shell. 37

van der Waals attractions *VAN-dur-walls a-TRAC-shuns* Dynamic attractions within or between molecules when oppositely charged regions draw near one another. 42

variable *VAIR-ee-ah-bul* A changeable factor. 17

variable number of tandem repeats *VARE-e-a-bull NUM-ber of TAN-dem re-PEETZ* Repetitions of a DNA sequence, used to map genes. 384

variably expressive *VARE-e-a-blee ex-PRESS-ive* A phenotype that varies in intensity in different individuals. 285

vas deferens *VAS DEF-er-enz* In the human male, a tube that comes from the epididymis and joins the urethra in the penis. 197

vasectomy *vas-EK-toe-me* Contraception by tying off the vas deferens. 260

vector *VEK-ter* A piece of DNA used to transfer DNA to another organism's cell. 375

vesicle *VESS-i-cal* A membrane-bound sac in a cell. 84

virus *VI-rus* An infectious particle consisting of a nucleic acid (DNA or RNA) wrapped in protein. 79

W

wavelength *WAYV-lenth* The distance a photon moves during a complete vibration. 146

white blood cell (leukocyte) *WITE BLUD SEL* A cell that helps fight infection.

wild type *WILD TYPE* The most common phenotype or allele for a certain gene in a population. 274

X

X inactivation *X IN-ak-tah-VA-shun* The turning off of one X chromosome in each cell of a female mammal at a certain point in prenatal development. 306

X-inactivation center *X IN-ak-tah-VA-shun SEN-ter* The part of the X chromosome that inactivates the X. 306

Z

zinc finger *ZEENK FING-er* A transcription factor motif. 349

zona pellucida *ZO-nah pel-LU-seh-dah* A thin, clear layer of proteins and sugars surrounding a secondary oocyte. 217

zygote *ZI-goat* In prenatal humans, the organism during the first two weeks of development. Also called a preembryo. 217

zygote intrafallopian transfer (ZIFT) *ZI-goat in-tra-fall-O-pee-on TRANS-fer* Insertion of an in vitro fertilized ovum into a woman's fallopian tube. 249

CREDITS

INDEX